NANOELECTRONICS

Micro & Nano Technologies Series

NANOELECTRONICS

Devices, Circuits and Systems

Edited by

Brajesh Kumar Kaushik

Indian Institute of Technology-Roorkee, Uttarakhand, India

ELSEVIER

Elsevier
Radarweg 29, PO Box 211, 1000 AE Amsterdam, Netherlands
The Boulevard, Langford Lane, Kidlington, Oxford OX5 1GB, United Kingdom
50 Hampshire Street, 5th Floor, Cambridge, MA 02139, United States

British Library Cataloguing-in-Publication Data
A catalogue record for this book is available from the British Library

Library of Congress Cataloging-in-Publication Data
A catalog record for this book is available from the Library of Congress

ISBN: 978-0-12-813353-8

For Information on all Elsevier publications
visit our website at https://www.elsevier.com/books-and-journals

Publisher: Matthew Deans
Acquisition Editor: Simon Holt
Editorial Project Manager: Joshua Mearns
Production Project Manager: R. Vijay Bharath
Cover Designer: Greg Harris

Typeset by MPS Limited, Chennai, India

Dedication

To my Father, Late Mr. Jai Prakash Kaushik, and my Mother, Mrs. Karuna Kaushik, for their affection and untiring efforts in my upbringing. Also dedicated to my wife, Supriya, and our loving son, Partha Kaushik. Life has been really enjoyable because of the efforts of all my friends and students. Dedicated to all of them.

—Brajesh Kumar Kaushik

Contents

List of Contributors

J. Ajayan Department of Electronics and Communication Engineering, SNS College of Technology, Coimbatore, India

Saurabh Chaudhury Department of Electrical Engineering, National Institute of Technology Silchar, Assam, India

Suprem R. Das Kansas State University, Manhattan, KS, United States

Prabhat Kumar Dubey Department of Electronics and Communication Engineering, Indian Institute of Technology Roorkee, Roorkee, Uttarakhand, India

Rosario D'Esposito IMS Laboratory, University of Bordeaux, Talence, France

Sébastien Frégonèse IMS Laboratory, University of Bordeaux, Talence, France

Navdeep Goyal Department of Physics, Panjab University, Chandigarh, India

Ravi S. Hegde Department of Electrical Engineering, Indian Institute of Technology, Gandhinagar, Gujarat, India

Rahul Kumar Jaiswal RF to THz Integrated Circuits, System & Sensors Laboratory, Department of Electronics and Communication Engineering, Indian Institute of Technology, Roorkee, Uttarakhand, India

Saumyakanti Khatua Department of Chemistry, Indian Institute of Technology, Gandhinagar, Gujarat, India

R.K. Kotnala CSIR-National Physical Laboratory, New Delhi, India

Yaochuan Mei DD Visions LLC, Austell, GA, United States

Sachin Mishra Biosensors and Bioelectronics Centre, IFM, Linköping University, Linköping, Sweden; Department of Physics, University of the Free State, Bloemfontein, South Africa

Sparsh Mittal IIT, Hyderabad, Telangana, India

D. Nirmal Electronics and Communication Engineering, Department of Electrical Sciences, Karunya Institute of Technology and Sciences, Coimbatore, India

Nidhi Pandit RF to THz Integrated Circuits, System & Sensors Laboratory, Department of Electronics and Communication Engineering, Indian Institute of Technology, Roorkee, Uttarakhand, India

Nagendra Prasad Pathak RF to THz Integrated Circuits, System & Sensors Laboratory, Department of Electronics and Communication Engineering, Indian Institute of Technology, Roorkee, Uttarakhand, India

Shibnath Pathak Research and Development, Lumentum Holdings Inc., San Jose, CA, United States

Bhargav Raval School of Nanoscience, Central University of Gujarat, Gandhinagar, India

Berardi Sensale-Rodriguez The University of Utah, Salt Lake City, UT, United States

Vaishali Shukla School of Nanoscience, Central University of Gujarat, Gandhinagar, India

Man Singh School of Chemical Sciences, Central University of Gujarat, Gandhinagar, India

Sanjeet Kumar Sinha Department of Electrical Engineering, National Institute of Technology Silchar, Assam, India

Kuldeep C. Verma Department of Physics, Panjab University, Chandigarh, India

Xinbo Wang The University of Utah, Salt Lake City, UT, United States

Thomas Zimmer IMS Laboratory, University of Bordeaux, Talence, France

About the Author

Dr. Brajesh Kumar Kaushik received Doctorate of Philosophy (PhD) in 2007 from Indian Institute of Technology, Roorkee, India. He joined in the Department of Electronics and Communication Engineering, Indian Institute of Technology, Roorkee, as an Assistant Professor in December 2009; and since April 2014 he has been an Associate Professor. He has served as General Chair, Technical Chair, and Keynote Speaker of many reputed international and national conferences. He is a Senior Member of IEEE and member of many expert committees constituted by government and nongovernment organizations. He is Editor of *IEEE Transactions on Electron Devices*; Associate Editor of *IET Circuits, Devices & Systems*; Editor of *Microelectronics Journal*, Elsevier; Editorial Board Member of *Journal of Engineering, Design and Technology*, Emerald; and Editor of *Journal of Electrical and Electronics Engineering Research*, Academic Journals. He also holds the position of Editor-in-Chief of *International Journal of VLSI Design & Communication Systems*, and *SciFed Journal of Spintronics & Quantum Electronics*. He has received many awards and recognitions from the International Biographical Center (IBC), Cambridge, United Kingdom. His name has been listed in Marquis Who's Who in Science and Engineering® and Marquis Who's Who in the World®. He has been conferred with Distinguished Lecturer award of IEEE Electron Devices Society (EDS) to offer EDS Chapters with quality lectures in his research domain. His research interests are in the areas of high-speed interconnects, low-power VLSI design, memory design, carbon nanotube–based designs, organic electronics, FinFET device circuit codesign, electronic design automation (EDA), spintronics-based devices, circuits and computing, image processing, and optics and photonics–based devices.

Preface

The nanoelectronics area has grown consistently over last few decades at all levels of design and implementation, let it be at material, device, circuit, and architectural. This trend is expected to continue in future also, but will require breakthroughs in newer technologies such as photonics, plasmonics, spintronics, and other areas. With the emerging materials and nanodevices, efforts are being made by researchers across the world to gain better physical insight of the device, circuit, and system characteristics. This book is a collection of chapters from such leading researchers in the respective areas of nanoelectronics, photonics, plasmonics, spintronics, nanostructured material, and other relevant areas. The 13 chapters of the book are categorized under five sections that cover modeling and applications of electronic, magnetic, and optical devices in various circuits and systems. The book eloquently presents some of the emerging nanodevices and materials with respect to existing challenges and future demands. The purpose of the book is to provide thorough understanding of the emerging areas in nanotechnology and is intended to serve as a reference for scientists and engineers involved in the interdisciplinary fields of nanotechnology.

With the device dimensions of the order of nanometers, the devices can no longer be described by simple models. The first section of this book primarily enhances knowledge base of the readers in the domain of device modeling and applications. Chapter 1, Tunnel FET: Devices and Circuits, of the book deals with detailed physical insight of the tunnel field effect transistor (TFET) devices wherein quantum tunneling phenomenon plays a vital role in operation of these devices. TFET has emerged as one of the most promising postcomplementary metal-oxide-semiconductor (CMOS) transistor technology. The latter part of this chapter discusses the usage of TFETs for integrated circuit applications. The performance improvement in silicon technology by incorporation of germanium in emitter-base regions of transistors has opened up new avenues to the device designer to improve device performance. In Chapter 2, Electrothermal Characterization, TCAD Simulations, and Physical Modeling of Advanced SiGe HBTs, physical modeling of these devices has been discussed. Using technology computer-aided design (TCAD) simulations, performance of these devices is predicted for future applications. Due to the strong relationship between electrical and thermal effects in silicon-germanium (SiGe) heterojunction bipolar transistors (HBTs), electro-thermal characterization of these devices is necessary and is examined through advanced simulation approaches. Chapter 3, InP-Based High-Electron-Mobility

Transistors for High-Frequency Applications, presents high-frequency applications of indium-phosphide (InP)—based high electron mobility transistors (HEMT). These are considered as promising candidates for high-frequency applications. Chapter 4, Organic Transistors, discusses the emerging field of organic transistor, their model, structure, and prospective applications.

Electronics is entering a new era of spintronics, where spin current plays a central role in driving devices. The emerging spin transfer torque random access memory (STT-RAM) promises many attractive features that makes it to one of the most promising candidates for next-generation logic and memory circuits. The readers can explore the field of spintronics in the second section of this book. The reliability improvement techniques for random access memories based on spin-transfer torque effect are elucidated in Chapter 5, Mitigating Read Disturbance Errors in STT-RAM Caches by Using Data Compression. Emerging materials multiferroics and diluted magnetic semiconductors (DMS) have created opportunities for improved device performance for variety of applications. Chapter 6, Multifunctionality of Spintronic Materials, explores the application of spintronics using multiferroics and DMS materials.

The integration of photonics and electronics components on a chip promises to provide a compact photonic integration platform for variety of applications with faster and better performance. The third section of the book discusses the technology development and applications over last few years in the field of optics and photonics. Chapter 7, Photonics Integrated Circuits, discusses about the potential of next-generation integrated photonic circuits. The integration of silicon photonics demands CMOS compatible materials that can make integrated optical systems adaptable for their use across variety of applications. With its remarkable optical and electrical properties, graphene has been explored as a multifunctional optoelectronic material to achieve a wide range of optoelectronic devices. Chapter 8, Graphene-Based Optical Interconnects, discusses graphene-based integrated optical interconnect and their performance that is benefited from the optical interaction of light with graphene.

The research area of plasmonics integrates nanophotonics with electronics. Plasmonics plays a dominant role in confinement, guiding, and strong local enhancement of waves that may be otherwise difficult by other means. The recent advancement in photonics and nanofabrication technology creates an access to flourishing development of plasmon-assisted devices. The fourth section of the book covers advances in the field of plasmonics. Plasmon-induced hot carrier formation is attracting research interest due to its potential for applications in variety of fields. Chapter 9, Hot Carrier Generation in Plasmonic Nanostructures: Physics and Device Applications, covers comprehensive theoretical description of the hot carrier generation process in plasmonic nanostructures and discusses novel devices capable of fully utilizing this principle. Metamaterials demonstrate better control for propagation of electromagnetic waves and thus showcase improved physical properties. The combination of plasmonic and metamaterial can lead to promising platform for exploring new possibilities. Chapter 10, Plasmonic Metamaterial-Based RF-THz Integrated Circuits: Design and Analysis, presents design and analysis of RF-THz–integrated circuits with emerging plasmonic metamaterials.

The last section of the book provides various possibilities of the devices and circuits with some emerging materials and addresses the challenges that still need to be addressed. The realization of nanodevices with such materials has extensively improved the overall performance. This part of the book discusses applications of indium antimonide (InSb) and indium arsenide (InAs) nanowire–based FET in Chapter 11, Advances in InSb and InAs Nanowire-Based Nanoelectronic Field-Effect Transistors, and carbon nanotubes (CNTs)–based devices in Chapter 12, Carbon Nanotube and Nanowires for Future Semiconductor Devices Applications. Recently nanocomposites-based memory has been gaining tremendous attention due to their superior memory effect. The advantages may enable nanocomposites to serve as a storage media. Chapter 13, Role of Nanocomposites in Future Nanoelectronic Information Storage Devices, presents the investigations and developments carried out in memory devices based on nanocomposites in recent past.

This book is a unique conglomeration of topics covering major aspects of developing nanotechnologies and intends to serve the readers with the recent advancements in the field of nanoelectronics, photonics, and other upcoming areas.

Brajesh Kumar Kaushik
Roorkee, India

Acknowledgments

This book is a collective work of many researchers committed to the diversified fields of nanotechnology. I would like to acknowledge all the authors with gratitude for spending their precious time in creation of this edition.

I wish to express my heartfelt gratitude to one and all who directly or indirectly contributed for the completion of this book.

Brajesh Kumar Kaushik

PART I

Device Modeling and Applications

CHAPTER 1

Tunnel FET: Devices and Circuits

Prabhat Kumar Dubey

Department of Electronics and Communication Engineering, Indian Institute of Technology Roorkee, Roorkee, Uttarakhand, India

1.1 CMOS Power Trends

The idea of scaling the transistors dimension has resulted in a faster and increased integration density of a chip. However, high-density chips are suffering with high-power dissipation. Therefore, there is a need to reduce the power dissipation. The power consumption of the devices is increasing with the subsequent technology nodes (Fig. 1.1) [2]. There are two kinds of power dissipations in complementary metal oxide semiconductor (CMOS) circuit design: static power dissipation and dynamic power dissipation. Typically, the dynamic power consumption is higher than the static power dissipation. Nevertheless, the situation changes with scaling down of the metal oxide semiconductor field-effect transistor (MOSFET) size. It means, the static power increases with every new technology nodes. The increase in the static power with scaling of the gate length of the device is depicted in Fig. 1.1. It indicates that the leakage power dissipation has been growing with scaling of the technology node and exceeds the dynamic power dissipation below 65 nm technology node.

To maintain the power performance and a high speed of the device, the power supply as well as the threshold voltage must be scale down in proportion to the channel length. However, a decrease in threshold voltage increases the OFF current (I_{OFF}) that increases the static power dissipation of the device up to an unacceptable level. Therefore, scaling of threshold voltage of the MOSFET device is limited by the fundamental minimum *SS* of 60 mV/dec [3]. Fig. 1.2 indicates that it is not possible to scale down the threshold voltage in proportion to supply voltage that reduces the overdrive voltage. Consequently, scaling of the supply voltage out-of-proportion to the threshold voltage reduces the drive current of device that results in a slower speed. Therefore, there are two major consequences of the nonscalability of the threshold voltage of MOSFETs:

1. Supply voltage scaling slowed down due to the reduction in overdrive voltage (Table 1.1).
2. Scaling of transistor dimensions without scaling the supply voltage increases the electric field inside the channel that enhances the leakage *current.*

Nanoelectronics. DOI: https://doi.org/10.1016/B978-0-12-813353-8.00002-6

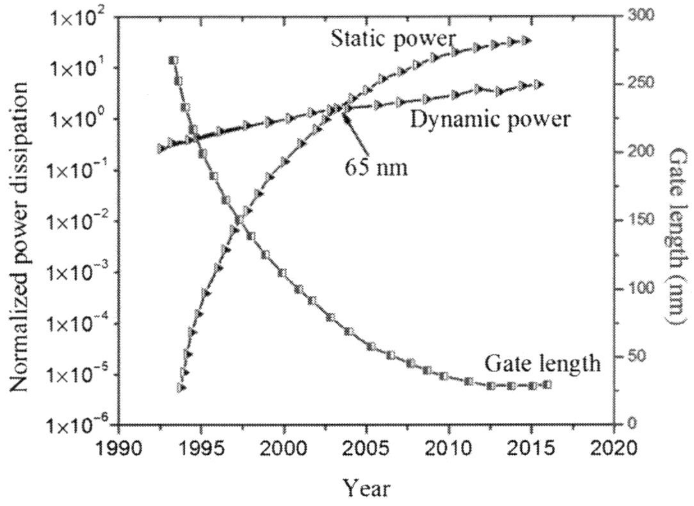

Figure 1.1
Power dissipation trend based on international technology roadmap for semiconductors [1].

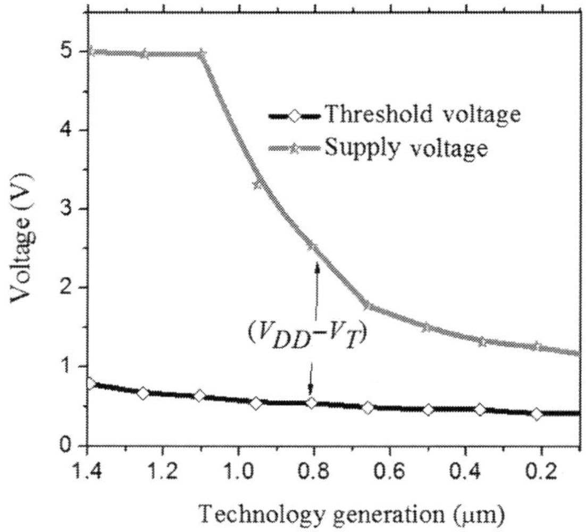

Figure 1.2
Supply voltage and threshold voltage scaling trends based on international technology roadmap for semiconductors [1].

Table 1.1: Scaling trend of supply voltage at different technology nodes [1].

Node (nm)	500	350	250	180	130	90	65	45
V_{DD} (V)	5.0	3.3	2.5	1.8	1.3	1.2	1.1	1.0

To overcome these problems, steep slope transistors attracted attention of device community researchers. A number of steep slope devices like impact ionization field-effect transistors (FETs), piezoelectric transistors, electromechanical FETs, feedback FETs, ferroelectric FETs, and tunneling field-effect transistors (TFETs) have been investigated. However, TFETs are the most promising device due to their MOSFET like device structures. TFETs work on the principle of band-to-band tunneling (BTBT) that is a quantum mechanical phenomenon. A brief description of quantum mechanical tunneling is presented in the next section.

1.2 Tunneling Phenomena

Tunneling of charge carriers through a potential barrier is a quantum mechanical phenomenon that occurs due to wave nature of the charge particle at small dimensions. Incidence of charge particle on a potential barrier of height more than the energy of the charge particle, it reflects or tunnel through it. The tunneling probability of the charge carrier depends on the potential barrier height, width, shape, and tunneling mass of the charge carrier [4]. The wave vector of the charge carrier is real in the incident and transmitted regions, but imaginary in the tunneling region. The imaginary wave vector in the tunneling region causes an exponential decay of the wave function amplitude and reduces the probability of transmission (Fig. 1.3). The tunneling of charge carriers is well understood in areas like tunneling of the electron across gate oxide, tunnel diodes, etc. This section focuses on the BTBT mechanism that describes the tunneling of the electrons from the valence band to the conduction band of a semiconductor.

The rate of BTBT across a potential barrier can be calculated using Kane's approach and Wentzel–Kramers–Brillouin (WKB) approach. Although, WKB approach is different from Kane's approach, it provides a similar result on simplification.

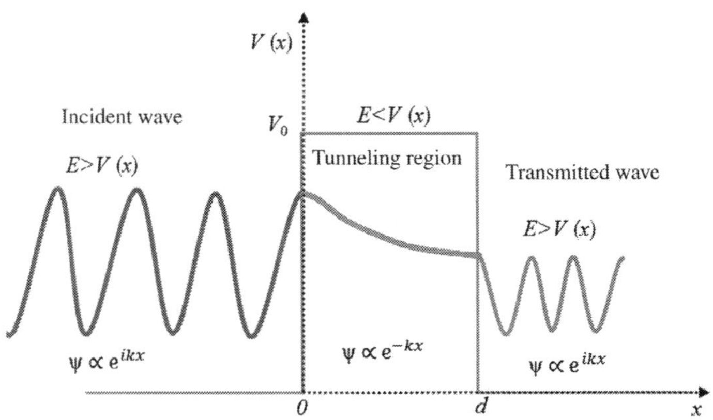

Figure 1.3
Tunneling through a rectangular potential barrier.

1.2.1 Kane's Formulation

Kane formulated the band-to-band generation rate ($cm^{-3}\,s^{-1}$) for the Zener tunneling from the valence band to the conduction band in a semiconductor [5]. The basic approach was based on the concept of the time-dependent perturbation theory, and Fermi's Golden rule to calculate the generation rate in the conduction band of semiconductors.

$$\text{i.e.} \quad G_{BTBT} = \frac{2\pi}{\hbar} \left| \langle \psi_1 | H | \psi_2 \rangle \right| \tag{1.1}$$

where the initial state $\langle \psi_1 |$ is the sum of the Bloch states in the valence band, final state $\langle \psi_2 |$is the sum of the Bloch states in the conduction band of the semiconductor, and H is the perturbation operator. Eq. (1.1) yields the expression of BTBT generation as follows:

$$G_{BTBT} = \frac{q^2 \sqrt{m^*} E^2}{18\ \pi \bar{h}^2 \sqrt{E_g}} \exp\left(-\frac{\pi \sqrt{m^*} E_g^{3/2}}{2\sqrt{2}\ q\ \hbar\ E} \right) \tag{1.2}$$

where q is unit electron charge, m^* is effective mass of the electron, E_g is effective energy band gap of the semiconductor, and E is electric field across the tunneling region. Eq. (1.2) indicates that the BTBT generation rate exponentially depends on the electric field.

1.2.2 WKB Approximation

The WKB approximation allows an alternative approach to calculate the BTBT generation rate. In this approach, a sum of all valence states with the momentum directed toward the tunneling barrier is considered for tunneling probability calculation. During the tunneling process, the energy and momentum of the charge carriers remain conserved. In one dimension, the tunneling probability can be expressed by the WKB approximation as [6]

$$T_{WKB} = \exp\left(2 \int_0^d \left| k(x) dx \right| \right) \tag{1.3}$$

where d is the tunneling barrier width, and $k(x)$ is the wave vector of the charge carrier. The wave vector $k(x)$ can be expressed as

$$k(x) = \sqrt{\frac{2m^*(E - V)}{\bar{h}^2}} \tag{1.4}$$

where E is the energy of the charge carrier, m^* is the effective mass of the electron, and V is the potential profile that can be given by the shape of the potential barrier.

1.3 Tunneling Field-Effect Transistors

The BTBT tunneling mechanism across a reverse biased junction realizes the steep slope current–voltage (I–V) characteristics. TFETs utilize this property of the reverse bias junction; it means current conduction in TFETs is based on the BTBT mechanism that improves SS. This makes TFETs a low-power device. Moreover, the process flow of TFETs is similar to MOSFETs and hence it can make the use of existing Si CMOS design and manufacturing infrastructure. This provides an opportunity to improve the integration density and performance of TFET-based integrated circuits (ICs). Furthermore, the structure of the TFET is similar to the MOSFET structure except the doping of source and drain is of opposite kind in TFETs. The schematic of an nTFET cross section and its band diagram in ON and OFF state is portrayed in Fig. 1.4. It consists of a highly doped p-type source, an intrinsic channel, and an n-type drain region [6].

A positive voltage at the drain terminal reverse biases the source–channel and drain–channel junctions. In the absence of gate voltage (V_{GS}), due to potential barrier neither thermionic nor band to band tunneling current flows. Therefore, the I_{OFF} of TFETs is very small that makes it an energy efficient device. In ON state, application of positive V_{GS} pushes the conduction band down in the channel region (Fig. 1.4B). It creates a tunneling path between source band and channel conduction bands. Therefore, TFETs operate on the principle of BTBT generation instead of the thermionic emission. Since the

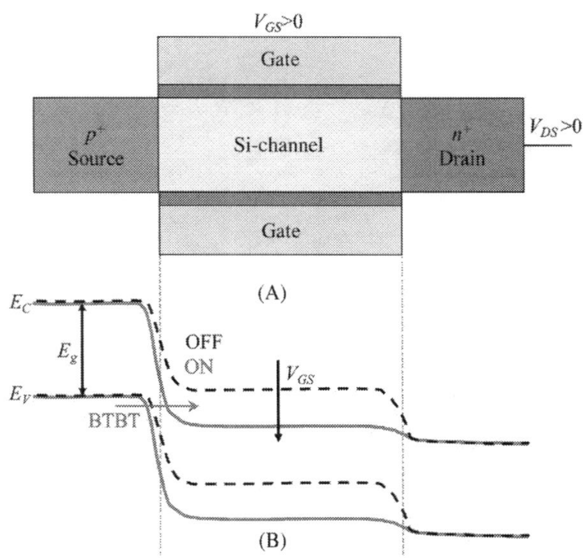

Figure 1.4
(A) Schematic of a nTFET cross section. (B) Band diagram of nTFET along the channel in ON and OFF states.

current flow in the TFETs is contributed by the valence band electrons of the source that contains lower energy, the TFETs are thermally cooled system and the *SS* of TFETs are not expected to be limited by the temperature [3].

1.3.1 Current–Voltage Characteristics

In the earlier sections, the steep slope property, device structure, and device physics of TFETs have been discussed. In this section, the *I–V* characteristics of the TFETs are described along with the physics. The *I–V* characteristics of a double gate TFET are depicted in Fig. 1.5. The transfer characteristics show a sub-60 *SS*. The value of drain current (I_{DS}) is much smaller than the expected to compete with the CMOS technology.

The application of positive V_{GS} pulls down the channel band profile that increases the charge carrier concentration in the channel region. Initially, no BTBT take place across the source–channel junction and the I_{DS} remains very small. However, as the V_{GS} pulls down the channel conduction band below the source valence band (Fig. 1.4B), BTBT starts across the source–channel junction and a sharp increase in I_{DS} is observed (Fig. 1.5B). Any further increase in V_{GS} continues pulling down the channel conduction band, until it aligns with the drain conduction band. After that, the channel conduction band does not show any significant shift with a further increase in V_{GS}, because the free charge carriers screen the effect of V_{GS} on the surface potential of the channel. Therefore, initially, the I_{DS} increases with V_{GS} sharply and then the effect of V_{GS}, on I_{DS}, reduces (see Fig. 1.5A) [7]. As a result, the *SS* of TFETs increases with V_{GS}.

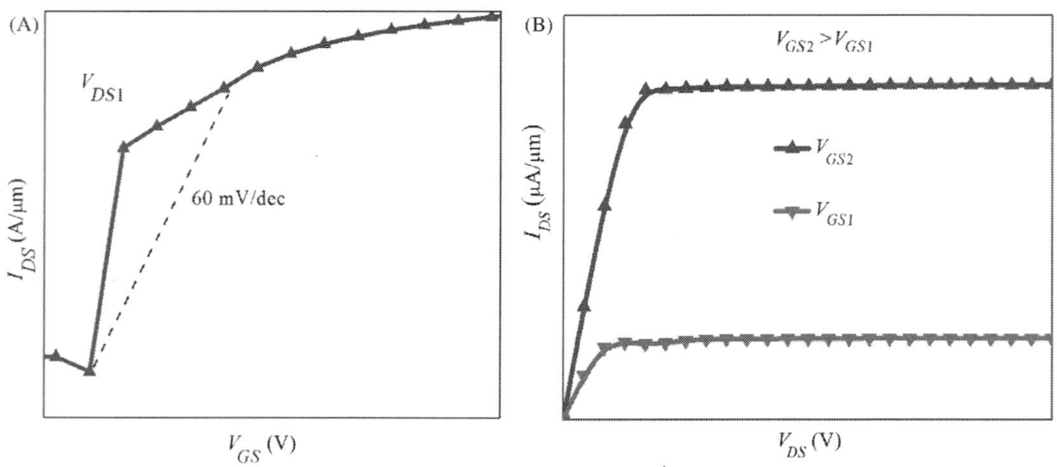

Figure 1.5
(A) Transfer characteristics of TFET. (B) Output characteristics of TFET.

An increase in the TFET drain voltage (V_{DS}) reduces the number of the free charge carriers in the channel. Consequently, it increases the channel potential profile and hence the electric field across the source/channel junction, at the same V_{GS}. Therefore, for small drain voltages, V_{DS} and V_{GS} both determine the tunnel junction status and hence I_{DS} increases with the V_{DS}. As the channel region becomes fully depleted, a further increase in V_{DS} does not make any change on surface potential and hence I_{DS} saturates. The saturation current in TFETs is determined by both V_{DS} and V_{GS}. Saturation current is high for higher V_{GS} [7].

1.3.2 Capacitance–Voltage Characteristics

This section discuss about the gate capacitance components in TFETs. The I_{ON} and switching speeds in ICs depend on the capacitances. An understanding of the gate capacitance components is important for TFET-based circuit design. Fig. 1.6 depicts the

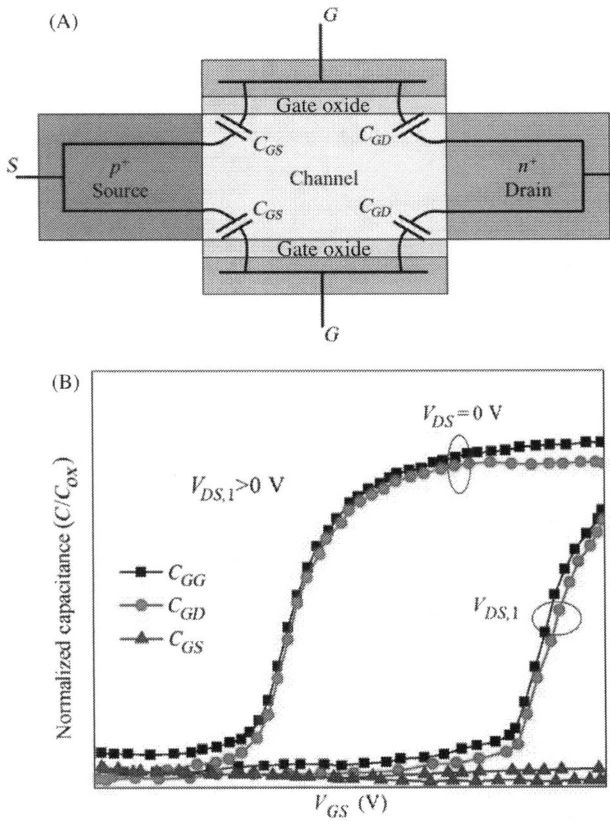

Figure 1.6

(A) Components of gate capacitances in TFETs. (B) Capacitance–voltage characteristics of TFETs, showing gate capacitance and its components as a function of the V_{GS}.

components of gate capacitance in TFETs and its capacitance–voltage characteristics [8]. Current conduction in TFETs is due to the BTBT mechanism across the source–channel junction. In OFF state ($V_{GS} = 0$ V, $V_{DS} = V_{DS,1}$), the tunneling probability is very low due to a wide depletion region at the source side junction that reduces the I_{OFF}. In ON state ($V_{GS} = 1$ V, $V_{DS} = V_{DS,1}$), the tunneling barrier thickness becomes small that allow the charge carriers to tunnel across the source–channel junction. Since the current conduction in TFETs is by the BTBT mechanism, the I_{ON} of TFETs is very low as compared to the MOSFETs. Therefore, in ON state ($V_{GS} = V_{DS} = 1$ V), a lower I_{ON} leads to reduced gate-to-source capacitance (C_{gs}) of TFETs [8]. Moreover, there is a very small potential drop at the drain–channel junction of TFETs as compared to the large reverse bias in MOSFETs (see Fig. 1.6B). This causes a comparatively larger gate-to-drain capacitance (C_{gd}) in TFETs [9]. It is interesting to note that the total gate capacitance C_{gg} is entirely dominated by the C_{gd} in TFETs in contrast to MOSFETs, where both C_{gs} and C_{gd} contribute significantly.

1.4 Challenges for TFETs

To outperform the CMOS technology, TFETs need to achieve the highI_{ON} (in the range of hundreds of mA), the lowest possible SS (<60 mV/dec), lowI_{OFF}, and supply voltage less than 0.5 V, simultaneously. However, TFETs are facing the problem of the lower I_{ON} due to BTBT mechanism of current conduction. A number of techniques have been proposed to improve the I_{ON} of the TFETs including different device structures as well as material systems. Other challenges for TFETs include its ambipolar behavior and high Millar capacitance.

1.4.1 ON Current Performance Boosters

As discussed earlier the I_{ON} of the TFETs is low due to BTBT current conduction mechanism. However, it can be improved in a number of ways like area scaled TFET design techniques, use of high-*k* dielectric gate oxides, using III–V semiconductors, and using the heterojunction-based TFET (HTFET) structures. Some important techniques to enhance I_{ON} is presented as follows.

1.4.1.1 High-k gate dielectric

The capacitance coupling between the gate and tunnel junction can be determined by the gate dielectric. High-*k* gate dielectric leads a better control of the gate on TFETs that improves the SS as well as I_{ON} of the device. The thickness of the gate dielectric affects the I_{ON} of the device with scaling of the gate oxide thickness results in an increase in I_{ON} [10]. In addition to improvement in I_{ON}, the scaling of gate oxide leads better control of the gate on turning OFF the device [10]. However, the scaling of the gate oxide thickness can increase the gate leakage current. Another drawback of high-*k* gate dielectric is that it can also create the defects at the high-*k* dielectric/semiconductor interface, if high-*k* gate dielectric is directly deposited on the semiconductor. These defects can degrade the *SS* of the device and can lead to trap assisted tunneling that shows a higher temperature dependence of the *SS*. To reduce the interface defects, an interfacial layer with better oxide/semiconductor interface is required between high-*k* dielectric and semiconductor. For example, Al_2O_3/HfO_2 bilayer gate oxide in InGaAs TFETs improves the *SS* in comparison to single HfO_2 gate oxide.

1.4.1.2 Area scaled devices

In conventional TFETs, the tunneling occurs within 1–2 nm of the inversion layer. The basic problem is that the tunneling area cannot be increased by changing the technology or process parameter. Therefore, a new class of TFETs, called as area scaled TFETs, have been proposed that resolve the problem of low I_{ON} by allowing a wider tunneling area [11,12]. Fig. 1.7 presents the area scaled device structure. In area scaled devices, the improvement in I_{ON} is due to an increase in the tunneling cross section area of the device.

The current conduction in an area scaled TFET is by the vertical/line tunneling mechanism. In line tunneling, tunneling occurs in a direction parallel to the line of the electric force. In conventional TFETs the current conduction is due to the point tunneling (tunneling in a direction perpendicular to the line of electric force) mechanism. In addition to improvement in I_{ON}, an area scaled TFET also leads to an improvement in the *SS* due to a sharp increase in I_{ON}.

1.4.1.3 III–V HTFETs

The I_{ON} of TFETs can also be improved by material system engineering. Si is an indirect band semiconductor that has a band gap of 1.12 eV. The indirect tunneling mechanism and a larger potential barrier height of Si material reduce the tunneling probability and hence I_{ON} in Si-based TFETs. Therefore, direct band gap materials like III–V semiconductors can improve the tunneling probability at the source–channel junctions. III–V semiconductors

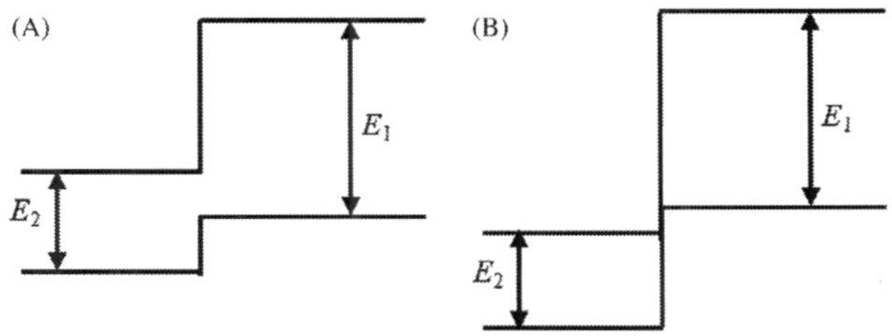

Figure 1.7
(A) A conventional TFET structure. (B) Area scaled TFET structure.

offer a wide range of materials for band gap engineering. A small band gap material can increase the I_{ON}, but it also increases the I_{OFF} that degrades the *SS* and ambipolar behavior of the device. These problems can be addressed by the application of heterojunction in TFETs. Heterojunction allows the band gap engineering in TFETs by using a combination of low band gap and high band gap materials. The use of a low band gap source material enhances the I_{ON} while the high band gap material in the channel and the drain region reduces the I_{OFF} and ambipolar current.

For example, GaSb/InAs HTFET shows a better I_{ON} in comparison to InAs and GaSb homojunction transistors [13]. One possible way to further increase the I_{ON} of HTFET without degrading its *SS* is to make the use of ternary alloys. This idea is based on the fact that in $Al_{x}Ga_{(1-x)}Sb$, the valance band edge varies linearly with change in the molar fraction, without much significant change in lattice constant [14]. Hence, the molar fraction grading of $Al_{x}Ga_{(1-x)}Sb$ can be used in the source region to minimize the source depletion region thickness [15]. The main problem of the molar-fraction-based device is the practical realization of the steep molar fraction grading. Therefore, there is a need of an alternate technique to increase I_{ON} of GaSb/InAs HTFETs. Another method to increase the I_{ON} without degrading the *SS* is the formation of GaSb quantum well in the InAs source region [16]. The operating principle of using the GaSb quantum well in InAs source is to create a significant hole density of state (DOS) at the source edge adjacent to the channel region. This enhancement in DOS helps in transmission of electron wave function in the channel region, and hence, increases the I_{ON}.

An $In_{0.53}Ga_{0.47}As$/InP heterojunction is also a very interesting material system for TFET application. The motivation behind $In_{0.53}Ga_{0.47}As$/InP HTFET is that the $In_{0.53}Ga_{0.47}As$ with a narrow energy band gap (0.73 eV) at source side increases the BTBT probability and hence I_{ON}. Moreover, InP with a wide band gap (1.35 eV) in the drain side of the channel reduces the ambipolar and I_{OFF} in the TFETs [17]. Another motivation toward this material system lies in the fact that InP is a material widely used in optical systems; application of InP in TFET can converge the electronics and photonics on a single platform. To further improve the I_{ON} without degrading the *SS* and ambipolar characteristics, $In_{0.53}Ga_{0.47}As$/$In_{0.53}Ga_{0.47}As$/InP/InP material system can be used. In this material system further insertion of $In_{0.53}Ga_{0.47}As$ at the source side channel reduces the tunneling path, and hence, enhances the I_{ON}.

Although the heterojunctions improve the performance of the TFETs, two types of heterojunctions are of specific importance: the staggered gap and the broken gap (Fig. 1.8). Under bias conditions, the staggered gap heterojunctions provide a reduced tunneling length as compared to homojunctions that leads to a better performance of HTFETs. For the broken-gap-type heterojunctions, there is no potential barrier region to pass through that offers a high transmission probability. However, the I_{OFF}

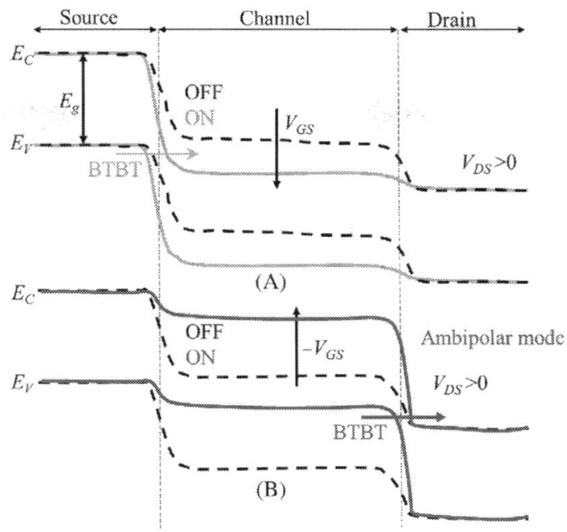

Figure 1.8
(A) Staggered band alignment heterojunction. (B) Broken band alignment heterojunction.

in broken gap HTFETs is comparatively high due to the absence of a significant potential barrier region.

The drawback of III–V semiconductors is its lower DOS that reduces the tunneling probability and hence I_{ON} of the device. Another drawback of III–V semiconductor is the difference in DOS of the conduction band and valence band [18]. This causes a difference in performance of *p*TFETs and *n*TFETs that creates a problem for complementary digital logic design. The low DOS of the conduction band in III–V semiconductor results in a large source doping degeneracy for *p*TFET. Heavily doping the semiconductor to increase I_{ON} exposes the exponential tail of Fermi–Dirac distribution that increases the *SS* value of TFET. Therefore, the optimum doping of *p*TFET is lower than *n*TFET. This difference in *p*TFET and *n*TFET causes problem in complementary implementation of TFET circuits. The n–n^+–p^+–i–*p*TFET structure can block the injection of a hot carrier from the source to channel by maintaining a high electric field at tunnel junction [19]. However, light doping *n*-region of the source can result in an increase in series resistance at the source side. An HTFET can also improve the performance of *p*TFETs, since hetero source material has a large conduction band DOS like silicon.

1.4.2 Ambipolarity

Symmetric TFETs shows ambipolar device characteristic [20]. It demonstrates *n*-type behavior with the electron as a dominant charge carrier for positive V_{GS}, and *p*-type

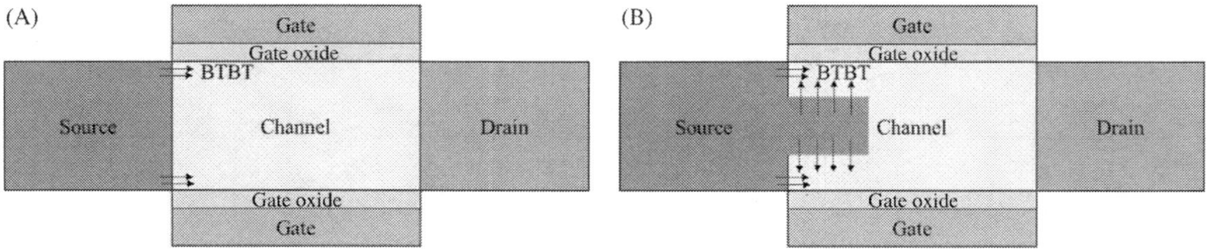

Figure 1.9
Band diagram of TFETs in ON, OFF, and ambipolar states.

behavior with hole as a dominant charge carrier for negative V_{GS}. For *p*-type behavior, the *n*-type drain works as a source region and *p*-type source behaves as a drain region (Fig. 1.9). As a result, in OFF state, the band diagram of a *p*TFET is similar to *n*TFET that suppress the BTBT and results in a small value of I_{OFF}. Application of negative V_{GS} uplifts the band in the channel region (Fig. 1.9). A BTBT starts as soon as the channel valence band (E_V) is lifted up to the conduction band (E_C) of the *n*-type drain region. This results in BTBT of the channel valence band electron in the drain conduction band region. As a result. there is current conduction in symmetric TFETs for both positive and negative V_{GS}. The ambipolarity of the TFETs is a major drawback in complimentary digital logic design using TFETs [20]. To overcome the problem of the ambipolarity, different device structure and device design technique have been proposed.

One method to reduce the ambipolar behavior of TFETs is asymmetric source–drain doping concentration. In this method, the doping concentration of the drain is kept lower than the source doping concentration that widens the barrier thickness at the drain–channel junction. It reduces the BTBT at the drain–channel junction for negative V_{GS}, as the BTBT reduces exponentially with an increase in tunneling barrier width. On the other hand, the I_{ON} remains unchanged due to a reduction in drain doping concentration, as it is controlled by the electron tunneling at the source–channel junction.

Other techniques to reduce the ambipolar current are gate–drain underlap structure [21], gate–drain overlap structure [22], hetero-gate-dielectric-based TFETs [23], etc. The I_{ON} of the TFETs depends on the source doping concentration and electron concentration of the channel near the source region. Therefore, it is possible to design a TFET with a resistance in the channel region near the drain without affecting the I_{ON} of device. The high resistance region near the drain can be created by the gate–drain underlap structure that can suppress the ambipolar current effectively. The gate–drain underlap structure reduces the impact of the gate on drain–channel junction and hence reduces ambipolar current. Similarly, in gate–drain overlap structure, application of negative V_{GS} depletes the *n*-type drain region near drain–channel junction that reduces the band bending at the junction for negative V_{GS} reducing the ambipolar current.

1.5 TFET Characteristics and Impact on the Circuit Design

In this section the challenges and advantages of TFETs from circuit design prospective has been discussed. On one hand, the TFET properties like a lower *SS*, lower threshold voltage, and low supply voltage operation provide an opportunity for low-power TFET-based circuit design. On the other hand, unidirectional conduction, delayed saturation, and a higher value of C_{gd} are the major challenges for TFET-based circuit designs. Therefore, the impact of properties of TFETs, like unidirectional conduction, delayed saturation, and higher Miller capacitances, on functionality of the circuits has been described.

1.5.1 Unidirectional Conduction

Unlike MOSFETs, the doping in the source and drain of TFETs is of the opposite kind that makes it operate as a reverse biased gated $p-i-n$ diode. This causes unidirectional current conduction characteristics of TFETs, in contrast to symmetric source–drain MOSFETs. Consequently, the device source–drain orientation needs to be addressed in TFET-based circuit design that creates complexity. The design of the pass transistor logic using a single transistor is not possible as it requires bidirectional conduction. Therefore, additional transistors are required to design pass transistor logic using TFETs. However, in some circuits (e.g., TFET radio frequency (RF) rectifiers) the unidirectional current conduction simultaneously improves the reverse leakage current and the power loss [24].

Fig. 1.10 shows a track-and-hold circuit. A simple track-and-hold circuit consists of a transistor switch and a holding capacitance. In high state of the clock signal, the switch remains in ON state and output follows (tracks) the input signal. As the clock signal becomes low, the switch turns OFF and the output holds a constant voltage. The issue with a single *n*TFET track-and-hold circuit is its asymmetric conduction. For example, suppose the output holds a voltage V_{hold} and the clock signal goes high, and if a new input signal (V_{input}) to be sampled is smaller than the V_{hold}, then the V_{DS} of the TFET will be negative and it will not conduct to track the input signal. Therefore, the single

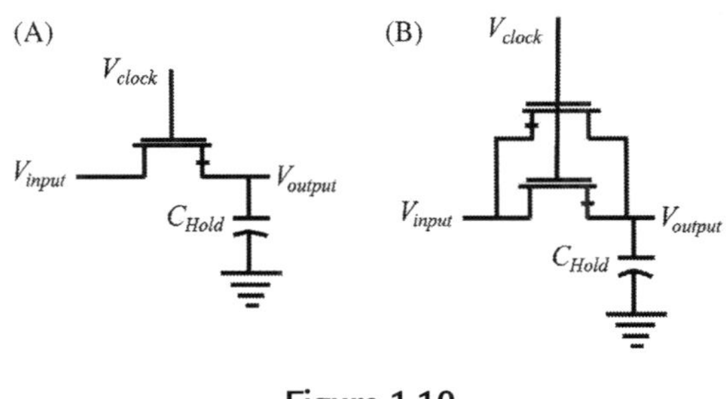

Figure 1.10
Track-and-hold circuit (A) using single *n*TFET and (B) using a pair of *n*TFET.

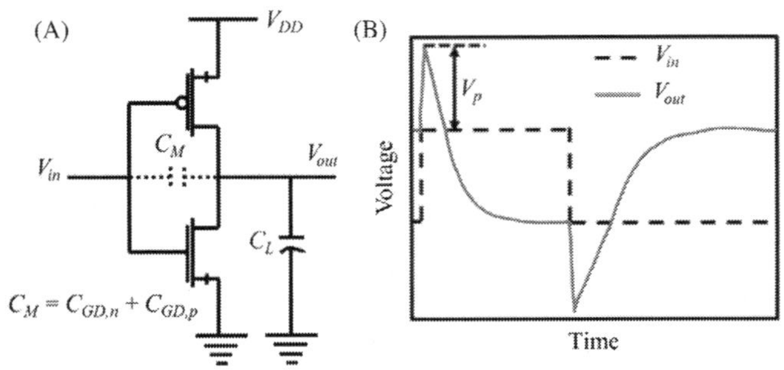

Figure 1.11
(A) A schematic of TFET-based inverter showing the effects of the Miller capacitance. (B) Effect of the Miller capacitance on the output waveform of the inverter.

*n*TFET-based T/H circuit will not work properly. To address this problem, a double *n*TFET transistor-based T/H circuit is required (Fig. 1.10B). The double *n*TFET combination of Fig. 1.10B makes a symmetric switch that ensures that the charging and discharging currents are similar [24].

1.5.2 Enhanced ON-State Miller Capacitance

As discussed in the previous sections, the distribution of total gate capacitance of TFETs is different from MOSFETs. Unlike MOSFETs, the C_{gd} of TFETs controls the total gate capacitance. This effect is known as the enhanced ON-state Miller capacitance effect. In ON-state MOSFETs, the contribution of C_{gs} to the C_{gg} is very large as compared to C_{gd}. In ON-state TFETs, however, due to an asymmetric charge distribution between the source–drain and presence of the source–channel barrier at the source side, the C_{gd} dominates the C_{gg}. This enhanced capacitive coupling between the gate and drain increases the Miller capacitance of TFETs. The enhanced Millar capacitance in TFETs results in a voltage "spike" during switching of the inverter (Fig. 1.11). This causes an increase in the circuit stabilizing time, delay, and dynamic power consumption during switching. Nevertheless, in static random-access memory (SRAM) design, higher Millar capacitances can improve the coupling between the data storage nodes and the support node recovery due to radiation that improves the SRAM soft error resilience [25]. The effect of enhanced Miller capacitance in TFETs can be reduced by (1) application of the large load capacitance, (2) using a low DOS semiconductor materials for TFETs, and (3) device engineering to reduce the coupling between the gate and drain.

1.6 Tunnel FET SRAM Design

Different techniques have been proposed to enable a low-voltage operation of CMOS digital logic circuits. A further requirement of the low-voltage operation of logic circuits led the subthreshold operation of the MOSFETs that increase the sensitivity of circuit parameters on device level variations. Moreover, it causes an exponential increase in delay, and leakage energy dominates the circuit operation. With the current scaling trend, SRAM bit cells can be prone to the additional sensitivity of transistors. Therefore, there is a need of robust SRAM circuit design. As TFETs are the suitable device for low-power operations, it can be used as an alternative for subthreshold MOSFET-based designs. This approach offers a more robust low-power SRAM design.

1.6.1 6T TFET SRAM Cell

A 6T SRAM consists of two cross-coupled inverter structures that are used to store the information. It consists of two pass gate transistors to access the information data storage point. In read operation, pre-charged bit line (*BL*)/bit bar line $(\overline{BL})$ start discharging through pass transistor according to information stored at Q. On the other hand, in write operation, one of the bit lines $(BL/\overline{BL})$ is pulled down to discharge the corresponding data node, according to the data to be written. Therefore, for a 6T SRAM structure, two cross-coupled robust inverters and two pass transistors that can conduct in both directions are required.

The 6T TFET SRAM structure (Fig. 1.12) suffers due to some basic properties of the TFETs. The TFETs are the devices that have very small threshold voltage. Therefore, it shows a delayed saturation in output characteristics. The delayed saturation of the TFET characteristics degrades the voltage transfer characteristics of the inverter and hence affects the performance of the SRAM. The unidirectional conduction of TFETs allows only read/write operation in a 6T TFET SRAM circuit (Fig. 1.12). As a result, the conventional 6T TFET SRAM cell needs modification to achieve both read and write operation.

1.6.2 8T TFET SRAM Cell

As discussed in the previous section that 6T TFET SRAM cell cannot perform both read and write operation simultaneously. To overcome this problem, an 8T TFET SRAM cell is proposed. An 8T TFET SRAM cell consists of two cross-coupled inverters along with four pass transistors to access data from data node (Fig. 1.12B). A pair of two pass transistors is connected in parallel to each other, such that they allow bidirectional current conduction and hence both read and write operation. The read operation is performed by enabling inward-facing TFET transistor using a word line (*WL*). Inward-facing TFET pass transistors discharge the bit lines according to the data stored at the data node that allow the reading operation. Both word lines needs to be enabled during write operation. The outward-facing transistor pulls down the data node to zero and the inward-facing transistor assist in writing by raising the voltage of the complementary data node. At low V_{DD}, a significant reduction

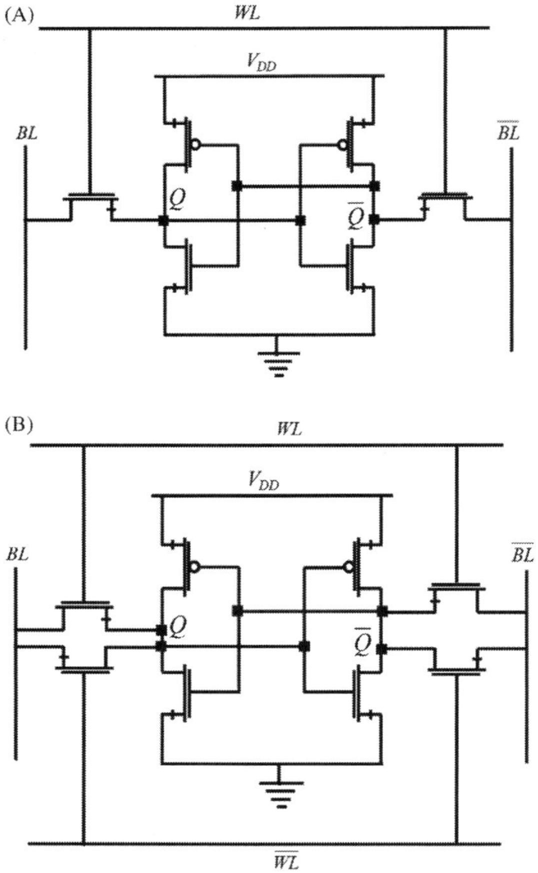

Figure 1.12
(A) 6T TFET SRAM cell. (B) 8T TFET SRAM cell.

in delay and dynamic power consumption can be achieved in TFET-based SRAM arrays as compared to MOSFET-based arrays [25]. It is due to steep switching of TFETs compared to subthreshold MOSFET operation.

1.7 TFET Analog/RF Application

Although, in digital applications, the scaling of transistors follows the Moore's law, in analog/RF design the scaling is not important and a mature technology is in use. It is due to the trade-offs in matching the performance parameters like gain, bandwidth, linearity, noise, and frequency for different analog applications. As the demand is increasing for a portable and battery-less energy harvesting systems, subthreshold CMOS-based ultra-low-power analog/RF circuit designs have been suggested for energy efficient applications.

However, the subthreshold CMOS design is still facing challenges due to trade-offs among speed, signal-to-noise ratio (SNR), robustness, and precision. In this section, analog/RF figure of merits (FOMs) of TFETs have been described for better understanding of the device performance in ultra-low-power applications [26].

Analog/RF FOMs can be expressed in terms of maximum oscillation frequency (f_{max}), cutoff frequency (f_T), transconductance (g_m), transconductance generation factor (TGF), and transconductance frequency product (TFP). These parameters can be determined by using a small signal at the input port and observing the voltage and current at the output port. Among these, g_m is an important parameter and plays an important role in achieving higher f_T as well as higher dc gain. The parameter g_m can be defined as change in the I_{DS} with respect to a change in V_{GS} at a constant V_{DS} (i.e., $g_m = \partial I_{DS}/\partial V_{GS}$). It is an important factor in analyzing the gain of the device and plays a vital role in analog/RF circuit design [27]. The optimum bias voltage for g_m should be high enough to accomplish a better high-frequency performance. In the subthreshold region the transconductance can be expressed as

$$g_m = \frac{\log(10) I_{DS}}{SS} \tag{1.5}$$

Unlike MOSFETs, the *SS* of the TFETs changes with an increase in I_{DS}. Therefore, g_m is very small for lower gate bias, and increases slowly with gate bias until the saturation region because of improvement in the current driving capability. Nevertheless, it decreases at higher gate bias because of the mobility degradations (or due to increase in *SS*). A high value of g_m ensures higher amplification and hence it can be preferred for analog applications [27]. As there in no channel length modulation at the drain terminal, the I_{DS} of TFETs is determined by the BTBT at the source–channel junction with a small dependence on the V_{DS}. Therefore, the output resistance (r_0) remains higher than the CMOS. The higher output resistance of TFETs leads to a higher intrinsic small signal voltage gain ($A_V = g_m r_0$).

Another important analog/RF parameter to check the RF performance of a device is cutoff frequency, defined as a frequency at which current gain becomes unity, and is given by $f_T = g_m/[2\pi(C_{gs} + C_{gd})]$. Cutoff frequency increases with gate bias due to an increase in g_m. After a certain V_{GS}, it falls with gate bias because of the combined effect of C_{gs} and C_{gd}, and limitation of g_m due to the mobility degradation. The maximum frequency of oscillation (f_{max}) is another key parameter that is defined as a frequency of unity power gain. The f_{max} corresponds to the transit frequency of the maximum available gain. It is a realistic parameter for the optimization of amplifiers and is given by

$$f_{\text{max}} = \sqrt{\frac{f_T}{8\pi R_{gd} C_{gd}}} \tag{1.6}$$

It is clear from Eq. (1.6) that $f_{\max}$ mainly depends on f_T and gate-to-drain parasitic capacitance (C_{gd}). C_{gd} is a dominant parameter for $f_{\max}$ analysis. It is notable that both the f_T and $f_{\max}$ depend on the geometrical parameters of the device. Furthermore, the parasitic capacitances need to be very small to achieve a higher f_T and $f_{\max}$ for fulfilling the desired RF requirements. Another important device design parameter is the TGF or transconductance to current ratio (g_m/I_{DS}) that is a measure of efficiency to convert current (power) into transconductance (speed) [27]. The maximum TGF can be obtained at small gate bias, and decreases as the V_{GS} increases. It is because of very small change in the I_{DS} in the saturation region (TGF is discussed in Section 1.7.1 in detail). The TFP is also an important FOM for high-frequency analysis, where TFP is basically the product of TGF and f_T, given by [27]

$$TFP = \left(\frac{g_m}{I_D}\right) f_T \tag{1.7}$$

It indicates a trade-off between bandwidth and power which is useful for moderate- to high-speed designs.

1.7.1 Transconductance Generation Factor (g_m/I_{DS})

TGF or transconductance to current ratio (g_m/I_{DS}) is an important parameter for low-power analog circuit designs. In MOSFETs, the highest value of g_m/I_{DS} can be calculated at the device subthreshold region as [25]

$$g_m = \frac{I_{D0}}{nV_T} \exp\left(\frac{V_{GS} - V_T}{nV_T}\right)\left(1 - \exp\left(\frac{-V_{DS}}{V_T}\right)\right) \tag{1.8}$$

$$\text{Or,} \quad \frac{g_m}{I_{DS}} \sim \frac{1}{nV_T} < \frac{1}{26\ \text{mV}} = 40\ \text{V}^{-1} \tag{1.9}$$

As a MOSFET can achieve a minimum SS of kT/q, the TGF of MOSFETs is limited to an upper limit of 40 V^{-1} at 300 K. In steep switching devices like TFETs, g_m/I_{DS} can overcome this limit:

$$\frac{g_m}{I_{DS}} = \frac{\partial I_{DS}}{\partial V_{GS}} \frac{1}{I_{DS}} = \frac{\partial \ln I_{DS}}{\partial V_{GS}} = \frac{\ln 10 \partial \log I_{DS}}{\partial V_{GS}} = \frac{\ln 10}{SS} \tag{1.10}$$

A trend of the g_m/I_{DS} versus I_{DS} curves of a TFET and MOSFET is depicted in Fig. 1.13. It can be observed that TFETs show an improvement in g_m/I_{DS} at low I_{DS} [25]. This offers a more energy efficient and ultra-low-power analog circuit design using TFETs. For instance, in an analog-to-digital converter (ADC) design, the power dissipation per sampling

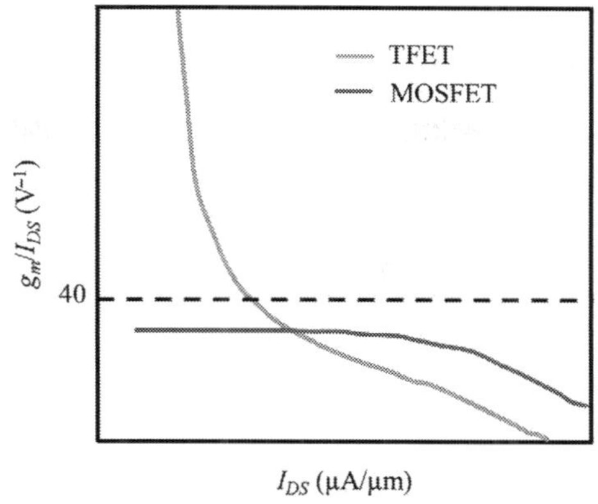

Fig. 1.13
g_m/I_{DS} versus I_{DS} comparison of TFET and MOSFET.

frequency is bounded by the g_m/I_{DS}, SNR, and V_{DD}. The SNR requirement limits the lower V_{DD} design and energy efficiency of ADCs [25]:

$$\frac{Power}{Sampling\,frequency} \sim \frac{kT \cdot SNR}{V_{CC}} \left(\frac{g_m}{I_{DS}}\right)^{-1} \tag{1.11}$$

Therefore, for a high g_m/I_{DS} in TFET, the V_{DD} can be scaled down by keeping the SNR ratio fix. This reduces the energy per conversion step of a low-power ADC design.

1.7.2 Linearity Performance

A linear system, with input $x(t)$ and output $y(t)$, has input–output relationship expressed as

$$y(t) = kx(t) \tag{1.12}$$

where k is a constant. In modern communication systems, a minimum distortion level is required for high-speed analog and RF applications. To achieve high linearity, transconductance should be constant over the desired input voltage. However, the transconductance of both MOSFET and TFET changes with the input voltage that represents nonlinear behavior. The linearity of an analog circuit can be determined by

$$g_m' = \frac{\partial^2 I_{DS}}{\partial V_{GS}^2} \tag{1.13}$$

$$g_m' = \frac{\partial^3 I_{DS}}{\partial V_{GS}^3} \tag{1.14}$$

The subthreshold operation increases the nonlinearity of the circuits due to a larger value of g'_m and g''_m. However, in a certain window of V_{GS}, TFETs can provide substantially lower nonlinearity in comparison to MOSFETs.

1.8 TFET-Based OTA

An operational transconductance amplifier (OTA) is an important component for low-power analog circuit designs such as neural amplifiers, a cellular neural network, $G_m - C$ filter synthesis, etc. In this section, application of TFET in OTA design has been described. The circuit diagram of a TFET OTA is depicted in Fig. 1.14. Transistors T_1 and T_2 generate the differential current in the circuit that is proportional to the differential input voltages. Current mirrors $T_3 - T_5$ and $T_4 - T_6$ are used to mirror the differential currents. These mirrored currents are added at the branch $T_6 - T_8$ and lead to a total load current. The improved output resistance of TFETs avoids cascading of the transistor at the output stage.

The performance parameters of the TFET-based OTA are described for a varying power conditions, and a comparison is made with equivalent MOSFET-OTA design. Power variation in an OTA design can be performed by changing the supply voltage (V_{DD}) or bias current (I_{OTA}). As in a system other analog components like biasing elements, floating switches, etc. limit the voltage scaling [26], the I_{OTA} is varied at a fixed V_{DD}.

The total transconductance of the OTA (G_m) can be given as [26]

$$G_m = g_{T1} M_R \tag{1.15}$$

where g_{T1} is the transconductance of the transistor T_1 (or T_2), and M_R is the current mirror ratio of the mirrors $T_3 - T_5$ (or $T_4 - T_6$) [26].

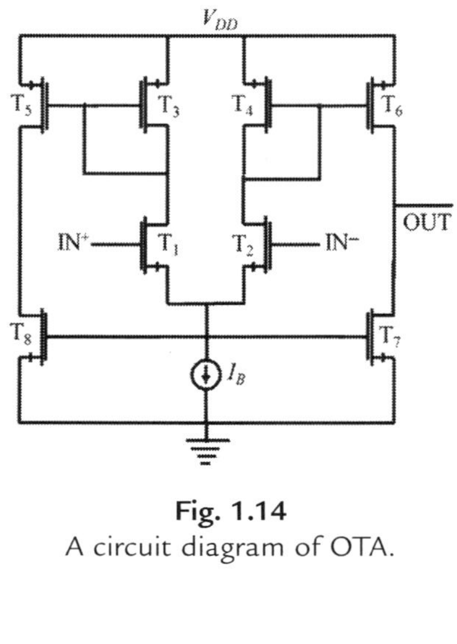

Fig. 1.14
A circuit diagram of OTA.

$$\text{Therefore,} \quad \frac{G_m}{I_{OTA}} = \frac{G_m}{I_B(M_R + 1)} = \frac{g_{T1}}{I_B}\frac{M_R}{M_R + 1} \tag{1.16}$$

As we reduce I_B, the dc gate–source bias of the transistor T_1 and T_2 reduces; as a result, it enhances the *SS* and g_{T1}/I_{DS} of the OTA. It can be observed that the value of G_m/I_{OTA} for a TFET-based OTA increases at lower I_B. In contrast, for a MOSFET-based OTA in the subthreshold region can be expressed as

$$\frac{g_{T1}}{I_{DS}} = \frac{1}{n}\frac{q}{kT} \tag{1.17}$$

where T is the temperature, q is the unit electronic charge, k is the Boltzmann constant, and n is a parameter related to its *SS*. Therefore, as we scale the I_B of MOSFET-based OTA, the G_m/I_{OTA} remains comparatively invariant. However, this ratio degrades with an increase in temperature according to Eq. (1.17). At very small bias currents, the drain induced barrier lowering does not allow the currents of branch T_5–T_7 and T_6–T_8 to follow the mirror ratio. Consequently, a MOSFET-based OTA design results in a higher leakage, and G_m/I_{OTA} ratio drops. Therefore, TFET-based design can operate in the ultra-low operating regime beyond achievable by the MOSFETs. However, at high-power conditions, MOSFETs outperform TFETs when TFET *SS* becomes higher than the MOSFET.

1.9 Summary

TFETs are introduced in this chapter. A brief description of the tunneling mechanism along with the tunneling rate calculation approach is presented. The operation of TFETs is described by explaining the current–voltage as well as capacitance–voltage characteristics. Issues associated with the performance of TFETs like ambipolarity and low I_{ON} are discussed, and techniques to resolve these problems like gate–drain underlap structure, gate–drain overlap structure, hetero-gate-dielectrics structure, and area scaled TFET structures are described. A detailed description on III–V TFETs are presented; it is demonstrated that III–V TFETs can resolve the low I_{ON} limitation. Moreover, a TFET-based circuit design approach is introduced and techniques to address the challenges related to delayed saturation, enhanced Miller capacitance, and unidirectional conduction are presented. Finally, TFET-based SRAM and OTA circuits are designed and different aspects of their performance parameters are discussed.

Acknowledgment

Part of the work presented in this chapter is supported by the INSPIRE program of Department of Science and Technology, Government of India.

References

[1] International Technology Roadmap for Semiconductors. <http://public.itrs.net>, 2006.

[2] J. Xue, T. Li, Y. Deng, Z. Yu, Full-chip leakage analysis for 65 nm CMOS technology and beyond, Integration VLSI J. 43 (2010) 353–364.

[3] A.M. Ionescu, H. Riel, Tunnel field-effect transistors as energy efficient electronic switches, Nature 479 (7373) (2011) 329–337.

[4] D.J. Griffiths, Introduction to Quantum Mechanics, Univ. Sci. Books, Sausalito, CA, 1995.

[5] E.O. Kane, Theory of tunneling, J. Appl. Phys. 32 (1) (1961) 83–91.

[6] A.C. Seabaugh, Q. Zhang, Low-voltage tunnel transistors for beyond CMOS logic, Proc. IEEE 98 (12) (2010) 2095–2110.

[7] A. Mallik, A. Chattopadhyay, Drain-dependence of tunnel field-effect transistor characteristics: the role of the channel, IEEE Trans. Electron Devices 58 (12) (2011) 4250–4257.

[8] Y. Yang, X. Tong, L.-T. Yang, P.-F. Guo, L. Fan, Y.-C. Yeo, Tunneling field-effect transistor: capacitance components and modeling, IEEE Electron Device Lett 31 (7) (2010) 752–754.

[9] S. Mookerjea, R. Krishnan, S. Datta, V. Narayanan, Effective capacitance and drive current for tunnel FET (TFET) CV/I estimation, IEEE Trans. Electron Devices 56 (9) (2009) 2092–2098.

[10] K. Boucart, A.M. Ionescu, Double-gate tunnel FET with high-k gate dielectric, IEEE Trans. Electron Devices 54 (7) (2007) 1725–1733.

[11] A. Rajoriya, M. Shrivastava, H. Gossner, T. Schulz, V.R. Rao, Sub 0.5 V operation of performance driven mobile systems based on area scaled tunnel FET devices, IEEE Trans. Electron Devices 60 (8) (2013) 2626–2633.

[12] K. Hemanjaneyulu, M. Shrivastava, Fin enabled area scaled tunnel FET, IEEE Trans. Electron Devices 62 (10) (2015) 3184–3191.

[13] J. Knoch, J. Appenzeller, Modeling of high-performance p-type III–V heterojunction tunnel FETs, IEEE Electron Device Lett. 31 (4) (2010) 305–307.

[14] I. Vurgaftman, J.R. Meyer, L.R. Ram-Mohan, Band parameters for III–V compound semiconductors and their alloys, J. Appl. Phys. 89 (11) (2001) 5815–5875.

[15] S. Brocard, M.G. Pala, D. Esseni, Large on-current enhancement in hetero-junction tunnel-FETs via molar fraction grading, IEEE Electron Device Lett. 35 (2) (2014) 184–186.

[16] M.G. Pala, S. Brocard, Exploiting hetero-junctions to improve the performance of III–V nanowire tunnel-FETs, IEEE J. Electron Devices Soc. 3 (3) (2015) 115–121.

[17] G. Zhou, Y. Lu, R. Li, Q. Zhang, Q. Liu, T. Vasen, et al., InGaAs/InP tunnel FETs with a subthreshold swing of 93 mV/dec and I_{ON}/I_{OFF} ratio near 10_6, IEEE Electron Device Lett. 33 (6) (2012) 782–784.

[18] A.S. Verhulst, D. Verreck, M.A. Pourghaderi, M. Van de Put, B. Soree, G. Groeseneken, et al., Can p-channel tunnel field-effect transistors perform as good as n-channel?, Appl. Phys. Lett. 105 (4) (2014) 043103.

[19] D. Verreck, A.S. Verhulst, B. Soree, N. Collaert, A. Mocuta, A. Thean, et al., Improved source design for p-type tunnel field effect transistors: towards truly complementary logic, Appl. Phys. Lett. 105 (24) (2014) 243506.

[20] A. Hraziia, C. Andrei, A. Vladimirescu, A. Amara, C. Anghel, An analysis on the ambipolar current in Si double-gate tunnel FETs, Solid-State Electron. 70 (2012) 67–72.

[21] A.S. Verhulst, W.G. Vandenberghe, K. Maex, G. Groeseneken, Tunnel field-effect transistor without gate–drain overlap, Appl. Phys. Lett. 91 (5) (2007) 053102–053103.

[22] D.B. Abdi, M.J. Kumar, Controlling ambipolar current in tunneling FETs using overlapping gate-on-drain, IEEE J. Electron Devices Soc 2 (6) (2014) 187–190.

[23] W.Y. Choi, W. Lee, Hetero-gate-dielectric tunneling field effect transistors, IEEE Trans. Electron Devices 57 (9) (2010) 2317–2319.

[24] B. Sedighi, X.S. Hu, H. Liu, J.J. Nahas, M. Niemier, Analog circuit design using tunnel-FETs, IEEE Trans. Circuits Syst. I, Reg. Papers 62 (1) (2015) 39–48.

[25] H. Liu, S. Datta, V. Narayanan, Steep switching tunnel FET: a promise to extend the energy efficient roadmap for post-CMOS digital and analog/RF applications, Proc. ISLPED (2013) 145–150.

[26] A.R. Trivedi, S. Carlo, S. Mukhopadhyay, Exploring tunnel-FET for ultra-low power analog applications: a case study on operational transconductance amplifier, Proc. ACM/IEEE Design Automation Conf. (2013) 1–6.

[27] P.N. Kondekar, K. Nigam, S. Pandey, D. Sharma, Design and analysis of polarity controlled electrically doped tunnel FET with bandgap engineering for analog/RF applications, IEEE Trans. on Electron Devices 64 (2) (2017).

CHAPTER 2

Electrothermal Characterization, TCAD Simulations, and Physical Modeling of Advanced SiGe HBTs

Rosario D'Esposito, Sébastien Frégonèse and Thomas Zimmer
IMS Laboratory, University of Bordeaux, Talence, France

2.1 SiGe HBT Technologies and Their Thermal Issues

2.1.1 THz Waves and Applications

The terahertz (THz) region is a part of the electromagnetic spectrum lying between microwaves and infrared light, and is typically referred to frequencies from 300 GHz to 3 THz [1], although there is no stated convention. THz waves can penetrate various materials (clothing, plastic, ceramics, etc.) as well as tissues with low water content (applications in millimeter wave dosimeters of human skin). Unlike X-rays, THz radiation is not ionizing, and, therefore, should cause no damage to live biological tissues. These characteristics allow a large number of new innovative applications [2] in the field of THz imaging and sensing for biology and medical sciences [3–5], non-destructive evaluation [6,7], and homeland security [8,9]. Moreover, ultra-high-speed communication systems at THz frequencies can be attractive for the professional as well as the consumer market [10,11]. However, with currently available and practical THz detectors and sources, measurements over distances larger than 20 m are very difficult [12], besides, the attenuation level in the earth atmosphere for THz waves is higher than for the neighboring bands [13]. Obtaining THz signals with sufficient output power is exceptionally challenging and the lack of suitable devices to generate, detect, or properly process THz signals is the so-called THz gap (Fig. 2.1). In fact, electronic sources, like transistors or multipliers, can go up to 1 THz at room temperature [14,15] and conventional laser diodes can easily generate signals at very high frequencies [16], but their use in the THz-gap region still appears impractical. For instance, free-electron lasers can generate electromagnetic radiation over a very wide range: from microwaves, through THz radiation to X-ray. However, apart from being extremely expensive, they are bulky and not suitable for consumer applications.

Nanoelectronics. DOI: https://doi.org/10.1016/B978-0-12-813353-8.00013-0

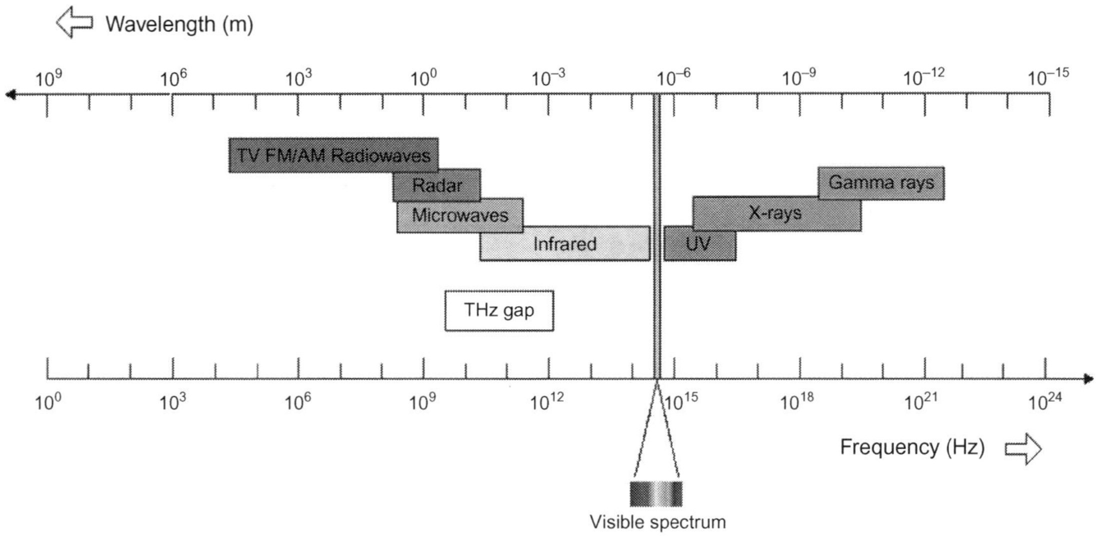

Figure 2.1
Overview of the electromagnetic spectrum and related applications.

2.1.2 SiGe BiCMOS Technologies

In millimeter wave applications, the III–V technologies have dominated the silicon technologies for many years. Performances of silicon (Si) technologies are still lower than other technologies in gallium arsenide (GaAs) or indium phosphide (InP). Devices fabricated in III–V technologies tend to have a superior performance compared to Si devices, due to higher carrier mobility; however, they have high fabrication costs and a poor reliability, making them not attractive for high volume applications. Moreover, due to their high thermal resistance, the integration level is much lower than for silicon. The power handling capability on Si is, respectively, 3.5 and 2.7 times as large as that on GaAs and InP substrates [17]. These arguments make silicon, even with lower performances, the favorable technology for large-scale applications. High-performance SiGe HBTs integrated in a bipolar complementary metal oxide semiconductor (BiCMOS) technology seem to be a promising solution.

The acronym HBT stands for heterostructure bipolar transistor, since in these devices one or both junctions are formed between dissimilar semiconductor materials, differently than the classical silicon-based bipolar transistor, in which we talk about homojunctions (e.g., the junctions are formed by similar semiconductor materials). In fact, the silicon (Si) and the germanium (Ge) have different energy bandgaps (E_g) between the conduction and the valence band ($E_{g,Si} = 1.12$ eV and $E_{g,Ge} = 0.66$ eV at 300 K). For this reason, the Si_xGe_{1-x} alloy has a smaller E_g than pure Si, which can be adjusted by tuning the percentage of Ge. The effect of this bandgap engineering is to ensure a high current gain (β), while at the same time maintaining a high-base doping [18], which in turn yields a reduced base resistance and therefore improved high-frequency characteristics, needed to design performing RF circuits.

Due to these technological innovations, SiGe HBTs provide a higher current and voltage driving capability, higher linearity, larger transconductance (g_m), lower 1/*f* noise property, excellent reliability, better device matching, and superior performance in comparison to standard silicon metal-oxide semiconductor field-effect transistors (MOSFETs). This makes them ideal devices for microwave power applications [19,20]. Therefore, several companies have chosen to integrate their high-performance SiGe HBTs in a lower cost CMOS process (BiCMOS) which allows an improved functionality, but at the same time increases process complexity, due to the HBT integration. Moreover, SiGe HBT technology continues to gain momentum over CMOS process nodes, as shown in Fig. 2.2; although the evolution of SiGe has been rather irregular, with missed technology nodes, a clear scaling law exists (as indicated by the trend in dashed line), which allows SiGe BiCMOS to retain an advantage of around two generations over CMOS in terms of transit frequency (f_T) and maximum oscillation frequency (f_{max}).

Optimized device structure along with aggressive scaling lead to comparable performance as III–V devices ($f_T = 300$ GHz, $f_{max} = 500$ GHz) [21]. The Seventh Framework

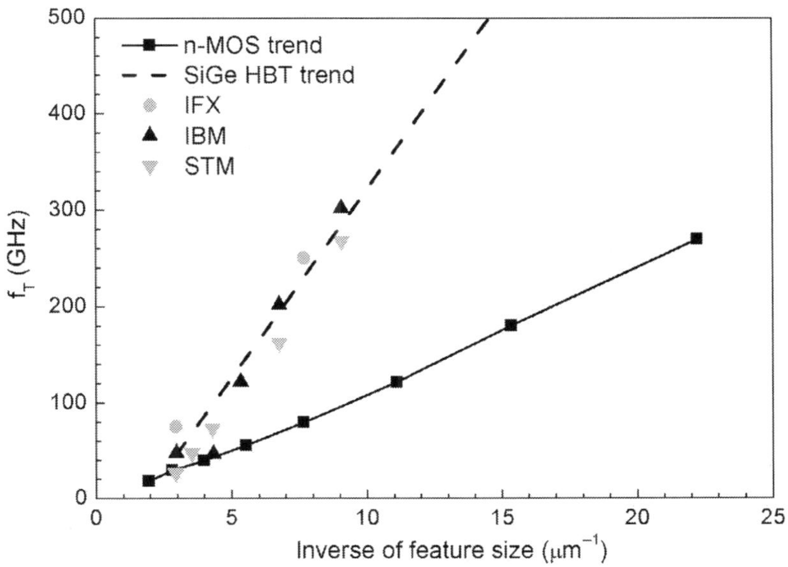

Figure 2.2
SiGe BiCMOS technology evolution over the inverse of the technology feature size. Comparison with CMOS technology trend.

Programme project DOTSEVEN has reached the development of SiGe:C HBTs technologies with f_{max} of around 700 GHz. This continuous improvement of the device performances allows the design of circuits for RF applications having outstanding performances. A power density of 18.5 mW/μm^2 at 1 dB compression was obtained at 94 GHz in [22], which can be attractive for designing power amplifiers in W-band. In [23], a record gate delay of 1.65 ps is demonstrated, which represents a record for bipolar transistor-based technologies. In [24], an 820-GHz transceiver chipset for THz active imaging applications was fabricated using IHP technology. This scenario opens the doors for THz applications [25] that can be mass-produced in a cost-effective way.

2.1.3 Thermal Issues in SiGe HBT Technology Nodes

In advanced SiGe technologies, higher device performances are mainly achieved with downscaling of the device dimensions and by improving the device architecture [26,27]. To allow device operation in the sub-THz range, the HBTs must be driven by critically high-current densities, which leads to a continuous increase of the power dissipation per unit area [28]. In these technologies, structural innovation has contributed to reduce the parasitic elements and to decrease the transit time by introducing shallow trench, deep trench, and other technological process steps [29]. The key improvements, (which also have a strong impact on the electrothermal behavior of the HBTs) are: (1) the introduction of the deep-trench isolation (DTI), which surrounds the whole device, thus helping to reduce the parasitic capacitances (see Fig. 2.3); the drawback, from a thermal point of view, is that the DTI is mainly made of silicon dioxide (SiO_2) having a very poor thermal conductivity and

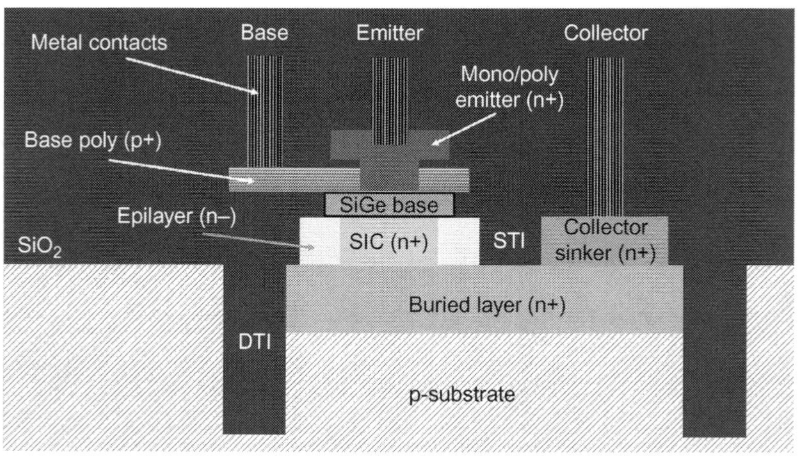

Figure 2.3
Cross section representing the schematic structure of a SiGe HBT with BEC arrangement for the contacts.

so its presence restricts the heat flow, resulting in larger thermal impedances (Z_{th}) [30–32]. (2) In order to boost the transit frequency (f_T), the base-collector (BC) junction and the epitaxial collector must be carefully designed. This permits to push the Kirk effect to higher current densities, leading to a higher f_T on the one hand, but on the other hand, the operating point is inevitably shifted to higher values of power dissipation [33]. Both (1) and (2) aspects result in a more pronounced self-heating effect, leading to higher device operating temperatures that affect the circuit performances, reliability, and packaging costs. Moreover, since the electrical behavior of SiGe HBTs is sensibly affected by the temperature reached by the component, an accurate electrothermal characterization and modeling of these phenomena is mandatory, in order to ensure realistic simulations during the circuit design phase.

In conventional approaches for the study of thermal phenomena arising in state-of-the-art SiGe HBTs, it is assumed that the main part of the power dissipation occurs at the BC space charge region, due to the intense electric fields and the high-current densities which are present in this region of the transistor. For thermal analysis at device level, it is possible to accurately estimate the spatial distribution of the power density from three-dimensional (3D) device simulations [34].

Due to the complex architecture of state-of-the-art components, which make use of doped materials, passivation layers, and metallization, having very different thermal properties, the study and modeling of the thermal diffusion mechanism within the transistor is a

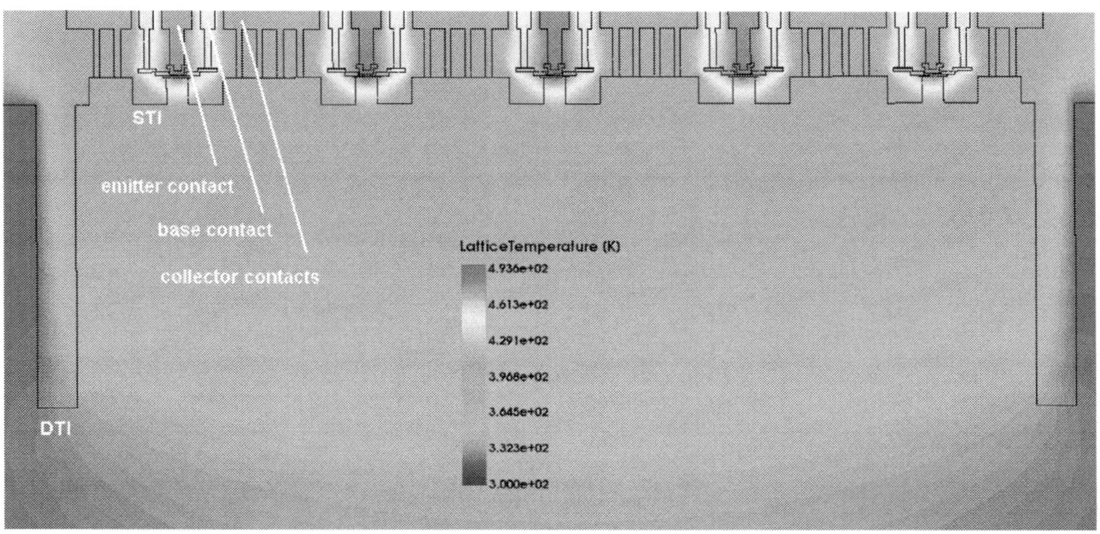

Figure 2.4
Temperature profile simulated in Sentaurus TCAD for a five-fingers SiGe HBT realized in ST B55 technology. The five heat sources are placed at the BC junction and the simulated 3D transistor structure includes different layers till the contacts at metal-1 level.

challenging topic of increasing interest. In Fig. 2.4, the temperature profile simulated in Sentaurus TCAD for a five-fingers transistor in ST B55 technology is shown, when planar heat sources are considered at the BC junctions: it is clear how the presence of different materials within the HBT yields thermal gradients; these temperature profiles propagate in different ways, according to the thermal conductivity of the different materials and thus an estimation of the temperature reached by the transistor requires a deep insight into the physics and the geometry of the component.

The determination of the temperature reached by the transistor is mandatory from a modeling point of view, since it impacts in a considerable way its electric behavior. When an HBT handles high levels of power dissipation, thermal instabilities can occur due to the positive feedback between the collector current and the device temperature [35,36]. In fact, HBTs can be considered unconditionally stable when driven by a constant base current; on the contrary, when biased with a constant base emitter voltage (V_{BE}), these devices show a positive electrothermal feedback. In this situation, in fact, the self-heating causes an increase in the collector current, which will determine an increase in the power dissipation, thus creating a positive feedback loop. This situation can cause the onset of the flyback behavior in the output characteristics, thus limiting the safe operating area (SOA) of the transistor in the region at high currents [37].

To mitigate the negative effects of self-heating, multifinger architectures can be used, which can have an increased power handling capability. These devices consist of multiple transistors in parallel, placed within the same DTI (as visible in Fig. 2.4). Due to the close proximity of the different fingers, thermal coupling effects can cause the central emitter to reach the highest temperature and thus drive more current than the other emitters, triggering a thermal instability mechanism. This situation can lead to an uneven distribution of the currents among the different emitters, till reaching the situation where one finger tends to conduct the whole current (current hogging effect) [38,39]. This unwanted focalization of the current in a single finger can lead to the generation of hot spots within the transistor and may eventually cause the device failure. To overcome these thermal instabilities, properly designed ballasting resistors can be added at the emitter nodes and can limit the rapid current increase and the current hogging phenomenon [40,41]. This solution, however, leads to degradation of the output power and of the power-added efficiency (PAE) of the transistor, so alternative transistor topologies can be helpful to extend the safe operating limits of the transistor without any loss in performance.

In general, due to the strong relationship between electrical and thermal effects in SiGe HBTs, an efficient characterization of the device is essential, in order to generate trustable compact models and thus allow precise electrical simulations for circuit designers. Consequently, a realistic modeling must provide an electrical representation, but also a precise and physics-based reproduction of the thermal characteristics of the HBTs, which can also be helpful for thermal stability and reliability analysis. Besides, due to the intrinsic nonlinearity of the transistor, third-order intermodulation products, generated near its thermal time constant, can trigger low-frequency variations of the junction temperature; these low-frequency temperature variations can affect the electrical behavior of the component [42], so a precise characterization of the thermal impedance is necessary and can be helpful in the design of predistortion linearizers [43]. It must be taken into account, though, that external parasitic elements, like the metallization in the back end of line (BEOL), can impact to a considerable extent the thermal impedance of the transistor [44,45] and so its impact should be analyzed and modeled.

2.2 Device Characterization in SiGe HBT Technologies

As already presented in the previous paragraph, state-of-the-art SiGe technologies are affected by serious thermal issues that can degrade the performances of the active components and eventually lead to their failure. A deep insight into the mechanisms of heat generation and diffusion at device level as well as an accurate characterization and modeling of these phenomena are mandatory, in order to allow a realistic circuit simulation during the design phase, but it can also be helpful to improve the performances and push the limits of a given technology.

This paragraph will be focused on the description of the measurement techniques adopted for the characterization of self-heating-related effects in state-of-the-art SiGe HBT technologies. First, the effect of self-heating on the electrical behavior of the transistor is shown by means of simulations for state-of-the-art technology; secondly, an overview of the methodologies to extract the static and dynamic thermal properties of the transistors is presented and finally the impact of thermal memory effect is discussed, with large signal two tones simulations, showing also the limitations in the use of a single-pole thermal network.

2.2.1 Modeling of Device Self-heating in HiCuM

The main goal of compact modeling is to enable circuit design. A compact model should be a physics-based mathematical description of the electrical behavior of a device for use in a circuit simulator. There is a large demand for fast and reliable compact models that are accurate for a great variety of technologies. A trustable compact model must provide high accuracy over a wide electrical and thermal range, geometrically scalable parameters calculation, numerical stability, and fast execution time. A compact model must reproduce the specific electrical behavior of the active component, while meeting the criteria of robustness, convergence of the simulations, and simulation time. It is the key element on which circuit designers rely and, therefore, the model should provide an accurate representation of the physical phenomena that may affect the transistor behavior, for all the possible bias conditions imposed by the circuit operation. Nowadays, the main models for silicon-based bipolar transistors are: Gummel-Poon (SGP), VBIC, MEXTRAM, and HiCuM, which are physics-based models. Among them, the HiCuM/L2 model is considered as a standard and is commonly adopted for the design of circuits using SiGe HBTs; this model has been chosen for all the simulation results reported in this chapter.

The HiCuM (high-current model) [46] is a scalable physics-based compact model for bipolar transistors, taking into account the high-current effects which are significant for certain high-speed transistor applications. The first version was introduced in 1986 by H.M. Rein and M. Schröter and it is now a Compact Model Council (CMC) standard model and integrated in a wide range of simulators. HiCuM is based on an extended and generalized integral charge−control relation (GICCR). However, in contrast to the (original) Gummel−Poon model, in HiCuM, the GICCR concept is applied without simplifications and additional fitting parameters (such as the Early voltages). Therefore, it allows a reliable design and circuit optimization. Detailed information about this compact model can be found in [46].

In a SiGe HBT, the heat is generated in all active areas, which results in an increase in the temperature of the component. The self-heating effect leads to a shift in the device electrical characteristics and it can be modeled with an equivalent electrical circuit. As shown in Fig. 2.5, in the HiCuM/L2 compact model, a dedicated sub-circuit allows the calculation of the increase in the junction temperature (ΔT_j); it consists of one thermal resistance R_{th} and one thermal capacitance C_{th} in parallel, forming a single-pole circuit, which represents the simplest model topology for calculating the junction temperature evolution.

The units for R_{th} and C_{th} are K/W and Ws/K, respectively. The voltage node in the thermal equivalent network corresponds to the temperature rise ΔT_j experienced by the transistor:

$$\Delta T_j = T_j - T_{amb} = P_d R_{th} \tag{2.1}$$

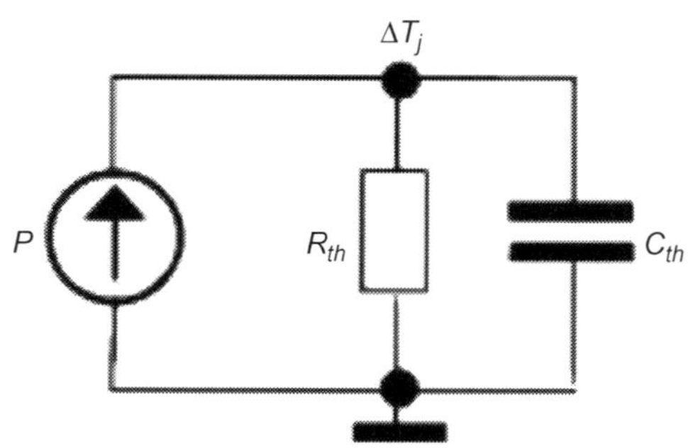

Figure 2.5
Thermal network used in HiCuM for calculating the HBT junction temperature.

where T_j and T_{amb} are the junction and ambient temperatures, respectively. The current source corresponds to the dissipated power P_d and represents the amount of heat that is generated by the device. In HiCuM/L2, the flag FLSH allows three different approximations for the calculation of P_d [46]:

- If FLSH = 0, the self-heating effect is not taken into account and the device temperature is constantly equal to the ambient temperature T_{amb}
- If FLSH = 1, only the most relevant dissipative elements are considered for the calculation of P_d, as follows:

$$P_d = |I_T V_{C'E'}| + |I_{AVL} V_{B'C'}| \tag{2.2}$$

- If FLSH = 2, P_d is calculated from all relevant dissipative elements in the equivalent circuit:

$$P_d = |I_T V_{C'E'}| + |I_{AVL} V_{B'C'}| + \sum \frac{|\Delta V_n^2|}{R_n} + |I_{TS} V_{B^*S}| \tag{2.3}$$

where $R_n = \{r_{Bi}, r_{Bx}, r_E, r_{Cx}\}$ are the series resistances connected to, respectively, the internal and external base, the emitter and the external collector; ΔV_n represents the corresponding voltage drop across these resistances [46].

The thermal network and the electrical part of the transistor model are solved together for each DC and dynamic operation. The temperature-dependent model parameters are calculated using the voltage at the temperature node. Furthermore, if the internal R_{th} and C_{th} are deactivated, the thermal node can be connected to more complex external thermal networks, in order to allow a more accurate simulation of specific thermal phenomena. For instance, this option is later used in Section 2.3.2 to consider the heat flow through the metal heat spreaders placed above the transistor.

2.2.2 Self-heating Effect on the Device DC and AC Characteristics

Measurements for device characterization are normally conducted on-wafer, due to the use of automatic or manual probe stations; the transistor test structures are usually realized in common emitter configuration and a ground-signal-ground (GSG) arrangement for the pads is preferred to avoid possible oscillations and allow RF measurements. The base and the collector are connected to the source and monitor units (SMU) of a semiconductor parameter analyzer, which applies and measures voltages and currents directly on the transistor (see Fig. 2.6). The parameter analyzer is connected to a GPIB interface (general purpose interface bus), and the desired voltage or current sweeps are controlled by dedicated software, allowing to collect and handle the measured data.

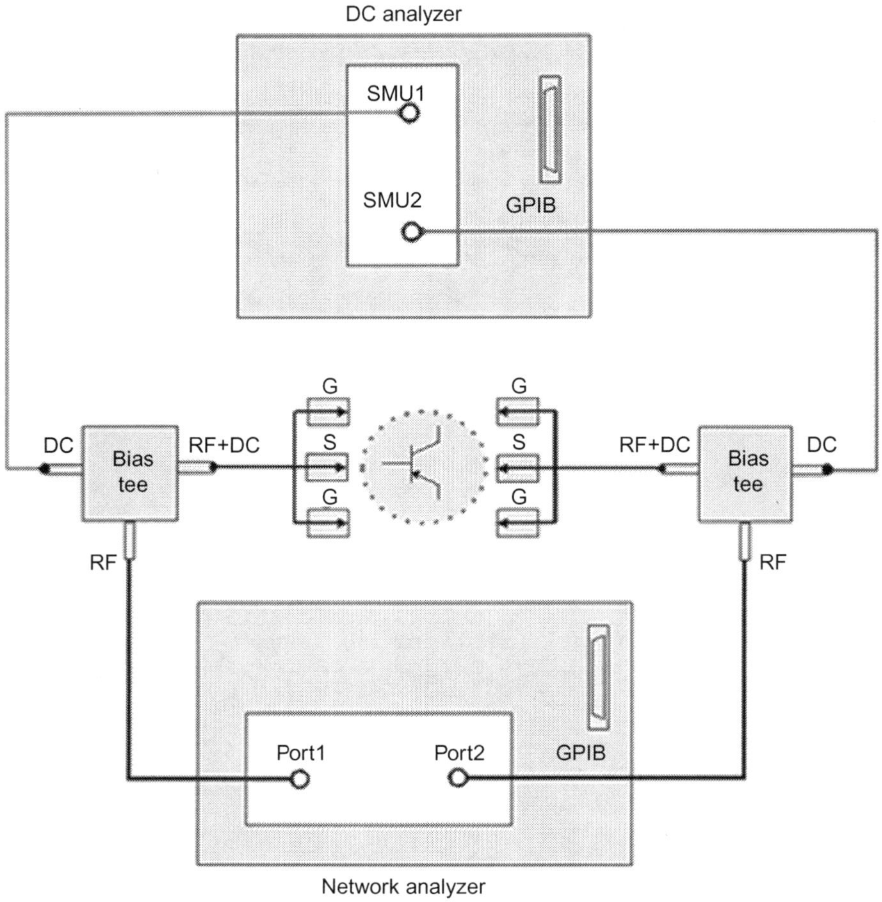

Figure 2.6
Block diagram showing a typical measurement setup for the DC and AC characterization of active components on-wafer.

The measurement of the $I(V)$ current–voltage relationships is the most basic form for a DC characterization of the transistors: in Fig. 2.7A the Gummel plot measured on an HBT realized in the Infineon B11HFC process (W_ExL_E = 0.22 × 5 μm^2) is reported, when a V_{BE} sweep is applied to the transistor for V_{CE} = 1.5 V. In this biasing condition, the collector current I_C increases with increasing temperature [35]; since the temperature reached by the component (T_j) is directly proportional to the power dissipated P_d [according to Eq. (2.1)] and the P_d is proportional to current flowing into the device [following Eq. (2.2) or Eq. (2.3)], it results that for this biasing condition, the I_C will experience an uncontrolled increase while the P_d increases (thermal runaway) [38], which can eventually lead to instability and device failure. In Fig. 2.7, the measurements are also compared to the results from simulation using the HiCuM/L2 compact model and a model card that has an R_{th}

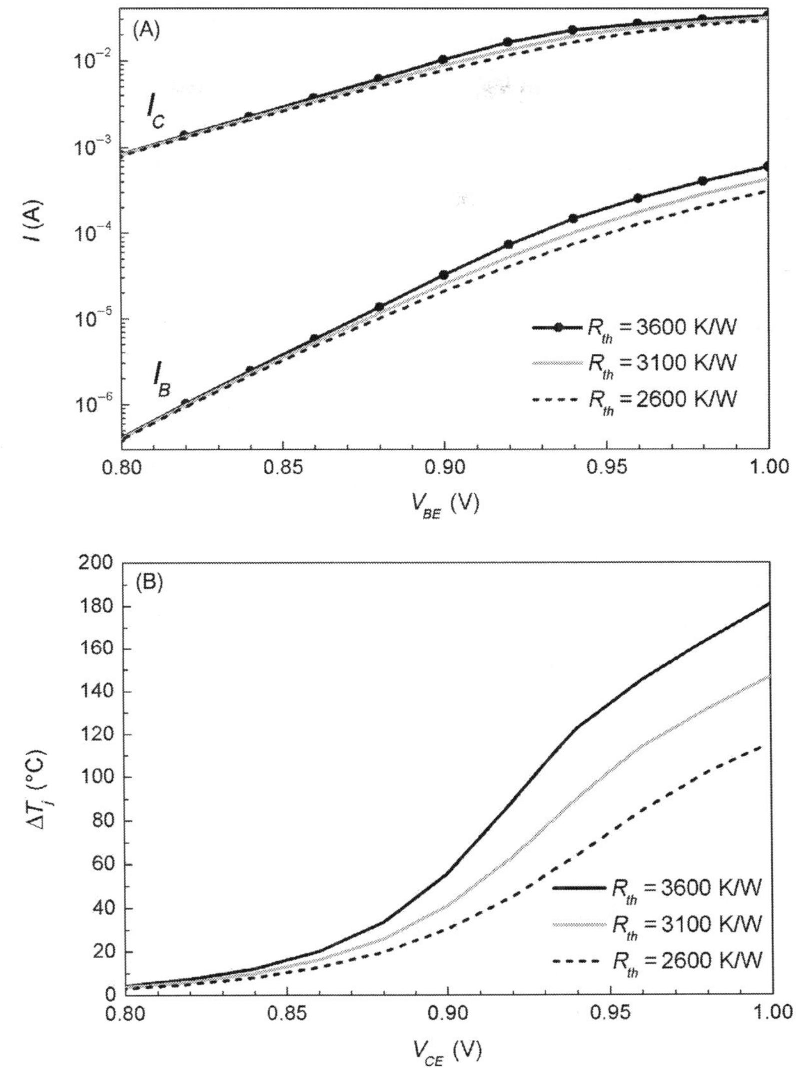

Figure 2.7

(A) Forward Gummel plot at constant $V_{CE} = 1.5$ V for an Infineon B11HFC CBEBC HBT with $A_E = 0.22 \times 5\ \mu m^2$: (B) corresponding temperature increase simulated in HiCuM. The reported curves correspond to different values of the R_{th}.

value of 3600 K/W. Two additional simulation results are added, using two lower values of the thermal resistance R_{th}: 3100 K/W and 2600 K/W, in order to evaluate the differences that can be expected in the electrical characteristics when the effects of self-heating are lower. It can be seen that, starting from a V_{BE} value of around 850 mV, as the R_{th} is set to smaller values, the simulated I_B decreases; concerning the simulated I_C instead, the maximum deviation is obtained for $V_{BE} = 925$ mV. This can appear strange, since observing the temperature increase predicted by the compact model (in Fig. 2.7B) it is clear that the temperature difference among three different model cards continues to increase for $V_{BE} > 925$ mV. It must be taken into account that for values of V_{BE} higher than 925 mV the increase in I_C is caused by a combination of other effects (mainly impact ionization, avalanche multiplication, and the feedback role played by the emitter resistance) rather than solely by self-heating and thus the difference among the I_C obtained for the three R_{th} considered starts to decrease.

The effect of the positive electrothermal feedback when forcing the device at constant V_{BE} is also visible on the output curves in Fig. 2.8A, and the same considerations made for the Gummel plot still hold: among the different R_{th} considered, the bias condition at which the highest variation of I_C is observed for $V_{BE} = 900$ mV; in fact for $V_{BE} = 850$ mV, the self-heating effects are less evident, since the temperature reached by the component is lower, whereas for $V_{BE} = 950$ mV the increase in I_C is caused by a combination of other effects, as previously discussed and thus the difference among the results obtained for the three R_{th} considered is lower. In Fig. 2.8B, another situation is presented, corresponding to the case where the transistor is driven by a current injected in the base terminal; in this scenario, instead, the self-heating causes the I_C current to drop when the high power dissipation region is reached and so the device is intrinsically stable.

For what concerns the AC characterization, one of the main figures of merit that is used for comparing the performances of a given transistor is the unity gain cutoff frequency (f_T), which is defined as the frequency where the extrapolated small-signal current-gain β_F equals 1 (Fig. 2.9A):

$$\left|\beta_F(f = f_T)\right| = \left|\frac{dI_C(f_T)}{dI_B(f_T)}\right|_{V_{CE}} = \left|H_{21}(f_T)\right| = 1 \tag{2.4}$$

$$f_T = f_{meas}\left|H_{21}(f_{meas})\right| = \frac{f_{meas}}{\Im\left(Y_{11}/Y_{21}\right)} \tag{2.5}$$

where f_{meas} has to be chosen as a frequency at which β_F is rolling-off at a slope of -20 dB/dec, In Eq. (2.5) two possible expressions to extract f_T are reported. Another commonly used figure of merit is the f_{max}, defined as the frequency at which the unilateral power-gain U (or Mason's invariant gain) becomes unity (Fig. 2.9B).

$$\left|U(f = f_{\max})\right| = \frac{|Y_{21} - Y_{12}|}{4(\Re(Y_{11})\Re(Y_{22}) - \Re(Y_{12})\Re(Y_{21}))} = 1 \tag{2.6}$$

$$f_{\max} = f_{meas}\sqrt{\left|U(f_{meas})\right|} \tag{2.7}$$

Eq. (2.7) is a commonly used expression to extract f_{max} based on the definition of U.

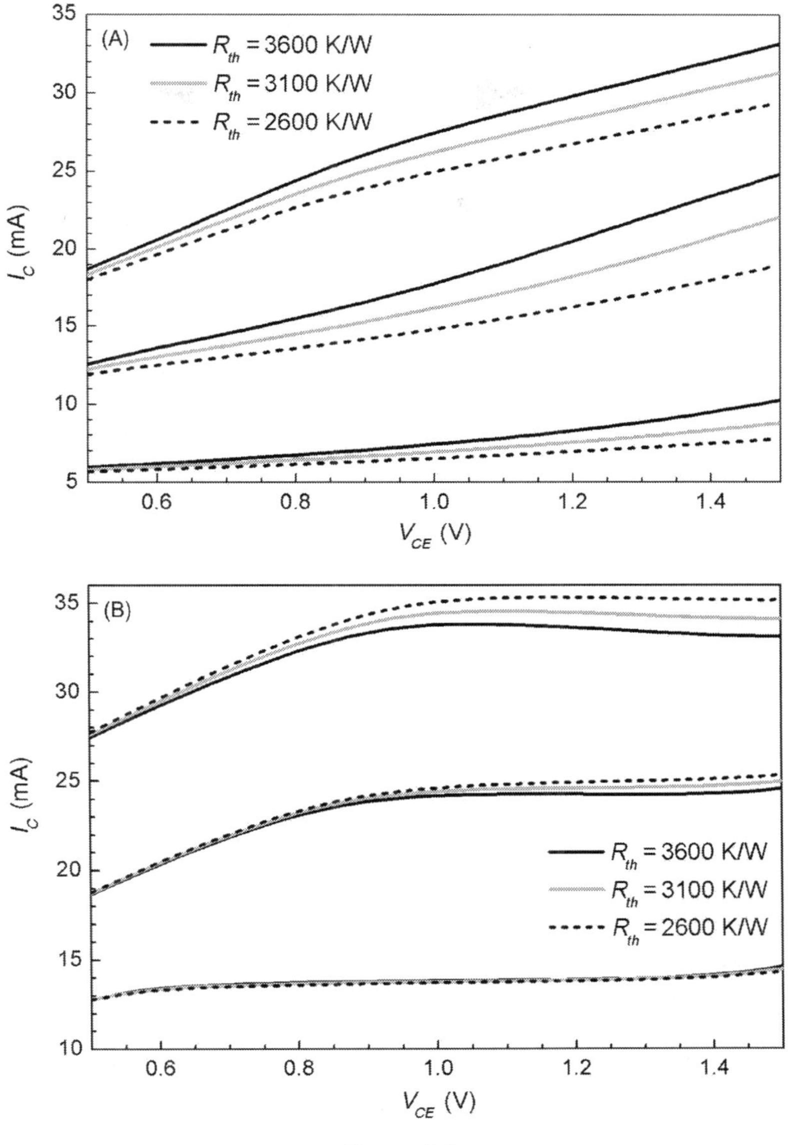

Figure 2.8
Simulated output curves for an Infineon B11HFC CBEBC HBT with $A_E = 0.22 \times 5\ \mu m^2$: (A) the forced V_{BE} ranges from 0.9 to 1 V; (B) the forced I_B ranges from 60 to 600 μA. Different values of the R_{th} are used in compact model simulations.

In Fig. 2.10, the simulated f_T and f_{max} are shown as function of the I_C for the same transistor and the same values of R_{th} discussed above; it can be noticed that varying the R_{th} (and thus changing the amount of self-heating effect within the transistor) impacts the f_T and f_{max} in the region around the peak and after.

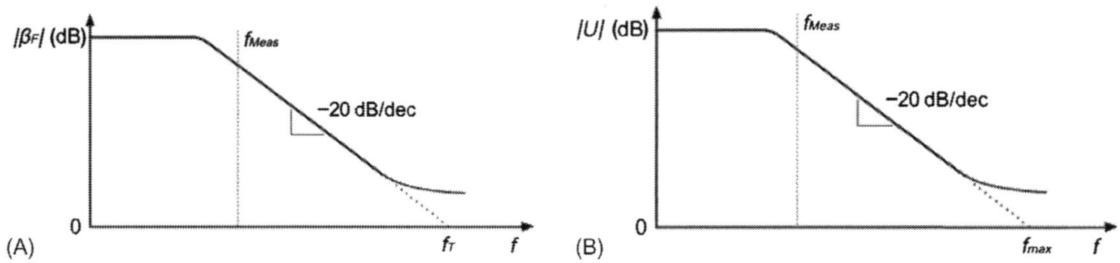

Figure 2.9
Illustration of the small-signal current-gain β_F (A) and Mason's gain (B) vs frequency; the f_T and f_{max} are extrapolated as the frequencies at which β_F and U, respectively, reach 0 dB.

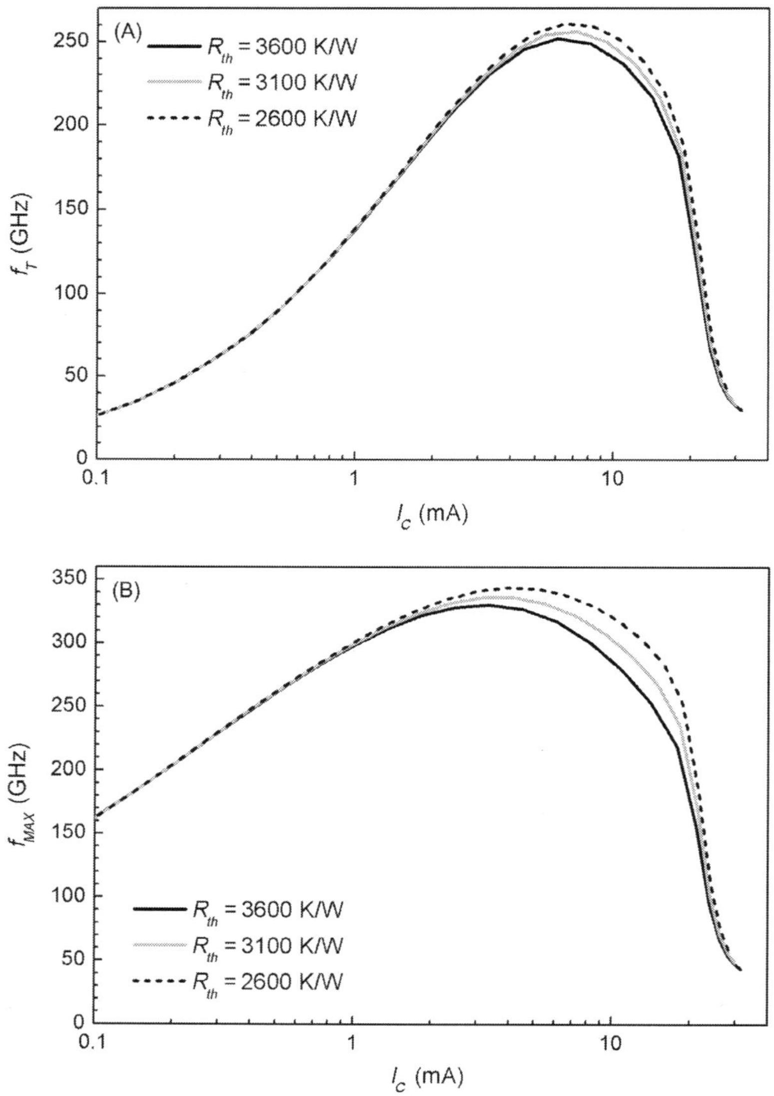

Figure 2.10
Simulated f_T (A) and f_{max} (B) plotted vs I_C for a constant V_{CE} of 1.5 V; different values of R_{th} are used in compact model simulations.

2.2.3 Extraction of the R_{th}

The extraction of R_{th} is one of the long lasting challenges encountered for device characterization and several different methodologies have been proposed in literature [47–51]. An alternative and very interesting R_{th} extraction technique is detailed in [52] and it is based on identifying the (V_{BE}, I_C) interception point on two Gummel plots measured at two different V_{CE} voltages and chuck temperatures. In [53], this extraction technique is compared to the other methodologies aforementioned, demonstrating that it is able to provide a superior accuracy for what concerns the temperature dependency of R_{th} at high junction temperatures; in fact, due to the differential approach adopted, any additional temperature dependence of the emitter and collector resistances is eliminated, thus extracting the sole temperature dependency of R_{th}.

In general, all the methods listed above consist of performing measurements at different chuck temperatures, in order to determine the thermal dependence of the electrical parameter which will be used as thermometer. The R_{th} extraction procedure used in this chapter is based on the theory given in [48] and the necessary steps are reported for the case of a five-fingers transistor realized in ST microelectronics B55 process ($0.18 \times 5\ \mu m^2$).

This procedure uses the base–emitter voltage as the temperature sensitive parameter and consists in measuring the V_{BE} as function of the ambient temperature (T_{amb}) for constant collector currents I_C and different V_{CE}; the approach used here is to measure I_C for swept V_{BE}, as done for Gummel plots and to determine V_{BE} as function of I_C by interpolation. In Fig. 2.11, the measured Gummel plots for different values of V_{CE} from 1 to 1.24 V and for four values of the chuck temperature (T_{chuck}) are shown, ranging from 300 to 380 K. Three different values of I_C (5, 10, and 20 mA) are chosen for the extrapolation, and the related V_{BE} which are extracted are then plotted as function of the chuck temperature, as reported in Fig. 2.12; this calibration curve shows a very linear relationship between V_{BE} and T_{chuck}.

As it can be seen in Fig. 2.12, for small values of I_C (e.g., at I_C chosen = 5 mA), where the power dissipation (and thus the self-heating) is negligible, the extracted V_{BE} does not vary too much for different V_{CE} values considered. However, if I_C is increased to 20 mA, it is observed that the V_{BE} values extracted start to vary if V_{CE} is swept; for this value of current, P_d is significantly higher, and thus self-heating can be considered responsible for the change of V_{BE} with V_{CE} varying. Next, different values of V_{BE} (from V_{BE1} to V_{BEn}) are chosen and the corresponding T_{chuck} are extracted by interpolation, these points correspond to the determination of the ambient temperature for which the transistor exhibits (for a given I_C value of 20 mA) the V_{BE} values which have been chosen.

These extracted chuck temperatures are now plotted vs their corresponding dissipated power dissipation (which is approximated here as $P_d = I_C\ V_{CE}$).

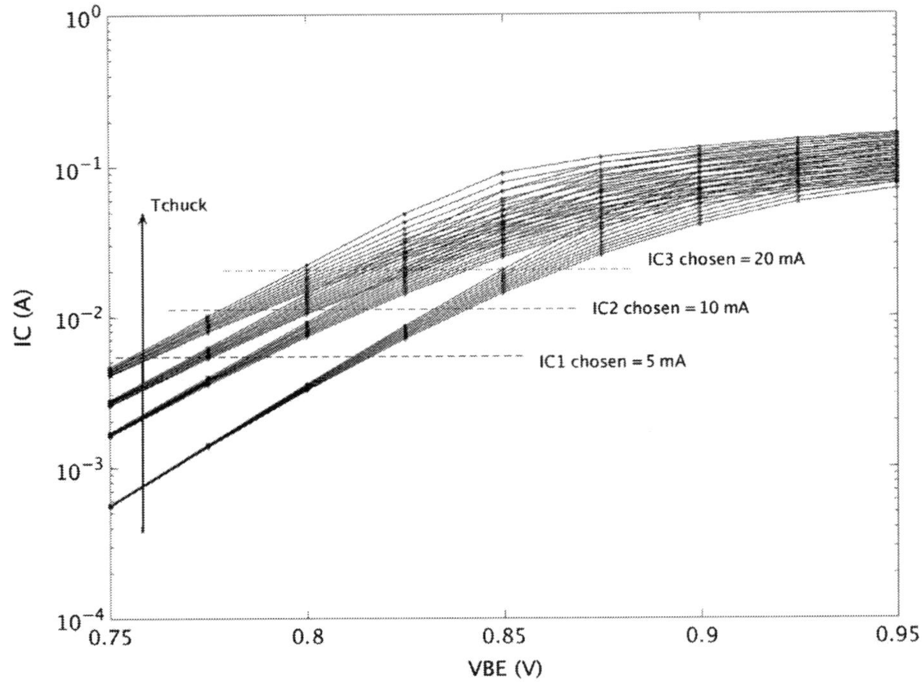

Figure 2.11
Gummel plots measured on a five-fingers HBT in ST B55 technology; the V_{CE} ranges from 1 to 1.24 V. The measurements are performed at different chuck temperatures ranging from 300 to 380 K. The values of I_C chosen for R_{th} extraction are also shown.

Since the junction temperature T_j can be expressed as the sum of the ambient temperature and the temperature increase in the device caused by self-heating:

$$T_j = R_{th}P_d + T_{chuck} \tag{2.8}$$

it follows that T_j can be determined by extrapolation of the T_{chuck} at $P_d = 0$, as marked in Fig. 2.13; moreover, the R_{th} can be estimated as the slope of the $T_{chuck}(P_d)$ curve. Repeating this procedure for all the values of V_{BE} which have been fixed leads to the results in Fig. 2.14A, which represent the extracted R_{th} as function of the junction temperature T_j. Since the T_j is extracted at fixed I_C, for every value of V_{BE}, the dV_{BE}/dT_j can also be determined using this technique (Fig. 2.14B); this parameter characterizes the technology and does not depend on the transistor layout.

2.2.4 *Extraction of the Z_{th}*

The characterization of dynamic electrothermal self-heating inside SiGe HBTs remains an essential task for model developers in order to determine the device operating temperature

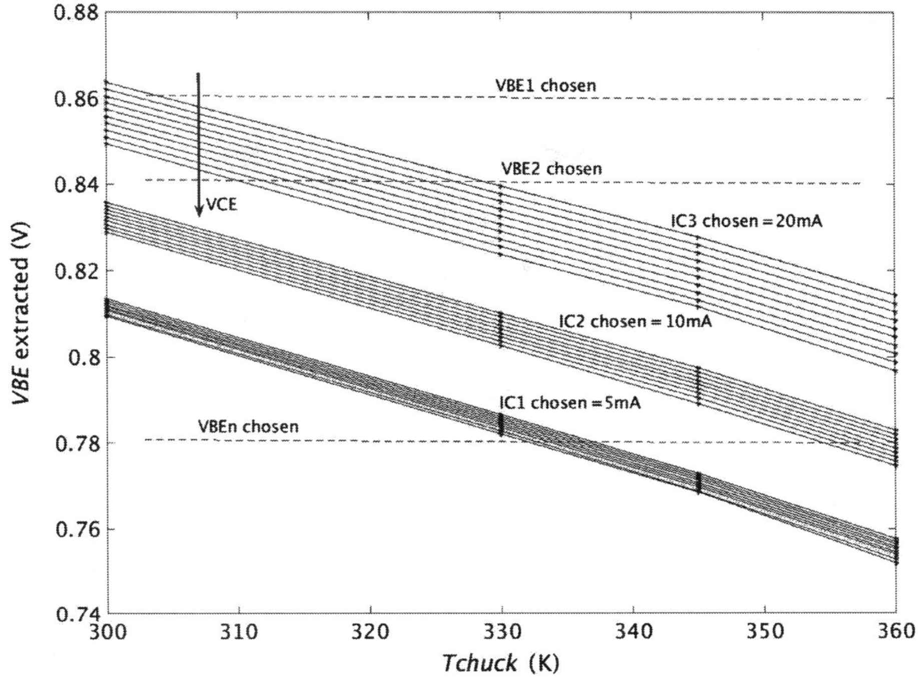

Figure 2.12

V_{BE} values extracted from the Gummel plots for the three values of I_C chosen. The V_{BE} are plotted vs T_{chuck} for different values of V_{CE} considered. The values of V_{BE} fixed to extract the corresponding T_{chuck} are also shown in the plot.

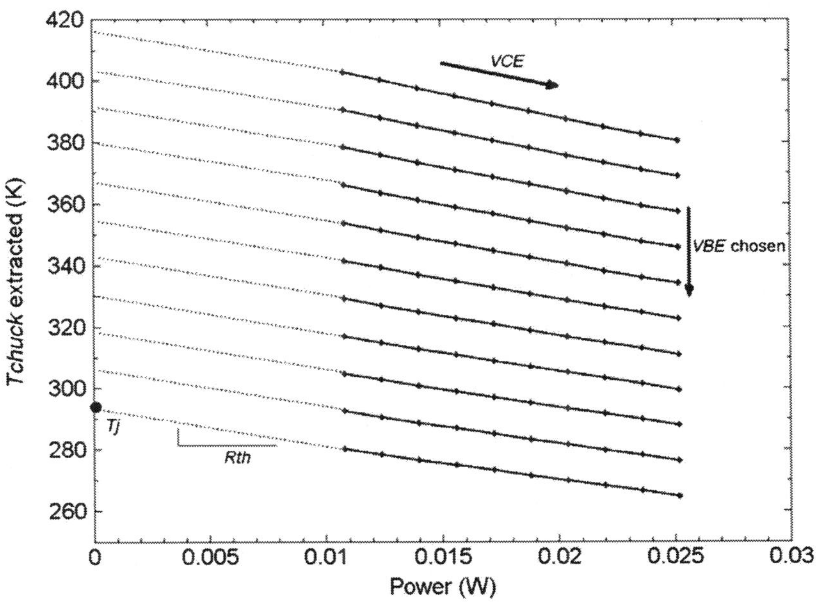

Figure 2.13

Chuck temperatures extracted for the values of V_{BE} which have been fixed, plotted as function of the corresponding P_d. This plot allows to extract, for each V_{BE} fixed, the corresponding values of the T_j and R_{th}.

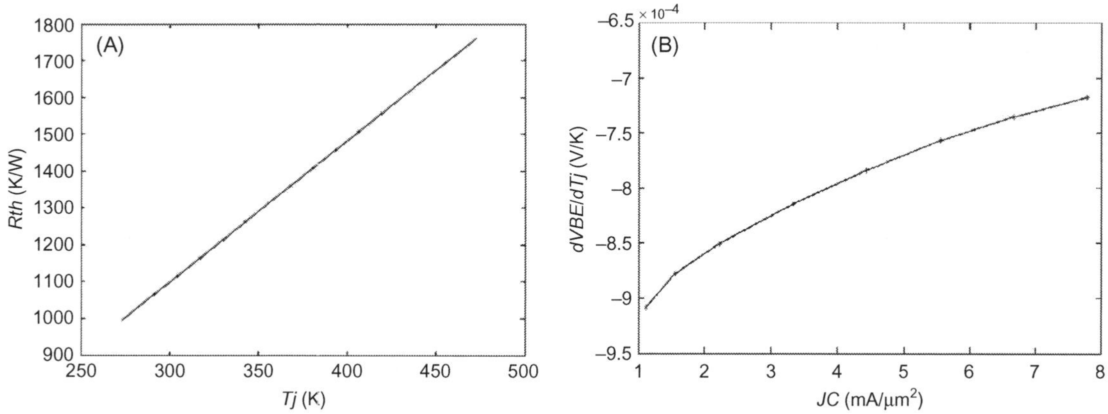

Figure 2.14
(A) R_{th} extracted as function of the T_j using the method described. (B) dV_{BE}/dT_j measured for the ST B55 technology.

in practical applications, where a signal having a certain frequency bandwidth is fed at the input of the transistor. The dynamic behavior can be described by the thermal impedance Z_{th} which can be seen as the parallel of R_{th} and C_{th} in the thermal subcircuit in Fig. 2.5.

The thermal response of SiGe HBTs ranges from about few kilohertz to about 200 MHz [54], hence their thermal impedance is limited to low frequencies; the Z_{th} can be extracted from low-frequency S-parameter measurements and it is a relatively simple and precise measurement procedure. Several approaches have been presented in literature to provide an expression to determine the Z_{th} from dedicated measurements at low frequency. For example, generalized formulations for thermal impedance from Y-parameters are proposed by Rinaldi [55] through a rigorous mathematical treatment. Also, dynamic characterization of thermal effect, based on H-parameters, has been proposed in [56,57]. The dispersive phenomenon presented by the device at low frequencies is caused by dynamic self-heating effects: performing S-parameter measurements, it is possible to observe a certain deviation of the S_{12} and S_{22} in the low-frequency range; in Fig. 2.15A, the magnitude of the S-parameters measured in the frequency range 10 kHz–3 GHz for the bias point $V_{BE} = 0.9$ V and $V_{CE} = 1.25$ V are reported, at which a significant amount of self-heating can be expected. The measurements are performed with the low-frequency vector network analyzer (VNA) Agilent E5061B (5 Hz-3 GHz), making use of bias tees particularly developed to cover this specific frequency range. The VNA was calibrated on-wafer using SOLT (short open load through) method. The measured results are de-embedded using open and short test structures.

The measurements refer to a five-fingers HBT fabricated in ST microelectronics B55 technology ($0.18 \times 5\ \mu m^2$): it is possible to notice that the S_{12} and S_{22} exhibit the largest variation among the four S-parameters. Analyzing these parameters for different bias

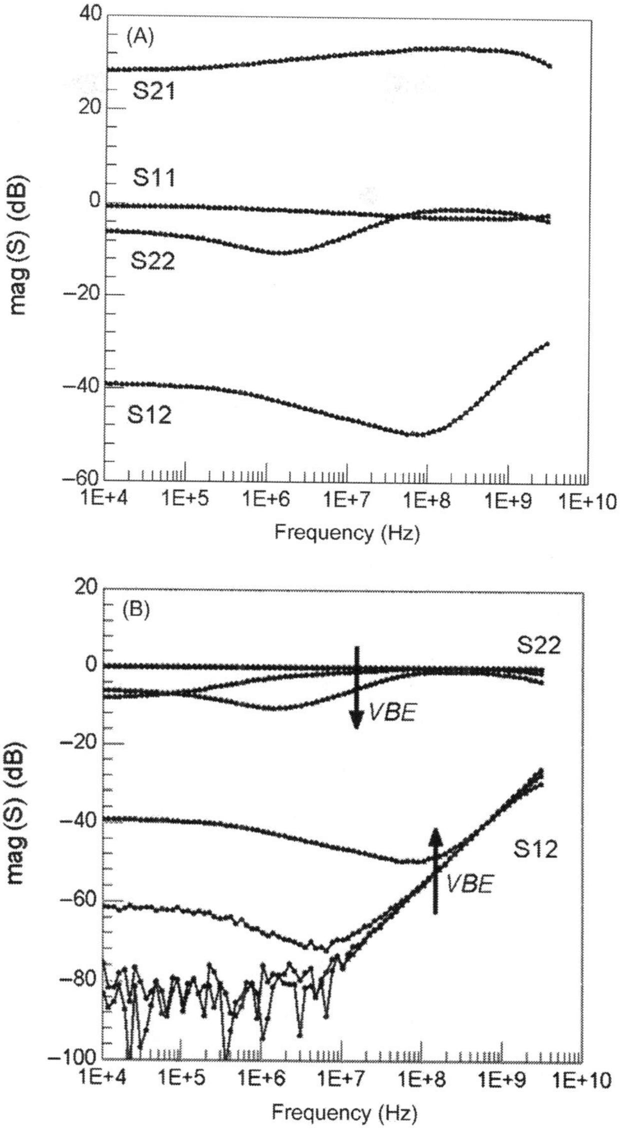

Figure 2.15

(A) Measured low-frequency S-parameters at $V_{BE} = 0.9$ V and $V_{CE} = 1.25$ V for the five-fingers HBT from ST B55 technology. (B) S_{12} and S_{22} for $V_{CE} = 1.25$ V and V_{BE} ranging from 0.75 to 0.9 V.

conditions, it is possible to identify the frequency range where dynamic self-heating effect is predominant; in Fig. 2.15B, S_{12} and S_{22} for different V_{BE} ranging from 0.75 to 0.9 V are plotted: it is shown that, as V_{BE} increases (and thus self-heating becomes more pronounced), it is possible to identify, for different values of V_{BE}, a deviation in the S-parameters in the low-frequency range (till around 200 MHz). At these low frequencies, in fact, the junction temperature is able to follow the variations of the power dissipation, thus affecting the S-parameters, whereas at higher frequencies, the junction temperature is not able to follow the evolution of power dissipation and thus just DC self-heating occurs, as will be detailed in the next section. The frequency that marks this difference in the thermal behavior will be referred to as thermal cutoff frequency. For the extraction of the Z_{th}, the S-parameters are converted to Y-parameters, as will be described below.

2.2.4.1 Theoretical formulation

Collector current I_C and base current I_B are functions of the three independent variables: (1) the base–emitter voltage V_{BE}, (2) the collector–emitter voltage V_{CE}, and (3) the junction temperature T_j. To generalize the following equations, V_{BE}, V_{CE}, I_B, and I_C are replaced by V_1, V_2, I_1, and I_2, respectively:

$$\begin{aligned} I_1 &= f_1(V_1, V_2, T_j) \\ I_2 &= f_2(V_1, V_2, T_j) \end{aligned} \tag{2.9}$$

A linearization valid for small signal excitations at a constant bias condition leads to:

$$\begin{aligned} i_1 &= Y_{11}v_1 + Y_{12}v_2 \\ i_2 &= Y_{21}v_1 + Y_{22}v_2 \end{aligned} \tag{2.10}$$

where each Y-parameter is defined as

$$Y_{mn} = \frac{\partial f_m}{\partial V_n} \quad \text{with} \quad m=1;2 \quad n=1;2. \tag{2.11}$$

Here, index 1 and 2 indicate the base and collector ports, respectively.

The following expression can be written for the Y-parameters [58]:

$$Y_{mn} = \left.\frac{dI_m}{dV_n}\right|_{V_m=const.} = \left.\frac{\partial f_m}{\partial V_n}\right|_{V_m,T_j=const.} + \left.\frac{\partial f_m}{\partial T_j}\frac{dT_j}{dV_n}\right|_{V_m=const.} = Y_{mn\_iso} + \left.\frac{\partial f_m}{\partial T_j}\frac{dT_j}{dV_n}\right|_{V_m=const} \tag{2.12}$$

where Y_{mn_iso} is temperature independent. The temperature dependence is approximated for small signal excitations with:

$$\alpha_m = \left.\frac{\partial f_m}{\partial T_j}\right|_{V_m=const.} \tag{2.13}$$

where α_m is the relative change of I_m per degree temperature change at a constant voltage V_m. The junction temperature T_j is defined as

$$T_j = T_{amb} + Z_{th}P_d \tag{2.14}$$

where T_{amb} is the ambient temperature (chuck temperature) and P_d is the dissipated power in the device that results from

$$P_d = V_1 I_1 + V_2 I_2 \tag{2.15}$$

The change in dissipated power with small signal excitation (dP_d) can be written as

$$dP_d = dV_1 I_1 + V_1 dI_1 + dV_2 I_2 + V_2 dI_2. \tag{2.16}$$

From Eq. (2.14) and Eq. (2.15), it follows that

$$\left.\frac{dT_j}{dV_n}\right|_{V_m=const.} = Z_{th}\left(I_n + V_n \left.\frac{\partial I_n}{\partial V_n}\right|_{V_m=const.} + V_k \left.\frac{\partial I_k}{\partial V_n}\right|_{V_m=const.}\right) \text{with} k = 3 - n \tag{2.17}$$

which can be transformed into

$$\left.\frac{dT_j}{dV_n}\right|_{V_m=const} = Z_{th}(I_n + V_n Y_{nn} + V_k Y_{kn}) \tag{2.18}$$

Using Eq. (2.12) and Eq. (2.18), the following expression for Z_{th} is obtained

$$Z_{th} = \frac{Y_{mn} - Y_{mn\_iso}}{(\partial f_m / \partial T_j)[I_n + V_n Y_{nn} + V_k Y_{kn}]} \tag{2.19}$$

In Eq. (2.12), the term dT_j/dV_n is greater for V_2 (V_{CE}) than for V_1 (V_{BE}), so Y_{12} and Y_{22} are the Y-parameters most affected by self-heating, and therefore, most suitable for the extraction of Z_{th}. Considering a formulation using Y_{22}, the resulting Z_{th} would be

$$Z_{th} = \frac{Y_{22} - Y_{22\_iso}}{\alpha_2[I_2 + V_2 Y_{22} + V_1 Y_{12}]} = \frac{Y_{22} - Y_{22\_iso}}{\alpha_{IC}[I_C + V_{CE} Y_{22} + V_{BE} Y_{12}]}. \tag{2.20}$$

As expressed in Eq. (2.20), it is assumed that the measured Y-parameters contain two parts: one is an isothermal part where only DC self-heating effects are visible and which has been measured at high frequency (above thermal cutoff frequency) and the other is the thermal part (below thermal cutoff frequency) where both dynamic and DC self-heating effects are equally visible.

In Fig. 2.16A, the measured Y_{22} parameters for different values of V_{BE} ranging from 0.75 to 0.95 V and $V_{CE} = 1.25$ V are plotted. It can be seen that, as V_{BE} (and consequently the amount of self-heating) increases, the variation of Y_{22} at low frequency becomes higher; in particular, the highest deviation is observed at $V_{BE} = 0.9$ V. From Fig. 2.16A, it looks evident the presence of two regions in the Y_{22} parameters: (1) the isothermal region (above 200 MHz) and (2) the dynamic self-heating affected region or non-isothermal

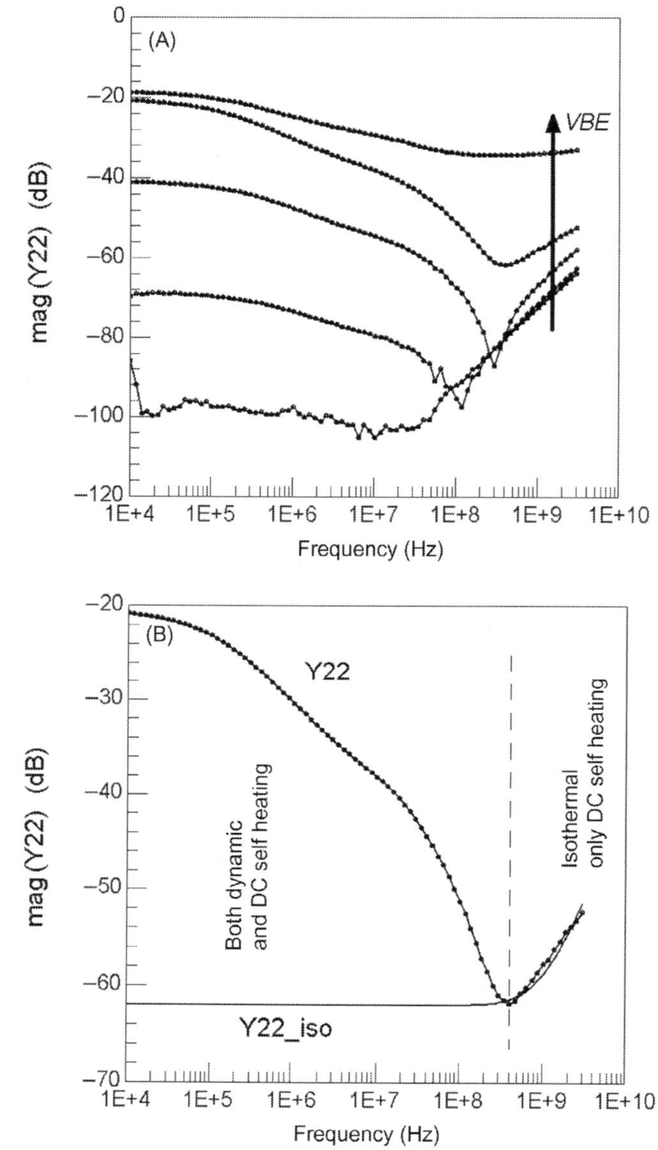

Figure 2.16

(A) Variation of the low-frequency Y_{22} parameter for $V_{CE} = 1.25$ V and V_{BE} ranging from 0.75 to 0.95 V. (B) Y_{22} for $V_{CE} = 1.25$ and $V_{BE} = 0.9$ V: the calculated Y_{22_iso} is shown.

region (below 200 MHz). Once the cutoff frequency is determined, the Y_{22_iso} necessary for the extraction of Z_{th} is calculated as an extrapolation of Y_{22}, starting from the highest frequency measured down till the cutoff frequency, as shown in Fig. 2.16B for the bias point $V_{BE} = 0.9$ V.

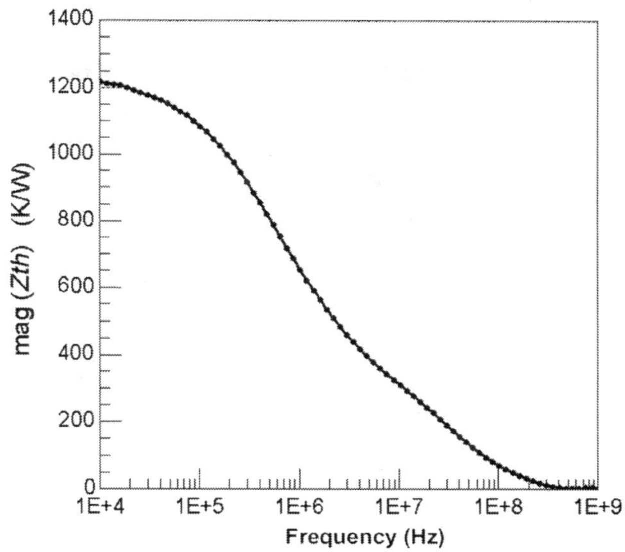

Figure 2.17
Extracted Z_{th} vs frequency in the 10 kHz–1 GHz range, using the extraction methodology described.

In Fig. 2.17, the corresponding Z_{th} extracted is reported, according to Eq. (2.20).

2.2.5 Recursive Thermal Network Models

As already detailed in Section 2.2.1, state-of-the-art compact models, like HiCuM/L2, normally adopt a single-pole network for the simulation of the junction temperature of the transistor (T_j). This network represents the easiest model possible to reproduce the evolution of T_j when a dynamic power dissipation is present within the transistor, but it can show its limitations in specific conditions of operation of the component. In fact, the heat diffusion phenomena from the heat source to the heat sink show a distributed nature, and thus multiple time constants can be needed to approximate this thermal behavior. In literature, different modeling approaches have been proposed to threat this problem, and mostly rely on the use of transmission-line equivalent circuits, where a distributed RC network is used to model the diffusion of the heat; more precisely, a Cauer-type network is normally chosen for this purpose, since it is more suitable as a thermal analogy from the physical point of view [59,60]. Using these distributed approaches, the transistor structure can be divided into multiple volume elements [61] and each segment can be associated to its corresponding lumped R_{th} and C_{th}, as shown in Fig. 2.18, making the following assumptions [61]:

- The graduation of layer thickness should be chosen so that progressively larger thermal time constants ($\tau_{th} = R_{th\ i}\ C_{th\ i}$) are produced in the direction of the heat propagation.

- If the dimensions of the heat source are smaller than the cross section of the heat conducting material, the heat spreading region can be assumed to be confined within a spreading angle α. For heat propagation in homogeneous media, a spreading angle α of about 40° can be considered [62].

In this chapter, a recursive thermal network will be mainly used, in which the lumped elements are scaled using two different recursive multiplication factors: $K_r(<1)$ and $K_c(>1)$, respectively, for the $R_{th\ i}$ and $C_{th\ i}$ as shown in Fig. 2.19. The elements of this network can be related to the different volumes in which the transistor structure is ideally subdivided according to the model in Fig. 2.18.

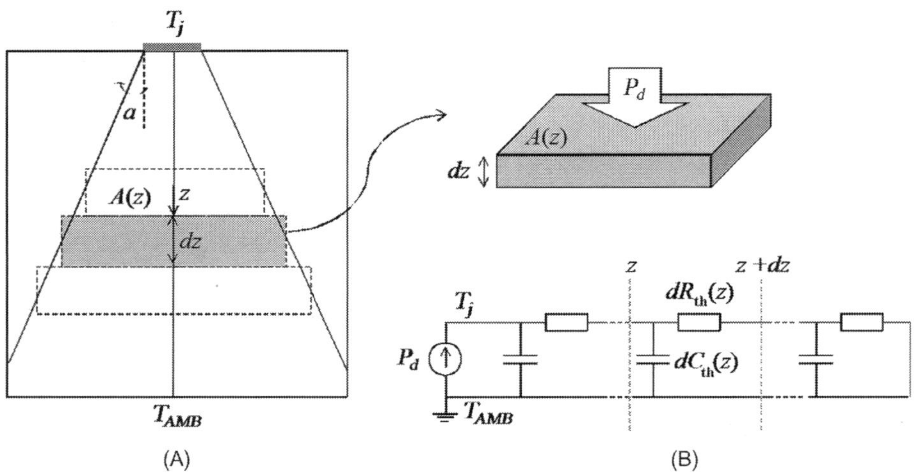

Figure 2.18
(A) Heat diffusion within spreading angle α in a simple structure and (B) electrical transmission line equivalent circuit diagram.

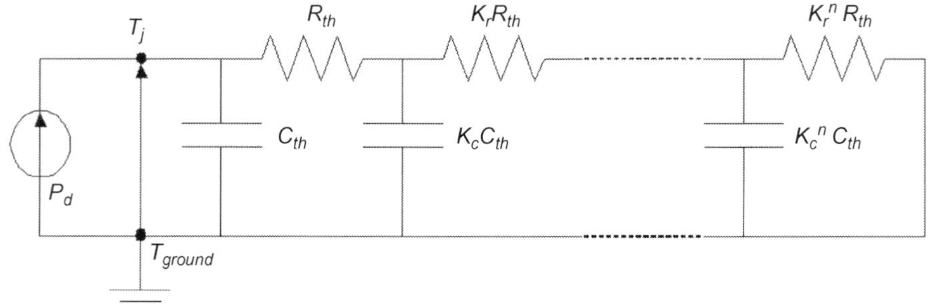

Figure 2.19
A Cauer-type recursive network for the Z_{th} representation (the different elements of the network are in recursive order).

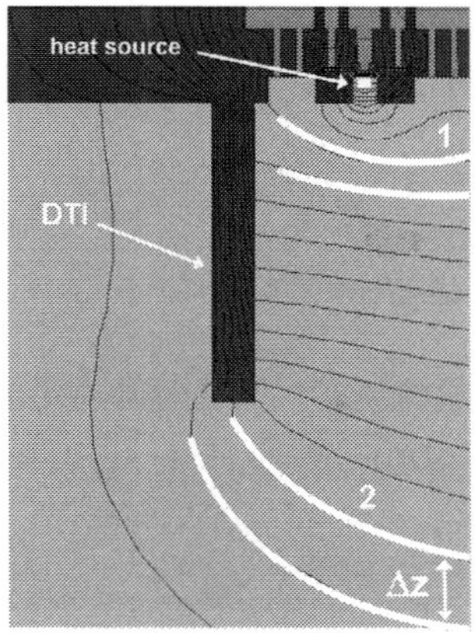

Figure 2.20
Thermal TCAD simulations for a five-fingers HBT in ST B55 technology. The image is zoomed on one finger to show the isothermal contours that identify the different volumes in which the structure can be subdivided.

Thermal TCAD simulations can help to identify the different volumes delimited by the isothermal contours created when a power source is applied to the transistor structure; in Fig. 2.20, the thermal profile simulated in TCAD Sentaurus for the five-fingers HBT realized in ST microelectronics B55 technology is reported for which the Z_{th} extraction has been previously demonstrated ($A_E = 0.18 \times 5\ \mu m^2$), in the case where a power dissipation of 30 mW per finger is applied. In the simulated structure, a zoom on just one finger is shown.

It can be observed in Fig. 2.20 that the isothermal contours identify multiple regions in the simulated structure and that the distance between the isothermal contours (and thus the volume) increases as the distance from the heat source increases. The R_{th} and C_{th} related to these volumes can be expressed as

$$\Delta R_{th} = \frac{\Delta z}{kA(z)} \tag{2.21}$$

$$\Delta C_{th} = \frac{k\,A(z)\,\Delta z}{\alpha} \tag{2.22}$$

where Δz represents the distance between two isothermal contours, $A(z)$ the related local cross section area, and k and α are, respectively, the thermal conductivity and the heat diffusion coefficient of the material. Comparing the ΔR_{th} related to the volume number 1 to the number 2 in Fig. 2.20, it will result $\Delta R_{th1} > \Delta R_{th2}$ (this assumption can be considered valid within the DTI enclosed area, where $A(z)$ is kept constant by the adiabatic wall created by the DTI), while the opposite will hold for the ΔC_{th}, being $\Delta C_{th1} < \Delta C_{th2}$. So, as the distance from the heat source increases, the R_{th} of the discrete volumes decreases, while the related C_{th} increases; the same trend is obtained in the recursive thermal network in Fig. 2.19, moving from the current source till the pole with index n.

This network architecture in Fig. 2.19 ensures a particular behavior as the frequency of the heat source is swept: at low frequency, the different C_{th} can be seen as open circuits and thus the Z_{th} can be approximated as the series of all the R_{th}; as the frequency increases, the high-valued capacitances located far from the current source start to shunt to ground, thus short-circuiting the related R_{th}. As the frequency further increases, the smaller capacitances located closer to the current source will start to shunt as well, thus yielding an equivalent Z_{th} which tends to 0 (compare to the Z_{th} vs frequency extracted in Fig. 2.17).

In other words, at low frequency, the whole volume of silicon below the heat source participates to determine the Z_{th}, and big-valued C_{th} are thermally charged and discharged; as the frequency increases this big volume cannot participate to the thermal sinusoidal variations, due to its intrinsic inertia and so the AC dynamic phenomena can be approximated to happen in a smaller volume, located close to the heat source. It is then evident that the Z_{th} has a physically distributed nature.

To check the inaccuracies that are introduced in compact models considering a simple single-pole thermal network, in Fig. 2.21, it is shown the measured Z_{th} on the five-fingers transistor under study and the results of simulation using a thermal network having, respectively, 1, 3, and 5 poles. The values of R_{th} and C_{th} are tuned to obtain the best fit

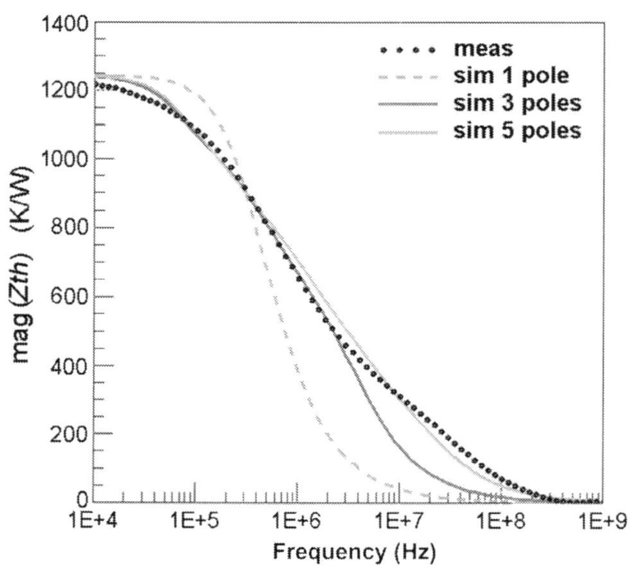

Figure 2.21
Comparison of the measured and simulated Z_{th} using a thermal network having 1, 3, and 5 poles.

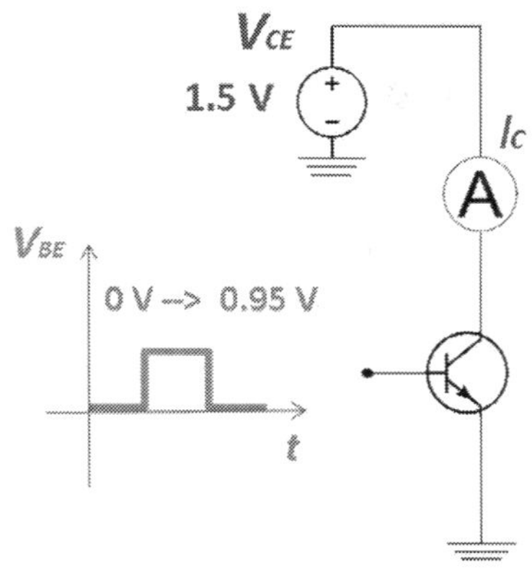

Figure 2.22
Measurement setup to characterize the thermal response of the transistor under pulsed operation.

and it is shown that three poles can lead to an acceptable accuracy in the measured frequency range.

An alternative way to characterize the distributed nature of the thermal impedance of state-of-the-art transistors is to perform pulsed measurements. When operated in pulsed conditions, SiGe HBTs show a transient variation of the junction temperature, which can lead to a slow response in the I_C evolution; in fact, if a pulsed V_{BE} is applied to the transistor (see Fig. 2.22), after a quick electrical response, the I_C will slowly reach its steady-state value, over a time proportional to the thermal time constant of the component [63–65].

In Fig. 2.23, it is shown the I_C measured on a HBT realized in the B11HFC technology from Infineon ($A_E = 0.22 \times 5\ \mu m^2$) for $V_{CE} = 1.5$ V, when a pulse of V_{BE} is applied, having an amplitude of 0.95 V. It can be seen that the I_C reaches a stable value in around 27 μs; this thermal evolution is not representable with a good accuracy using a single-pole thermal network (represented in the plot by a dotted line), whereas 3 poles recursive network can provide a more realistic simulation of the pulsed I_C. The simulations using a single-pole network can significantly deviate from the measurements in correspondence of the instant where the V_{BE} pulse is applied, especially in the case of large emitter area transistors, where the I_C can reach higher values [65].

2.2.6 Behavior of the Transistor Under Two Tones Excitation

As shown in Section 2.2.4, the thermal impedance of state-of-the-art SiGe HBTs has a cutoff frequency in the order of few hundreds of MHz; in other words, if a sinusoid at a given frequency f_{IN} is fed at the input of the transistor (and thus a sinusoidal power is dissipated within the component), the junction temperature T_j will experience a sinusoidal

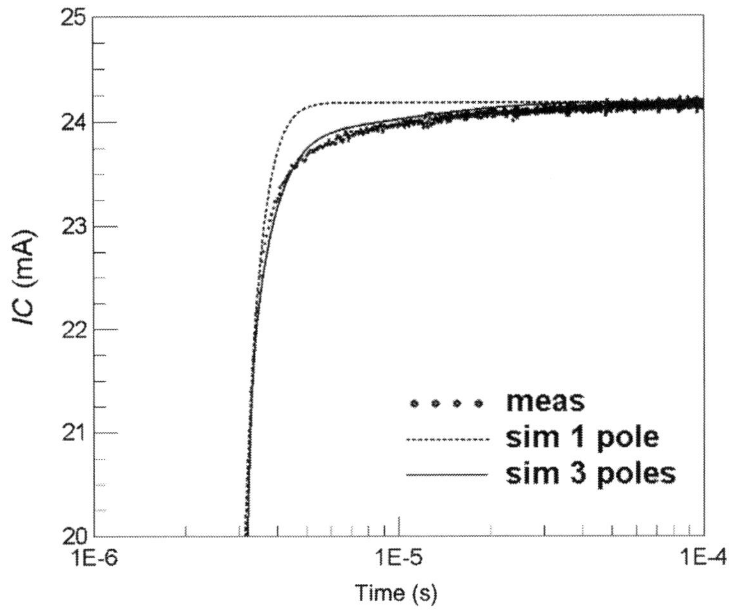

Figure 2.23
Measured and simulated I_C in pulsed conditions using a thermal network having 1 and 3 poles on an HBT in B11HFC technology.

variation, synchronized with the time evolution of the power dissipation. This will happen till f_{IN} is lower than the thermal cutoff frequency of the transistor, whereas for higher frequencies the T_j will remain stable, due to the thermal inertia of the component [66].

Even if the thermal impedance of these components is mainly located in the low-frequency range, it can affect the transistor behavior also when it is operated at much higher frequencies; in fact, when a large-amplitude signal or a modulated input signal is put at the input of the transistor, due to its intrinsic nonlinearities, intermodulation products are generated [67]. In power amplifier design, an important figure of merit used to characterize linearity is the intermodulation distortion (IMD), caused when multiple frequencies are mixed to form sum and difference products of the original signals; these signals occur as integer multiples of the input frequencies. It is possible to identify IMD of different orders, according to the frequencies which are mixed; however, the third order IMD (IMD3) is of particular importance in narrowband signals because its distortion is often in the signal band. In Fig. 2.24, the output spectrum of a transistor affected by IMD products is represented, when two frequency tones (f_1 and f_2) are present at its input.

In general, these nonlinear phenomena can be caused by electrical effects, but can also come from thermal self-heating. The mechanism of IMD3 generation for a two-tone input signal is shown in Fig. 2.25; there are two processes which lead to the generation of IMD3

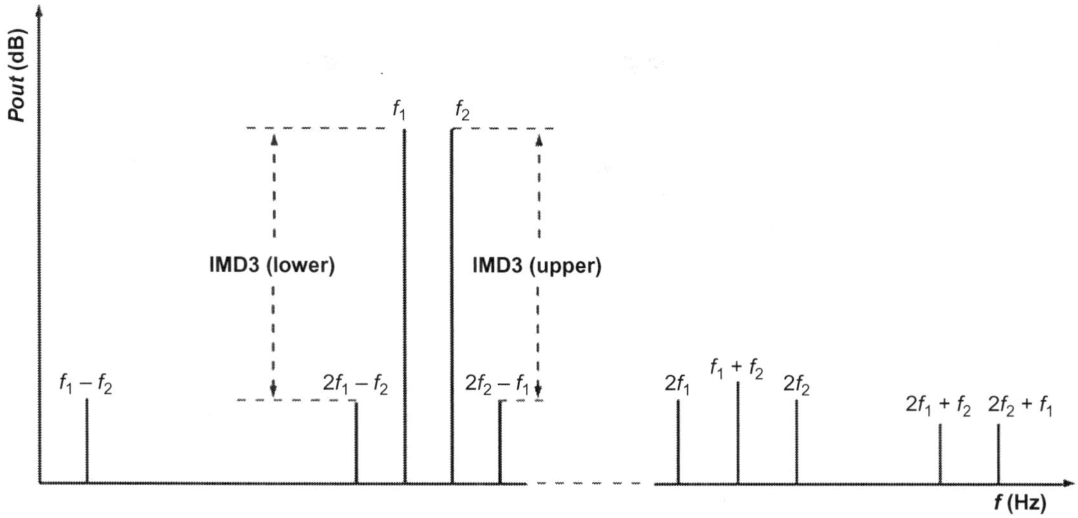

Figure 2.24
Schematic representation of different intermodulation products that can be generated when nonlinearities arise in a transistor.

at the output: the first is related to the third-order nonlinear transfer function that directly generates IMD3 (the products identified on the (A) path in Fig. 2.25). The second is related to the nonlinear transfer functions of second-order named (B) and (C) in Fig. 2.25: it has to be considered, in fact, that a nonlinear component, such as a transistor, generally has more than one nonlinear transfer function; the IMD3 can thus be also generated when a signal passes two times through second-order nonlinear transfer functions. This second process is called the memory effect [68]. Due to the memory effect, the baseband signals ($\omega_1 - \omega_2$ and $\omega_2 - \omega_1$), and the second-order higher harmonic signals ($2\omega_1$ and $2\omega_2$) interact with the fundamental frequency tones (ω_1 and ω_2), as shown in Fig. 2.25. Since the spectra of the dynamic self-heating phenomena can be very close to the baseband components $\omega_1 - \omega_2$ and $\omega_2 - \omega_1$, these dynamic temperature variations modulate the electrical memory effect. This is the origin of the thermal memory effect [42,66,68,69]. The frequency interactions leading to output intermodulation products affected by thermal memory effect are highlighted with a thick gray line in Fig. 2.25.

Due to the complex nature of these phenomena, in order to test the linearity of power amplifiers, it is useful to drive them with a two-tones test signal with varying tone spacing and amplitudes, to check how IMD mechanisms affect the transistor behavior. If the amplifier behaves properly for a wide range of tone spacings, it can be expected to behave correctly also in the case of wideband signals.

In Fig. 2.26, it is shown an ADS setup that can be used to simulate the effect of tone spacing on the time evolution of T_j predicted by the compact model. The transistor is biased

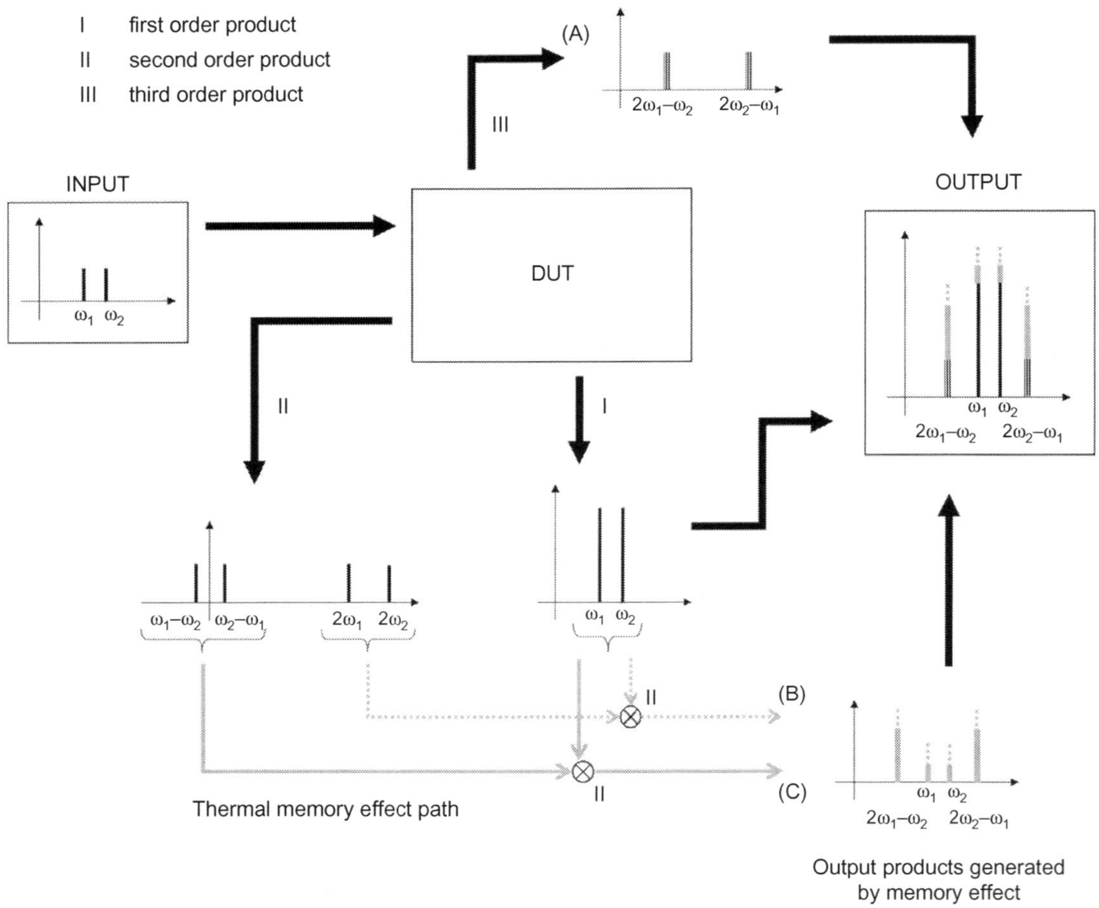

Figure 2.25

Mechanisms of IMD3 generation for a two-tone input signal. The frequency interactions that lead to output intermodulation products affected by thermal memory effect are highlighted with a thick gray line.

at a $V_{BE} = 0.925$ V and $V_{CE} = 1.5$ V, at the input two sinusoidal tones are applied (f_1 and f_2), both having an amplitude of -20 dBm. A three-poles network is used for these simulations, which has been validated upon low-frequency measurements (Fig. 2.23 shows the measured and simulated Z_{th} using the three-poles network and a single-pole network for the transistor under study).

In Fig. 2.27, the evolution of temperature for a single frequency of operation is reported: the f_2 tone is switched off, while the frequency of f_1 is set equal to 1 and 100 MHz; it is evident that, at low frequency, the T_j (in gray) is synchronized to the power dissipated in the transistor (in black) and a considerable temperature variation of around 15 °C is predicted. If the frequency of the input signal is set to 10 MHz instead, due to the reduced

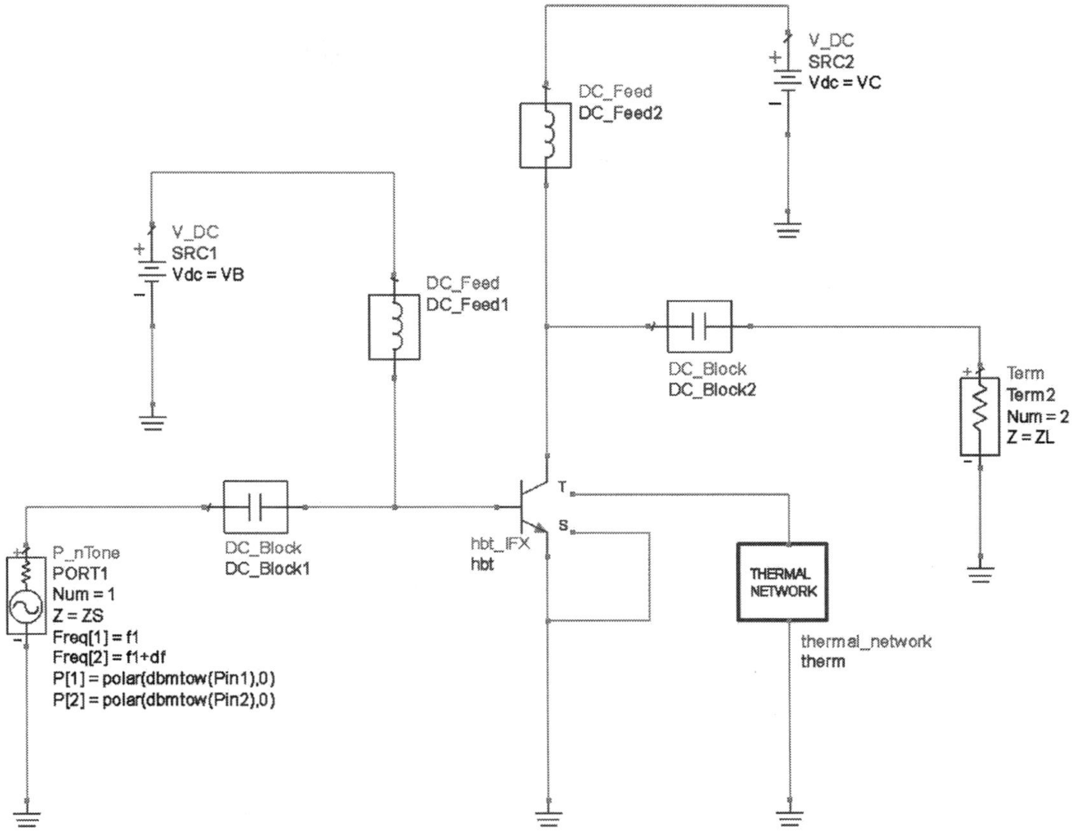

Figure 2.26
ADS schematic used to simulate the behavior of the transistor under a two-tones large signal excitation.

Z_{th}, the T_j responds to a lower extent to the power dissipation variations, resulting in an almost constant T_j. At this frequency of operation, the power dissipated in the transistor appears distorted due to a phase shift between the I_C and V_{CE} waveforms. If then a two-tones signal is be applied, it must be taken into account that, when driven by a two-tone input signal (ω_1 and ω_2), the baseband signals $\omega_1 - \omega_2$ and $\omega_2 - \omega_1$ generated by nonlinearity (see Fig. 2.25), act as a modulating signal on the carrier frequencies, and thus the transistor's current and voltage will follow the envelope of $\omega_1 - \omega_2$ and $\omega_2 - \omega_1$, as shown in Fig. 2.28.

In Fig. 2.29, the temperature evolution is shown, when the transistor is stimulated with a two-tones signal composed of $f_1 = 1$ GHz and $f_2 = 1$ GHz + df, where df represents the frequency shift between the two sinusoids, which is set to 1 MHz in Fig. 2.29A and 100 MHz in Fig. 2.29B. It can be seen that when $df = 1$ MHz, the T_j evolves following the

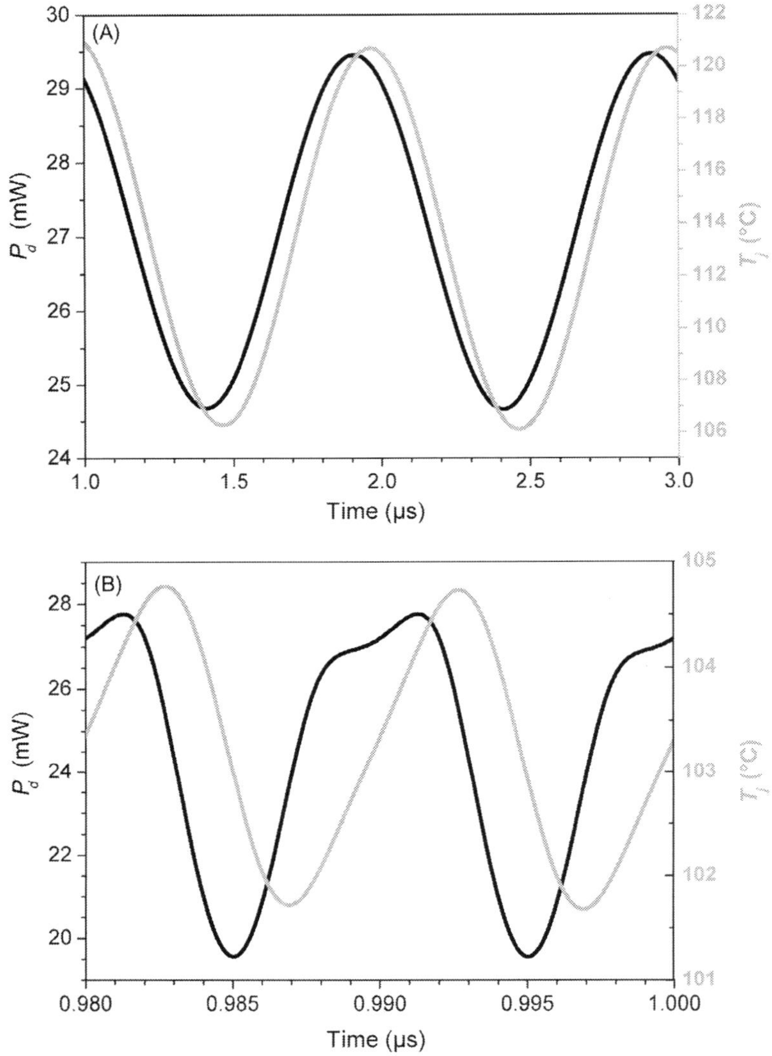

Figure 2.27
Simulated power dissipation and junction temperature for a single frequency excitation at a bias point of $V_{BE} = 0.925$ V and $V_{CE} = 1.5$ V. The frequency is set to 1 MHz (A) and 100 MHz (B).

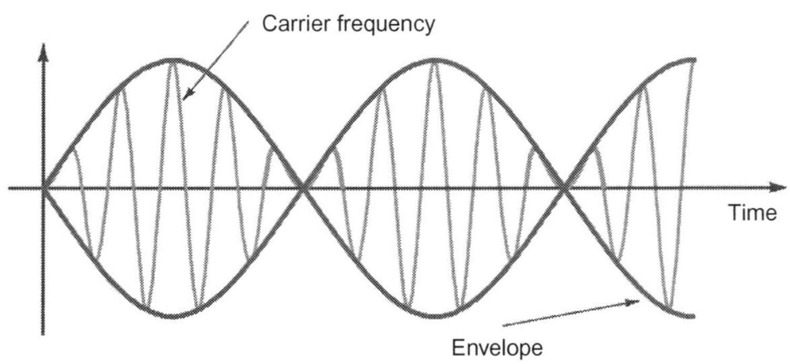

Figure 2.28
Representation of the time evolution of a high-frequency carrier modulated by a low-frequency envelope.

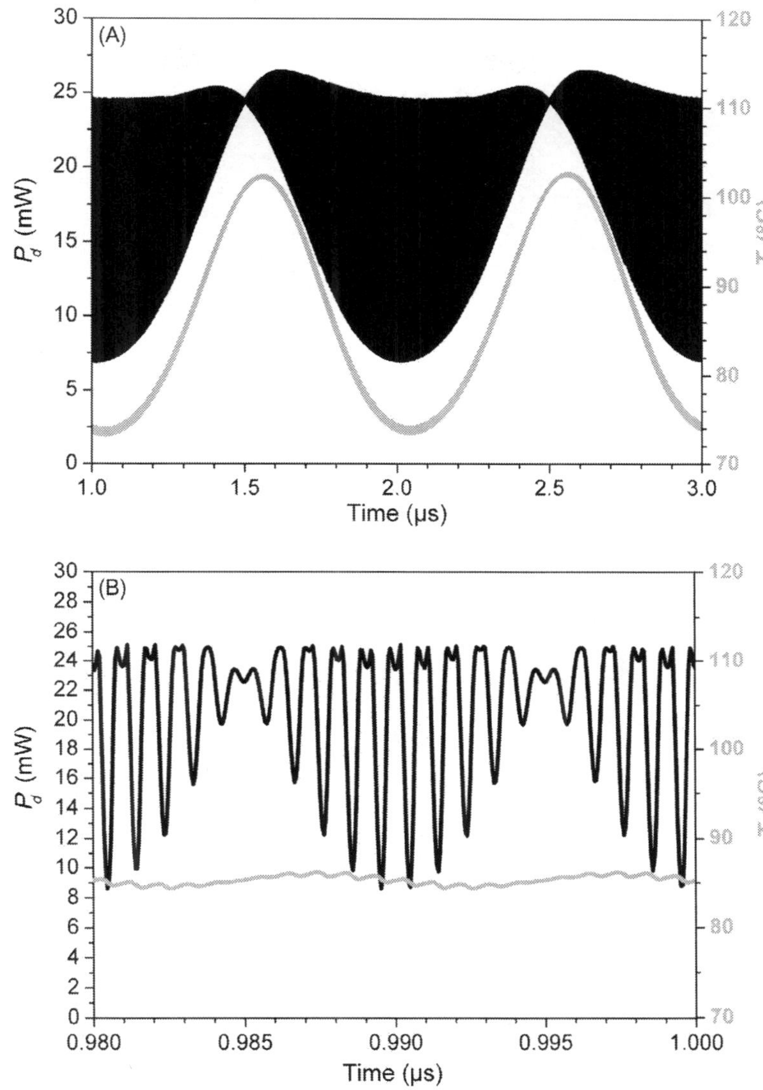

Figure 2.29
Simulated power dissipation and junction temperature for a two-tones excitation at a bias point of $V_{BE} = 0.925$ V and $V_{CE} = 1.5$ V. The frequency spacing is set to 1 MHz (A) and 100 MHz (B).

envelope of the power dissipated in the transistor, with a peak-to-peak simulated temperature of 30 °C. In the other case, where the envelope of the power dissipation has a frequency of 100 MHz, the T_j can be approximated to be stable to 85 °C, since its variations are negligible.

To conclude, in Fig. 2.30, a comparison of the temperature evolution is reported that has been obtained for a tone spacing of 1 and 100 MHz using the three-poles network

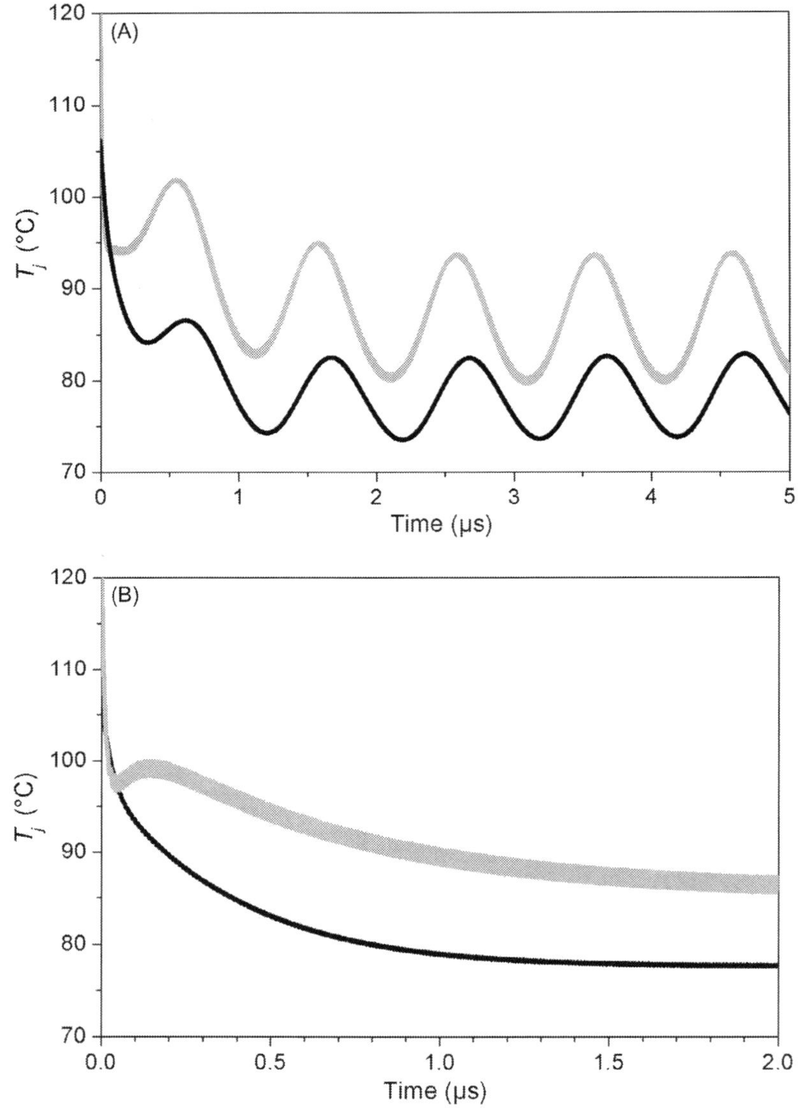

Figure 2.30

Simulated power dissipation and junction temperature for a two-tones excitation, comparison of the results obtained using a thermal network having a single pole (black line) or three poles (grey line). The frequency spacing is set to 1 MHz (A) and 100 MHz (B).

(grey line) and the single-pole network (black line) which have been adopted in the simulations in Fig. 2.23. It can be observed that a different thermal evolution is predicted by the compact model: in particular, the temperature assessment follows a different evolution at the beginning of the simulation, corresponding to the moment at which the transistor is turned on. Also the sinusoidal temperature variation is different in the two cases, since the two models present two different dynamic behaviors. More precisely, the single-pole network exhibits a lower thermal cutoff frequency and so its thermal capacitance shunts to ground earlier in frequency (see Fig. 2.21), thus attenuating the dynamic self-heating; this is the reason for which the sinusoidal temperature variations predicted by the single pole network are smaller.

The temperature variations triggered by the thermal memory effect should be correctly taken into account, since they can hugely impact the electrical behavior of transistor; at circuit level, it can be important to dispose of a transistor model that can accurately simulate these effects, in order to design pre-distortion linearizers [43,68,70].

2.3 Electrothermal Impact of the BEOL Metallization in SiGe HBTs

In the previous paragraph measurement and modeling techniques have been presented that provide accurate information about the static and dynamic self-heating effects inside SiGe HBTs. It has been discussed that the BC junction is thermally modeled as a heat source and it must be pointed out that the heat generated actually flows downward toward the back of the wafer and is also conveyed upward, in the BEOL region, whose dimensions are not negligible compared to the active part of the device. This region is composed of metal and SiO_2 layers, having very different thermal properties, so its presence has a certain thermal impact [45]. In different works, like [71,72], it has been demonstrated that the heat can find a preferential path through the metallization of the BEOL, and thus a thermal gradient can be observed through the metal layers, due to the higher thermal conductivity of copper compared to that of SiO_2. In [63,73], different techniques have been proposed to extract and model the BEOL contributions for thermal resistance and impedance, whereas in [74] and [75] it is demonstrated that an accurate design of the metal connections can lead to improved SOA specifications and better RF performances.

This paragraph presents an analysis of the thermal impact of the metal in the BEOL and it is also shown that a careful design of the metallization above the transistor can help improving the thermal handling capability of SiGe HBTs of the most advanced technologies. For these purposes, dedicated transistor structures are described and characterized; the HBTs are realized in a state-of-the-art SiGe BiCMOS technology from Infineon (B11HFC) having an $f_t = 250$ GHz, $f_{max} = 370$ GHz and six levels of metallization [76]. Different sets of test structures are presented and consist of a transistor cell containing specially designed metal stacks in the BEOL, which act as heat spreaders. Transistors having two different emitter widths are characterized: $L_E = 0.34$ μm and 0.22 μm. The first one has the capability of driving higher currents, whereas the latter ensures less parasitic resistances and capacitances; the choice of the geometry thus depends on the application. A complete DC and RF electrical characterization at small and large signal, as well as a thermal characterization are performed on the proposed test structures, in order to underline the performance improvements that are possible to achieve using these innovative topologies. A compact model is proposed to take into account the impact of the BEOL metallization on the thermal impedance; the model is validated for the $0.22 \times 5\ \mu m^2$ geometry (which is mainly used in RF design) and is based on a recursive RC network, which can be connected to the thermal node of the HiCuM model. For this set of test structures, the variation of the thermal behavior with the doping concentration is also analyzed by means of TCAD simulations in which both static and dynamic power dissipations are applied and the results are compared to the experimental data obtained from measurements.

The approach of exploiting metal layers to help evacuate the heat generated by active devices has also been used in other state-of-the-art technologies, like 3D-integrated circuits, which are also affected by serious self-heating and mutual-heating issues, due to a very high density of power dissipating components. To overcome these issues, that highly degrade the performances and reliability, thermal vias (made of copper or other materials with high thermal conductivity) are designed and realized through the oxide, thus allowing a better handling and extraction of the power dissipated [77–79].

The same principle of exploiting the metallization of the BEOL has been applied in different published works, like [45,72] that demonstrated a significant influence on the thermal behavior of SiGe HBTs of the most advanced technologies. In these works, the impact of the BEOL on the thermal properties of the HBTs has been analyzed by means of TCAD simulations. On the optimization of these BEOL stacks, limited information is available in the literature. In [80], for instance, some innovative test structures are proposed to investigate the impact of the strain created by metal dummies on the bandgap inside the base of the transistors. The metal stacks are placed above the active part of the transistor and the result is a sensible increment of the I_C as well a higher transit frequency (f_T). This improved electrical behavior is caused by a variation in the base bandgap structure, whereas no sensible improvement of the thermal properties has been reported. This is probably due to the fact that the metal dummies are not connected by vias to the emitter contact, which reaches high temperatures when the device is operated at high power dissipation [81].

2.3.1 Electrothermal Characterization of Dedicated HBT Test Structures

The test structures under study consist of a single transistor having a CBEBC arrangement for the contacts and are connected in common emitter configuration. The reference test structure is an HBT that has only metal-1 above all the contacts. For all the test structures analyzed in this section, the drawn emitter window is $5 \times 0.34\ \mu m^2$; this transistor geometry yields high values of power dissipation and the addition of metal stacks above the heat source can thus show some sensible variation in the electrothermal behavior.

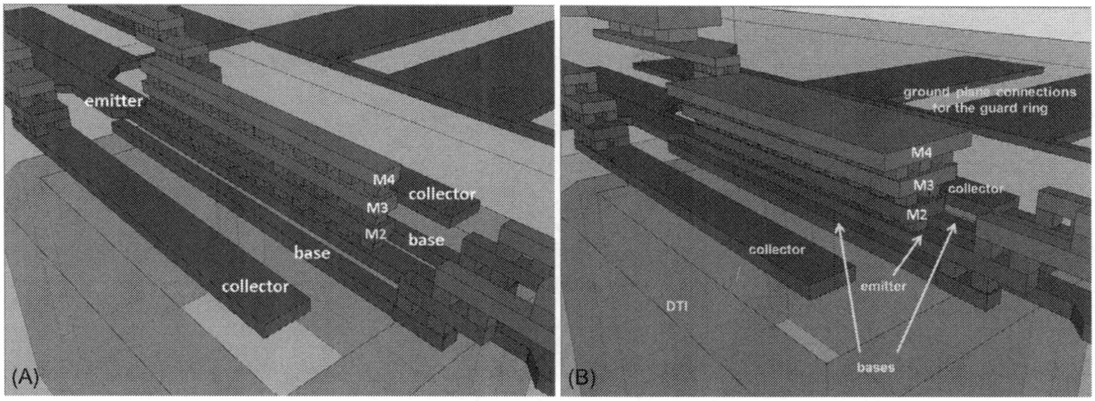

Figure 2.31
3D representation of the $E4_{narr}$ (A) and $E4_{wide}$ (B) test structures.

Different approaches are used to design the BEOL metallization upon the active part of the component. For the first set of test structures, the metal bars are stacked one on top of the other and connected by vias; all the metal layers have the same length and width (0.34 μm). These structures are named $E2_{narr}$, $E3_{narr}$, and $E4_{narr}$, where the number stands for the level of the last metal stack that is added and E indicates that the metal dummies are placed above the emitter contact (refer to Fig. 2.31A). A second set is similar to the first, but this time the additional metal dummies have a gradually increasing width, as shown in Fig. 2.31B: for this set of structures, the overall volume of added metal is higher than the previously discussed set; these structures are named $E2_{wide}$, $E3_{wide}$, and $E4_{wide}$ and the width of the metal bars upon the emitter is gradually increased till reaching 1.52 μm for the $E4_{wide}$ structure. Finally, a last set is characterized with metal stacks above the two base contacts, as shown in Fig. 2.32; two versions are realized, named B2 and B3, whereas the reference structure will be referred to as B1E1 (in this case, as already mentioned, just metal-1 is present on both the base and emitter contacts).

The added metal stacks are designed in such a way that they do not impact the access lines compared to the reference structure B1E1; the same access resistance is in fact guaranteed for all the test structures under study by contacting the transistor nodes of the different structures using similar metal interconnections (Fig. 2.32).

In Fig. 2.33, a schematic representation of the thermal resistances is sketched that are distributed along the BEOL in a transistor test structure containing metal dummies connected upon the emitter (like in the case of $E4_{narr}$ or $E4_{wide}$). Since the BC junction is thermally modeled as a heat source, in the equivalent electrical circuit it is represented with a current source. In this schematic representation, the heat actually flows downward toward the back of the wafer through R_{th}FEOL, which models the thermal properties of the front-end-of-line part, and is also conveyed upward through the metallization, which can be seen

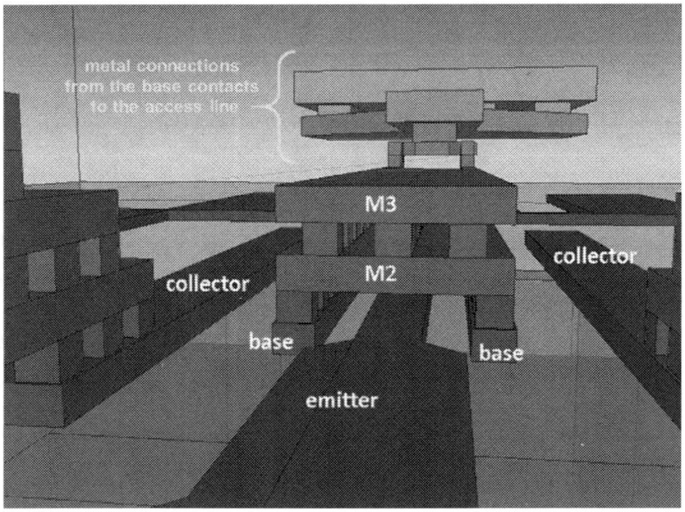

Figure 2.32
Schematic 3D representation of the B3 test structure.

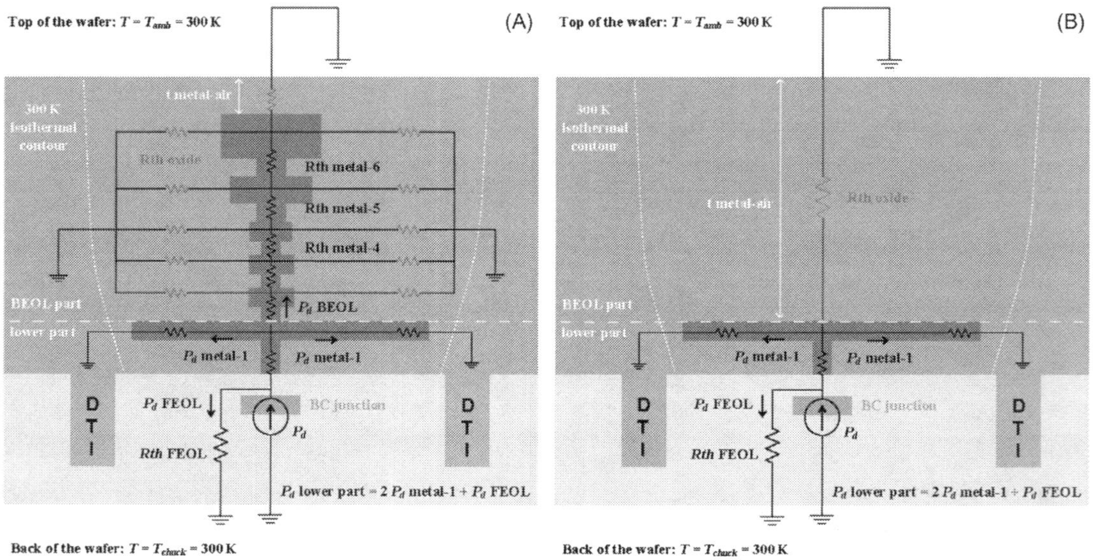

Figure 2.33
Schematic representation of the thermal resistances distributed across the transistor structure in the case where metal heat spreaders are present in the BEOL (A) and in the case they are absent (B).

as a preferential path for the heat flux, due to the higher thermal conductivity of copper compared to that of SiO_2 [45]. It has been shown in [45] by means of TCAD simulations that at a certain lateral distance from the metal stacks (that reach a temperature T_{metal}, higher than T_{amb}) the structure presents an isothermal contour at ambient temperature; for this reason adding metal stacks results in adding T-networks (formed by $R_{th\ metal}$ and $R_{th\ oxide}$), thus allowing alternate paths for the heat flow to the thermal ground (T_{amb}). At the same time, as more metal levels are stacked, the distance $t_{metal-\mathrm{air}}$ decreases, thus decreasing the corresponding R_{th} and contributing further to the diminution of the overall R_{th} of the component. Instead, to represent a thermal model for a test structure (like B1E1) which does not contain metal dummies, since the metal stacks from metal-2 to metal-4 are absent, the thermal resistances of the metal layers in the region named "BEOL part" should not be considered.

2.3.1.1 DC electrical characterization

A DC characterization of the structures has been done using a Keysight E5270B DC analyzer; the measurements are performed on a probe station equipped with a thermal chuck.

Fig. 2.34 shows the output curves measured at a chuck temperature $T_{chuck} = 300$ K; for sake of clarity, within each set of test structures only the ones reaching the highest levels of

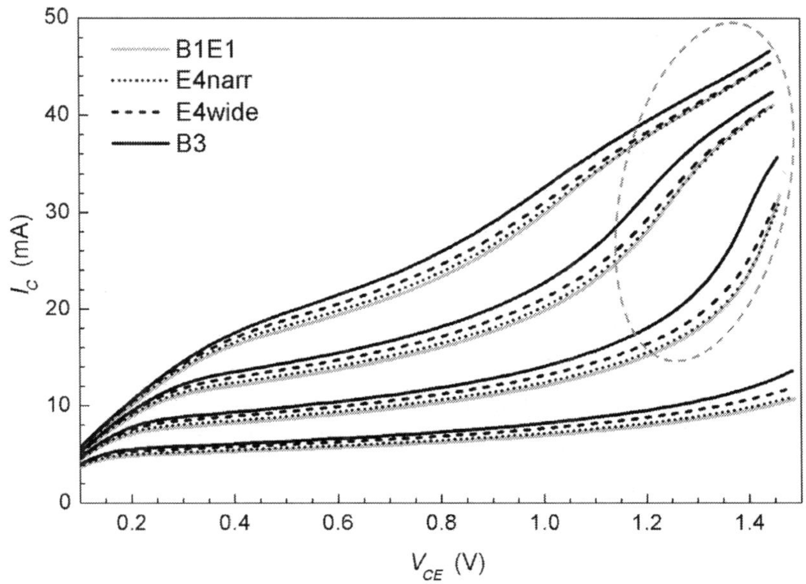

Figure 2.34
Output curves measured for different HBT test structures realized in the Infineon B11HFC process; V_{BE} ranges from 830 to 890 mV.

Table 2.1: I_C variation at $V_{BE} = 0.83$ V and $V_{CE} = 0.5$ V

Test structures	ΔI_C %
B1E1	0.0
$E2_{narr}$	1.8
$E3_{narr}$	2.5
$E4_{narr}$	3.4
$E2_{wide}$	5.5
$E3_{wide}$	9.0
$E4_{wide}$	10.0
B2	10.2
B3	15.8

metal heat spreaders are shown. It can be noticed in Fig. 2.34 that in the low V_{CE} region of the output curves (where the power dissipation is quite low), the test structures that have metal dummies above the transistor drain a slightly higher I_C than the reference B1E1 (the percentage variations observed at $V_{BE} = 0.83$ V and $V_{CE} = 0.5$ V are shown in Table 2.1). This variation is due to the mechanical strain effects of the metal dummies that lead to a change in the bandgap structure of the HBT, as described in [80].

As reported in Table 2.1, as the level of metal stacks and the width of metal dummies increases, the percentage increase in I_C ($\Delta I_C\%$) tends to increase, in accordance to the results published in [80]. Furthermore, it is noticeable that adding metal dummies on the base contacts results in a more pronounced strain effect and thus a higher $\Delta I_C\%$. From Fig. 2.34, it is observed that the I_C trend of the compared structures starts to be different getting closer to the regions of high power dissipation (P_d) highlighted with a dashed circle, where the total P_d is estimated to range between 20 and 65 mW. The different behavior of the test structures in the high power dissipation region is instead mostly related to a variation in the R_{th}.

In Fig. 2.35, the output conductance $g_{out} = \partial I_C / \partial V_{CE}$ is plotted as a function of the measured current in each structure. It is evident that, for a given I_C, higher than 15 mA, the structures containing metal dummies exhibit a lower g_{out} than the reference structure. It has been verified, by means of simulations using a HiCuM model of the transistors under study, that the decrease in g_{out} is caused by the combination of two different effects: (1) the increase in the saturation current (which is induced by the BEOL mechanical stress) and (2) the decrease in the R_{th}. Regarding the latter, it must be pointed out that the self-heating effect (among other effects like avalanche multiplication) is responsible for the sudden increase in the I_C in the high power dissipation region (clearly visible in Fig. 2.34 for $V_{BE} = 0.85$ V).

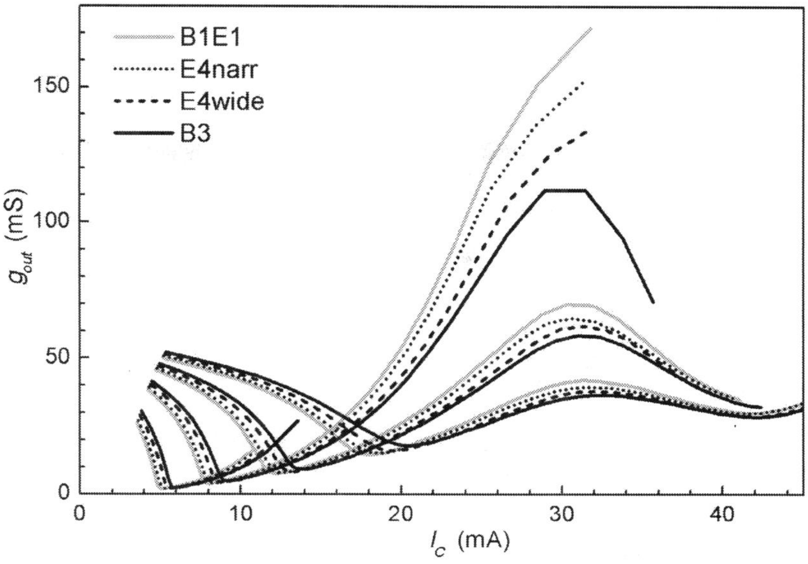

Figure 2.35
Output conductance for the same bias points and test structures shown in Fig. 2.34.

As shown in Fig. 2.35, the peak of g_{out} (or, equivalently, the maximum slope of I_C in the $I_C(V_{CE})$ plane in Fig. 2.34) is reached for all the test structures at an I_C value of around 30 mA. For this value of I_C, at $V_{BE} = 0.85$ V, a maximum g_{out} variation of around 54% is found between the reference structure B1E1 and structure B3. For I_C values higher than 30 mA, the g_{out} starts to decrease; this phenomenon is caused by the extrinsic emitter resistance of the HBT, which plays a feedback role, limiting the internal base–emitter voltage for high values of I_C and thus counterbalancing the I_C increase caused by the self-heating effect. It must be pointed out that a careful design of metal heat spreaders in the BEOL can not only improve the DC electrothermal performances of the HBTs, but even lead to an improvement of their SOA specifications; in fact, as it has been demonstrated in [74], the decrease in g_{out} ensures that the transistor can withstand higher V_{CE} voltages (or drain higher I_C currents) and thus its SOA is sensibly extended.

2.3.1.2 Thermal characterization

As already discussed, it should be expected that the addition of metal stacks above the active part of the transistor allows an alternate path for the heat flow, thus lowering its R_{th}. An R_{th} extraction is realized performing DC measurements at constant I_B for different values of T_{chuck}, to calibrate the V_{BE} as a function of the ambient temperature. Then, using an approach similar to the extraction method proposed in [48], the R_{th} is calculated and is plotted in Fig. 2.36 as a function of the T_j for some of the test structures under study. It is possible to see that all the test structures present an R_{th} that

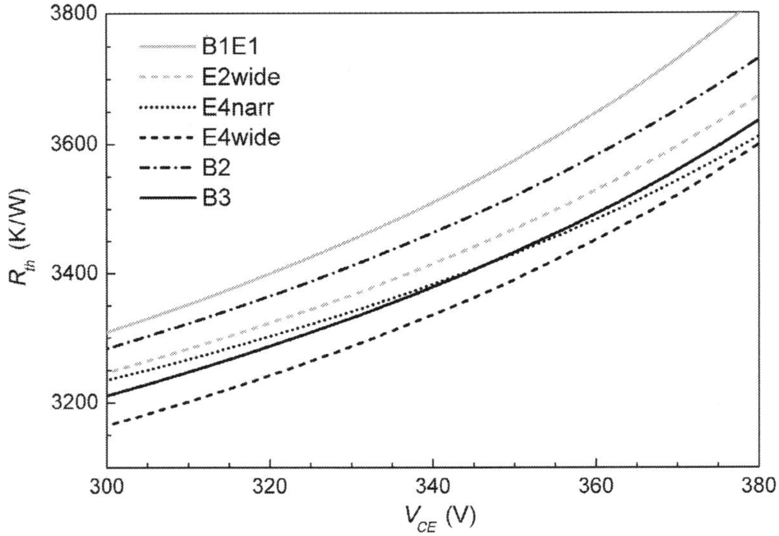

Figure 2.36
Extracted R_{th} vs T_j for some of the structures presented.

is lower than B1E1 and the best thermal behavior is realized for the test structure $E4_{wide}$, which presents an R_{th} around 5% lower than B1E1. Comparing the R_{th} of test structure $E4_{narr}$ and $E4_{wide}$, it can be stated as well that increasing the width of the metal dummies yields an enhanced thermal behavior. A sensible R_{th} decrease is obtained also for the set of structures B2 and B3.

2.3.1.3 Small signal RF characterization

To continue the analysis of the test structures, a small signal RF characterization is proposed and the f_T and f_{MAX} of the transistors under study are extracted. For this purpose, S-parameters are measured at 35 GHz using a ZVA67 Rohde-Schwarz VNA and de-embedded using the same open and short structures for all the transistors under study. Also in this case, for sake of clarity, only the results for the most representative test structures having the highest level of metallization are reported and commented.

From the 3D images in Figs. 2.31 and 2.32, it is evident that the metal dummies that have been added are electrically connected to the emitter contact or to the base contact and could therefore add potential parasitic elements at high frequency, impacting in a negative way the AC figures of merit like the f_T and f_{MAX}.

The extracted f_T, evaluated at an applied V_{CE} of 1.5 V is plotted vs the measured I_C in Fig. 2.37A and demonstrates instead that the presence of dummies does not degrade this figure of merit. On the contrary, a sensible increment of the peak f_T is noticeable for the test structures $E4_{wide}$ and B3. For I_C higher than 10–15 mA the structures containing

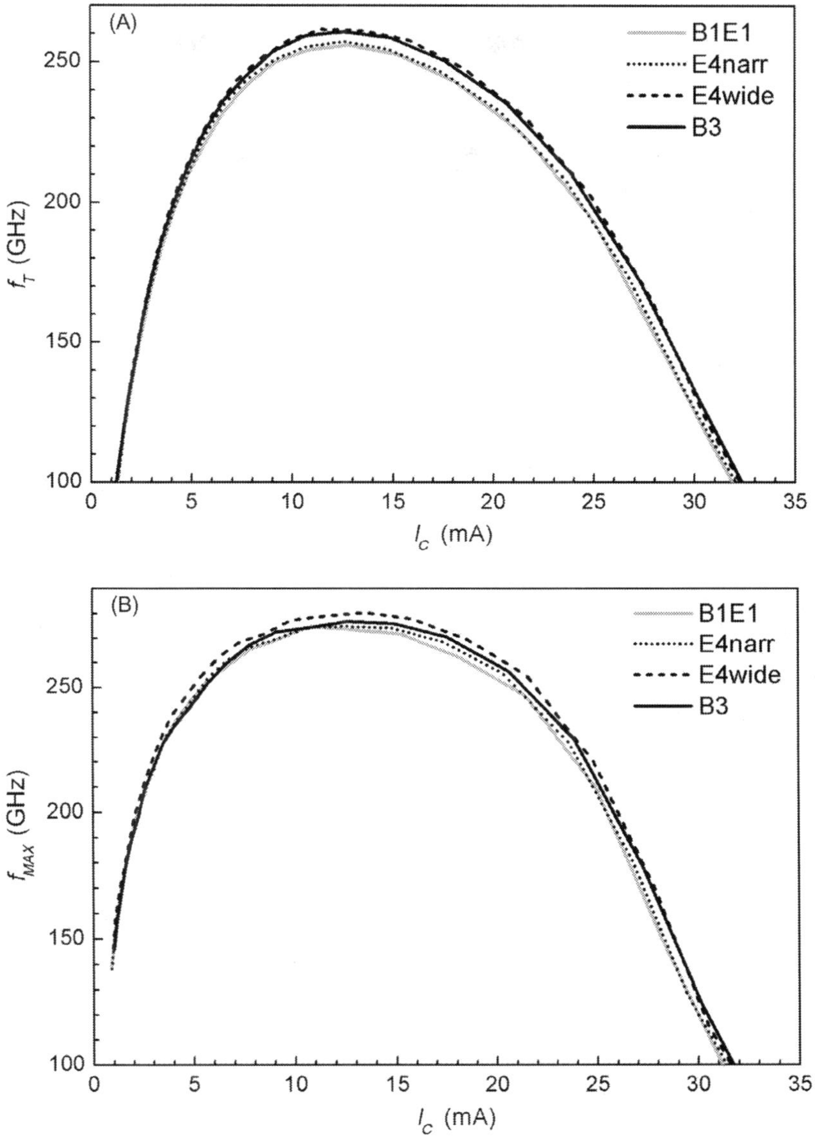

Figure 2.37
f_T (A) and f_{MAX} (B) evaluated at V_{CE} of 1.5 V.

metal dummies perform better in terms of f_T, probably due to the effect of the mechanical stress and to the reduction of R_{th}, which plays a positive role in the electrothermal behavior of the transistor when high levels of P_d are reached. The same improvement is detectable on the f_{MAX}, which is extracted at the same bias conditions as the f_T and also shown in Fig. 2.37B.

2.3.1.4 Large signal RF measurements

The behavior of the different transistor structures is finally tested under large signal conditions at 77 GHz on a single-tone load-pull measurement system. The passive load and source tuners (FOCUS CCMT) present in the measurement setup allow generating complex impedances between 57 and 90 GHz. The Rohde & Schwarz SMF100A is used as a signal generator in combination with the SMZ90 frequency multiplier. A directional coupler divides the input signal into two parts: one is guided to the power meter for measuring the power delivered to the DUT and the other one is the actual input signal. The isolator prevents the reflected signal from returning in the coupler and the power meter at the output port allows to measure the P_{out} (Fig. 2.38).

In Fig. 2.39, the output power (P_{out}) and output efficiency ($\eta = P_{out} / I_C V_{CE}$) contours obtained from a sweep of the output load over 70 complex impedances on test structures

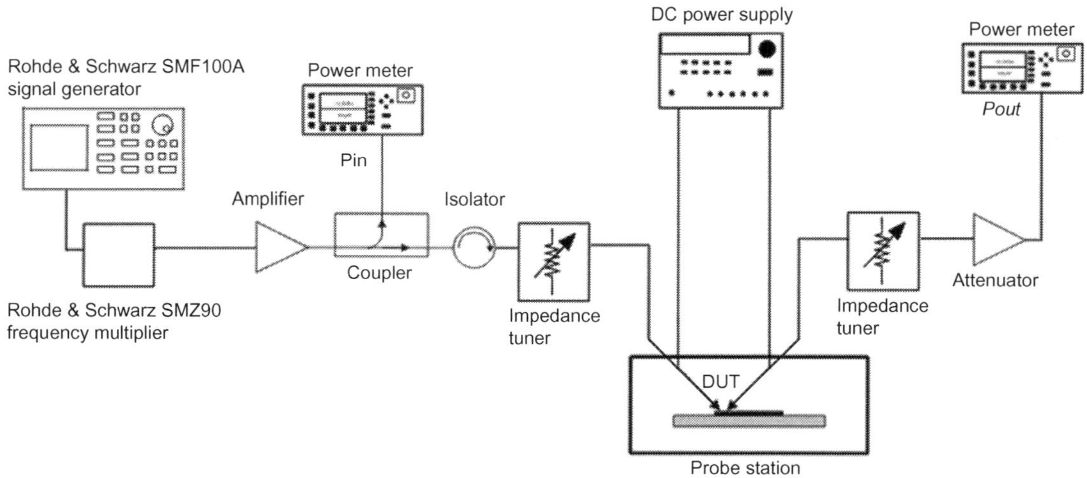

Figure 2.38
Schematic representation of the load pull measurement setup.

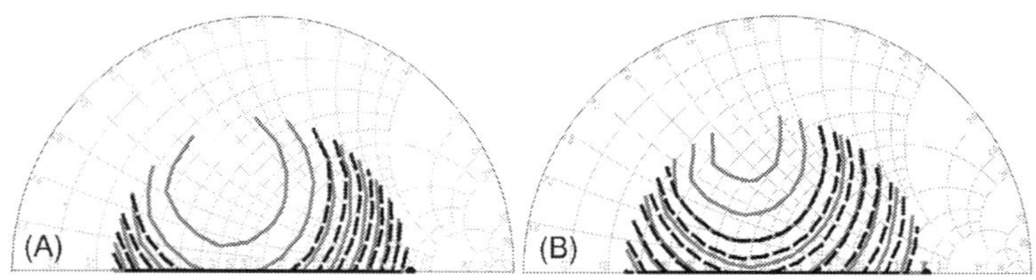

Figure 2.39
P_{out} (A) and η (B) contours for the bias point $V_{BE} = 835$ mV and $V_{CE} = 1.25$ V when a P_{avs} of 0 dBm is applied. The contours for B1E1 are shown in grey, whereas in black are reported the results for E4$_{\text{wide}}$. The P_{out} contours range from 1.2 to 4.5 dBm, whereas the η contours range from 6.1% to 18.2%.

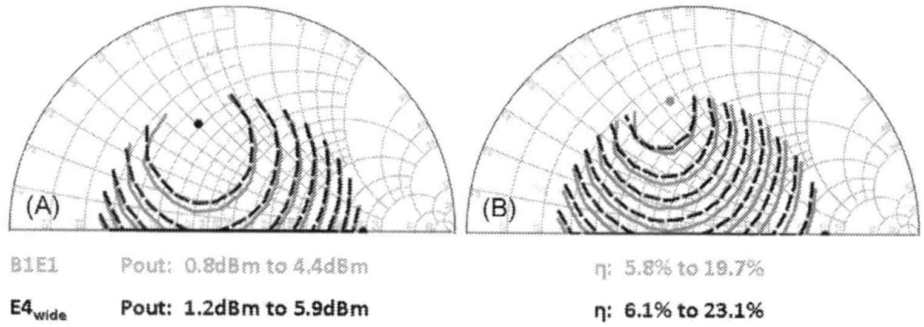

Figure 2.40
P_{out} (A) and η (B) contours for the bias point $V_{BE} = 835$ mV and $V_{CE} = 1.25$ V when a P_{avs} of 0 dBm is applied. The contours for B1E1 are shown in grey, whereas in black are reported the results for $E4_{wide}$. The ranges of the P_{out} and η contours are different than in Fig. 2.39 and are indicated for the two test structures.

B1E1 and $E4_{wide}$ are reported; in this plot, the measured P_{out} and η are reported using the same ranges for the two structures compared (from 1.2 to 4.5 dBm for the P_{out} and from 6% to 18% for the η). Although in this representation the B1E1 and $E4_{wide}$ contours seem to be different, it is to be noticed that the loads yielding the maximum values of P_{out} and η are located in the same regions of the Smith chart for the two test structures under comparison; this is shown in Fig. 2.40, where the range limits for P_{out} and η are adapted to the minimum and maximum values measured on each of the two transistors. For all the Smith chart contours shown, the power available from the source (P_{avs}) is 0 dBm, whereas the base and collector voltages are set, respectively, to 835 mV and 1.25 V, which corresponds to the bias conditions of peak f_T.

Next, the input impedance Z_S is fixed at 50 Ω, whereas the output impedance Z_L is set, respectively, to 50 Ω and to the load ensuring the highest η, which is found to be (20.7 + j37.6)Ω; the power available from the source (P_{avs}) is swept from −12 to 7 dBm. The results reported here are for the same bias point previously considered: $V_{BE} = 835$ mV and $V_{CE} = 1.25$ V.

It is possible to see in Fig. 2.41 that at $Z_L = 50$ Ω the structure $E4_{wide}$ is able to deliver around 0.5 dBm of additional power to the output over the whole range of swept P_{avs}; when the output impedance is (20.7 + j37.6)Ω this difference can reach 2 dBm, showing that the impact of the BEOL at large signal can be significant and that it is also function of the load conditions applied to the transistor. Fig. 2.42 shows the output efficiency for the three structures under study; it has been measured an η increase till 2.5% and 8.2% passing from test structure B1E1 to $E4_{wide}$ for a Z_L of, respectively, 50 Ω and (20.7 + j37.6)Ω.

To explain this enhancement, it is helpful to investigate the evolution of I_C as the input power increases (Fig. 2.43): it is shown that for low P_{avs} it is the structure B3 that drains the highest I_C, due to the effect of mechanical stress already reported in Table 2.1. As P_{avs}

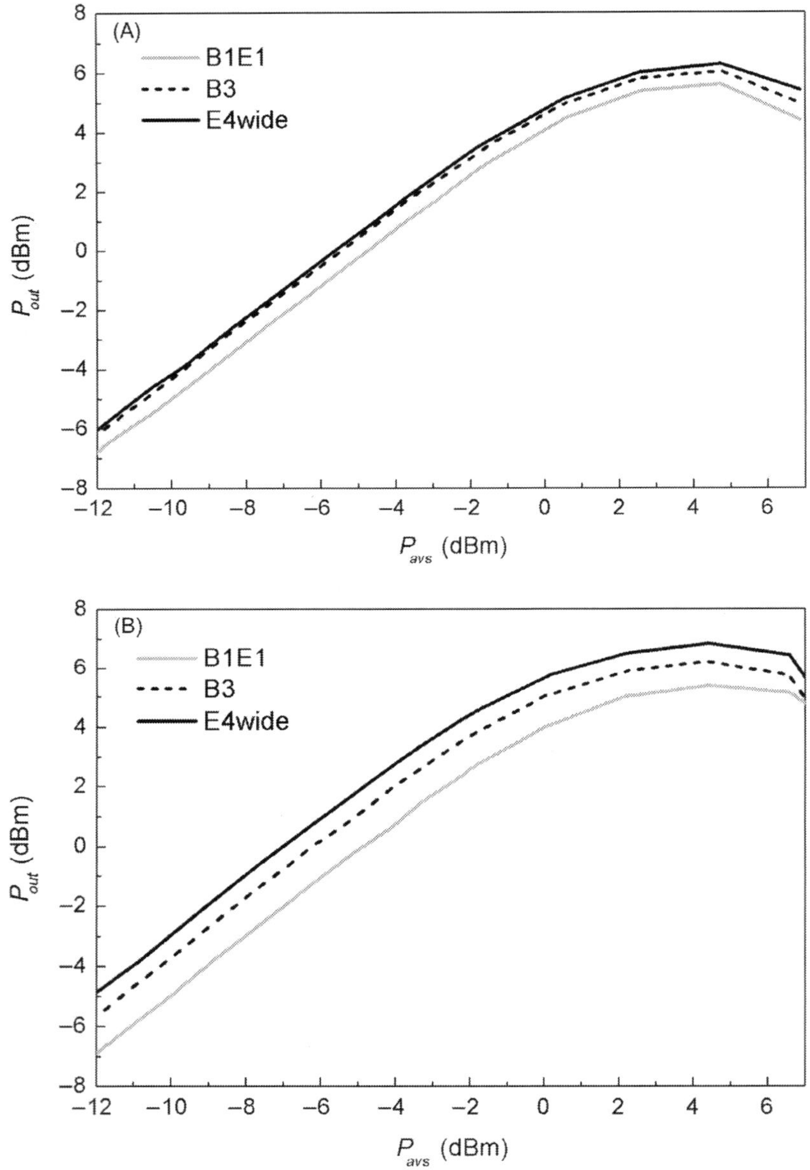

Figure 2.41
Output power plotted vs P_{avs} for $V_{BE} = 835$ mV and $V_{CE} = 1.25$ V; for a load impedance of 50 Ω (A) and for the load yielding maximum η: $(20.7 + j37.6)\Omega$ (B).

increases, though, it is $E4_{wide}$ that drains the lowest current, since it has the lowest R_{th} among the other structures. This lower I_C yields lower power dissipation and, together with the slightly higher P_{out} (Fig. 2.41) implies a better η figure of merit. Large signal measurements at different bias points show a similar behavior and are not reported.

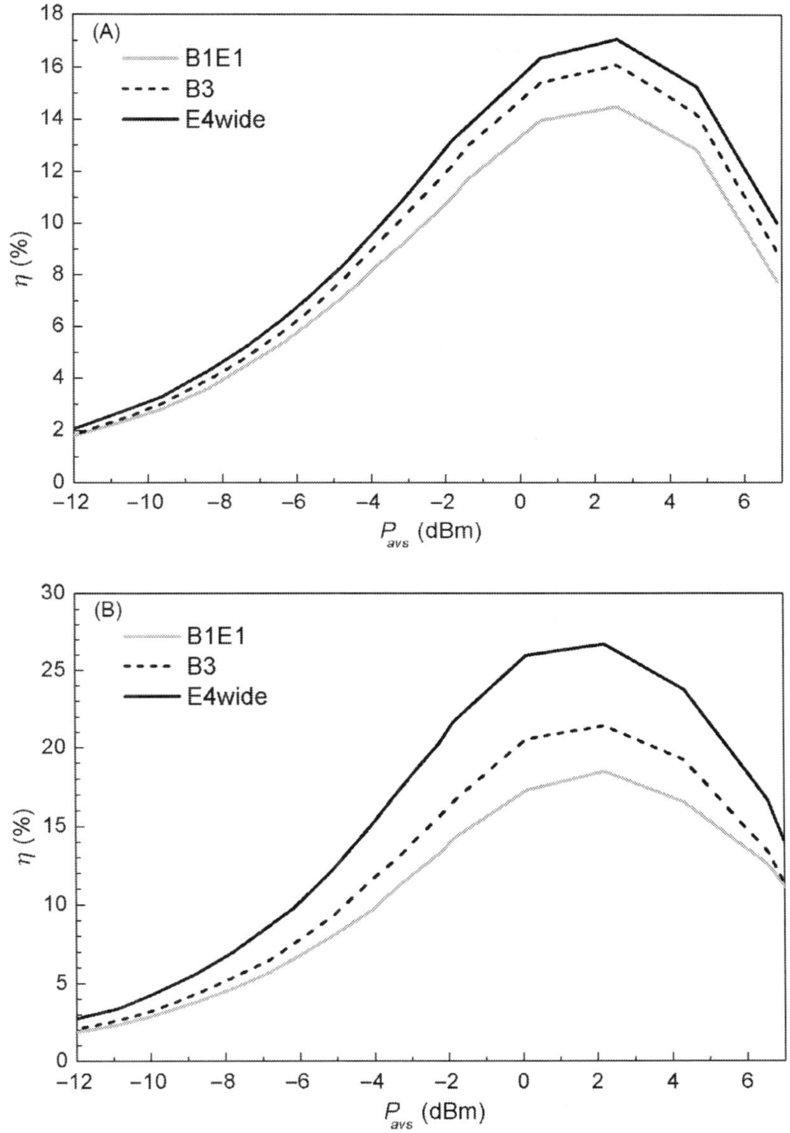

Figure 2.42
Output efficiency vs P_{avs} for $V_{BE} = 835$ mV and $V_{CE} = 1.25$ V; for a load impedance of 50 Ω (A) and for the load yielding maximum η: $(20.7 + j37.6)\Omega$ (B).

2.3.2 Compact Modeling of the BEOL Thermal Impact

After having presented in the previous sections, a complete electrical characterization of the BEOL impact, the focus will be hereby on the possibility to take into account its thermal impact into compact models. Now the focus will be on test structures realized using a

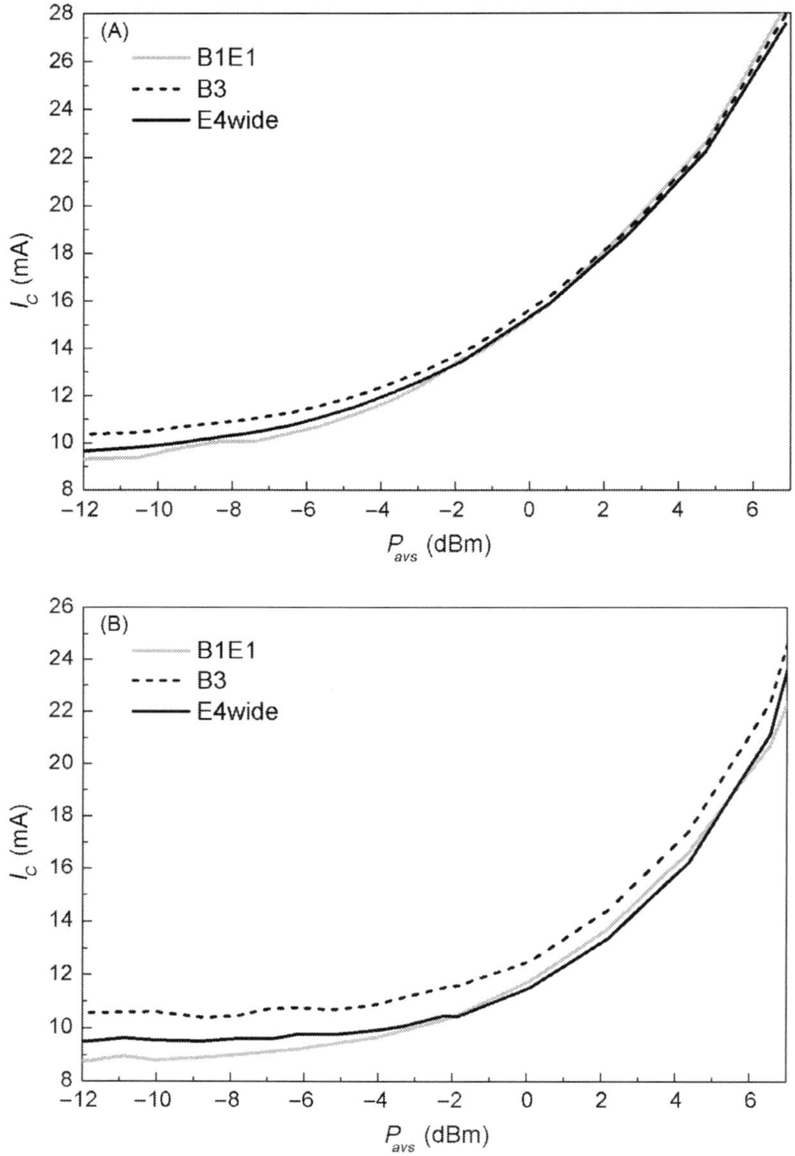

Figure 2.43

Collector current vs P_{avs} for $V_{BE} = 835$ mV and $V_{CE} = 1.25$ V; for a load impedance of 50 Ω (A) and for the load yielding maximum η: $(20.7 + j37.6)\Omega$ (B).

$0.22 \times 5\ \mu m^2$ transistor geometry and again a CBEBC configuration for the contacts; this geometry ensures a lower R_B and C_{BC}, which result in a higher f_{MAX}, making it a good choice in circuit design where RF performances are important.

The test structures consist of a single transistor cell in common emitter configuration and are realized on a different wafer than the ones presented in Section 2.3.1. As it is possible

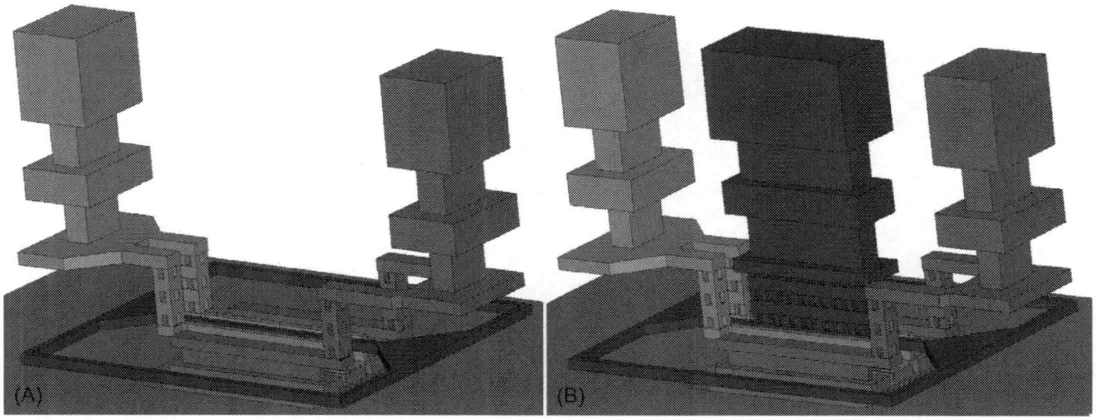

Figure 2.44
3D images of the standard test-structure M1 (A) and test-structure M6 with metal stacks above the emitter (B). These test structures contain HBTs designed using the Infineon B11HFC process and having a drawn emitter area of $0.22 \times 5\ \mu m^2$.

to see in Fig. 2.44B, metal stacks are placed above the emitter contact till metal-6. The colors of the metallization are: blue for the base connections, green for the collector, brown for the emitter; the active part of the device is placed below the BEOL, and is partially visible (in yellow), while silicon is colored in orange. At the metal-6 level, the base and collector are then connected to transmission lines (not shown), while the emitter is connected to the guard ring, and then to a ground plane (not shown). This structure will be named M6 (Fig. 2.44B) and is compared to a reference one, named M1, where just metal-1 is present above the emitter, to electrically connect it to ground, without other metal layers above (Fig. 2.44A).

2.3.2.1 Thermal modeling of the BEOL metallization

In Figs. 2.45 and 2.46, an equivalent electrical model to simulate the thermal behavior of the two test structures is proposed. Since the thermal impedances of these devices show a distributed nature [45,54,82], three-poles Cauer-type recursive networks are used. For structure M1 in Fig. 2.45, just one network is used to model the lower part, which takes into account the thermal behavior of the device from the bottom of the wafer up to metal-1; then in the case of the M6 test structure, a second network is added in parallel to the previous one, to simulate the presence of the BEOL part (Fig. 2.46). In the two proposed models, the temperature dependency of R_{th} is also taken into account.

2.3.2.2 DC electrical characterization

To evaluate the thermal behavior of the two test structures under study, first of all a DC characterization is made on a probe station equipped with a thermal chuck (that sets

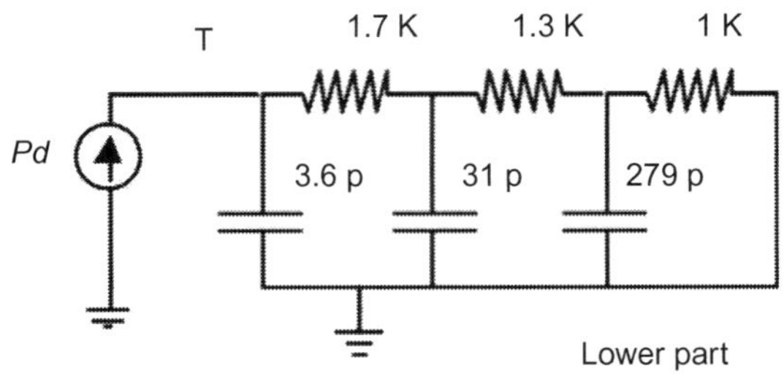

Figure 2.45
Cauer-type recursive network used to model the thermal impedance of the M1 structure; this network models just the thermal behavior of the lower part of the device up to metal-1.

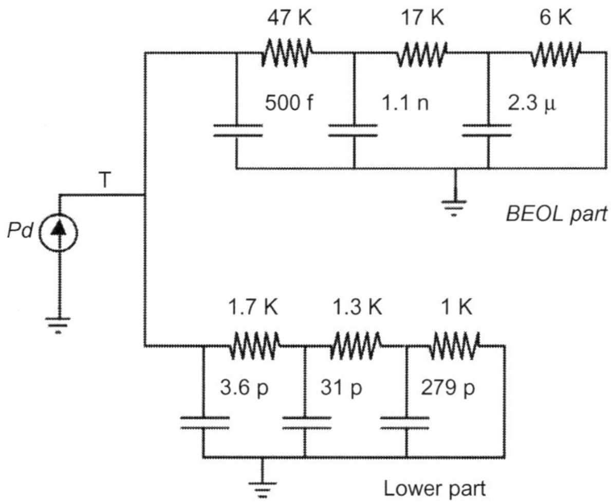

Figure 2.46
Cauer-type recursive network used to model the thermal impedance of the M6 structure. The dissipated power P_d flows in the lower part (that takes into account the thermal behavior of the device up to metal-1) and in the BEOL part (which models the thermal behavior of the metal stacks from metal-2 to metal-6).

$T_{amb} = 300$ K): as expected for the output characteristics (measured with an Agilent 4155 A DC analyzer), the transistor in the M6 structure drains less collector current than M1 in the bias regions where high self-heating occurs (Fig. 2.47); the bias point where this ΔI_C is maximum for $V_{BE} = 0.925$ V and $V_{CE} = 1.5$ V, for which passing from structure M1 to M6 gives an I_C drop from 18.5 to 16.7 mA (around 10% decrease, which is not negligible). The two thermal networks presented in the previous section are connected to the temperature node of a HiCuM model and the values of the different R_{th} are adjusted to fit the measured data; the output curves obtained from the simulations are represented with solid lines in Fig. 2.47. For the highest power dissipation, the T_j predicted by the simulator is 427 K for M1 and 412 K for M6. The equivalent R_{th} used in the thermal models drops from 3950 to

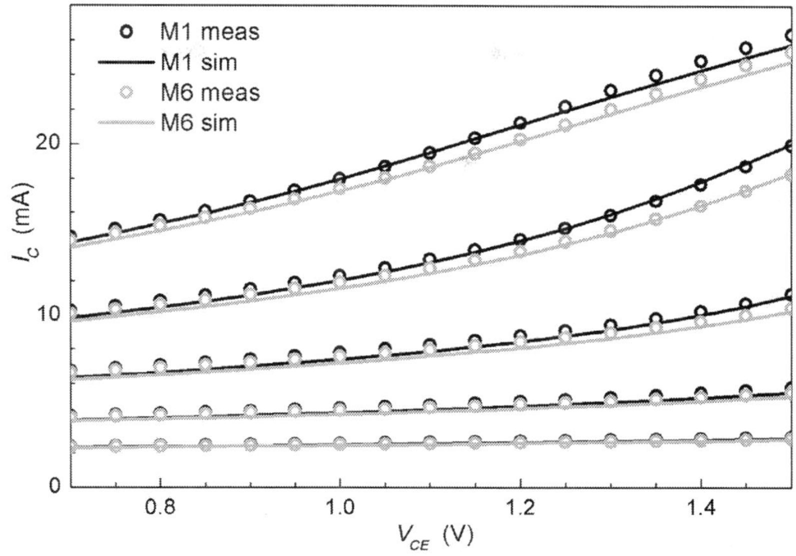

Figure 2.47
I_C (V_{CE}) for different V_{BE} values: zoom on the output curve in the region where self-heating is more present; V_{BE} ranges from 850 to 950 mV.

3739 K/W (around 5%) passing from structure M1 to M6, in agreement with the results in [45]. According to the model shown in Fig. 2.46, with a total P_d of 40 mW, the amount of power that is dissipated in the lower part is 36 mW, whereas 4 mW will be dissipated in the BEOL part.

2.3.2.3 Low-frequency measurements

To evaluate the impact of the BEOL on the C_{th}, the device must be stimulated in dynamic operation. To extract the thermal cutoff frequency of these components, which is in the range of few MHz [54], the evolution vs frequency of the Y- or H-parameters are commonly used as temperature sensitive parameters and therefore used in the Z_{th} calculation [54,55,83]. In this case, Y_{22} is found to be among the most sensible temperature parameters and therefore chosen to tune the elements of the proposed three poles recursive networks (this parameter can be demonstrated to be directly proportional to Z_{th}, as detailed in Section 2.2.4), obtaining all the values for R_{th} and C_{th} presented in Figs. 2.45 and 2.46.

In Fig. 2.48, the magnitude vs frequency of Y_{22} for the two test-structures under study at the DC bias point is shown, where the maximum variation of I_C was found. Observing the decay of the measured $|Y_{22}|$ vs frequency, it looks clear that a single-pole thermal network (which is normally used in compact models) is not sufficient to reproduce the distributed $|Y_{22}|$ trend vs frequency. It is therefore strongly recommended to use at least a three-poles network to accurately take into account dynamic thermal effects in the low-frequency

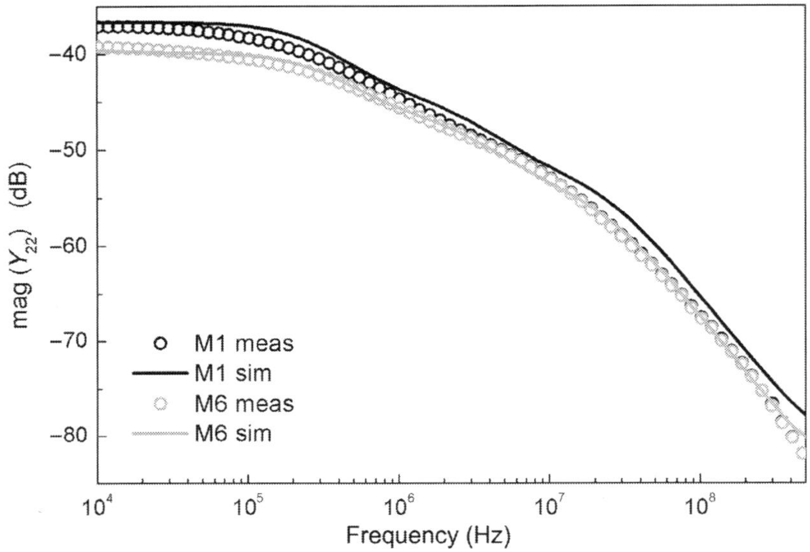

Figure 2.48
Magnitude vs frequency of the temperature sensitive Y_{22} parameter for $V_{BE} = 0.9$ V and $V_{CE} = 1.5$ V.

range; this is particularly important in circuit designs where IMD3 products arise within the thermal bandwidth of the HBTs. It is moreover evident that the added metal stacks lower $|Y_{22}|$ (and thus $|Z_{th}|$) in the frequency range up to around 3 MHz; at higher frequencies the $|Y_{22}|$ trend is almost the same for the two test structures. It looks evident that structure M6 has a lower cutoff frequency, so it can be assumed that some big thermal capacitances are added due to the presence of the metal stacks in the BEOL. It is in fact observed in Fig. 2.46 that the C_{th} of the BEOL part are much higher than those related to the lower part.

2.3.2.4 Pulsed measurements

Time-domain measurements are also performed, to further verify the proposed model: the device is constantly biased on the collector, while on the base a voltage pulse is generated using a Keithley 4200 containing a 4225-SCS module. Fig. 2.49 shows the I_C that is measured on test-structure M1 and M6 for different amplitudes of V_{BE}. At the instant, when the pulse is applied, I_C exhibits a quick increase (corresponding to an initial fast electrical response) followed by a slow increment due to the much slower thermal time constant. It can also be noticed that for test-structure M6 the I_C increase is much slower than M1, due to the big C_{th} that are added; besides the steady-state I_C value is lower for M6 due to less self-heating (lower R_{th}); the simulation provides a good reproduction of these thermal phenomena. The corresponding temperature evolution that is simulated using the HiCuM model connected to the proposed thermal network is shown in Fig. 2.50. It can be observed

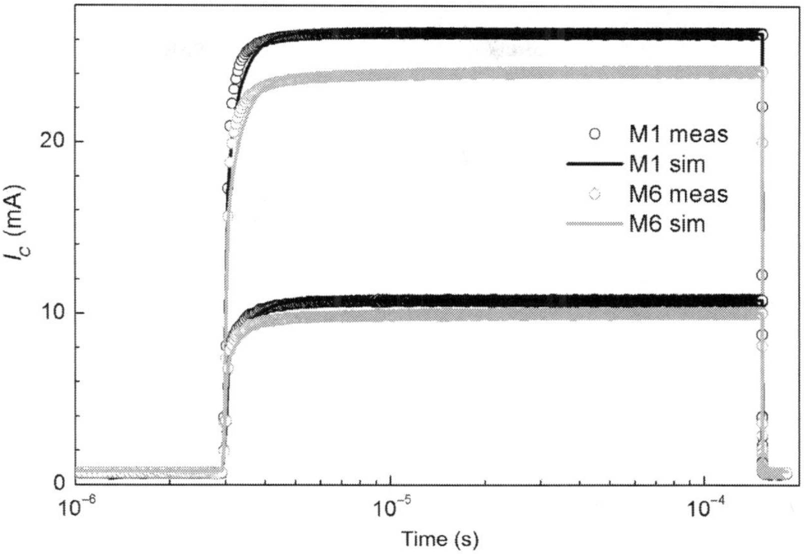

Figure 2.49
Measured and simulated I_C vs time for $V_{CE} = 1.5$ V; on-voltages of the pulse signal V_{BE} are 0.9 and 0.95 V; *x*-axis is in log scale to better evidence the evolution of I_C vs time.

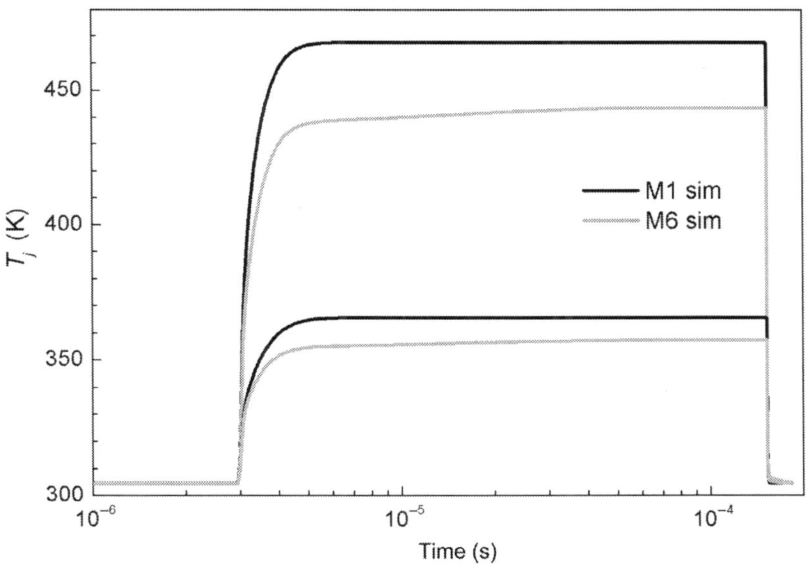

Figure 2.50
Resulting temperature evolution from the simulation of the HiCuM model using the proposed thermal networks in the same bias conditions detailed in Fig. 2.49.

that the simulated T_j follows the same evolution of the I_C, since it is the device temperature itself which is responsible for the slow variation of I_C in pulsed operation.

2.3.2.5 Large-signal two-tones simulations

Finally, once the thermal models have been calibrated upon dedicated measurements, they are used to simulate the behavior of the two transistor structures under a two-tones excitation; the setup that is considered is the same which has been presented in Section 2.2.6, where it has been discussed about how the low-frequency envelope of the power dissipation can modulate the T_j of the transistor. The bias conditions are: $V_{BE} = 925$ mV and $V_{CE} = 1.5$ V, the two input frequencies are set to $f_1 = 1$ GHz and $f_2 = 1$ GHz + df and the power of both is -20 dB.

In Fig. 2.51, the temperature evolution simulated by the HiCuM model for both M1 and M6 is shown, when the frequency spacing between the tones (df) is of 100 kHz and 100 MHz. As it is expected, the temperature of M6 will be lower, due to the effect of the R_{th} decrease; moreover, it can be noticed that, for $df = 100$ kHz, the temperature excursion (the difference between the maximum and minimum temperatures reached) caused by the envelope of the power dissipation is higher for M1 than for M6 (30 °C vs 22 °C, respectively); in the case of $df = 100$ MHz, instead, the temperature excursion appears the same for both structures (around 2 °C of excursion). This effect is justified by the fact that, when the structures are stimulated by a power dissipation envelope having a low frequency, the big thermal capacitances that are added due to the BEOL contribution help to keep the temperature of the transistor more stable; when instead the df increases, these high-value C_{th} do not contribute any more to the temperature variations. In this situation, the dynamic thermal behavior in the two structures becomes comparable and a much smaller thermal variation of 2 °C is simulated for both transistors; this is also confirmed by a comparison of the Z_{th} measured on the two transistors in Fig. 2.55.

2.3.3 Static and Dynamic 3D TCAD Thermal Simulations

To evaluate the heat diffusion mechanisms in the test structures M1 and M6 presented in Section 2.3.2, 3D numerical simulations are performed using Sentaurus device simulator. The temperature distributions due to device self-heating are calculated by solving the heat flow equation:

$$C_L \frac{\partial T}{\partial t} - \nabla \cdot k \nabla T = q \tag{2.23}$$

where T represents the temperature and q is the power generated by the heat source. In the proposed simulations, the heat source is placed at the BC junction of the device (the 3D TCAD simulator only allows the use of planar heat sources).

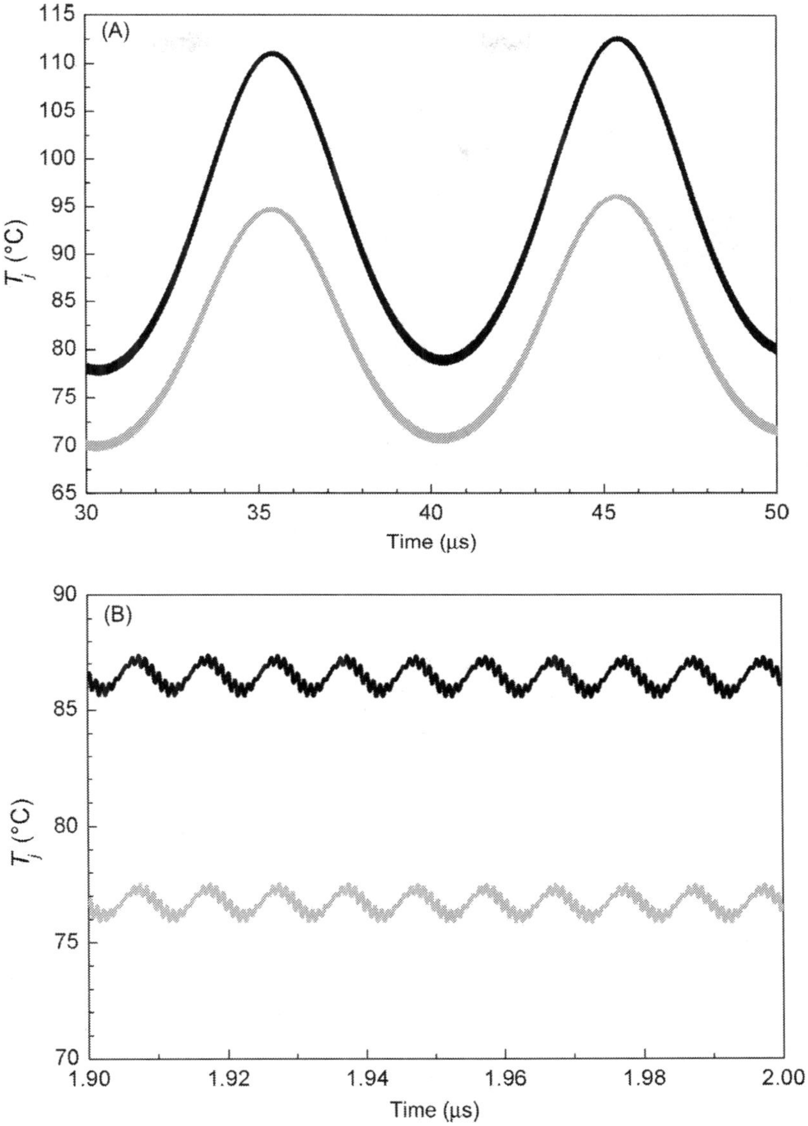

Figure 2.51
Two-tones large-signal simulations on the M1 (black line) and M6 (grey line) transistors for two different values of the tone spacing: (A) 100 kHz and (B) 100 MHz.

The simulated transistor structure is placed on a silicon substrate which is 300 μm thick; since the whole structure is symmetric with respect to the *xz* plane, only half of the device is considered (see Fig. 2.52 and refer to Fig. 2.44 for a more detailed reproduction of the metallization in the two test structures). To reproduce the same thermal conditions

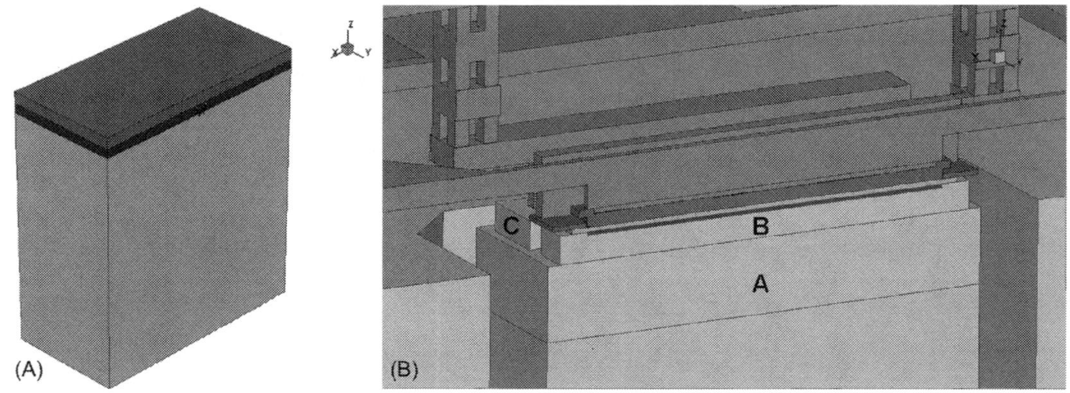

Figure 2.52

(A) The structure simulated in Sentaurus TCAD environment: silicon is represented in pink, SiO_2 in brown, copper in orange, and the air layer is colored in blue. (B) Zoom-in the active part of the transistor: polysilicon is colored in magenta, SiGe in green, and the heat source is highlighted in red, whereas SiO_2 is hidden for better visibility; the buried layer is indicated with the letter A, the selectively implanted collector (SIC) with B and the collector sinker with C.

than during the measurements, the boundary conditions defined in the simulation environment are as follows:

- The boundary surfaces of the silicon block are adiabatic, i.e., there is no heat flow to the surroundings.
- The wafer back side is isothermal and initially kept at ambient temperature (300 K). This is considered as the thermal ground.
- A thin air layer (15 μm thick) is present above the oxide passivation. Within this layer, there is no particle movement (laminar sub-layer) due to the fluid viscosity and the convection mechanisms can be neglected, considering just thermal conduction [84]. Due to the very low k of air, this boundary condition is very close to be adiabatic.
- An isothermal boundary condition is considered above the thin air layer; the temperature is set to 300 K.

2.3.3.1 Thermal parameters and doping dependence

Thermal simulations are usually performed by setting the thermal conductivities k of the materials equal to the values measured from bulk undoped samples (listed in Table 2.2).

However, in state-of-the-art technologies, high values of doping are reached in some well-defined regions within the transistor and this can yield a local change of the thermal properties of the material, especially the thermal conductivity k [71]. It must be taken into account that the doping profile information is often protected by strict confidentiality and that few experimental data are available in scientific literature about the change of the

Table 2.2: Thermal properties of the materials used for TCAD simulations in the case when the impact of doping is not considered

Material	Thermal conductivity (K) (W/cm K)	Heat capacity (Cp,v) (J/cm^3 K)
SiO_2	0.014	1.67
Silicon	1.54 (at 300 K)	1.63
Copper	3.85	3.42
Polysilicon	1.50	1.63
SiGe	Si:1.54; Ge:0.6	Si:1.63; Ge:1.67

Table 2.3: Thermal conductivity used in the regions of the transistor where a high-doping concentration is present

Region	Material	Doping considered (atoms/cm^3)	Thermal conductivity (k) (W/cm K)
Y-shaped emitter	Polysilicon	$1 * 10^{21}$ (Arsenic)	0.15
Base contact	Polysilicon	$6 * 10^{20}$ (Boron)	0.5
Base	$Si_{0.8}Ge_{0.2}$	$6 * 10^{20}$ (Boron) for Si	0.08
SIC	Silicon	$1 * 10^{19}$ (Arsenic)	1.1
Buried layer	Silicon	$1 * 10^{20} - 5 * 10^{19}$ (Arsenic)	0.45–0.75
Collector sinker	Silicon	$1 * 10^{21}$ (Arsenic)	0.15

thermal properties with the doping concentration. For this reason, a standard but realistic doping profile is assumed for the technology under study and the values of thermal conductivity k are modified according to the data published in [85] for what concerns silicon and in [86] for what concerns polysilicon; furthermore, the same temperature dependency of k than undoped silicon is assumed in the regions where a lower k is considered due to the high doping concentration. At this point, to evaluate the impact of the doping profile on the thermal behavior of the transistor, three different cases are studied and compared to the measured results: (1) the default k used in Sentaurus environment is employed for all the materials, without taking into account the thermal impact of the doping (as reported in Table 2.2); (2) the k is changed in the regions of the transistor where a high level of doping concentration is present, according to the values shown in Table 2.3 and considering $k = 75$ W/cmK for the buried layer; (3) the k in the different regions is changed according to the values shown in Table 2.3, using $k = 45$ W/cm K for the buried layer. The different regions of the transistor considered can be identified in Fig. 2.52.

2.3.3.2 Static thermal analysis and 3D simulations

In Fig. 2.53, the temperature profile and the heat flux in the two structures are reported, when a static power dissipation of 40 mW is applied to the heat source (a value of $k = 45$ W/cm K is considered for the buried layer). It is clear how the metal stacks placed

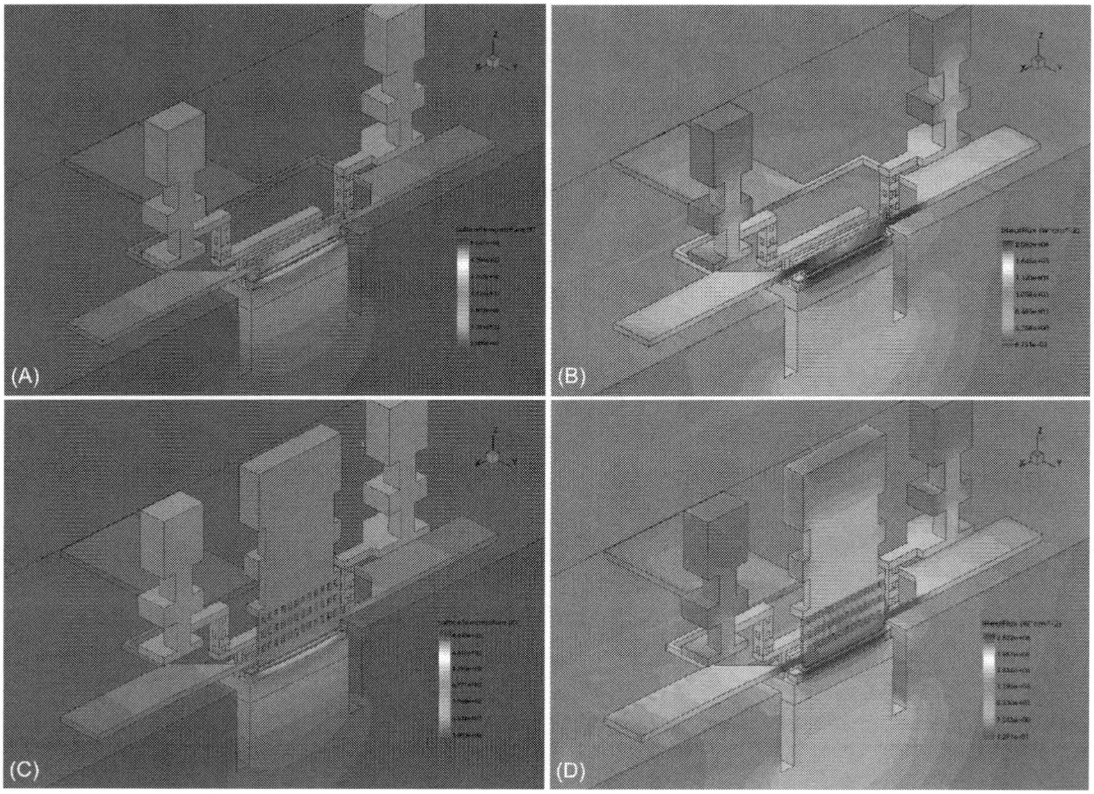

Figure 2.53
(A) thermal profile and (B) heat flux simulated for the M1 structure; (C) thermal profile and (D) heat flux simulated for the M6 structure. In all the simulations, $P_d = 40$ mW and the effect of the doping concentration on the thermal conductivity k is taken into account, considering $k = 45$ W/cm K for the buried layer.

upon the transistor play an active role in its thermal behavior, allowing an alternative path for the heat flow and thus lowering the temperature that is reached in the active region.

Following the procedure described in Section 2.2.3, dedicated on-wafer measurements are performed on the M1 and M6 transistors; the resulting R_{th} are plotted in Fig. 2.54. It is shown that, for the two HBTs under study, the addition of metal-6 on the emitter contact leads to a decrease in the transistor R_{th} of around 10%. In Fig. 2.54, the measured R_{th} are compared to the results extracted from the TCAD simulations for test structure M1 and M6. The R_{th} is obtained by the simple expression P_d/T_j; although the most obvious solution may seem to consider as T_j the temperature at the BC junction, where the heat source is present, it must be pointed out that it is instead the temperature reached by the BE junction which influences the electrical behavior of the transistor. A rigorous extraction of the R_{th} from TCAD simulations, then, should consider that the T_j is evaluated as an average temperature

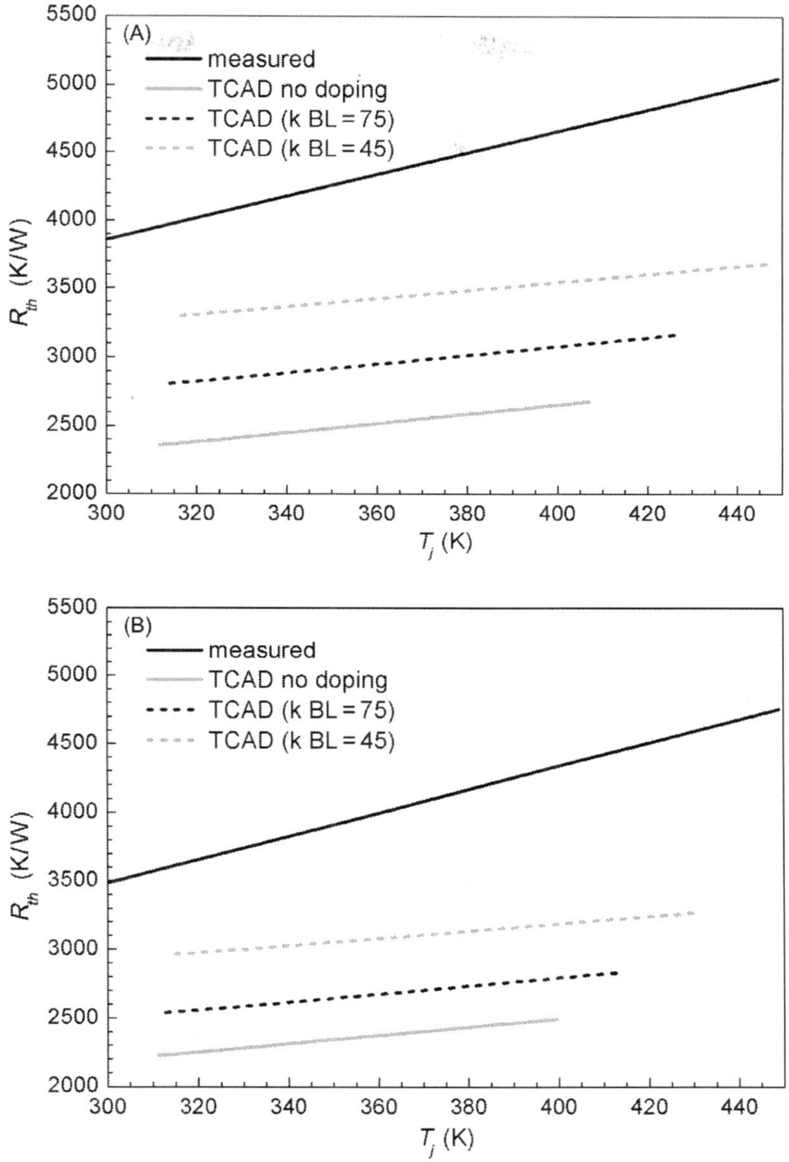

Figure 2.54

Thermal resistance Rth for the test structures M1 (A) and M6 (B); comparison between measurements and TCAD simulations, for different assumptions about the thermal impact of the doping profile.

reached by the BE junction. Considering that the BC and BE junctions are separated by the SiGe layer, which has a very low value of k, the temperature simulated at the two junctions can be quite different, and thus the error introduced in the R_{th} calculation can be considerable.

It is evident how neglecting the impact of the k decrease in the regions with high doping concentration leads to an underestimation of the R_{th}; furthermore, a certain mismatch is observed in the slope of the R_{th} vs T_j, which can be attributed to a change in the temperature dependency of k in the regions of the device which are highly doped. No experimental data seem to be available in literature about this phenomenon.

2.3.3.3 Dynamic thermal analysis and 3D simulations

Regarding the dynamic thermal characterization, instead, the thermal impedance Z_{th} is extracted from low-frequency S-parameters measured in the range from 10 kHz to 1 GHz and is reported in Fig. 2.55; it is shown that the Z_{th} of test structure M6 is lower in the low frequency range, but as the frequency increases, the difference among the two test structures tends to evanish, till around 3 MHz, from where the Z_{th} shows the same trend for both M1 and M6.

To understand why the two HBTs under study present a different Z_{th} at low frequencies and an identical decay at higher frequencies, it is necessary to have an insight in the way the heat spreads in a transistor structure excited by a sinusoidal power dissipation and introduce the concept of thermal penetration depth. In fact, the induced temperature oscillations will be distributed in a different way within the transistor, according to the frequency of the power stimulus. To explain this, let us assume as an example the simplified case of a semi-infinite block of silicon having an initial uniform temperature T_0; when a uniformly distributed power dissipation is applied at one side, eventually the whole block of material will tend to reach a higher temperature T_1. But before this happens, a thermal gradient can

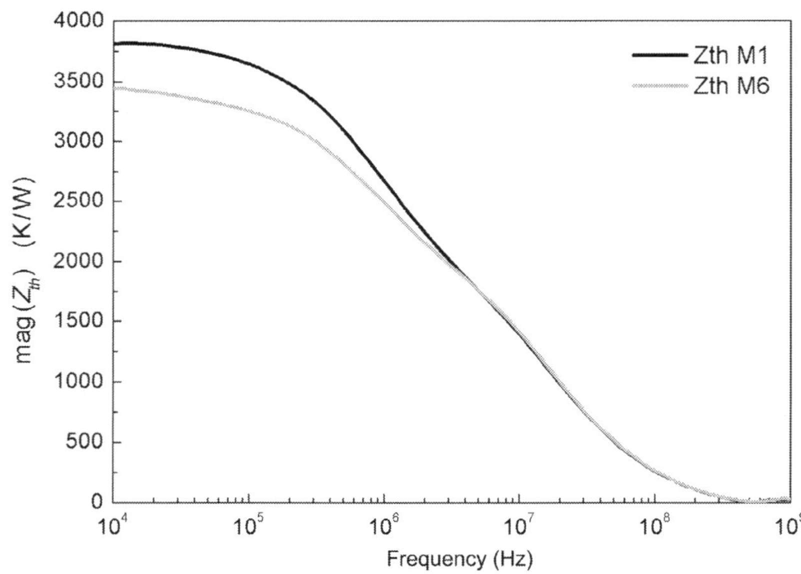

Figure 2.55
Measured thermal impedance Z_{th} in the range 10 kHz–1 GHz for the test structures M1 and M6.

be observed within a certain distance from the heat source, while beyond this boundary the temperature of the block is still T_0. The thermal penetration depth (δ) is defined as the distance that the heat diffuses through the material (under a sinusoidal stimulus) during a time $1/f$ and is approximated with the expression [87]:

$$\delta = \sqrt{\frac{k}{\rho\, C_L\, \pi f}} \tag{2.24}$$

where κ, ρ and C_L are, respectively, the thermal conductivity, density and lattice heat capacity of the material.

Note that Eq. (2.24) is only valid in 1D heat transfer by conduction; moreover, it must be considered that a state-of-the-art SiGe HBT has a complex structure, made of multiple material layers, having very different thermal properties. Nevertheless, Eq. (2.24) indicates the general trend that should be expected; in particular, since δ is inversely proportional to the square root of f, the volume of silicon that effectively contributes to Z_{th} reduces if the frequency increases. This means that, at high frequency of operation, the thermal oscillations take place in a very small volume around the heat source, whereas at lower frequencies, the temperature variations can significantly penetrate the structure, both downwards (toward the back of the wafer) and upward, till eventually reaching the higher levels of the metal in the BEOL. This explains why sweeping the frequency axis in Fig. 2.55 from the highest towards the lowest frequencies, we observe a deviation in the Z_{th} trend of the two structures at around 3 MHz. It can be in fact assumed that, at this frequency, the thermal penetration depth has reached the abscissa corresponding to metal-1 [44]. At frequencies, lower than 3 MHz the δ will penetrate further in the metallization, and since M6 has heat spreaders above metal-1 it will present a lower Z_{th} than M1.

From a physical modeling point of view, as shown in Fig. 2.46, the presence of metal upon the emitter contact can be modeled with the addition of thermal capacitances of high value in parallel to the R_{th} and C_{th} that model the thermal behavior of the lower part of the transistor (which is the same for both the M1 and M6 test structures) [63]. These added capacitances are thus responsible for the thermal behavior at low frequency till around 3 MHz, but increasing the frequency of the stimulus (represented, in this analysis, by the power dissipated in the component) the metal in the BEOL does no more respond thermally, it is rather the silicon in the region of the FEOL surrounding the heat source that determines the Z_{th}. This justifies the behavior observed when a two-tones signal is applied to the two transistors (refer to the simulations in Fig. 2.51). At 100 kHz in fact, a significant difference is observed among the Z_{th} of M1 and M6, whereas at 100 MHz their thermal impedances are the same; it follows that the amplitude of the T_j oscillations will be the same in the two HBTs at high frequencies, whereas at low frequencies they will be higher for M1, due to its higher Z_{th}.

To show how the δ reaches different distances according to the frequency of operation, 3D thermal simulations have been performed on the transistor structure M1, when sinusoidal P_d

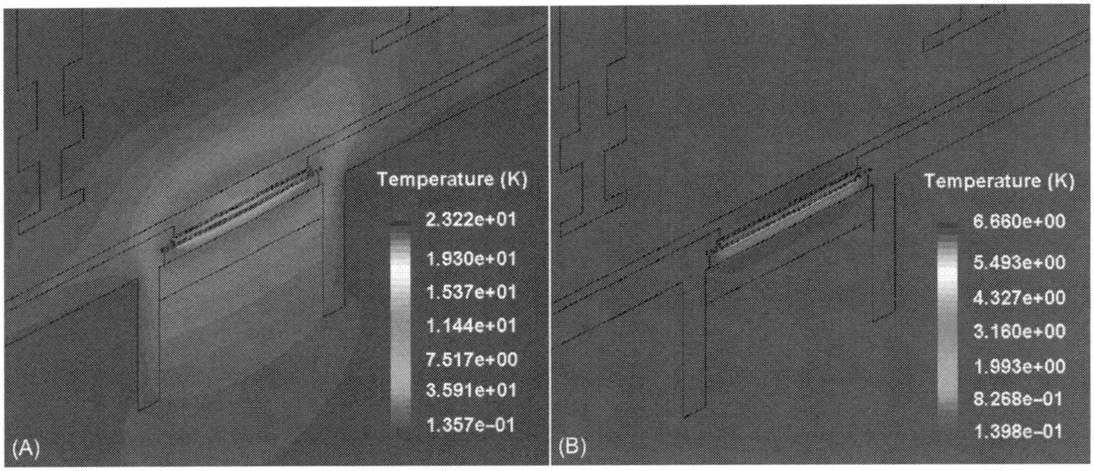

Figure 2.56
Simulated 3D structure of the B11HFC HBT (zoom on the transistor), showing the thermal penetration depth (δ) profile. A sinusoidal power having a frequency of 50 kHz (A) and 100 MHz (B) are applied at the BC junction. In both cases, P_dDC = 40 mW, P_dAC = 2 mW.

at 50 kHz and 100 MHz are applied at the BC junction. The simulation results in Fig. 2.56 represent the amplitude of the temperature oscillations, which gives a direct image of the spatial distribution of δ. It is evident how, at 100 MHz, the thermal variations are only located in a small volume around the BC junction, whereas at 50 MHz they are visible till the highest levels of metal in the BEOL.

Dedicated transistor structures having metal heat spreaders in the BEOL can thus help to identify the frequencies at which the temperature oscillations imposed by a sinusoidal P_d reach a certain distance from the heat source [44]; this characterization can be helpful in the development of physical thermal models for the Z_{th} based on distributed thermal networks.

As shown in this section, the configuration of materials in the BEOL, although far from the BC junction, where a high amount of power dissipation can be present, has a non-negligible thermal impact. In particular, the presence of metal connections can considerably affect the Z_{th} decay at low frequency, but it must be pointed out that instead, at frequencies starting from around 1–3 MHz, the same Z_{th} trend has to be expected, regardless of any modifications of the HBT structure.

References

[1] P.H. Siegel, Terahertz technology, IEEE Transactions on Microwave Theory and Techniques 50 (2002) 910–928. Available from: https://doi.org/10.1109/22.989974.

[2] M. Tonouchi, Cutting-edge terahertz technology, Nat Photon 1 (2007) 97–105. Available from: https://doi.org/10.1038/nphoton.2007.3.

[3] B. Fischer, M. Hoffmann, H. Helm, R. Wilk, F. Rutz, T. Kleine-Ostmann, M. Koch, P. Jepsen, Terahertz time-domain spectroscopy and imaging of artificial RNA, Opt. Express 13 (2005) 5205–5215. Available from: https://doi.org/10.1364/OPEX.13.005205.

[4] M. Nagel, P. Haring Bolivar, M. Brucherseifer, H. Kurz, A. Bosserhoff, R. Büttner, Integrated THz technology for label-free genetic diagnostics, Appl. Phys. Lett. 80 (2002) 154. Available from: https://doi.org/10.1063/1.1428619.

[5] R.M. Woodward, B.E. Cole, V.P. Wallace, R.J. Pye, D.D. Arnone, E.H. Linfield, M. Pepper, Terahertz pulse imaging in reflection geometry of human skin cancer and skin tissue, Phys. Med. Biol. 47 (2002) 3853–3863. Available from: https://doi.org/10.1088/0031-9155/47/21/325.

[6] Hua Zhong, Jingzhou Xu, Xu Xie, Tao Yuan, Reightler, R. Madaras, E. Xi-Cheng Zhang, Nondestructive defect identification with terahertz time-of-flight tomography, IEEE Sensors Journal 5 (2005) 203–208. Available from: https://doi.org/10.1109/JSEN.2004.841341.

[7] D.M. Mittleman, J. Cunningham, M.C. Nuss, M. Geva, Noncontact semiconductor wafer characterization with the terahertz Hall effect, Appl. Phys. Lett. 71 (1997) 16. Available from: https://doi.org/10.1063/1.119456.

[8] K. Kawase, Y. Ogawa, Y. Watanabe, H. Inoue, Non-destructive terahertz imaging of illicit drugs using spectral fingerprints, Opt Express 11 (2003) 2549–2554.

[9] D.L. Woolard, J.O. Jensen, R.J. Hwu, Terahertz Science and Technology for Military and Security Applications, World Scientific, 2007.

[10] Kao-Cheng Huang, Zhaocheng Wang, Terahertz terabit wireless communication, IEEE Microwave Magazine 12 (2011) 108–116. Available from: https://doi.org/10.1109/MMM.2011.940596.

[11] N. Kukutsu, A. Hirata, M. Yaita, K. Ajito, H. Takahashi, T. Kosugi, H.-J. Song, A. Wakatsuki, Y. Muramoto, T. Nagatsuma, Y. Kado, Toward practical applications over 100 GHz, in: Microwave Symposium Digest (MTT), 2010 IEEE MTT-S International. Presented at the Microwave Symposium Digest (MTT), 2010 IEEE MTT-S International, IEEE, 2010 pp. 1134–1137. https://doi.org/10.1109/MWSYM.2010.5515913.

[12] R. Piesiewicz, T. Kleine-Ostmann, N. Krumbholz, D. Mittleman, M. Koch, J. Schoebel, T. Kurner, Short-range ultra-broadband terahertz communications: concepts and perspectives, IEEE Antennas and Propagation Magazine 49 (2007) 24–39. Available from: https://doi.org/10.1109/MAP.2007.4455844.

[13] U.R. Pfeiffer, E. Öjefors, Terahertz imaging with CMOS/BiCMOS process technologies, in: ESSCIRC, 2010 Proceedings of The. Presented at the ESSCIRC, 2010 Proceedings of the, IEEE, 2010. pp. 52–60. https://doi.org/10.1109/ESSCIRC.2010.5619773.

[14] Eunyoung Seok, Dongha Shim, Chuying Mao, Sankaran Ruonan Han, S. Changhua Cao, W. Knap, K. O. Kenneth, Progress and challenges towards terahertz CMOS integrated circuits, IEEE Journal of Solid-State Circuits 45 (2010) 1554–1564. Available from: https://doi.org/10.1109/JSSC.2010.2049793.

[15] J.V. Siles, J. Grajal, Physics-based design and optimization of Schottky diode frequency multipliers for terahertz applications, IEEE Transactions on Microwave Theory and Techniques 58 (2010) 1933–1942. Available from: https://doi.org/10.1109/TMTT.2010.2050103.

[16] Q. Hu, B.S. Williams, S. Kumar, H. Callebaut, S. Kohen, J.L. Reno, Resonant-phonon-assisted THz quantum-cascade lasers with metal–metal waveguides, Semicond. Sci. Technol. 20 (2005) S228–S236. Available from: https://doi.org/10.1088/0268-1242/20/7/013.

[17] G.-B. Gao, M.-Z. Wang, X. Gui, H. Morkoc, Thermal design studies of high-power heterojunction bipolar transistors, IEEE Transactions on Electron Devices 36 (1989) 854–863. Available from: https://doi.org/10.1109/16.299666.

[18] J.D. Cressler, G. Niu, Silicon-germanium Heterojunction Bipolar Transistors, Artech House, 2003.

[19] P. Chevalier, T. Lacave, E. Canderle, A. Pottrain, Y. Carminati, J. Rosa, F. Pourchon, N. Derrier, G. Avenier, A. Montagne, A. Balteanu, E. Dacquay, I. Sarkas, D. Celi, D. Gloria, C. Gaquiere, S.P. Voinigescu, A. Chantre, Oct. scaling of SiGe BiCMOS technologies for applications above 100 GHz, in:

2012 IEEE Compound Semiconductor Integrated Circuit Symposium (CSICS). Presented at the 2012 IEEE Compound Semiconductor Integrated Circuit Symposium (CSICS), pp. 1–4. doi: 10.1109/CSICS.2012.6340083.

[20] H. Rücker, B. Heinemann, A. Fox, SiGe BiCMOS Technologies for Applications above 100 GHz, in: 2012 IEEE Compound Semiconductor Integrated Circuit Symposium (CSICS). Presented at the 2012 IEEE Compound Semiconductor Integrated Circuit Symposium (CSICS), 2012. pp. 1–4. https://doi.org/10.1109/CSICS.2012.6340061.

[21] B. Heinemann, R. Barth, D. Bolze, J. Drews, G.G. Fischer, A. Fox, O. Fursenko, T. Grabolla, U. Haak, D. Knoll, R. Kurps, M. Lisker, S. Marschmeyer, H. Rèücker, D. Schmidt, J. Schmidt, M.A. Schubert, B. Tillack, C. Wipf, D. Wolansky, Y. Yamamoto, SiGe HBT technology with fT/fmax of 300GHz/500GHz and 2.0 ps CML gate delay, in: Electron Devices Meeting (IEDM), 2010 IEEE International. Presented at the Electron Devices Meeting (IEDM), 2010 IEEE International, IEEE, 2010. p. 30.5.1-30.5.4. https://doi.org/10.1109/IEDM.2010.5703452.

[22] A. Pottrain, T. Lacave, D. Ducatteau, D. Gloria, P. Chevalier, C. Gaquiere, High power density performances of SiGe HBT from BiCMOS technology at W-Band, Electron Device Letters, IEEE 33 (2012) 182–184. Available from: https://doi.org/10.1109/LED.2011.2177631.

[23] M. Weiss, C. Majek, A.K. Sahoo, C. Maneux, O. Mazouffre, P. Chevalier, A. Chantre, T. Zimmer, Optimized ring oscillator with 1.65-ps gate delay in a SiGe:C HBT technology, IEEE Electron Device Letters 34 (2013) 1214–1216. Available from: https://doi.org/10.1109/LED.2013.2277550.

[24] E. Öjefors, J. Grzyb, Yan Zhao, B. Heinemann, B. Tillack, U.R. Pfeiffer, A 820GHz SiGe chipset for terahertz active imaging applications, in: Solid-State Circuits Conference Digest of Technical Papers (ISSCC), 2011 IEEE International. Presented at the Solid-State Circuits Conference Digest of Technical Papers (ISSCC), 2011 IEEE International, IEEE, 2011. pp. 224–226. https://doi.org/10.1109/ISSCC.2011.5746294.

[25] U.R. Pfeiffer, E. Ojefors, A. Lisauskas, H.G. Roskos, Opportunities for silicon at mmWave and terahertz frequencies, in: 2008 IEEE Bipolar/BiCMOS Circuits and Technology Meeting. Presented at the 2008 IEEE Bipolar/BiCMOS Circuits and Technology Meeting - BCTM, Monterey, CA, 2008. 149–156. https://doi.org/10.1109/BIPOL.2008.4662734.

[26] P. Chevalier, T.F. Meister, B. Heinemann, S. Van Huylenbroeck, W. Liebl, A. Fox, A. Sibaja-Hernandez, A. Chantre, Towards THz SiGe HBTs, in: Bipolar/BiCMOS Circuits and Technology Meeting (BCTM), 2011 IEEE. 2011. pp. 57–65. doi:10.1109/BCTM.2011.6082749.

[27] A. Chantre, P. Chevalier, T. Lacave, G. Avenier, M. Buczko, Y. Campidelli, L. Depoyan, L. Berthier, C. Gacquière, Pushing conventional SiGe HBT technology towards "Dotfive" terahertz, in: Microwave Integrated Circuits Conference (EuMIC), 2010 European. Presented at the Microwave Integrated Circuits Conference (EuMIC), 2010 European, IEEE, 2010. pp. 21–24.

[28] D. Vasileska, K. Raleva, S.M. Goodnick, Modeling heating effects in nanoscale devices: the present and the future, J Comput Electron 7 (2008) 66–93. Available from: https://doi.org/10.1007/s10825-008-0254-y.

[29] D.L. Harame, J.H. Comfort, J.D. Cressler, E.F. Crabbe, J.Y.-C. Sun, B.S. Meyerson, T. Tice, Si/SiGe epitaxial-base transistors. I. Materials, physics, and circuits, IEEE Transactions on Electron Devices 42 (1995) 455–468. Available from: https://doi.org/10.1109/16.368039.

[30] I. Marano, d'V. Alessandro, N. Rinaldi, Effectively modeling the thermal behavior of trench-isolated bipolar transistors, in: EuroSimE 2008—International Conference on Thermal, Mechanical and Multi-Physics Simulation and Experiments in Microelectronics and Micro-Systems. Presented at the Multi-Physics simulation and Experiments in Microelectronics, Freiburg im Breisgau, Germany, 2008. pp. 1–8. https://doi.org/10.1109/ESIME.2008.4525060.

[31] J.-S. Rieh, J. Johnson, S. Furkay, D. Greenberg, G. Freeman, S. Subbanna, Structural dependence of the thermal resistance of trench-isolated bipolar transistors, in: Bipolar/BiCMOS Circuits and Technology Meeting, 2002. Proceedings of the 2002. Presented at the Bipolar/BiCMOS Circuits and Technology Meeting, 2002. Proceedings of the 2002, pp. 100–103. https://doi.org/10.1109/BIPOL.2002.1042896.

[32] D.J. Walkey, T.J. Smy, C. Reimer, M. Schröter, H. Tran, Marchesan David, Modeling thermal resistance in trench-isolated bipolar technologies including trench heat flow, Solid-State Electronics 46 (2002) 7–17. Available from: https://doi.org/10.1016/S0038-1101(01)00305-7.

[33] G. Freeman, J.-S. Rieh, Z. Yang, F. Guarin, Reliability and performance scaling of very high speed SiGe HBTs, Microelectronics Reliability 44 (2004) 397–410. Available from: https://doi.org/10.1016/j.microrel.2003.11.003.

[34] A. Magnani, G. Sasso, V. d'Alessandro, L. Codecasa, N. Rinaldi, K. Aufinger, Advanced thermal resistance simulation of SiGe HBTs including backend cooling effect, in: 2015 21st International Workshop on Thermal Investigations of ICs and Systems (THERMINIC). Presented at the 2015 21st International Workshop on Thermal Investigations of ICs and Systems (THERMINIC), 2015, pp. 1–5. https://doi.org/10.1109/THERMINIC.2015.7389631.

[35] W. Liu, Thermal coupling in 2-finger heterojunction bipolar transistors, IEEE Transactions on Electron Devices 42 (1995) 1033–1038. Available from: https://doi.org/10.1109/16.387234.

[36] Winkler, R.H., May. Thermal properties of high-power transistors. IEEE Transactions on Electron Devices 14, 260–263. https://doi.org/10.1109/T-ED.1967.15939.

[37] N. Rinaldi, V. D'Alessandro, Theory of electrothermal behavior of bipolar transistors: Part I -single-finger devices, IEEE Transactions on Electron Devices 52 (2005) 2009–2021. Available from: https://doi.org/10.1109/TED.2005.854274.

[38] L. La Spina, V. D'Alessandro, S. Russo, N. Rinaldi, L.K. Nanver, Influence of concurrent electrothermal and avalanche effects on the safe operating area of multifinger bipolar transistors, IEEE Transactions on Electron Devices 56 (2009) 483–491. Available from: https://doi.org/10.1109/TED.2008.2011574.

[39] F.L. Traversa, F. Cappelluti, F. Bonani, G. Ghione, Assessment of thermal instabilities and oscillations in multifinger heterojunction bipolar transistors through a harmonic-balance-based CAD-oriented dynamic stability analysis technique, IEEE Transactions on Microwave Theory and Techniques 57 (2009) 3461–3468. Available from: https://doi.org/10.1109/TMTT.2009.2034229.

[40] W. Liu, A. Khatibzadeh, The collapse of current gain in multi-finger heterojunction bipolar transistors: its substrate temperature dependence, instability criteria, and modeling, IEEE Transactions on Electron Devices 41 (1994) 1698–1707. Available from: https://doi.org/10.1109/16.324577.

[41] W. Liu, S. Nelson, D.G. Hill, A. Khatibzadeh, Current gain collapse in microwave multifinger heterojunction bipolar transistors operated at very high power densities, IEEE Transactions on Electron Devices 40 (1993) 1917–1927. Available from: https://doi.org/10.1109/16.239729.

[42] R. Ishikawa, J. Kimura, K. Honjo, Analytical design method for a low-distortion microwave InGaP/GaAs HBT amplifier based on transient thermal behavior in a GaAs substrate, IEEE Transactions on Components, Packaging and Manufacturing Technology 3 (2013) 1705–1712. Available from: https://doi.org/10.1109/TCPMT.2013.2262504.

[43] Takahashi, Y., Ishikawa, R., Honjo, K., 2006. Precise modeling of thermal memory effect for power amplifier using multi-stage thermal RC-ladder network, in: Microwave Conference, 2006. APMC 2006. Asia-Pacific. pp. 287–290. https://doi.org/10.1109/APMC.2006.4429424.

[44] R. D'Esposito, S. Balanethiram, J.L. Battaglia, S. Frégonèse, T. Zimmer, Thermal penetration depth analysis and impact of the BEOL metals on the thermal impedance of SiGe HBTs, IEEE Electron Device Letters (2017) 1. Available from: https://doi.org/10.1109/LED.2017.2743043. 1.

[45] Sahoo, A.K., Fregonese, S., Weis, M., Maneux, C., Malbert, N., Zimmer, T., 2013. Impact of back-end-of-line on thermal impedance in SiGe HBTs, in: 2013 International Conference on Simulation of Semiconductor Processes and Devices (SISPAD). pp. 188–191. https://doi.org/10.1109/SISPAD.2013.6650606.

[46] M. Schröter, A. Chakravorty, Compact Hierarchical Bipolar Transistor Modeling with HiCuM, World Scientific, 2010.

[47] R. Menozzi, J. Barrett, P. Ersland, A new method to extract HBT thermal resistance and its temperature and power dependence, IEEE Transactions on Device and Materials Reliability 5 (2005) 595–601. Available from: https://doi.org/10.1109/TDMR.2005.854210.

[48] M. Pfost, V. Kubrak, P. Brenner, A practical method to extract the thermal resistance for heterojunction bipolar transistors, in: 33rd Conference on European Solid-State Device Research, 2003. ESSDERC '03. 2003, pp. 335–338. https://doi.org/10.1109/ESSDERC.2003.1256882.

[49] J.-S. Rieh, D. Greenberg, B. Jagannathan, G. Freeman, S. Subbanna, Measurement and modeling of thermal resistance of high speed SiGe heterojunction bipolar transistors, in: 2001 Topical Meeting on Silicon Monolithic Integrated Circuits in RF Systems, 2001. Digest of Papers. 2001, pp. 110–113. https://doi.org/10.1109/SMIC.2001.942350.

[50] H. Tran, M. Schroter, D.J. Walkey, D. Marchesan, T.J. Smy, Simultaneous extraction of thermal and emitter series resistances in bipolar transistors, in: Bipolar/BiCMOS Circuits and Technology Meeting, 1997. Proceedings of The 1997, pp. 170–173.https://doi.org/10.1109/BIPOL.1997.647427.

[51] T. Vanhoucke, H.M.J. Boots, W.D. van Noort, Revised method for extraction of the thermal resistance applied to bulk and SOI SiGe HBTs, IEEE Electron Device Letters 25 (2004) 150–152. Available from: https://doi.org/10.1109/LED.2004.824242.

[52] J. Berkner, Extraction of thermal reistance and its temperature dependence using DC methods, in: HiCuM Workshop 2007.

[53] S. Balanethiram, J. Berkner, R. D'Esposito, S. Fregonese, D. Celi, T. Zimmer, Extracting the temperature dependence of thermal resistance from temperature-controlled DC measurements of SiGe HBTs, in: Bipolar/BiCMOS Circuits and Technology Meeting (BCTM), 2017a IEEE.

[54] A. El Rafei, A. Saleh, R. Sommet, J.M. Nebus, R. Quere, Experimental Characterization and Modeling of the Thermal Behavior of SiGe HBTs, IEEE Transactions on Electron Devices 59 (2012) 1921–1927. Available from: https://doi.org/10.1109/TED.2012.2196765.

[55] N. Rinaldi, Small-signal operation of semiconductor devices including self-heating, with application to thermal characterization and instability analysis, IEEE Transactions on Electron Devices 48 (2001) 323–331. Available from: https://doi.org/10.1109/16.902734.

[56] Lonac, J.A., Santarelli, A., Melczarsky, I., Filicori, F., 2005. A simple technique for measuring the thermal impedance and the thermal resistance of HBTs, in: Gallium Arsenide and Other Semiconductor Application Symposium, 2005. EGAAS 2005. European. pp. 197–200.

[57] A.E. Rafei, R. Sommet, R. Quere, Electrical Measurement of the Thermal Impedance of Bipolar Transistors, IEEE Electron Device Letters 31 (2010) 939–941. Available from: https://doi.org/10.1109/LED.2010.2052232.

[58] E.K. Mueller, Internal thermal feedback in four-poles especially in transistors, Proceedings of the IEEE 52 (1964) 924–930. Available from: https://doi.org/10.1109/PROC.1964.3183.

[59] Igic, P.M., Mawby, P.A., Towers, M.S., Batcup, S., 2001. Dynamic electro-thermal physically based compact models of the power devices for device and circuit simulations, in: Semiconductor Thermal Measurement and Management, 2001. Seventeenth Annual IEEE Symposium. pp. 35–42. https://doi.org/10.1109/STHERM.2001.915142.

[60] B. Vermeersch, G.D. Mey, A fixed-angle heat spreading model for dynamic thermal characterization of rear-cooled substrates, in: Twenty-Third Annual IEEE Semiconductor Thermal Measurement and Management Symposium. 2007. pp. 95–101. https://doi.org/10.1109/STHERM.2007.352393.

[61] D.M. März, P. Nance, Thermal Modeling of Power-electronic System, Fraunhofer Institut Integrierte Schaltungen, 2000.

[62] A.K. Sahoo, Electro-thermal characterizations, compact modeling and TCAD based device simulations of advanced SiGe: C BiCMOS HBTs and of nanometric CMOS FET. Bordeaux 1 2012.

[63] R. D'Esposito, S. Fregonese, T. Zimmer, A. Chakravorty, Dedicated test-structures for investigation of the thermal impact of the BEOL in advanced SiGe HBTs in time and frequency domain, in: 2016 International Conference on Microelectronic Test Structures (ICMTS). 2016b, pp. 28–31. https://doi.org/10.1109/ICMTS.2016.7476168.

[64] R. D'Esposito, M. Weis, A.K. Sahoo, S. Fregonese, T. Zimmer, A study on transient intra-device thermal coupling in multifinger SiGe HBTs, in: 2014 IEEE Bipolar/BiCMOS Circuits and Technology Meeting (BCTM). 2014, pp. 179–182. https://doi.org/10.1109/BCTM.2014.6981309.

[65] A.K. Sahoo, S. Fregonese, M. Weiss, N. Malbert, T. Zimmer, Electro-thermal characterization of Si-Ge HBTs with pulse measurement and transient simulation, in: Solid-State Device Research Conference (ESSDERC), 2011 Proceedings of the European. 2011, pp. 239–242. https://doi.org/10.1109/ESSDERC.2011.6044190.

[66] V. Camarchia, F. Cappelluti, M. Pirola, S.D. Guerrieri, G. Ghione, Self-consistent electrothermal modeling of class A, AB, and B Power GaN HEMTs under modulated RF excitation, IEEE Transactions on Microwave Theory and Techniques 55 (2007) 1824–1831. Available from: https://doi.org/10.1109/TMTT.2007.903839.

[67] S.C. Cripps, RF Power Amplifiers for Wireless Communications, Artech House, 1999.

[68] J.H.K. Vuolevi, T. Rahkonen, J.P.A. Manninen, Measurement technique for characterizing memory effects in RF power amplifiers, IEEE Transactions on Microwave Theory and Techniques 49 (2001) 1383–1389. Available from: https://doi.org/10.1109/22.939917.

[69] R. Ishikawa, J. Kimura, K. Honjo, Analytic parameter determination for thermal memory effect compensation circuit in microwave InGaP/GaAs HBT power amplifiers, in: Microwave Conference Proceedings (APMC), 2011 Asia-Pacific. 2011, pp. 315–318.

[70] W. Bosch, G. Gatti, Measurement and simulation of memory effects in predistortion linearizers, IEEE Transactions on Microwave Theory and Techniques 37 (1989) 1885–1890. Available from: https://doi.org/10.1109/22.44098.

[71] V. d'Alessandro, A. Magnani, L. Codecasa, N. Rinaldi, K. Aufinger, Advanced thermal simulation of SiGe:C HBTs including back-end-of-line, Microelectronics Reliability 67 (2016) 38–45. Available from: https://doi.org/10.1016/j.microrel.2016.06.005.

[72] A.D.D. Dwivedi, A. Chakravorty, R. D'Esposito, A.K. Sahoo, S. Fregonese, T. Zimmer, Effects of BEOL on self-heating and thermal coupling in SiGe multi-finger HBTs under real operating condition, Solid-State Electronics 115 (2016) 1–6. Available from: https://doi.org/10.1016/j.sse.2015.09.016. Part A.

[73] S. Balanethiram, R. D'Esposito, A. Chakravorty, S. Fregonese, T. Zimmer, Extraction of BEOL contributions for thermal resistance in SiGe HBTs, IEEE Transactions on Electron Devices 64 (2017) 1380–1384. Available from: https://doi.org/10.1109/TED.2016.2645615.

[74] R. D'Esposito, S. Frégonèse, A. Chakravorty, P. Chevalier, D. Céli, T. Zimmer, Innovative SiGe HBT topologies with improved electrothermal behavior, IEEE Transactions on Electron Devices 63 (2016) 2677–2683. Available from: https://doi.org/10.1109/TED.2016.2570601.

[75] R. D'Esposito, M.D. Matos, S. Fregonese, S. Balanethiram, A. Chakravorty, K. Aufinger, T. Zimmer, Influence of the BEOL metallization design on the overall performances of SiGe HBTs, in: 2016 13th IEEE International Conference on Solid-State and Integrated Circuit Technology (ICSICT). 2016c. pp. 358–360. doi: 10.1109/ICMTS.2016.7476168.

[76] J. Böck, K. Aufinger, S. Boguth, C. Dahl, H. Knapp, W. Liebl, D. Manger, T.F. Meister, A. Pribil, J. Wursthorn, R. Lachner, B. Heinemann, H. Rücker, A. Fox, R. Barth, G. Fischer, S. Marschmeyer, D. Schmidt, A. Trusch, C. Wipf, SiGe HBT and BiCMOS process integration optimization within the DOTSEVEN project, in: 2015 IEEE Bipolar/BiCMOS Circuits and Technology Meeting - BCTM. 2015. pp. 121–124. https://doi.org/10.1109/BCTM.2015.7340549.

[77] M.S. Kim, S. Cho, J. Min, M.R. Pulugurtha, N. Huang, S. Sitaraman, V. Sundaram, M. Velez, A. Ravindran, Y. Joshi, R. Tummala, Modeling, design and demonstration of ultra-miniaturized glass PA modules with efficient thermal dissipation, in: Electronic Components and Technology Conference (ECTC), 2015 IEEE 65th. 2015. pp. 1163–1167. https://doi.org/10.1109/ECTC.2015.7159742.

[78] H.-C. Tseng, J.-L. Lin, Enhanced thermal performance of InGaP/GaAs collector-Up HBTs with a miniaturized backside heat-dissipation structure, IEEE Transactions on Components, Packaging and Manufacturing Technology 2 (2012) 2040–2043. Available from: https://doi.org/10.1109/TCPMT.2012.2219622.

[79] M. Turowski, S. Dooley, P. Wilkerson, A. Raman, M. Casto, Full-chip to device level 3D thermal analysis of RF integrated circuits, in: 11th Intersociety Conference on Thermal and Thermomechanical Phenomena in Electronic Systems, 2008. ITHERM 2008. pp. 315–324. https://doi.org/10.1109/ITHERM.2008.4544286.

[80] E. Canderle, P. Chevalier, G. Avenier, N. Derrier, D. Celi, C. Gaquiere, Impact of BEOL stress on BiCMOS9MW HBTs, in: 2013 IEEE Bipolar/BiCMOS Circuits and Technology Meeting (BCTM). 2013. pp. 223–226. https://doi.org/10.1109/BCTM.2013.6798181.

[81] E. Canderle, Etudes et developpement de transistors bipolaires Si/SiGe:C rapides dans un noeud BiCMOS 55nm, PhD Thesis, Université de Lille, France. ed. 2014.

[82] O. Sevimli, A.E. Parker, A.P. Fattorini, S.J. Mahon, Measurement and Modeling of Thermal Behavior in InGaP/GaAs HBTs, IEEE Transactions on Electron Devices 60 (2013) 1632–1639. Available from: https://doi.org/10.1109/TED.2013.2254117.

[83] R. Sommet, A.A.L. de Souza, A. Xiong, J. Obregon, J.C. Nallatamby, M. Prigent, R. Quere, On the determination of the thermal impedance of microwave bipolar transistors, in: 2010 12th IEEE Intersociety Conference on Thermal and Thermomechanical Phenomena in Electronic Systems (ITherm). 2010. pp. 1–8. https://doi.org/10.1109/ITHERM.2010.5501346.

[84] S. Kaka, Y. Yener, Convective Heat Transfer, Second Edition, CRC Press, 1995.

[85] W. Liu, K. Etessam-Yazdani, R. Hussin, M. Asheghi, Modeling and data for thermal conductivity of ultrathin single-crystal SOI layers at high temperature, IEEE Transactions on Electron Devices 53 (2006) 1868–1876. Available from: https://doi.org/10.1109/TED.2006.877874.

[86] A.D. McConnell, S. Uma, K.E. Goodson, Thermal conductivity of doped polysilicon layers, Journal of Microelectromechanical Systems 10 (2001) 360–369. Available from: https://doi.org/10.1109/84.946782.

[87] A. Salazar, Energy propagation of thermal waves, Eur. J. Phys. 27 (2006) 1349. Available from: https://doi.org/10.1088/0143-0807/27/6/009.

CHAPTER 3

InP-Based High-Electron-Mobility Transistors for High-Frequency Applications

D. Nirmal[1] and J. Ajayan[2]

[1]Electronics and Communication Engineering, Department of Electrical Sciences, Karunya Institute of Technology and Sciences, Coimbatore, India [2]Department of Electronics and Communication Engineering, SNS College of Technology, Coimbatore, India

3.1 History and Background of HEMT

In the mid-1970s, the molecular beam epitaxy (MBE) growth technique was developed and modulation doping technique was demonstrated. In this period of time, Takashi Mimura and his collaborators were working on a project based on GaAs metal oxide field effect transistors (MOSFETs) at Fujitsu Lab, Japan. The first high-electron-mobility transistor (HEMT) was a depletion-mode device (D-HEMT) which was invented by Mimura and his co-workers at Fujitsu Lab, Japan, in 1980 [1]. This D-HEMT was demonstrated using the concept of modulation-doped heterojunction superlattice, which was first demonstrated in AlGaAs/GaAs material system by Dingle and his colleagues at Bell Laboratory in 1978 [2]. The modulation-doped heterojunction superlattice leads to the formation of two-dimensional electron gas (2DEG) at the heterointerface between two semiconductors of different bandgaps. Since the 2DEG is separated from the dopants, the electrons in the quantum well (QW) has higher mobility than conventional field effect transistor (FET) structures. The first enhancement-mode HEMT (E-HEMT) was also developed in 1980 by Takashi Mimura [1] and his collaborators at Fujitsu Lab, Japan (Fig. 3.1).

To allow conduction, semiconductors are doped with impurities which donate either mobile electrons or holes. However, these electrons are slowed down through collisions with the impurities (dopants) used to generate them in the first place. HEMTs avoid this through the use of high mobility electrons generated using the heterojunction of a highly doped wide-bandgap n-type donor-supply layer (AlGaAs in our example) and a non-doped narrow-bandgap channel layer with no dopant impurities (GaAs in this case).

Nanoelectronics. DOI: https://doi.org/10.1016/B978-0-12-813353-8.00012-9

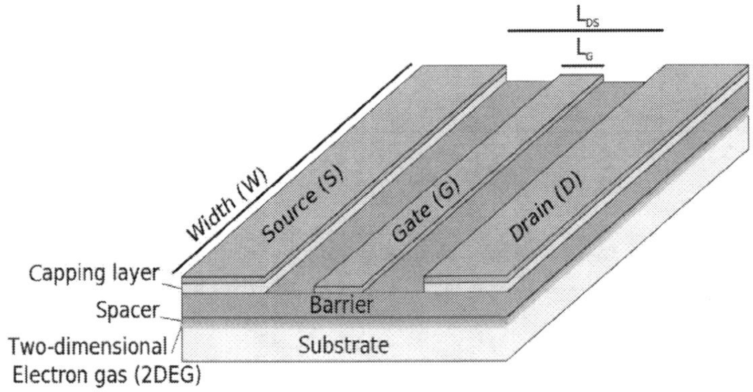

Figure 3.1
Structure of basic HEMT.

The electrons generated in the thin n-type AlGaAs layer drop completely into the GaAs layer to form a depleted AlGaAs layer, because the heterojunction created by different bandgap materials forms a QW (a steep canyon) in the conduction band on the GaAs side where the electrons can move quickly without colliding with any impurities due to undoped GaAs layer, and from which they cannot escape. The effect of this creates a very thin layer of highly mobile conducting electrons with very high concentration, giving the channel very low resistivity (or to put it another way, "high electron mobility").

3.2 Applications

Applications (e.g., for AlGaAs on GaAs) are similar to those of metal semiconductor field effect transistor (MESFETs)−microwave- and millimeter-wave communications, imaging, radar, and radio astronomy—any application where high gain and low noise at high frequencies are required. HEMTs have shown current gain to frequencies greater than 600 GHz and power gain to frequencies greater than 1 THz [3]. (Heterojunction bipolar transistors were demonstrated at current gain frequencies over 600 GHz in April 2005.) Numerous companies worldwide develop and manufacture HEMT-based devices. These can be discrete transistors but are more usually in the form of a "monolithic microwave-integrated circuit" (MMIC). HEMTs are found in many types of equipment ranging from cellphones and digital broadcast satellite (DBS) receivers to electronic warfare systems such as radar and for radio astronomy.

Furthermore, gallium nitride HEMTs on silicon substrates are used as power switching transistors for voltage converter applications. Compared to silicon power transistors gallium nitride HEMTs feature low on-state resistances, and low switching losses due to the wide bandgap properties. Gallium nitride power HEMTs are commercially available up to voltages of 200−600 V.

3.3 Working Principle

Modulation-doped heterojunction superlattice along with bandgap engineering techniques have been used in HEMTs to create 2DEG at the heterointerface between electron supply layer and channel layers. The modulation-doping technique was first demonstrated by Ray Dingle at Bell Labs. Since the 2DEG is separated from the dopants, the electrons in the QW has higher mobility than conventional FET structures.

3.3.1 Two-Dimensional Electron Gas in HEMT

In conventional MOSFETs, when electrons move from source to drain the mobility of electrons decreases through collisions with the impurities (dopants) in the channel. But HEMTs overcome this mobility degradation due to collision with impurities through the use of high mobility electrons generated using the heterojunction of a heavily doped wide bandgap n-type donor (supply layer e.g., InAlAs) and an undoped narrow bandgap channel layer with no dopant impurities (e.g., InGaAs). The mobile electrons generated in the supply layer (e.g., InAlAs) drop completely into an undoped channel layer (e.g., InGaAs) which results in the depletion of InAlAs donor layer, because the heterojunction formed by semiconductors with different bandgaps creates a QW in the conduction band on the InGaAs material, where the mobile electrons are able to move freely and quickly by avoiding collision with the dopants since the InGaAs channel layer is undoped, and also these conducting electrons cannot escape from the QW. This results in the formation of a very thin layer of high mobility electrons with very high density, in other words "high-electron mobility." The conductivity of the channel layer in the HEMTs can be altered by applying a potential at its gate terminal-like other conventional FET devices. The formation of 2DEG in AlGaAs/GaAs material system is illustrated in Fig. 3.2.

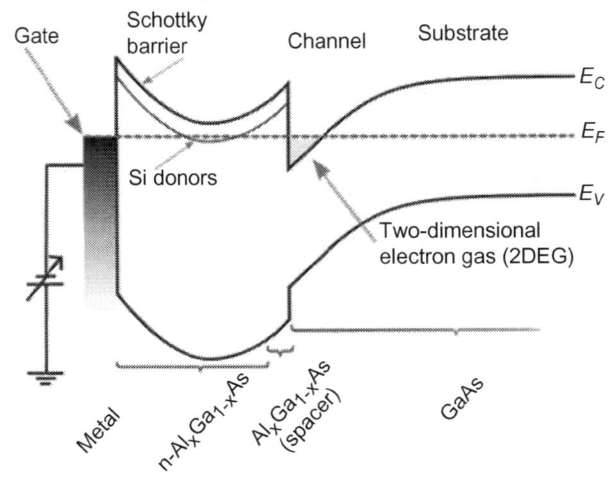

Figure 3.2
Illustration of 2-DEG in AlGaAs/GaAs HEMT.

3.4 Materials and its Properties—(InP/GaAs)

Indium phosphide (InP) and gallium arsenide (GaAs) are widely used as substrates for electronic and optic devices such as heterojunction bipolar transistors (HBTs), FETs, HEMTs, light emitting diodes (LEDs), etc. InP is a binary semiconductor formed by combining indium and phosphorus which has a face-centered cubic crystal structure, called "zinc blende." InP can be prepared from the chemical reaction of white phosphorus and indium iodide at 400 °C. InP and GaAs semiconductors are widely used as substrate and channel materials for low-power high-frequency-integrated circuit applications due to their excellent electron velocities and outstanding thermal stability characteristics in comparison with the conventional semiconductors such as silicon and germanium. The direct bandgap characteristics of these semiconductors make them suitable for optoelectronics devices like LEDs and laser diodes.

GaAs is also a III–V compound semiconductor formed of the elements gallium (III-group) and arsenic (V-group) which is widely used to fabricate ICs at microwave frequencies, infrared LEDs, laser diodes, solar cells, etc. AlGaAs and InGaAs can be grown by using GaAs as a substrate material. GaAs is also a direct bandgap semiconductor. Recent works on HEMTs by various research groups indicate that InP HEMTs show superior transistor gain, cut off frequency (f_T), maximum oscillation frequency (f_{max}), and low-noise behavior compared with GaN and GaAs HEMTs. These outstanding performances of InP HEMTs are attributed to the excellent electron mobility, electron saturation velocity, and 2D sheet charge densities (2DEG) of the InGaAs or InAs channel materials and also the aggressive scaling of geometrical dimensions of the device and reduction of parasitics by the proper selection of device architecture plays a vital role in enhancing the device performance. But, in the present scenario, GaAs-substrate-based HEMTs, MMICs, and terahertz monolithic integrated circuits (TMICs) have huge demand over InP-substrate-based devices due to their requirements in high data rate, low-noise commercial optical, and wireless communication systems. This is mainly because of the large cost of production for InP devices and ICs.

InP-substrate-based HEMTs, MMICs, and TMICs do not experience that much demand in the semiconductor industry compared to its alternative, i.e., GaAs-based devices. The demand for GaAs HEMTs is increasing day by day for the production of submillimeter, millimeter, and terahertz MMICs and TMICs with high speed, low power, low operating voltage, and low-noise applications, since GaAs as a substrate provides many advantages over devices on InP-substrate-like lower substrate cost because of large volume manufacturing, higher yield, better mechanical strength, and well-developed device fabrication technology with 6-in and 8-in GaAs substrates. Most importantly, increasing the amount of indium in the channel of HEMTs on InP substrate significantly affects the on-state and off-state breakdown voltages, and it also introduces serious kink effects in the drain current characteristics. The properties of InP and GaAs semiconductor materials are given in Table 3.1.

Table 3.1: Properties of InP and GaAs semiconductors

Parameter	GaAs	InP
Crystal structure	Zinc blende	Zinc blende
Lattice constant (Å)	5.653	5.869
Energy gap (eV)	1.424	1.344
Melting point (°C)	1240	1060
Electron mobility (cm^2/Vs)	8500	5400
Hole mobility (cm^2/Vs)	400	200
Intrinsic carrier concentration (cm^{-3})	2.1×10^6	1.3×10^7
Intrinsic resistivity ($\Omega \cdot cm$)	3.3×10^8	8.6×10^7

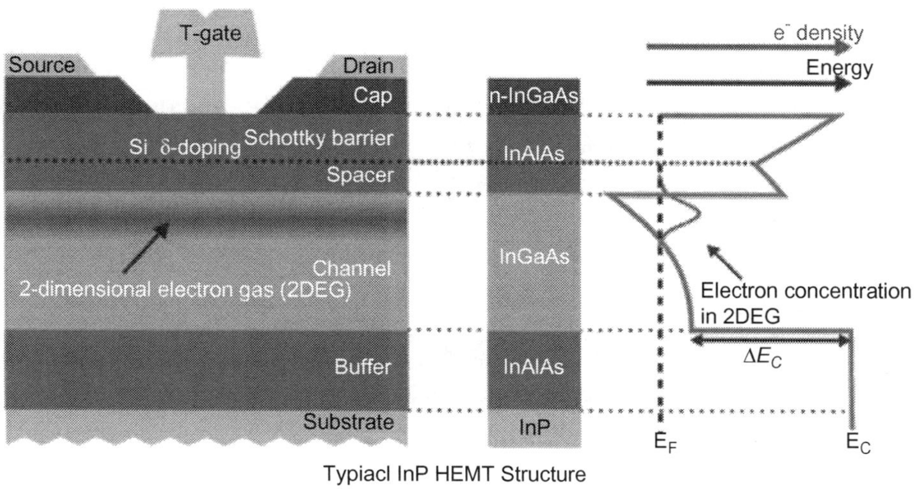

Figure 3.3
Layer structure of a typical InP/InAlAs/InGaAs HEMT.

3.5 General Structure of Inp HEMT

The typical epitaxial layer structure of InP/InAlAs/InGaAs HEMT is shown in Fig. 3.3. It consists of an InP substrate, InAlAs-buffer, InGaAs channel, InAlAs spacer, InAlAs Schottky barrier layer, and a heavily doped InGaAs cap layer. In conventional MOSFETs, a single layer of silicon forms the source, drain, and channel, with the exception that the source and drain regions are heavily doped with impurities which donate free electrons (or holes) to the lightly doped channel. In these conventional MOSFETs, when electrons move from source to drain, the mobility of electrons decreases through collisions with the impurities (dopants) in the channel. But HEMTs overcome this mobility degradation due to collision with impurities through the use of high mobility electrons generated using the heterojunction of a heavily doped wide bandgap n-type donor (supply layer e.g., InAlAs) and an undoped narrow bandgap channel layer with no dopant impurities (e.g., InGaAs).

The mobile electrons generated in the supply layer (e.g., InAlAs) drop completely into an undoped channel layer (e.g., InGaAs) which results in the depletion of InAlAs donor layer. The heterojunction formed by semiconductors with different bandgaps creates a QW in the conduction band on the InGaAs material where the mobile electrons are able to move freely and quickly by avoiding collision with the dopants as the InGaAs channel layer is undoped and these conducting electrons cannot escape from the QW. This results in the formation of a very thin layer of high mobility electrons with very high density, in other words "high electron mobility." The conductivity of the channel layer in the HEMTs can be altered by applying a potential at its gate terminal-like other conventional FET devices.

The buffer layer is used to isolate the defects of substrate from the channel. The spacer layer isolates the mobile electrons from the positively charged silicon donor atoms, which helps to minimize the impurity scattering which is the primary cause for mobility degradation in conventional MOSFETs. The separation of electrons from the donor atoms by spacer layer leads to the enhancement of mobility of electrons in the QW. The access resistance can be minimized by using a heavily doped n-type cap layer.

3.6 DC and Microwave Characteristics of HEMT

The transconductance (g_m) of the HEMTs can be defined as the ratio of change in drain current to change in V_{GS} for a fixed V_{DS} and can be mathematically expressed as

$$g_m = \left[\frac{\partial I_{DS}}{\partial V_{GS}}\right]_{V_{DS}}, \tag{3.1}$$

where the drain current for InAlAs/InGaAs HEMTs can be computed by considering the DD model and gradual channel effects and the expression for computing the drain current is given below).

$$I_{DS} = \mu_{eff}.W.V_{th}.\frac{\partial Q_{ch}}{\partial x} - \mu_{eff}\cdot W\cdot Q_{ch}\cdot\frac{\partial \Psi}{\partial x}, \tag{3.2}$$

where μ_{eff} is the effective electron mobility, W is the channel width, V_{th} is the threshold voltage, Ψ is the surface potential, and Q_{ch} is the channel charge.

The channel charge can be computed as

$$Q_{ch} = q\cdot n_s(x). \tag{3.3}$$

For below 25-nm gate-length regime, short channel effects (SCEs) are a major area of concern and it should be minimized to improve the peak transconductance and drive current. One way to suppress the SCEs is to reduce the gap between the gate foot and 2DEG in the channel which is possible by the reduction in spacer and barrier layer

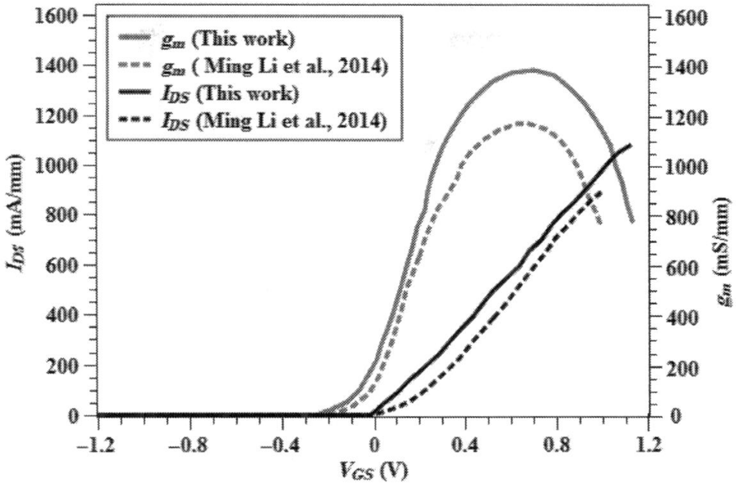

Figure 3.4
The variation of g_m and I_{DS} with the gate-to-source voltage for $L_g = 50$ nm $In_{0.52}Al_{0.48}As/In_{0.7}Ga_{0.3}As/InAs/In_{0.7}Ga_{0.3}As$ mHEMT on GaAs substrate.

thicknesses. The insertion of Pt metal into the barrier layer can also suppress the SCEs. In order to validate the performance of HEMT shown in Fig. 3.4, the simulated results of $L_g = 50$ nm mHEMT is compared with the experimental results given in [4]. For $L_g = 50$ nm, the maximum transconductance obtained for the proposed mHEMT is 1378 mS/mm, whereas Ming Li et al. reported the maximum transconductance as 1186 mS/mm. That is the proposed device shows 14% improvement in transconductance. The variation of transconductance (g_m) and drain current (I_{DS}) with gate-to-source voltage (V_{GS}) for $L_g = 50$ nm HEMT devices is shown in Fig. 3.5. For $L_g = 20$ nm, the proposed device exhibits a maximum transconductance of 2180 mS/mm. The V_{GS} dependence of g_m for $L_g = 20$ nm proposed device is shown in Fig. 3.4. For the proposed device structure, when the gate length is reduced from 50 to 20 nm, the transconductance is increased by 37%, which indicates that reducing the gate length is essential for improving the transconductance of the HEMTs (Fig. 3.5).

The variation of transconductance (g_m) and drain current (I_{DS}) with gate-to-source voltage (V_{GS}) for $L_g = 20$ nm proposed device is shown in Fig. 3.6.

3.7 Drain Current Characteristics

The drain current for ultra-short gate-length HEMTs greatly depends on the velocity saturation of electrons. The drift velocity of electrons in the channel is directly proportional

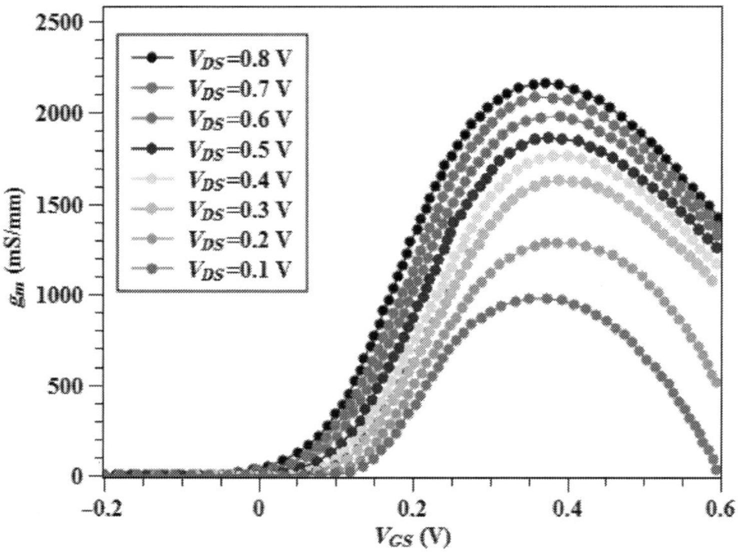

Figure 3.5
The g_m vs V_{GS} characteristics for $L_g = 20$ nm, $W = 2 \times 10$ μm $In_{0.52}Al_{0.48}As/In_{0.7}Ga_{0.3}As/InAs/In_{0.7}Ga_{0.3}As$ mHEMT on GaAs substrate.

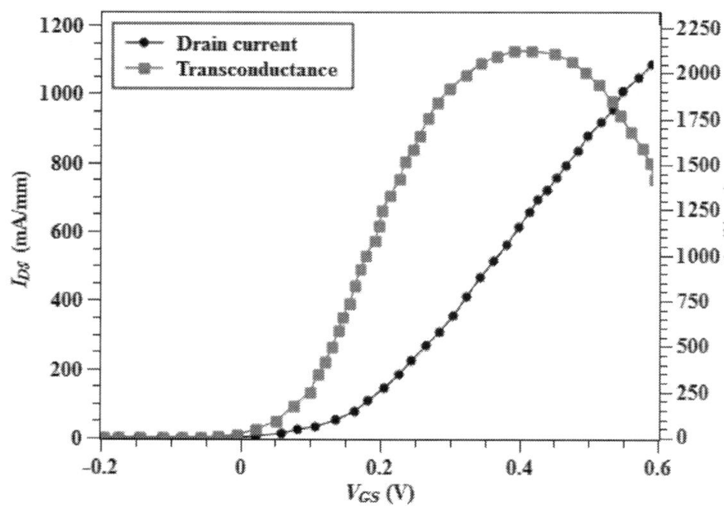

Figure 3.6
The variation of g_m and I_{DS} with the gate-to-source voltage for $L_g = 20$ nm $In_{0.52}Al_{0.48}As/In_{0.7}Ga_{0.3}As/InAs/In_{0.7}Ga_{0.3}As$ mHEMT on GaAs substrate.

to the electric field under the gate. The relationship between drift velocity of electrons in the channel and electric field under the gate can be mathematically expressed as

$$V(x) = \mu_0 \cdot \left(\frac{E(x)}{1 + (E(x)/E_C)} \right) \quad \text{for } E < E_C, \tag{3.4}$$

$$V(x) = V_{sat} = \mu_0 . E_C \quad \text{for } E > E_C, \tag{3.5}$$

where μ_0 is the low-field mobility of electrons, V_{sat} is the saturation velocity of electrons, and E_C is the reference electric field.

$$E(x) = \frac{dV(x)}{dx}. \tag{3.6}$$

$E(x)$ and $V(x)$ are the electric field in the channel and drift velocity of electrons in the channel, respectively.

For short gate-length HEMTs, based on velocity saturation model, the drain current can be computed as

$$I_{DS} = q \cdot w \cdot n_s(x) \cdot V(x), \tag{3.7}$$

where w is the gate width and $n_s(x)$ is the charge density in the channel. Equation 3.7 indicates that HEMTs with larger gate width are capable of driving large current. The drain current is also directly proportional to the sheet charge density in the channel. Therefore in the proposed HEMT structure, double delta-doping planes were used to achieve large drain current. The buried Pt technology improves the drift velocity of electrons under the gate. The drain current characteristics for $L_g = 50$ nm mHEMTs are shown in Fig. 3.8. The maximum drain current obtained for $L_g = 50$ nm proposed device is 904 mA/mm. Wang Li-Dan et al. [4] reported the maximum drain current as 786 mA/mm for $L_g = 50$ nm. For $L_g = 50$ nm, the proposed device shows 13% improvement in drain current. The I_{DS} vs V_{DS} characteristics for $L_g = 20$ nm and $W = 2 \times 10$ μm $In_{0.52}Al_{0.48}As/In_{0.7}Ga_{0.3}As/InAs/In_{0.7}Ga_{0.3}As$ mHEMT on GaAs substrate is shown in Fig. 3.9. The maximum drain current obtained for the proposed device with $L_g = 20$ nm and $W = 2 \times 10$ μm is 1100 mA/mm. When the gate length is reduced from 50 to 20 nm, the drain current is improved by 18%. The threshold voltage obtained for $L_g = 20$ nm and $W = 2 \times 10$ μm of the proposed device is 0.09 V. The positive threshold voltage ensures enhancement-mode operation of the device. The reason for drain current improvement in the proposed device is the high electron mobility obtained due to the use of InAs channel along with the reduction in source/drain parasitic resistance due to the use of heavily multilayer cap (Figs. 3.7 and 3.8).

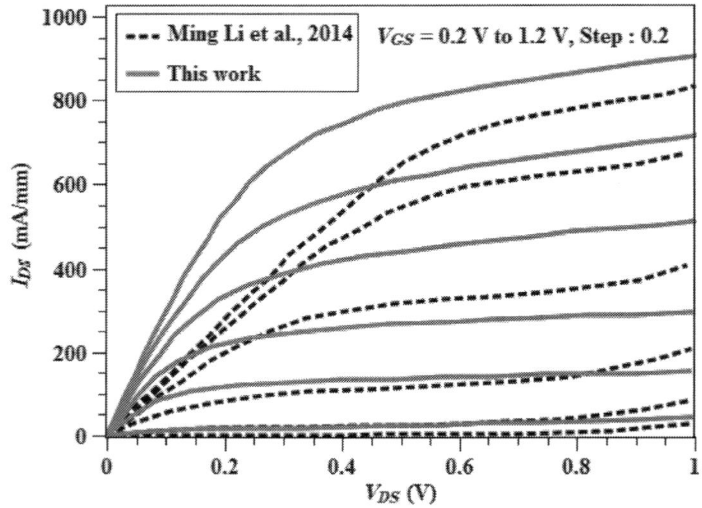

Figure 3.7
I_{DS} vs V_{DS} characteristics for $L_g = 50$ nm mHEMTs.

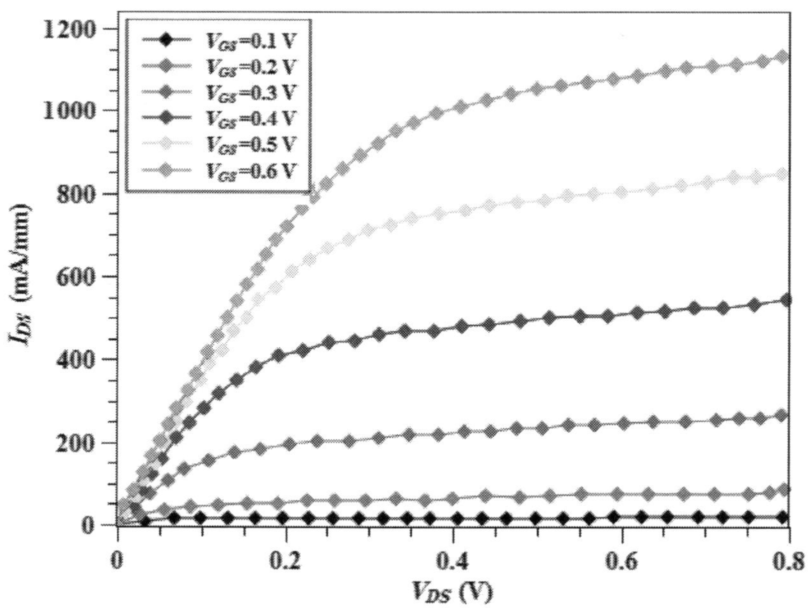

Figure 3.8
I_{DS} vs V_{DS} characteristics for $L_g = 20$ nm, $W = 2 \times 10$ μm $In_{0.52}Al_{0.48}As/In_{0.7}Ga_{0.3}As/InAs/In_{0.7}Ga_{0.3}As$ mHEMT on GaAs substrate.

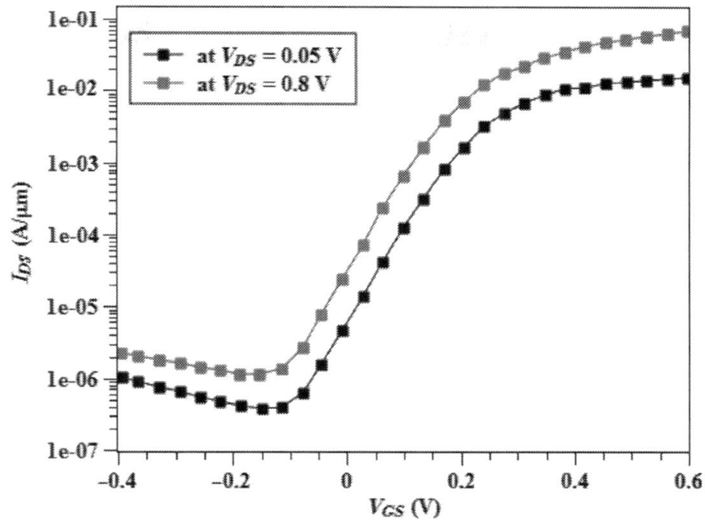

Figure 3.9
The subthreshold characteristics for $L_g = 20$ nm, $W = 2 \times 10$ μm $In_{0.52}Al_{0.48}As/In_{0.7}Ga_{0.3}As/InAs/In_{0.7}Ga_{0.3}As$ mHEMT on GaAs substra.

3.8 Subthreshold and Gate Leakage Characteristics

The variation of subthreshold current as a function of gate-to-source voltage (V_{GS}) for $L_g = 20$ nm, $W = 2 \times 10$ μm proposed mHEMT is shown in Fig. 3.4. The subthreshold current variation is plotted for drain-to-source voltages, $V_{DS} = 0.8$ V and $V_{DS} = 0.05$ V. The variation of gate leakage current as a function of gate-to-source voltage (V_{GS}) for $L_g = 20$ nm, $W = 2 \times 10$ μm proposed mHEMT is shown in Fig. 3.11. The gate leakage current is also plotted for $V_{DS} = 0.8$ V and $V_{DS} = 0.05$ V. From Fig. 3.10, it is clear that increase in drain-to-source voltage for a fixed gate-to-source voltage results in the increase in subthreshold current. From Fig. 3.11, it is evident that increase in drain-to-source voltage for a fixed gate-to-source voltage results in reduced gate leakage current (Fig. 3.9).

3.9 Measurement of DC and RF Performance of the Device

In this chapter, the DC and RF performance of the HEMT on InP substrate without doped source/drain regions is analyzed using Sentaurus technology computer-aided design (TCAD) tool. Physics-based drift-diffusion carrier transport model is used for TCAD simulation. The TCAD simulations were performed at room temperature. The important parameters adopted in the device simulations are dielectric constant (ϵ_r), elastic constant, thermal expansion, lattice constants, effective electron mass (m_e), effective hole mass (m_h), electron affinity (χ), energy gap (E_g), electron mobility (μ_n), hole mobility (μ_p), ionization

Table 3.2: Parameters for GaAs and InP semiconductors used for TCAD simulation (at room temperature)

Parameter	GaAs	InP
Dielectric constant (static)	12.9	12.5
Dielectric constant (at high frequency)	10.89	9.61
Effective electron mass (m_e)	0.063 m_0	0.08 m_0
Effective hole mass (m_h)	0.51 m_0	0.6 m_0
Electron affinity (χ)	4.07 eV	4.38 Ev
Lattice constant	5.653 Å	5.868 Å
Electron mobility (μ_n)	8500 cm^2/Vs	5400 cm^2/Vs
Hole mobility (μ_p)	400 cm^2/Vs	200 cm^2/Vs
Bandgap (E_g)	1.424 eV	1.344 eV
Intrinsic carrier concentration	$2.1 \times 10^6/cm^3$	$1.3 \times 10^7/cm^3$
N_C	$4.7 \times 10^{17}/cm^3$	$5.7 \times 10^{17}/cm^3$
N_V	$9 \times 10^{18}/cm^3$	$1.1 \times 10^{19}/cm^3$
Diffusion coefficient of electron (D_n)	200 cm^2/s	130 cm^2/S
Diffusion coefficient of holes (D_p)	10 cm^2/s	5 cm^2/S
Radiative recombination coefficient	7.2×10^{-10} cm^3/s	1.2×10^{-10} cm^3/S
Auger recombination coefficient	10^{-30} cm^6/s	9×10^{-31} cm^6/S
Thermal expansion constant	5.73×10^{-6} $°C^{-1}$	4.60×10^{-6} $°C^{-1}$
Intrinsic resistivity	3.3×10^8 $\Omega \cdot cm$	8.6×10^7 $\Omega \cdot cm$
Refractive index	3.3	3.1

parameters, effective conduction band and valence band density of states (Nc and Nv).The models used are concentration, field dependent, mobility models, Shockley–Read–Hall (SRH) recombination, Auger recombination, radiative recombination models and quantum effect model. The DC and RF performance of the device has been analyzed by measuring key parameters of the HEMTs such as transconductance, drain current, gate and subthreshold leakage currents, cut off and maximum oscillation frequencies. The values of some of the important parameters of GaAs and InP materials used for TCAD simulation are given in Table 3.2. The elastic constants for GaAs used in the TCAD simulation are $C_{11} = 11.90 \times 10^{11}$ dyn/cm^2, $C_{12} = 5.34 \times 10^{11}$ dyn/cm^2 and $C_{44} = 5.96 \times 10^{11}$ dyn/cm^2. Also the elastic constants for InP used in the TCAD simulation are $C_{11} = 10.11 \times 10^{11}$ dyn/cm^2, $C_{12} = 5.61 \times 10^{11}$ dyn/cm^2, and $C_{44} = 4.56 \times 10^{11}$ dyn/cm^2.

3.10 Transconductance Characteristics

The transconductance (g_m) of the HEMTs can be defined as the ratio of change in drain current to change in V_{GS} for a fixed V_{DS} and can be mathematically expressed as

$$g_m = \left[\frac{\partial I_{DS}}{\partial V_{GS}}\right]_{V_{DS}}, \tag{3.8}$$

where the drain current for InAlAs/InGaAs HEMTs can be computed by considering the DD model and gradual channel effects and the expression for computing the drain current is given below (K. Sourabh et al. 2012).

$$I_{DS} = \mu_{eff} \cdot W \cdot V_{th} \cdot \frac{\partial Q_{ch}}{\partial x} - \mu_{eff} \cdot W \cdot Q_{ch} \cdot \frac{\partial \Psi}{\partial x}, \tag{3.9}$$

where μ_{eff} is the effective electron mobility, W is the channel width, V_{th} is the threshold voltage, Ψ is the surface potential, and Q_{ch} is the channel charge.

The channel charge can be computed as

$$Q_{ch} = q \cdot n_s(x). \tag{3.10}$$

The 2DEG density in the channel also varies with gate bias which can be mathematically expressed as

$$n_S(x) = \frac{\varepsilon}{q}\left(\frac{V_{gs} - V_{th} - V_C(x)}{t_{g-c}}\right), \tag{3.11}$$

where t_{g-c} is the spacing between gate and 2DEG which is the sum of barrier thickness and spacer thickness and the distance of 2DEG from the spacer–channel interface. The relationship between drift velocity of electrons in the channel ($V_C(x)$) and electric field under the gate can be mathematically expressed as

$$V_C(x) = \mu_0 \cdot \left(\frac{E(x)}{1 + (E(x)/E_C)}\right) \quad \text{for } E < E_C, \tag{3.12}$$

$$V_C(x) = V_{sat} = \mu_0 \cdot E_C \quad \text{for } E > E_C, \tag{3.13}$$

where μ_0 is the low-field mobility of electrons, V_{sat} is the saturation velocity of electrons, and E_C is the reference electric field.

$$E(x) = \frac{dV(x)}{dx}. \tag{3.14}$$

$E(x)$ and $V(x)$ are the electric field in the channel and drift velocity of electrons in the channel, respectively. The Pt sinking used in the proposed mHEMT structure shifts the threshold voltage (V_{th}) toward the positive region. The variation of threshold voltage as a function of delta doping density $\delta(n)$ and the separation between 2DEG and the foot of the T-gate metal (t_{g-c}) can be mathematically expressed as

$$V_{th} = \Phi_B - \frac{\Delta E_C}{q} + \frac{E_{F0}}{q} - \frac{q \cdot \delta(n) \cdot t_{g-c}}{\varepsilon}, \tag{3.15}$$

where E_{F0} is the Fermi level at zero $n_s(x)$, and ϕ_B is the effective Schottky barrier height. The transconductance of the HEMT can be approximated as

$$g_m = \frac{\varepsilon_s \cdot V_{sat} \cdot W_g}{t_{g-c}}. \tag{3.16}$$

For improving the transconductance gate width (W_g) should be as large as possible and the physical gap between the gate and channel should be as small as possible according to the relationship given in Eq.(3.16) ϵ_s is the dielectric constant of the semiconductor material placed between gate and channel. The velocity of saturation can be mathematically expressed as a function of saturation electric field (E_{sat}) which is given in Eq.(3.17)

$$V_{sat} = \mu_n E_{sat}. \tag{3.17}$$

Therefore, Eq.(3.16) can be rewritten as

$$g_m = \frac{W_g.\mu_n.E_{sat}.\varepsilon_s}{t_{g-c}} \propto \mu_n. \tag{3.18}$$

The relationship between mobility of electron, scattering time constant (τ) and its effective mass is expressed in Eq.(3.19)

$$\mu_n \propto \frac{\tau}{m_e^*}. \tag{3.19}$$

For validating the performance of the proposed HEMT on InP substrate, the simulations were performed for $L_g = 30$ nm and $L_g = 20$ nm and the simulated DC and RF performance of $L_g = 30$ nm proposed HEMT on InP substrate is compared with the experimental results given in [5]. For $L_g = 30$ nm, the proposed HEMT on InP substrate exhibits a transconductance of 2265 mS/mm, whereas the transconductance reported by Kim et al. for the same gate-length device is 1830 mS/mm. The variation of transconductance with gate-to-source voltage for $L_g = 30$ nm InP HEMTs is plotted in Fig. 3.10. That is for $L_g = 30$ nm, the proposed device shows 19% improvement in transconductance compared with [5]. The variation of g_m and I_{DS} with V_{GS} for $L_g = 20$ nm and $W = 2 \times 10$ μm proposed HEMT on InP substrate is plotted in Fig. 3.11. Also the variation of g_m with V_{GS} and V_{DS} for $L_g = 20$ nm and $W = 2 \times 10$ μm proposed HEMT on InP substrate is plotted in Fig. 3.12. The proposed HEMT on InP substrate with $L_g = 20$ nm and $W = 2 \times 10$ μm exhibits a transconductance of 2630 mS/mm. The threshold voltage obtained for the proposed HEMT on InP substrate with $L_g = 20$ nm and $W = 2 \times 10$ μm is 0.07 V (Figs. 3.10 and 3.11).

The improved transconductance for 20-nm-gate-length device comes mainly from the reduced gate length and reduction of parasitics which is analyzed in the following sections (Fig. 3.12).

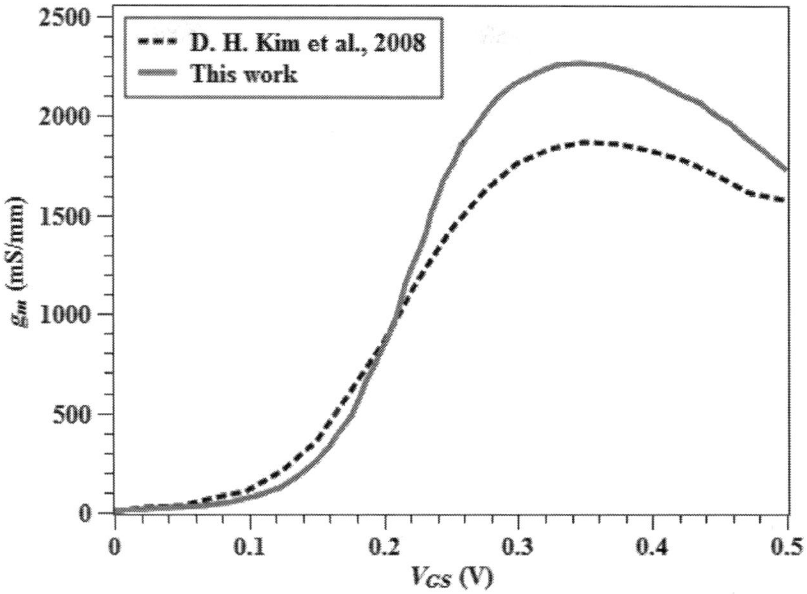

Figure 3.10
g_m vs V_{GS} characteristics for $L_g = 30$ nm, composite channel HEMT on InP-substrate without doped S/D regions.

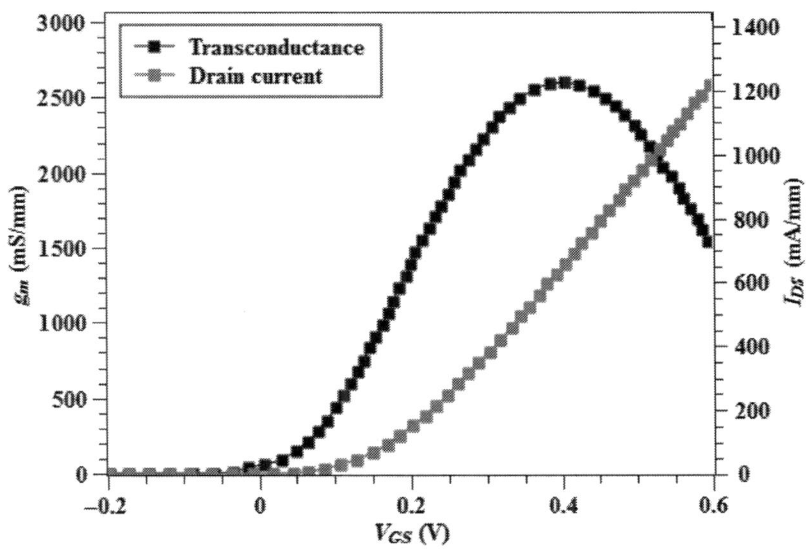

Figure 3.11
The variation of g_m and I_{DS} with V_{GS} for $L_g = 20$ nm and $W = 2 \times 10$ μm, composite channel HEMT on InP substrate without doped S/D regions.

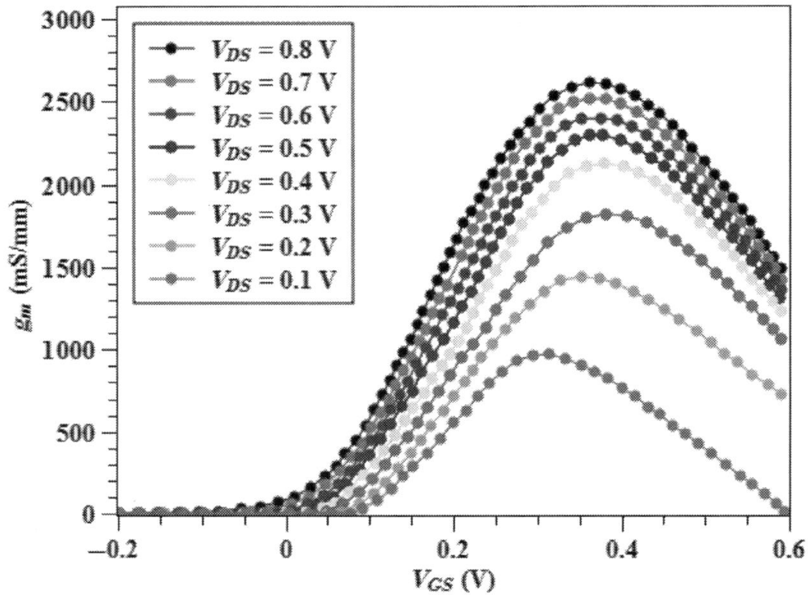

Figure 3.12
The variation of g_m with V_{GS} and V_{DS} for $L_g = 20$ nm and $W = 2 \times 10$ μm, composite channel HEMT on InP substrate without doped S/D regions.

3.11 Drain Current Characteristics

The output characteristics for $L_g = 30$ nm lattice-matched HEMTs on InP substrates using composite channel and platinum sinking technologies are shown in Fig. 3.13. The maximum drain current obtained for $L_g = 30$ nm proposed device is 798 mA/mm. Kim and del Alamo [5] reported the maximum drain current as 638 mA/mm for $L_g = 30$ nm. For $L_g = 30$ nm, the proposed device shows 20% improvement in drain current. The I_{DS} vs V_{DS} characteristics for $L_g = 20$ nm and $W = 2 \times 10$ μm $In_{0.52}Al_{0.48}As/In_{0.7}Ga_{0.3}As/InAs/In_{0.7}Ga_{0.3}As$ HEMT on InP substrate without heavily doped source/drain regions is shown in Fig. 3.14. The maximum drain current obtained for $L_g = 20$ nm and $W = 2 \times 10$ μm proposed device is 1272 mA/mm. The threshold voltage obtained for $L_g = 20$ nm and $W = 2 \times 10$ μm proposed device is 0.07 V. The positive threshold voltage ensures enhancement-mode operation of the device. The reason for drain current improvement in the proposed device is that the high electron mobility obtained due to the use of InAs channel along with the reduction in source/drain parasitic resistance due to the use of multilayer cap regions (Figs. 3.13 and 3.14).

3.12 Subthreshold and Gate Leakage Characteristics

For InAlAs/InGaAs HEMTs, the on-state gate leakage current increases with the increase in V_{DS} and the gate leakage characteristic plot contain a typical bell-shaped curve. At high

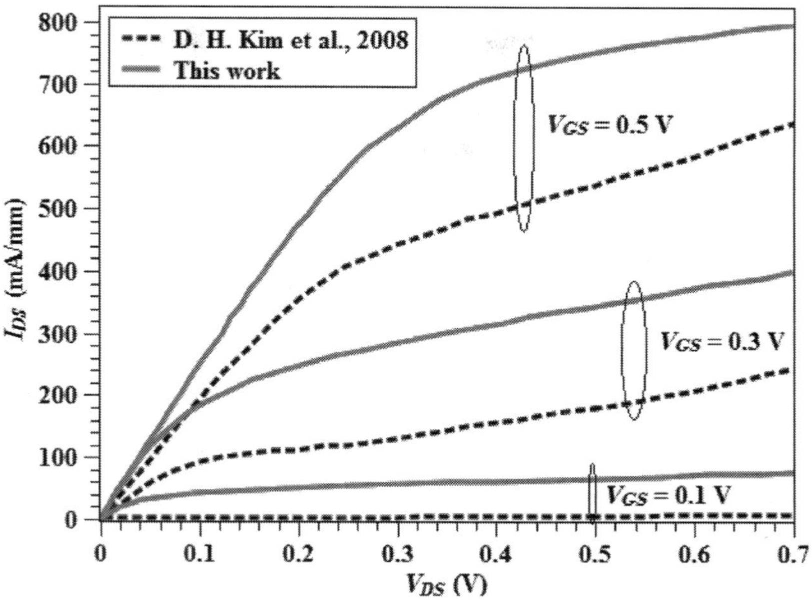

Figure 3.13
The output characteristics for $L_g = 30$ nm, composite channel HEMT on InP substrate without doped S/D regions.

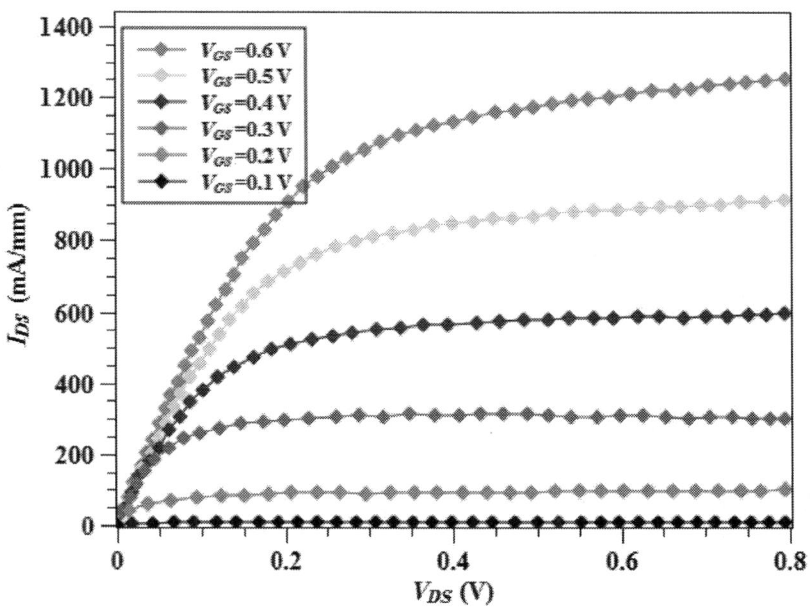

Figure 3.14
The output characteristics for $L_g = 20$ nm and $W = 2 \times 10$ μm, composite channel HEMT on InP substrate without doped S/D regions.

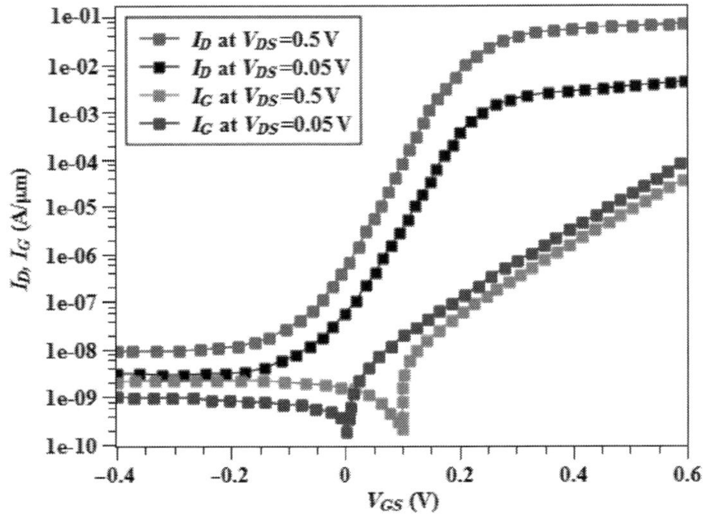

Figure 3.15
Subthreshold and gate leakage current characteristics for $L_g = 20$ nm and $W = 2 \times 10$ μm, HEMT on InP substrate without doped S/D regions.

V_{DS}, the gate leakage current will be high due to impact ionization mechanism. As V_{DS} increases, the hot electrons from the 2DEG acquire enough energy to create new electron–hole pairs based on a process called impact ionization. These newly created hot electrons are collected at the drain terminal which results in the sudden increase of drain current and the holes accumulated in the QW will move horizontally with respect to the 2DEG and recombines with electrons at the source terminal. At very high electric field, some of the newly created holes will be able to tunnel through the heterojunction barrier and reach the gate and cause large increase in the gate current. The net leakage current in the HEMTs increases with the decrease in L_g which is mainly due to the lowering of barrier height. The subthreshold leakage current can be reduced by minimizing the physical gap between the 2DEG and the gate foot. That is, for reducing the subthreshold leakage, the most effective way is reducing the thickness of spacer and barrier layers under the gate (Fig. 3.15).

The variation of subthreshold current and gate leakage current as a function of V_{GS} for $L_g = 20$ nm, $W = 2 \times 10$ μm proposed HEMT on InP substrate without doped source/drain regions is shown in Fig. 3.15. The subthreshold current variation is plotted for drain-to-source voltages, $V_{DS} = 0.5$ V and $V_{DS} = 0.05$ V. The gate leakage current is also plotted for $V_{DS} = 0.5$ V and $V_{DS} = 0.05$ V. From Fig. 3.15 it is clear that increase in drain-to-source voltage for a fixed gate-to-source voltage results in the increase in subthreshold current. Also from Fig. 3.15, it is evident that increase in drain-to-source voltage for a fixed gate-to-source voltage results in reduced gate leakage current. The gate leakage current depends mainly on the gate-to-channel separation. Reducing the gate-to-channel separation, results in increased leakage currents. In this work, the gate-to-channel separation was kept at 3 nm and this was possible with the use of platinum sinking technology.

3.13 Future Scope

The continued growth in millimeter- and sub-millimeter wave-frequency applications such as high data-rate communication systems, space-based millimeter wave receivers and exciters, low-noise amplifiers, and power amplifiers has propelled the demand for high-frequency transistors such as HBTs, HEMTs, and MOSFETs. Among these devices, HEMTs have been considered as the most preferable candidates for future THz applications due to their high-speed and low-noise characteristics. Conventional silicon-based transistors are incapable of meeting the demands arising from the consumer electronics, as well as military, space, and defence applications. New semiconductor materials, fabrication technologies, and device structures are required for future high-speed, low-power applications.

Outstanding DC and RF performance has been demonstrated by InAs inserted InAlAs/InGaAs/InAlAs HEMTs on InP or GaAs substrates. The mobility and saturation velocity of electrons in such HEMTs can be improved by increasing the indium content in the channel, which makes InAs-inserted InGaAs channel-based HEMTs the most attractive candidates for future low-power high-speed logic applications.

Although the RF performance of HEMT is being vigorously pushed toward terahertz region, novel device structures such as metal-oxide semiconductor high-electron-mobility transistors (MOSHEMT) are needed to be explored for future ultra-high-speed frequency applications especially when it comes to THz frequencies.

Despite the high-frequency performances, the gate leakage current in HEMTs must be reduced to meet the future requirements of international technology roadmap for semiconductors (ITRS). Therefore the effects of high-k gate dielectrics on the performance of MOSHEMTs must be explored for future high-speed logic applications.

References

[1] T. Mimura, S. Hiyamizu, T. Fujii, K. Nanbu, A new field-effect transistor with selectively doped GaAs/n−$Al_xGa_{1-x}As$ heterojunctions, Japanese Journal of Applied Physics 19 (5) (1980) L225−L227.

[2] R. Dingle, H.L. Störmer, A.C. Gossard, W. Wiegmann, Electron mobilities in modulation-doped semiconductor heterojunction superlattices, Applied Physics Letters 33 (7) (1978) 665−667.

[3] Takashi Mimura, The HEMT-Fujitsu First, FUJITSU 36 (4) (1985) 346−354.

[4] Wang Li-Dan, Ding Peng, Su Yong-Bo, Chen Jiao, Zhang Bi-Chan, Jin Zhi, 100-nm T-gate InAlAs/InGaAs InP-based HEMTs with fT = 249 GHz and fmax = 415 GHz, Chinese Physics B 23 (3) (2014) 038501. 038501.

[5] D.H. Kim, Jesus A. del Alamo, 30-nm E-mode InAs PHEMTs for THz and future logic applications, IEDM Technical Digest (2008) 719−722.

Further Reading

Takashi Mimura, Status of the GaAs metal-oxide-semi-conductor technology, IEEE Transactions on Electron Devices 27 (6) (1980) 1147–1149.

T. Mimura, S. Hiyamizu, H. Hashimoto, and H. Ishikawa, (1980) "An enhancement-mode high electron mobility transistor for VLSI", in Proc. 12th Solid-State Devices Conf., Tokyo, Japan, Aug. 1980, p. 364.

E. Yi Chang, Kuo Chien-I, InAs thin channel high electron mobility transistors with very high current gain cut-off frequency for emerging submillimeter wave applications, Applied Physics Express 6 (2013) 034001. 1.

Xiaobing Mei, Wayne Yoshida, Mike Lange, Jane Lee, Joe Zhou, Po-Hsin Liu, et al., First demonstration of amplification at 1 THz using 25-nm InP high electron mobility transistor process, IEEE Electron Device Letters 36 (4) (2015) 327–329.

T. Zimmer, D.O. Bodi, J.M. Dumas, N. Labat, A. Touboul, Y. Danto, Kink effect in HEMT structures; a trap related semi-quantitative model and an empirical approach for SPICE simulation, Solid-State Electronics 35 (10) (1992) 1543–1548.

B.J. Vasallo, H. Rodilla, T. Gonzalez, G. Moschetti, J. Grahn, J. Mateos, Monte Carlo study of kink effects in isolated gate InAs/AlSb high electron mobility transistors, Journal of Applied Physics 108 (2010) 094505. 1.

D.H. Kim, J.A. Alamo, J.H. Lee, K.S. Seo, Beyond CMOS: impact of side recess spacing on the logic performance of 50 nm $In_{0.7}Ga_{0.3}As$ HEMTs, Journal of Semiconductor Technology and Science 6 (3) (2006) 146–153.

R. Lai, X.B. Mei, W.R. Deal, W. Yoshida, Y.M. Kim, P.H. Liu, et al., Sub-50 nm InP HEMT device with fmax greater than 1 THz, IEEE International Electron Devices Meeting December (2007) 609–612.

K.J. Chen, T. Enoki, K. Maezawa, K. Arai, M. Yamamoto, High performance InP based enhancement mode HEMTs using non-alloyed ohmic contacts and platinum based buried gate technologies, IEEE Transactions on Electron Devices 43 (2) (1996) 252–257.

Y. Yamashita, A. Endoh, Pseudomorphic $In_{0.52}Al_{0.48}As/In_{0.7}Ga_{0.3}As$ HEMTs with an Ultra High f_T of 562 GHz, IEEE Electron Device Letters 23 (10) (2002) 573–575.

L.D. Nguyen, A.S. Brown, M.A. Thompson, L.M. Jelloian, 50-nm self-aligned-gate pseudomorphic AlInAs/GaInAs high electron mobility transistors, IEEE Transactions on Electron Devices 39 (1992) 2007–2014.

T. Enoki, M. Tomizawa, Y. Umeda, Y. Ishii, 0.05-μm-gate InAlAs/InGaAs high electron mobility transistor and reduction of its short channel effects, Japanese Journal of Applied Physics 33 (1994) 798–803.

T. Suemitsu, T. Ishii, H. Yokoyama, T. Enoki, Y. Ishii, T. Tamamura, 30-nm-gate InP-based lattice-matched high electron mobility transistors with 350 GHz cut-off frequency, Japanese Journal of Applied Physics 38 (1999) L154–L156.

T. Suemitsu, T. Ishii, Y. Ishii, Gate and recess engineering for ultrahigh-speed InP-based HEMTs, IEICE Transactions on Electronics E84-C (2001) 1283–1288.

CHAPTER 4

Organic Transistor- Device Structure, Model and Applications

Yaochuan Mei

DD Visions LLC, Austell, GA, United States

4.1 Organic Electronics: Low-Cost, Large-Area, and Flexible

Electronics based on inorganic semiconductors (predominantly silicon) currently address a huge market, but unfortunately they are ill suited for the emerging class of large-area, ultra-low cost, and flexible electronic products. This limitation is due to their mechanical properties (these semiconductors and the substrates compatible with them are rugged and relatively high cost as a result of complex processing steps). Organic electronics technologies, on the other hand, are based on semiconductors that can be manufactured at low cost and in large volumes: such devices address new applications and products, which are off-limits for conventional electronics, or offer improved economic benefit. Examples include, but are not limited to active electronic anticounterfeiting and tracking devices embedded into banknotes (Fig. 4.1A) [1], flexible and rollable displays (Fig. 4.1B) [2], sensors (Fig. 4.1C) [3], bio-electronics (sensing stamps, imaging arrays, implanted devices, etc.) (Fig. 4.1D) [4], radio-frequency identification (RFID) tags (Fig. 4.1E) [5], thermoelectrics (Fig. 4.1F) [6], disposable and wearable electronics [7–11], and so on.

Because organic semiconductors can be manufactured at moderate temperatures, they could be deposited onto a wide range of surfaces, from rigid silicon wafers to flexible plastics, even skin, paper, or textiles. This class of semiconductors offers an attractive platform for future electronics developments and applications, and it will be the focus of this chapter.

4.2 Field-Effect Transistors Structure

The layers of a field-effect transistor (FET) include three electrodes (the source, the drain, and the gate), a dielectric layer, and a semiconductor layer. Gate voltage establishes an electrical field in the dielectric layer and induces charge into the semiconductor: negative (positive) gate voltage is required to induce holes (electrons) in the channel. The source and

Nanoelectronics. DOI: https://doi.org/10.1016/B978-0-12-813353-8.00010-5

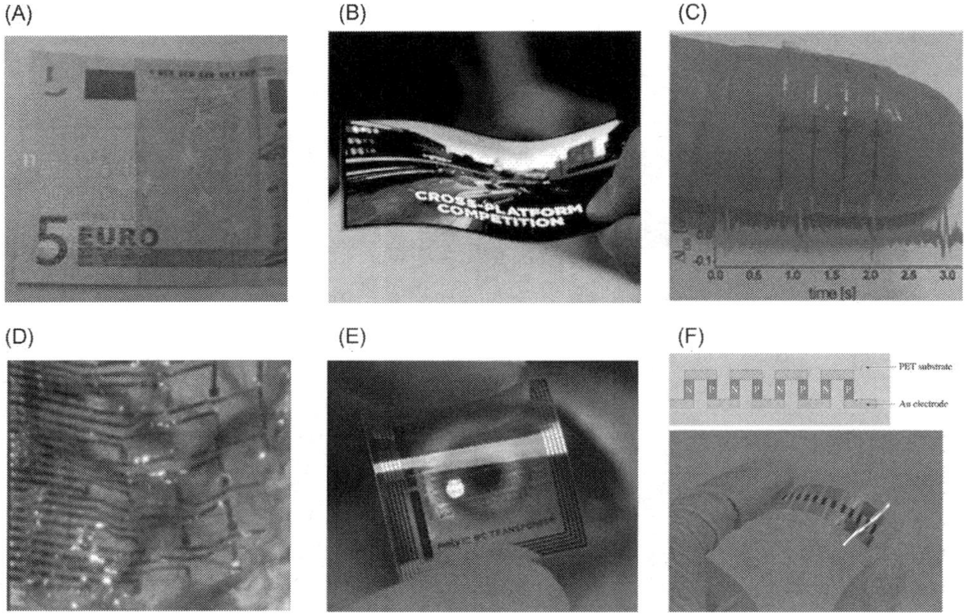

Figure 4.1
Applications based on organic semiconductors. (A) Flexible devices on banknote (Ref. [1]), (B) flexible display (Ref. [2]), (C) biosensor (Ref. [3]), (D) bio-integrated electronics (Ref. [4]), (E) RFID (Ref. [5]), and (F) thermoelectrics (Ref. [6]).

the drain electrodes are in direct contact with the semiconductor layer, and they are responsible for charge injection and collection through the conducting channel.

Several device structures can be built up based on the order in which the layers are deposited, as shown in Fig. 4.2.

Bottom-gate bottom-contact structure (A) offers the advantage of easy and fast measurement of semiconductor characteristics, as all layers beside semiconductor are prefabricated. The architectures (B) and (C) have the advantage of having less contact resistance compared with coplanar structure [8]. The top-gate dielectric (C), (D) is, in general, an organic film, because the deposition of inorganics could degrade the semiconductor layer. The nature of the dielectric material can impact the device performance by effects like polaron coupling [24] and thermal strain. Top-gate structures have an added advantage on unstable semiconductors, as here the dielectric layer directly encapsulates semiconductor after deposition.

Three typical methods were used to obtain the organic semiconductor thin film: drop-casting, spin-coating, and spray-coating (Fig. 4.3). Organic semiconductor solution was obtained by dissolving powder into room temperature organic solvents.

For drop-casting fabrication method (Fig. 4.3A), drops of solution on substrate takes slowly evaporate and obtain the crystalline film. The long-time drying of the active layer generates

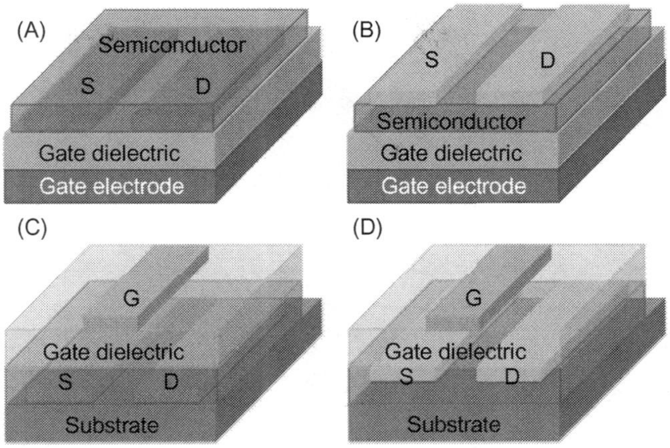

Figure 4.2
Different FET device geometries including (A) bottom-gate bottom-contact, (B) bottom-gate top-contact with air gap, (C) top-gate bottom-contact, and (D) top-gate top-contact. *G*: gate, *S*: source, *D*: drain.

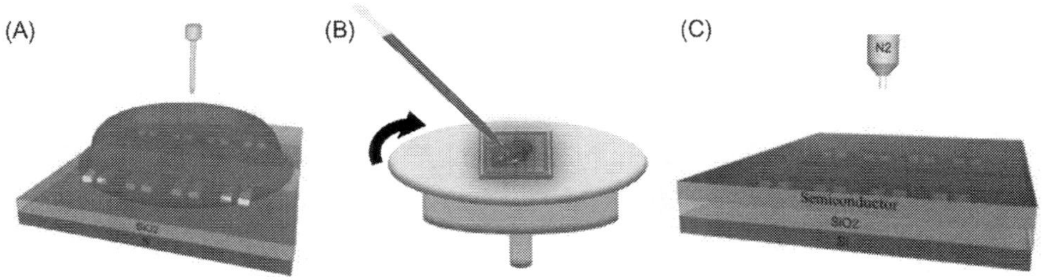

Figure 4.3
FET solution-processing methods: (A) drop-casting, (B) spin-coating, and (C) spray-coating.

large grains that can even be single crystals, which generally lead to higher mobility. Nevertheless, this method generally yields large variations in device performance.

In spin-coating (Fig. 4.3B), the organic solution was spin-coated over the substrates at a high speed. In spray-coating (Fig. 4.3C), argon or nitrogen are used to drive a more diluted solution onto substrate. Fast deposition methods like spin-coating and spray-coating could generally provide more uniform film.

4.3 Field-Effect Transistors Characterization

To test the electrical performance of the organic field-effect transistors (OFETs), we sweep source–gate voltage and source–drain voltage, respectively, and measure the transfer and

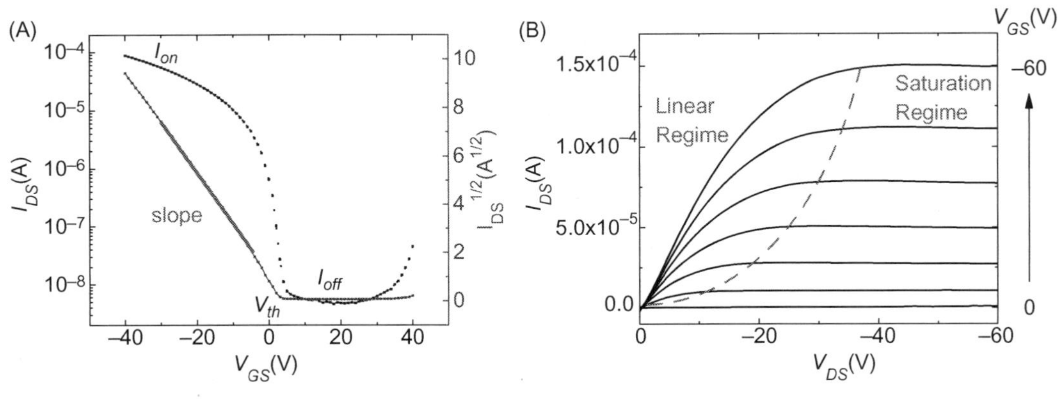

Figure 4.4
Typical FET characteristics: (A) transfer and (B) transport curves of a p-channel device.

transport currents (curves shown in Fig. 4.4). Important quantities for the measurement are as follows:

I_{DS}: source–drain current
W: channel width
L: channel length
μ: field – effect mobility
C_i: gate unit area capacitance
V_{DS} : source – drain bias
V_{GS}: source – gate bias
V_T: threshold voltage.

The transfer curve measurement was taken at two different source–drain voltages (V_{DS}): one in the device's linear regime and another in the device's saturation regime. Depending on the sign of V_{DS} and V_{GS}: positive or negative electrons or holes could be accumulated, respectively. Depending on the gate voltage relative to threshold voltage, the drain current shows transitions from the "off" to the "on" state. At low V_{DS}, the drain current increases linearly with the applied drain-to-source voltage, following equation:

$$I_{DS} = \frac{W\mu C_i}{L}(V_{GS} - V_T)V_{DS.} \tag{4.1}$$

At large V_{DS} (equal to or larger than gate-to-source voltage minus the threshold voltage), the drain current reaches a maximum value. Then the device reaches saturation regime, where saturated drain current follows [8]:

$$I_{DS} = \frac{W}{2L}\mu C_i(V_{GS} - V_T)^2 \tag{4.2}$$

Charge-carrier mobility in the saturation regime is an important figure of merit, as it shows intrinsic charge property with minimal contact effects.

4.4 Organic Semiconductors Selection

Organic electronics utilize carbon-based small molecules and polymers as the active electronic layers. Different materials can be adopted as semiconductors, dielectrics (insulators), or even injecting electrodes (conductors) [12–14]. Organic semiconductors combine functional tunability with the low cost per unit area offered by their production in large volumes, using solution-based approaches, but unfortunately organic devices suffer from insufficient performance for most of the applications that are envisioned for these materials. Classical examples of high performance organic semiconductors include polycyclic aromatic compounds such as pentacene, anthracene, rubrene, and their derivatives [15–21]. Fig. 4.5 depicts the chemical structure of pentacene (A) and rubrene (B), respectively. In Fig. 4.5C and D, we show their crystalline packing, while in Fig. 4.5E and F we include examples of field-effect devices fabricated at their surface. These materials can exhibit high charge-carrier mobilities (e.g., around 40 $cm^2V^{-1}s^{-1}$ for pentacene and around 20 $cm^2V^{-1}s^{-1}$ for rubrene) [22–24]. While they are a great platform for fundamental studies, their processing still adopts vacuum and/or high-temperature processes, which preclude rapid, low-cost fabrication of devices based on such materials.

Silylethyne substitution of high-performance semiconductors and their derivatives was introduced as an efficient strategy to enhance their solubility and allowed their deposition from solution-based methods [25–26]. Over the past decade, development of a large number of soluble small-molecule organic semiconductors following this route yielded significant improvements in the performance of OFETs and organic photovoltaic (OPVs) devices based on these coumpounds [27–34]. Triisopropylsilylethynyl pentacene (TIPS pentacene, Fig. 4.6A) and 2,8-difluoro-5,11-bis(triethyl silylethynyl) anthradithiophene (diF-TES ADT, Fig. 4.6B), in particular, generated much interest due to their high mobilities both as single phase and when blended with polymers [37–40]. Devices processed from dilute solution reach high field-effect mobilities (over 2 $cm^2V^{-1}s^{-1}$ for both TIPS pentacene [41] and diF-TES ADT [42]) as upon optimization of the film formation, and demonstrate the potential of OFET for incorporation in large-area, low-cost electronic applications. Crystal structure of these compounds adopts a 2D π-stacking motif (Fig. 4.6C and D).

Modifications on silylethyne side group and on backbone were shown to change the molecular packing, i.e., distance along the long and short axis between adjacent molecules, as well as the areal molecular overlap [43]. These variations in molecular arrangement could improve or disrupt charge transport properties, because of the changes that they introduce in the overlap of the molecular backbone. This is to be expected, as it was

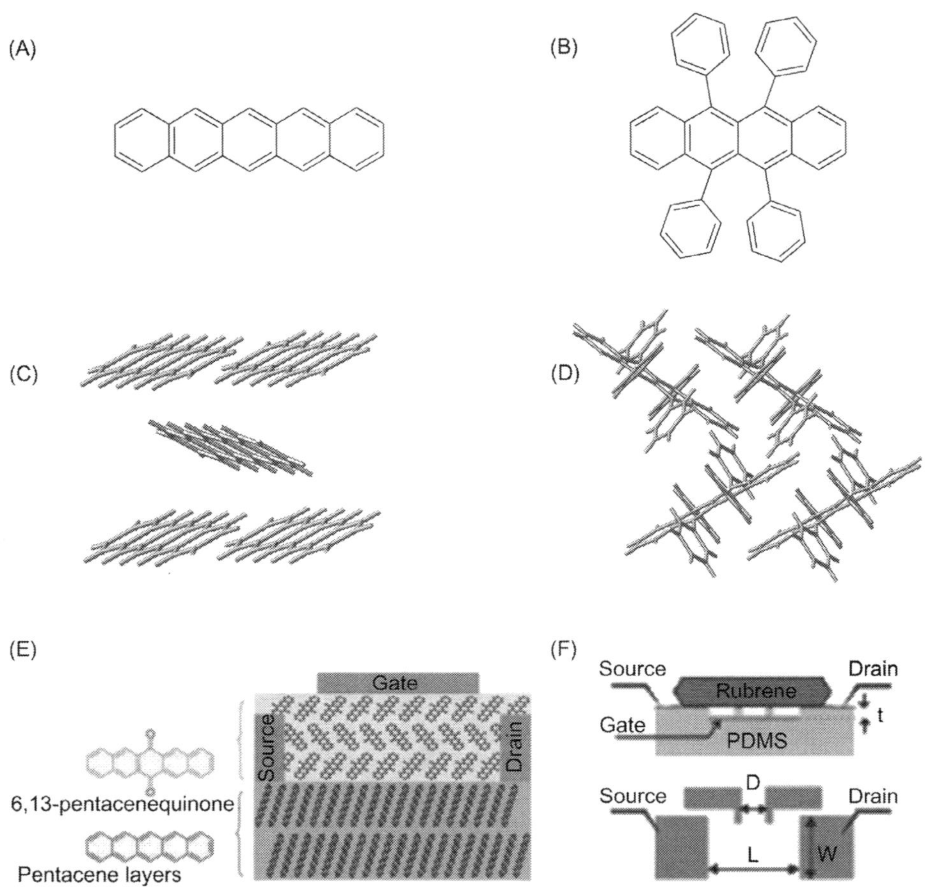

Figure 4.5

Molecular structure (A and B), crystal packing (C and D), and device structure with high performance ((E) from Ref. [22] and (F) from Ref. [23]) for pentacene and rubrene, respectively.

predicted by theoretical studies showing that such differences can strongly affect the intermolecular electronic couplings [44]. As an example, calculations on tetracene showing these variations in charge transfer are shown in Fig. 4.7. This work shows that the electronic coupling, quantified in charge transfer integral, *t*, is directly determined by molecular packing. From these calculations, charge transfer between molecules increases with decreasing intermolecular spacing. Subtle changes in the coplanar molecular overlap also influence the coupling. Achieving the desired crystalline packing, which can yield high charge-carrier mobility is, however, challenging because the structure of these compounds is highly sensitive to the details of processing as a result of the weak, van der Waals intermolecular bonds present in the solid. Fluorination of the fused-aromatic backbone in such structures was introduced to enhance crystallization via halogen interactions and improve chemical stability, but even for these compounds, the prediction and, consequently,

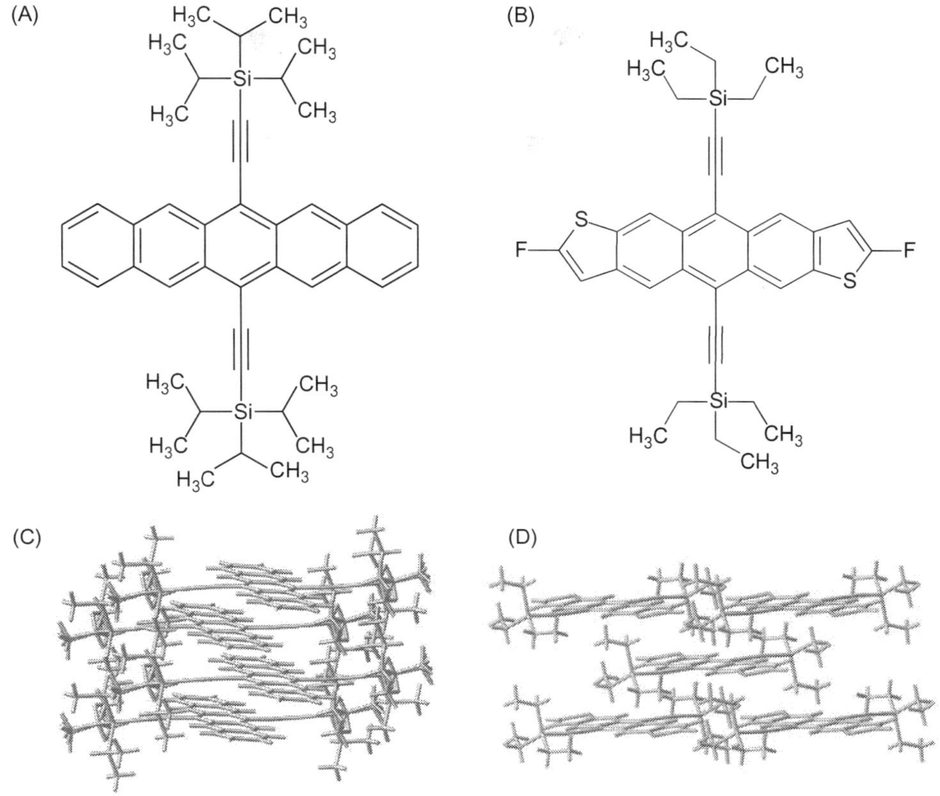

Figure 4.6
Soluble pentacene derivatives. (A) TIPS pentacene. (B) Difluoro-triethylsilylethynyl anthradithiophene (diF-TES ADT). (C) and (D) are their corresponding molecular packing.

achievement of a desired crystalline packing is not yet possible to be achieved in a reliable way [37,45]. Efforts have been made in fine-tuning processing and fabrication details to achieve improved crystalline packing.

4.5 Interfacial Engineering in Field-Effect Transistors

The performance of organic devices is generally modest, compared with the inorganic counterparts. This, however, is not necessary an intrinsic property of the organic semiconductor layer, but it may originate from extrinsic effects such as the presence of impurities, thin-film microstructure, traps at the organic semiconductor–dielectric interface, contact effects, etc. Self-assembled monolayers (SAMs) are often being used to enhance the electrical performance of OFETs by improving the organic semiconductor film morphology and/or lowering the injection barrier at the organic semiconductor/metal interface [46–49].

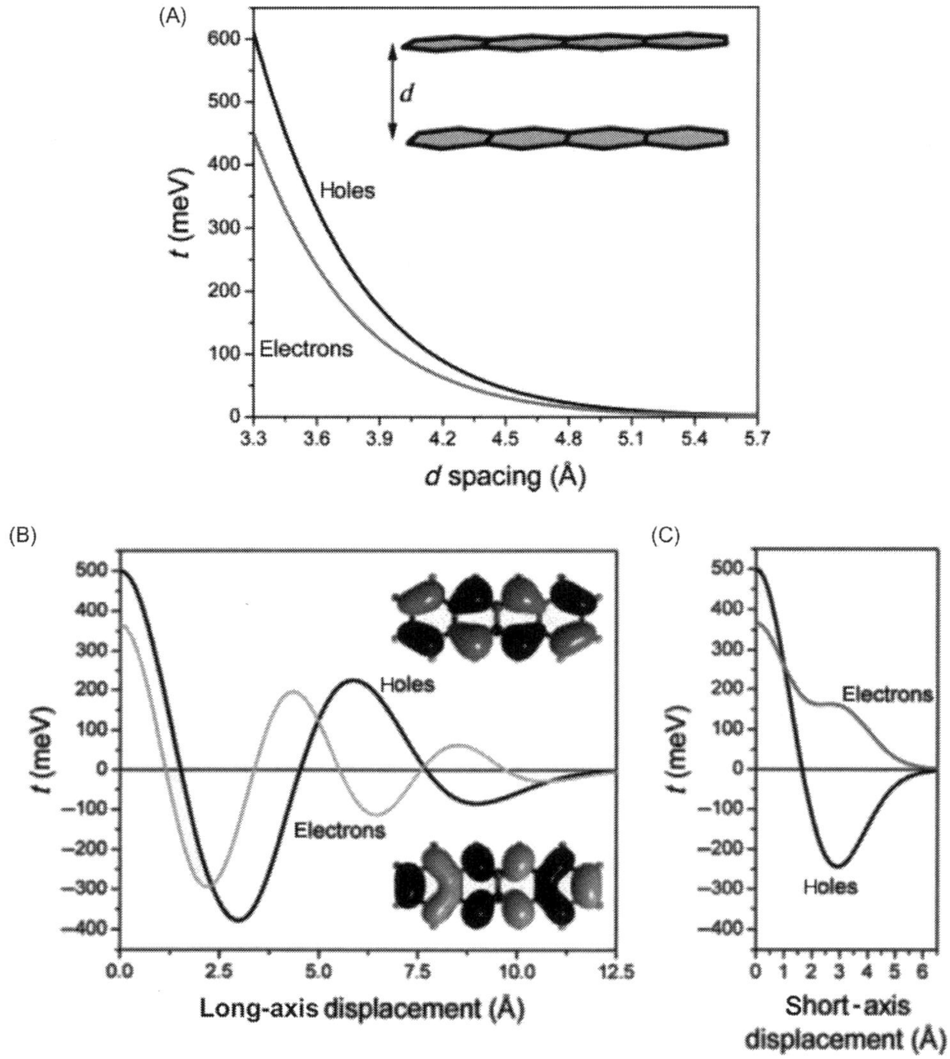

Figure 4.7

Calculated transfer integrals for electrons and holes in a tetracene cofacial dimer, as a function of (A) the intermolecular distance, (B) translation of one molecule along its long axis, and (C) short axis. Source: *Adapted from V. Coropceanu, et al., Charge transport in organic semiconductors, Chem. Rev. 107 (2007) 926–952.*

As shown in Fig. 4.8A, SAMs-treated dielectrics improve FET performance by reducing the interfacial trap densities and charge scattering [35]. From Fig. 4.8B, SAMs-treated contacts enhance FET performance by improving work function, thus charge injection [36].

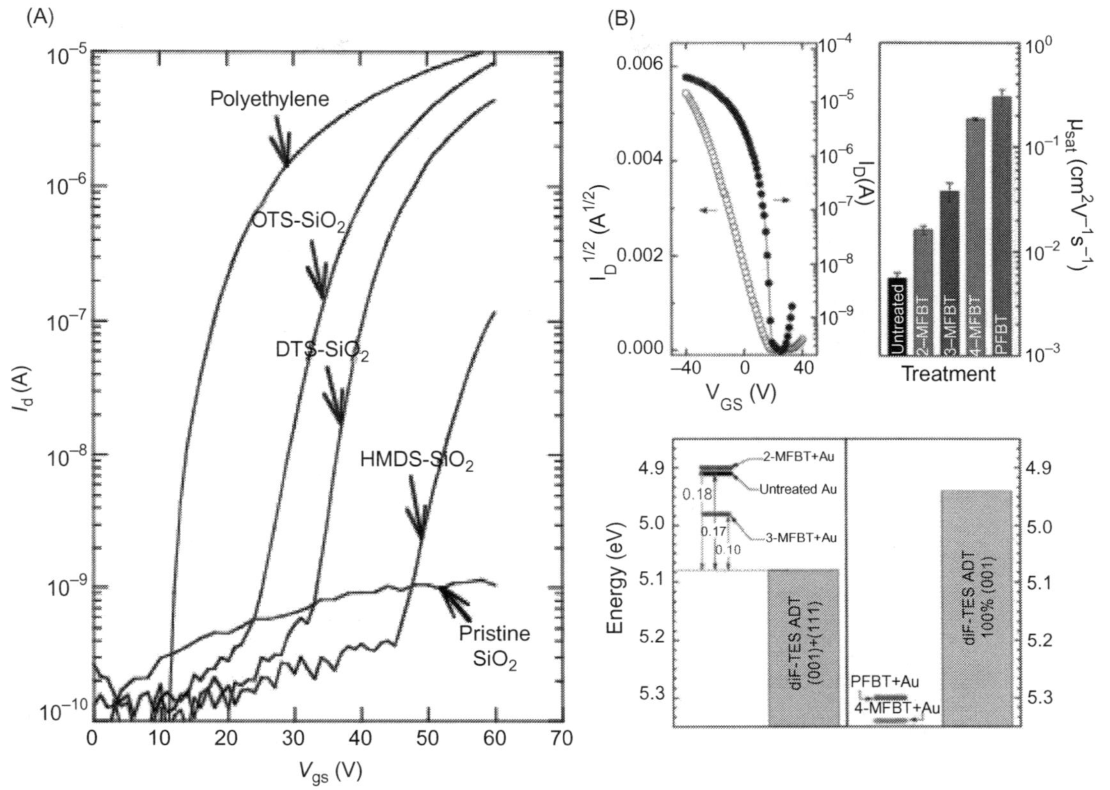

Figure 4.8

Various device performances based on self-assembled monolayers applied on (A) dielectric. Source: *Adapted from L.-L. Chua, et al., General observation of n-type field-effect behaviour in organic semiconductors. Nature 434 (2005) 194–199. (B) (Contact) adapted from J.W. Ward, et al., Tailored interfaces for self-patterning organic thin-film transistors, J. Mater. Chem. 22 (2012) 19047–19053.*

4.5.1 Changes in Surface Energy as a Result of SAM Treatment

Thin-films of organic semiconductors consist of grains separated by grain boundaries. Grain boundaries affect electrical properties by introducing resistances and voids [50–52]. Grain growth in thin-film is governed by interfacial surface energy and can be modeled by using the Kardar–Parisi–Zhang (KPZ) equation [53]:

$$\frac{\partial h}{\partial t} = \nu \nabla^2 h + \frac{\lambda}{2}(\nabla h)^2 + \eta. \tag{4.3}$$

Here, thickness (h) is a function of deposition time (t), smoothening factor ν is correlated to surface tension, λ is the growth front form depending on growth orientation, and η is noise in deposition. Assuming that grains grow isotropically without noise, film formation from large and small surface energy substrates could be modeled and shown in Fig. 4.9.

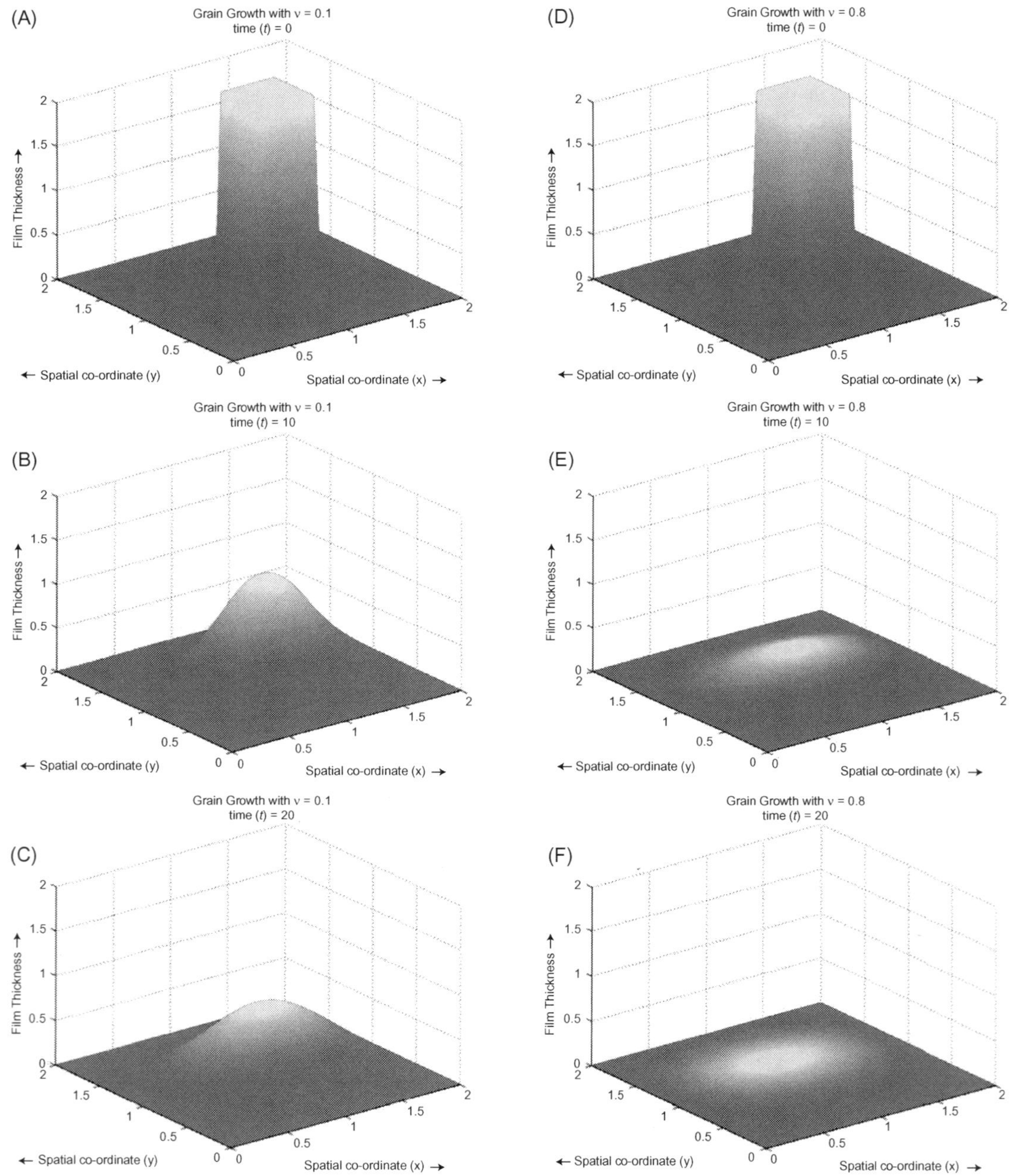

Figure 4.9

Film formation at each time step on small surface energy substrate (A–C) and large surface energy substrate (D–F).

Here, we assume organic material being originally deposited to a same 0.5*0.5 square in (A) and (D). Smaller surface energy leads to smaller ν value in (A–C). After same deposition time, organic material can be observed more concentrated with fewer grain formation sites, leading to larger grains in the semiconductor film shown in (C) compared with (F).

The total surface energy of the substrate (γ_S^{Total}) consists of two components:

$$\gamma_S^{Total} = \gamma_S^{LW} + \gamma_S^{AB}, \tag{4.4}$$

where γ_S^{LW} is the nonpolar Lifshitz–Van der Waals interaction and γ_S^{AB} is the polar acid–base interaction, and γ_S^{AB} can be defined as the geometric mean of the electron acceptor or Lewis acid (γ_S^+) and the electron donor or Lewis base (γ_S^-) components. The contact angle value is related to the surface tension from the following equation [1]:

$$\gamma_L \cos\theta = -\gamma_L + 2\left(\sqrt{\gamma_L^{LW}\gamma_S^{LW}} + \sqrt{\gamma_L^-\gamma_S^-} + \sqrt{\gamma_L^+\gamma_S^+}\right). \tag{4.5}$$

Here γ_L is the surface tension of the liquid and γ_L^{LW}, γ_L^-, and γ_L^+ are extracted from the literature. The contact angles for three different liquids on the same type of surface were measured by Rame–Hart goniometer. By creating a system of equations with the values obtained for the three liquids, we could solve for the surface energy and components of the substrate (γ_S^{Total}, γ_S^{LW},γ_S^{AB}).

4.5.2 Work Function Shift

Work function is a material property that reflects the energy needed to remove an electron from surface to vacuum. This property affects charge injection in FETs. Frisbie et al. used metal–molecule–metal junctions and found the presence of a barrier to charge transport at the metal molecule interfaces [54]. The barrier height decreases with increasing metal work function, leading to decreased contact resistance as shown in Fig. 4.10.

Work function of contacts and the shift in the Au work function as a result of SAM treatment could be easily determined by electrostatic voltmeter in Kelvin probe. The contact potential difference between the treated surface and the probe could be measured and the absolute work function, ϕ, is calculated from [36]

$$\phi = \mathrm{e}\left(V_{sample} - V_{probe}\right) + \phi_{ref}, \tag{4.6}$$

where V_{sample} is the signal from sample and V_{probe} is the signal from the reference surface.

The interfacial trap density (N_T) was characterized from the FET electrical characteristics. At constant temperature T, this quantity is a function of the dielectric capacitance, Ci, and

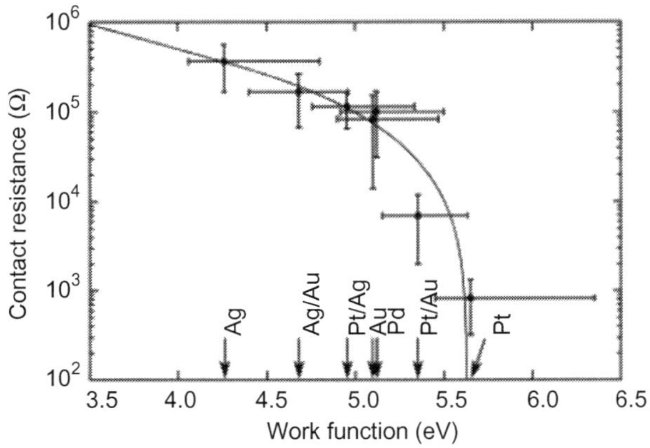

Figure 4.10
Contact resistance in metal–molecule–metal junctions as a function of different metal work functions. Source: *Adapted from J.M. Beebe, V.B. Engelkes, L.L. Miller, C.D. Frisbie, Contact resistance in metalmolecule – metal junctions based on aliphatic SAMs: effects of surface linker and metal work function, J. Am. Chem. Soc. 124 (2002) 11268 – 11269.*

the subthreshold swing S (the least gate voltage required to change the drain current by one order of magnitude) following the expression:

$$N_T = \frac{C_i}{e^2}\left(\frac{eS}{ln10 \cdot k_B T} - 1\right).^{54} \quad (4.7)$$

Here, e is the elementary charge and k_B is Boltzmann constant.

4.5.3 Contact Resistance

Transmission line method is commonly applied to measure contact resistance in OFETs [55]. The total resistance of the OFET is $R_{\text{on}} = \frac{\partial V_{DS}}{\partial I_D}$, which was determined from the linear regime of device operation (small V_{DS}) as the inverse slope of I_D versus V_{DS} graph, see Fig. 4.3B. This results from the added series resistance of the contacts (R_c) and the channel (R_{ch}):

$$R_{on} = R_{ch}(L) + 2R_c. \quad (4.8)$$

Thus, by plotting R_{on} as a function of channel length, L, and extrapolating this curve to $L = 0$, we could extract the value of R_c.

As an example, in Fig. 4.11, width normalized contact resistance R_cW for a set of organic thin-film transistors (TFTs) is calculated by extending the plot to $L = 0$, $R_cW \sim 200\ \Omega\text{m}$ [56].

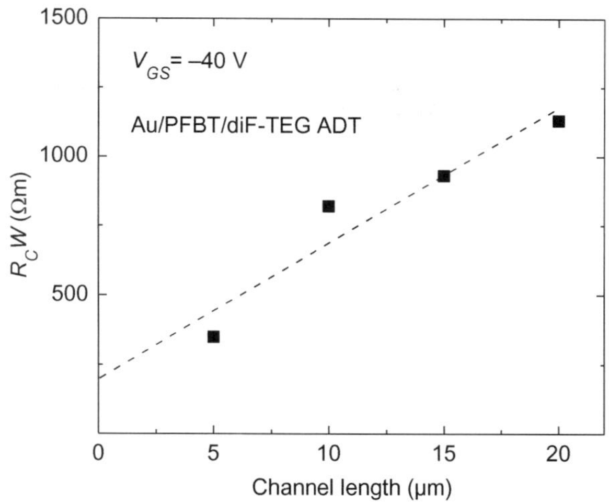

Figure 4.11
Plot of R_cW versus channel length for a series of diF-TEG ADT devices spun-coated on Au/PFBT contacts, at an applied gate voltage of $V_{GS} = -40$ V.

References

[1] U. Zschieschang, et al., Organic electronics on banknotes, Advanced Materials 23 (2011) 654–658.

[2] L. Li, et al., Advances and challenges for flexible energy storage and conversion devices and systems, Energy Environ. Sci. 7 (2014) 2101–2122.

[3] A. Campana, T. Cramer, D.T. Simon, M. Berggren, F. Biscarini, Electrocardiographic Recording with Conformable Organic Electrochemical Transistor Fabricated on Resorbable Bioscaffold, Advanced Materials 26 (2014) 3874–3878.

[4] D.-H. Kim, et al., Dissolvable films of silk fibroin for ultrathin conformal bio-integrated electronics, Nat Mater 9 (2010) 511–517.

[5] T.N. Jackson, Organic Semiconductors: Beyond Moore's Law, Nat Mater 4 (2005) 581–582.

[6] F. Jiao, et al., Inkjet-printed flexible organic thin-film thermoelectric devices based on p- and n-type poly (metal 1,1,2,2-ethenetetrathiolate)s/polymer composites through ball-milling, Philosophical Transactions of the Royal Society of London A: Mathematical, Physical and Engineering Sciences 372 (2014).

[7] A.C. Arias, J.D. MacKenzie, I. McCulloch, J. Rivnay, A. Salleo, Materials and Applications for Large Area Electronics: Solution-Based Approaches, Chemical Reviews 110 (2010) 3–24.

[8] H. Klauk, Organic thin-film transistors, Chemical Society Reviews 39 (2010) 2643–2666.

[9] H. Yan, et al., A high-mobility electron-transporting polymer for printed transistors, Nature 457 (2009) 679. U1.

[10] T. Sekitani, T. Someya, Stretchable, Large-area Organic Electronics, Advanced Materials 22 (2010) 2228–2246.

[11] S.C.B. Mannsfeld, et al., Highly sensitive flexible pressure sensors with microstructured rubber dielectric layers, Nature Materials 9 (2010) 859–864.

[12] K.P. Goetz, et al., Charge-transfer complexes: new perspectives on an old class of compounds, Journal of Materials Chemistry C 2 (2014) 3065–3076.

[13] H. Shirakawa, E.J. Louis, A.G. MacDiarmid, C.K. Chiang, A.J. Heeger, Synthesis of electrically conducting organic polymers: halogen derivatives of polyacetylene, (CH), Journal of the Chemical Society, Chemical Communications (1977) 578–580.

[14] A.F. Garito, A.J. Heeger, Design and synthesis of organic metals, Accounts of Chemical Research 7 (1974) 232–240.

[15] F. Würthner, R. Schmidt, Electronic and Crystal Engineering of Acenes for Solution-Processible Self-Assembling Organic Semiconductors, Chem Phys Chem 7 (2006) 793–797.

[16] J.A. Lim, H.S. Lee, W.H. Lee, K. Cho, Control of the Morphology and Structural Development of Solution-Processed Functionalized Acenes for High-Performance Organic Transistors, Advanced Functional Materials 19 (2009) 1515–1525.

[17] K.P. Goetz, et al., Effect of Acene Length on Electronic Properties in 5-, 6-, and 7-Ringed Heteroacenes, Advanced Materials 23 (2011) 3698–3703.

[18] M.L. Tang, T. Okamoto, Z. Bao, High-Performance Organic Semiconductors: Asymmetric Linear Acenes Containing Sulphur, Journal of the American Chemical Society 128 (2006) 16002–16003.

[19] C. Sutton, C. Risko, J.-L. Bredas, Non-Covalent Intermolecular Interactions in Organic Electronic Materials: Implications for the Molecular Packing vs. Electronic Properties of Acenes, Chemistry of Materials, acs.chemmater (2015) 5b03266. acs.chemmater.5b03266.

[20] M.M. Payne, S.R. Parkin, J.E. Anthony, Functionalized Higher Acenes: Hexacene and Heptacene, Journal of the American Chemical Society 127 (2005) 8028–8029.

[21] D.J. Gundlach, et al., Contact-induced crystallinity for high-performance soluble acene-based transistors and circuits, Nat Mater 7 (2008) 216–221.

[22] O.D. Jurchescu, M. Popinciuc, B.J. van Wees, T.T.M. Palstra, Interface-controlled, high-mobility organic transistors, Advanced Materials 19 (2007) 688–689.

[23] V. Podzorov, et al., Intrinsic charge transport on the surface of organic semiconductors, Physical Review Letters 93 (2004).

[24] I.N. Hulea, et al., Tunable Frohlich polarons in organic single-crystal transistors, Nature Materials 5 (2006) 982–986.

[25] J.E. Anthony, J.S. Brooks, D.L. Eaton, S.R. Parkin, Functionalized pentacene: Improved electronic properties from control of solid-state order, Journal of the American Chemical Society 123 (2001) 9482–9483.

[26] Y. Mei, M.A. Loth, M. Payne, W. Zhang, J. Smith, C.S. Day, S.R. Parkin, M. Heeney, I. McCulloch, T.D. Anthopoulos, J.E. Anthony, O.D. Jurchescu, High mobility field-effect transistors with versatile processing from a small-molecule organic semiconductor, Adv. Mater. 25 (31) (2013) 4352–4357.

[27] S.K. Park, J.E. Anthony, T.N. Jackson, Solution-processed TIPS-pentacene organic thin-film-transistor circuits, Ieee Electron Device Letters 28 (2007) 877–879.

[28] M.M. Payne, S.R. Parkin, J.E. Anthony, C.C. Kuo, T.N. Jackson, Organic field-effect transistors from solution-deposited functionalized acenes with mobilities as high as 1 cm(2)/V-s, Journal of the American Chemical Society 127 (2005) 4986–4987.

[29] C.D. Sheraw, T.N. Jackson, D.L. Eaton, J.E. Anthony, Functionalized pentacene active layer organic thin-film transistors, Advanced Materials 15 (2003) 2009–2011.

[30] D. Isik, et al., Ambipolar organic thin film transistors based on a soluble pentacene derivative, Applied Physics Letters 99 (2011).

[31] M.L. Tang, A.D. Reichardt, T. Okamoto, N. Miyaki, Z. Bao, Functionalized asymmetric linear acenes for high-performance organic semiconductors, Advanced Functional Materials 18 (2008) 1579–1585.

[32] M.L. Tang, A.D. Reichardt, N. Miyaki, R.M. Stoltenberg, Z. Bao, Ambipolar, high performance, acene-based organic thin film transistors, Journal of the American Chemical Society 130 (2008) 6064.

[33] M.T. Lloyd, J.E. Anthony, G.G. Malliaras, Photovoltaics from soluble small molecules, Materials Today 10 (2007) 34–41.

[34] M.T. Lloyd, et al., Photovoltaic cells from a soluble pentacene derivative, Organic Electronics 7 (2006) 243–248.

[35] L.-L. Chua, et al., General observation of n-type field-effect behaviour in organic semiconductors, Nature 434 (2005) 194–199.

[36] J.W. Ward, et al., Tailored interfaces for self-patterning organic thin-film transistors, Journal of Materials Chemistry 22 (2012) 19047–19053.

[37] S. Subramanian, et al., Chromophore fluorination enhances crystallization and stability of soluble anthradithiophene semiconductors, Journal of the American Chemical Society 130 (2008) 2706–2707.

[38] O.D. Jurchescu, et al., Organic Single-Crystal Field-Effect Transistors of a Soluble Anthradithiophene, Chemistry of Materials 20 (2008) 6733–6737.

[39] J. Smith, et al., Solution-Processed Small Molecule-Polymer Blend Organic Thin-Film Transistors with Hole Mobility Greater than 5 cm2/Vs, Advanced Materials 24 (2012) 2441–2446.

[40] R. Hamilton, et al., High-Performance Polymer-Small Molecule Blend Organic Transistors, Advanced Materials 21 (2009) 1166–1171.

[41] G. Giri, S. Park, M. Vosgueritchian, M.M. Shulaker, Z. Bao, High-Mobility, Aligned Crystalline Domains of TIPS-Pentacene with Metastable Polymorphs Through Lateral Confinement of Crystal Growth, Advanced Materials 26 (2014) 487–493.

[42] P.J. Diemer, et al., Quantitative analysis of the density of trap states at the semiconductor-dielectric interface in organic field-effect transistors, Applied Physics Letters 107 (2015) 103303. 103303.

[43] J.E. Anthony, Functionalized acenes and heteroacenes for organic electronics, Chemical Reviews 106 (2006) 5028–5048.

[44] V. Coropceanu, et al., Charge Transport in Organic Semiconductors, Chemical Reviews 107 (2007) 926–952.

[45] S.K. Park, D.A. Mourey, S. Subramanian, J.E. Anthony, T.N. Jackson, High-mobility spin-cast organic thin film transistors, Applied Physics Letters 93 (2008) 043301.

[46] D.J. Gundlach, et al., Contact-induced crystallinity for high-performance soluble acene- based transistors and circuits, Nat Mater 7 (2008) 216–221.

[47] J.C. Love, L.A. Estroff, J.K. Kriebel, R.G. Nuzzo, G.M. Whitesides, Self-Assembled Monolayers of Thiolates on Metals as a Form of Nanotechnology, Chemical Reviews 105 (2005) 1103–1170.

[48] Y. Mei, P.J. Diemer, M.R. Niazi, R.K. Hallani, K. Jarolimek, C.S. Day, C. Risko, J.E. Anthony, A. Amassian, O.D. Jurchescu, Crossover from band-like to thermally activated charge transport in organic transistors due to strain-induced traps, Proc. Natl. Acad. Sci. USA 114 (2017) E6739–E6748.

[49] Y. Mei, D. Fogel, J. Chen, J.W. Ward, M.M. Payne, J.E. Anthony, O.D. Jurchescu, Interface engineering to enhance charge injection and transport in solution-deposited organic transistors, Org. Electron 50 (2017) 100–105.

[50] G. Horowitz, M.E. Hajlaoui, Grain size dependent mobility in polycrystalline organic field-effect transistors, Synthetic Metals 122 (2001) 185–189.

[51] T.W. Kelley, C.D. Frisbie, Gate voltage dependent resistance of a single organic semiconductor grain boundary, Journal of Physical Chemistry B 105 (2001) 4538–4540.

[52] A. Di Carlo, F. Piacenza, A. Bolognesi, B. Stadlober, H. Maresch, Influence of grain sizes on the mobility of organic thin-film transistors, Applied Physics Letters 86 (2005) 263501. 263501.

[53] J.A. Venables, Introduction to surface and thin film processes, (Cambridge University Press, New York, 2003.

[54] J.M. Beebe, V.B. Engelkes, L.L. Miller, C.D. Frisbie, Contact resistance in metalmolecule-metal junctions based on aliphatic SAMs: Effects of surface linker and metal work function, Journal of the American Chemical Society 124 (2002) 11268–11269.

[55] D.J. Gundlach, et al., An experimental study of contact effects in organic thin film transistors, Journal of Applied Physics 100 (2006).

[56] S.M. Baier, M.S. Shur, K. Lee, N.C. Cirillo, S.A. Hanka, FET Characterization using gated-TLM structure, IEEE Transactions on Electron Devices 32 (1985) 2824–2829.

Further Reading

Sze, S. M. & Ng, K. K. Physics of semiconductor device (2007).

PART II

Spintronics

CHAPTER 5

Mitigating Read Disturbance Errors in STT-RAM Caches by Using Data Compression

Sparsh Mittal

IIT, Hyderabad, Telangana, India

5.1 Introduction

Recent trends of increasing core counts and LLC capacity have motivated researchers to explore low leakage alternatives of SRAM for designing large LLCs. Due to their high density and near-zero leakage power consumption, nonvolatile memories such as STT-RAM and ReRAM (resistive RAM) have received significant attention in recent years [1,2]. Of these, STT-RAM is considered especially suitable for designing LLCs due to its close to SRAM read latency and high write endurance [3].

Since the write latency/energy of STT-RAM are higher than those of SRAM, previous work has mainly focused on addressing this overhead to enable use of STT-RAM for designing on-chip caches [4]. However, a more severe issue of "read disturbance error" (RDE) in STT-RAM caches has not been adequately addressed. Specifically, with feature size scaling, the write current reduces; however, read current does not reduce as much [5]. In fact, for sub-32nm feature size, the magnitude of read current becomes so close to the write current that a read operation is likely to modify the stored data and this is referred to as RDE (refer Section 5.2 for more details). Since reads happen on critical access path, RDE is likely to severely affect performance and reliability. In fact, it is expected that with ongoing process scaling, *readability* and *not writability* will become the most crucial bottleneck for STT-RAM [6–9].

Contributions: In this chapter, we present a technique, named "SHIELD," which shields STT-RAM caches against both RDE and high write latency/energy overhead. SHIELD works by using data compression to reduce the number of bits written to cache (Section 5.3). First, SHIELD does not write any bits to STT-RAM block for all zero data, since during next read

Nanoelectronics. DOI: https://doi.org/10.1016/B978-0-12-813353-8.00001-4

operation, such data can be reconstructed from compression encoding bits. This avoids RDE and high write latency/energy issue for all zero data. Second, SHIELD keeps two copies of data in the block if the data have compressed width (*CW*) of at most 32B (assuming 64B block size). On the next read operation, one copy gets RDE which is not corrected since the second copy still remains free of RDE. This avoids one restore operation for such narrow data ($0 < CW \leq 32B$). On any future read to this block, the single error-free copy of data is read and then restored. Third, for any data with $CW > 32B$, including an uncompressed data-item, a single copy is written which is restored after each read operation.

Salient Features: By virtue of writing compressed data, SHIELD reduces the bits written in restore operation and saves write energy. Thus, SHIELD addresses both RDE and write overhead issue for compressed data, whereas previous techniques [10,11] reduce some restore operations only and do not reduce the overhead of read and write operations (Section 5.4). SHIELD forgoes the capacity advantage of compression, and thus stores only one cache line in a block. Hence, SHIELD does not suffer from overheads, which generally accompany compression techniques [12] such as extra tags, compaction, fragmentation, changes in cache replacement policy, etc. Further, some techniques for reducing soft error vulnerability in SRAM caches [13] work by keeping two or three copies of an entire cache block (64B) in a cache set. This, however, leads to sharp degradation in cache capacity. By comparison, SHIELD duplicates compressed data of at most 32B size within a 64B cache block itself, and thus SHIELD does not degrade cache capacity. While SHIELD can work with any compression technique, in this chapter, we demonstrate SHIELD with base-delta-immediate (BDI) compression algorithm [14].

Evaluation and Results: We have performed microarchitectural simulation with an X86 full-system simulator (Section 5.5). Also, we have compared SHIELD with two mechanisms for addressing RDE, namely restore-after-read (also called HCRR) and low current long latency read (LCLL) (refer Section 5.5.4). The results show that SHIELD provides higher performance and energy efficiency compared to these techniques (Section 5.6). Compared to an ideal RDE-free STT-RAM LLC, for single and dual-core configurations, SHIELD provides *energy saving* of 1.7% and 3.0%, respectively, whereas LCLL (which on average performs better than HCRR) incurs *energy loss* of 5.1% and 5.2%, respectively. Also, the relative performance with SHIELD is $0.98\times$ and $0.97\times$, whereas that with LCLL is $0.96\times$ and $0.96\times$, respectively. Additional experiments show that SHIELD works well for different cache sizes, and selective replication used in SHIELD contributes positively to its effectiveness.

5.2 Background

We now provide a brief background and refer the reader to previous work for more details [5,15,16].

5.2.1 Motivation for Using Nonvolatile Memories

Conventionally, SRAM has been used for designing on-chip caches. However, SRAM has low density and high leakage power due to which large SRAM caches contribute greatly to the overall power consumption. Although several architectural techniques have been proposed to manage the power consumption of SRAM caches such as cache reconfiguration, near-threshold computing, etc., the power budget targets of next-generation systems require magnitude order higher energy efficiency. This has motivated the researchers to explore alternatives of SRAM such as eDRAM, STT-RAM, etc. [17].

5.2.2 Working of STT-RAM

STT-RAM utilizes a magnetic tunnel junction (MTJ) as the memory storage. An MTJ contains two ferromagnetic layers separated by an oxide barrier layer. The magnetization direction of one ferromagnetic layer is fixed while that of the other ferromagnetic layer can be altered by passing a current. The resistance of the MTJ is determined by the relative magnetization direction of these two layers. If the two layers have different directions, the resistance of the MTJ is high and vice versa. Using this property, a binary value is stored in an STT-RAM cell [4]. Another characteristic of STT-RAM is that the read and write operations to it are asymmetric, since a write operation consumes larger time and energy than a read operation.

Although STT-RAM has lower density than PCM and RRAM and higher write latency and energy than SRAM, it has been widely used for designing caches due to its high write endurance. Another advantage of STT-RAM is that its nonvolatility can be traded to improve its write energy and latency. Based on the application characteristic and the level of cache hierarchy, a designer can choose a suitable value of retention period.

5.2.3 Origin of Read Disturbance Error

STT-RAM stores data in MTJ [3]. For performing both read and write operations to MTJ, a voltage is applied between bit line and source line. The only difference between read and write operations is that read operations use smaller voltage than write operations [10]. With decreasing feature size, there is an exponential reduction in MTJ area [10] such that halving the feature size leads to 25% MTJ area and 75% write current reduction. However, read current does not scale well with feature size scaling since sensing correct data using low current ($< 20\ \mu A$) is challenging [5,11]. At large feature size (e.g., 130 nm), read current is much lower than the write current; however, at small feature size (e.g., 32 nm), read and write current magnitudes become so close that a read may inadvertently write (i.e., disturb) the cell being read. This is referred to as RDE.

5.2.4 Characteristics of Read Disturbance Error

RDE is data-dependent, e.g., using read current from BL to SL disturbs cells storing "1" only and not "0". In STT-RAM, reads and writes cannot be independently optimized, e.g., reducing the MTJ thermal stability for improving switching performance increases probability of read disturbance, and thus RDE mitigation involves tradeoffs. For these reasons, RDE is expected to become the most severe bottleneck in STT-RAM scaling and performance.

It is interesting to note the difference between the RDE in STT-RAM and the write disturbance error (WDE) in PCM [6,18]. While RDE affects the same cell which is read, WDE affects the nearby cells. Also, RDE happens on a read operation, whereas WDE happens on a write operation.

5.2.5 Strategies for Addressing RDE

There are several possible strategies for addressing RDE:

1. One possible approach for reducing RDEs is to increase STT-RAM write current, which increases the margin between read and write currents. However, since STT-RAM write current and energy values are already high, increasing them further will make STT-RAM unsuitable for use in designing on-chip caches.
2. Conventionally, error-correcting codes are used for mitigating errors in caches [19,20], and along similar lines ECC can be used for mitigating RDE in STT-RAM [21]. However, at small feature sizes, RDE causes very high error rates, such that even with strong error-correcting codes (e.g., 5EC6ED which can correct five errors and detect six errors), the error rate remains higher than that acceptable for on-chip caches [10].
3. Once the data value is sensed by sense amplifiers (SAs), it remains locked in SAs, and thus it remains correct (free of RDE) [11]. Hence, after a read operation, writing the stored data back to cells (called "restore" operation) can avoid RDE. This scheme, known as restore-after-read (or HCRR) has been used in a recent STT-RAM prototype [5]. However, this scheme fails to leverage the properties of data value and cache access behavior and hence incurs large energy and performance penalty (refer Section 5.6.1).
4. To address RDE, low current SAs can be used which have larger sensing period. This approach does not create RDE but may increase sensing latency by three times [11,22], which harms performance.
5. Other researchers have proposed architectural management techniques such as phased access of MRU and non-MRU ways [23], avoiding restore operations by using LCLL reads [11], data compression [24], data duplication [24], by exploiting cache/register file access properties [10,25] and by using an SRAM buffer to absorb reads for

minimizing reads from STT-RAM [25]. By comparison, some techniques postpone restores by scheduling them at idle times.

6. A few other device-level techniques have been proposed for mitigating RDEs [9,15,26–29], however, these techniques impose latency, area and/or power overheads [27], and their integration into product system may require addressing several unforeseen challenges.

Clearly, architecture-level techniques for mitigating RDEs such as SHIELD can complement device-level techniques and can be especially useful at small feature sizes.

5.2.6 Cache Properties

A cache has a set associative structure, where a newly arrived cache block may reside in any of the cache ways, as decided by the cache replacement policy. The optimization target for designing each level of cache is different. The first-level cache (L1) is accessed very frequently and hence it should have high speed and write endurance, even if it has a small size or high leakage power. The LLC (L2 in this chapter) is designed to reduce off-chip accesses and hence it must have large capacity. A high latency of LLC can be usually tolerated using techniques such as instruction-level parallelism. In general, based on their latency and write endurance values along with capacity advantage, STT-RAM is more suitable for the L2 than the L1. Further, based on the latency tolerance property of L2 cache, we use data compression in L2 cache and not in L1 cache.

5.3 SHIELD: Key Idea and Architecture

Before describing the main idea and working of SHIELD in detail, we briefly discuss the compression algorithm used. We also define a metric which is helpful in understanding the effectiveness of any RDE-mitigation technique.

5.3.1 Compression Algorithm

SHIELD uses data compression approach for reducing the amount of data written to cache. While SHIELD can work with any compression technique, in this chapter, we illustrate its working using BDI compression algorithm [14] due to its low decompression latency and reasonable compression ratio.

It is well known that for many cache blocks, the difference between the values stored in the blocks is small [12]. Based on this, BDI represents a block with a base and an array of differences, which reduces the total block size compared to the original size. To compress blocks with two different dynamic ranges, BDI uses two bases where one base is always zero and the second base is taken from actual data contents [14]. BDI algorithm first tries to compress every element using a zero base. Then, for any uncompressed block, compression is attempted by taking the first uncompressed element as the base. A 1-bit mask is used to record whether the corresponding base is zero.

A BDI-compressed block is represented as $B_p\Delta_q$, where p and q represent the size of base and *Delta* (difference), respectively. BDI uses six such patterns, viz., $B_8\Delta_1$, $B_8\Delta_2$, $B_8\Delta_4$, $B_4\Delta_1$, $B_4\Delta_2$, and $B_2\Delta_1$. BDI also checks if the block has repeated eight-byte values or is all zero [14]. Thus, after attempting to compress the block data using these eight compression states, the one providing the smallest compressed size is finally chosen.

5.3.2 Defining Consecutive Reads

For any cache block, we define a metric *consecutive read* (CRead), which measures the average number of consecutive reads (i.e., no intermediate writes) seen by it during its residency in LLC. Also, for any application, the CRead is defined as the average CRead for all its cache blocks. Note that CRead is defined differently than the "reuse-count" of a block which measures both read and write operations in one generation of a block. For example, in Fig. 5.1, CRead for the block is 2 (=(2 + 1 + 3)/3), whereas reuse-count is 8.

To see the significance of CRead, assume a hypothetical application where every LLC read is followed by a write operation, and thus CRead = 1. Since an actual write operation automatically corrects any RDE-affected data, an oracle scheme can completely avoid restore operations for this application, and thus this application does not need restore operations. As another example, if every write is followed by 10 reads on average, then CRead = 10. In this case, automatic correction through write operations happens infrequently and larger number of restore operations are required. Clearly, higher the value of CRead, higher is the restore requirement of an application and vice versa.

5.3.3 SHIELD: Key Idea

Caches work on the temporal locality principle, which states that a data item which is referenced at one point of time will be referenced again in the near future. Such data item is stored in caches to avoid accessing main memory. For an RDE-affected STT-RAM cache, data are written to cache on both actual write operations and restore operations. To reduce the data written to STT-RAM during both these operations, SHIELD writes data in

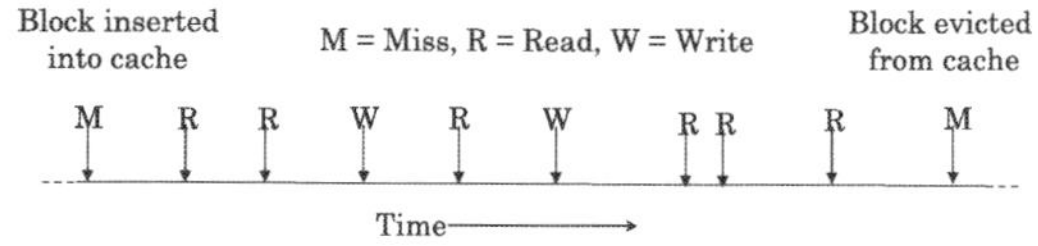

Figure 5.1
One generation of a cache block (for defining CRead).

compressed form using BDI algorithm. This reduces the STT-RAM cache write energy. We perform two optimizations to BDI algorithm [24,30] for avoiding restore operations and further reducing the amount of data written to cache.

First and importantly, for data with all zeros, no bit is stored in the STT-RAM block since the original uncompressed data can be easily recovered based on compression state encoding bits. Second, the first delta in BDI algorithm is always zero since it is the difference between the nonzero base with itself. Hence, we do not store this delta. The corresponding element is generated from the nonzero base.

Based on our optimizations and LLC access pattern, SHIELD uses the following insights to reduce restore operations.

1. For all zero data ($CW = 0$), no bits are written to or read from STT-RAM blocks, and thus RDE does not occur for such data. Hence, *all* the restore operations to such data are completely avoided.
2. It is well known that due to filtering from L1 cache, LLC sees much smaller data locality than L1 cache [31]. Hence, most LLC blocks are read only few times. In fact, our experiments have shown that average value of CRead for single and dual-core configurations is 1.61 and 1.32, respectively (refer Figs. 5.3(c) and 5.4(c)). Clearly, after any write operation, cache blocks see less than two read operations. Based on this, SHIELD keeps two copies of the data with $0 < CW \leq 32B$ within the same block. On a later read operation, one copy gets RDE, whereas the second copy still remains free from RDE. Thus, keeping two copies is equivalent to restoring the data at the time of original write operation to avoid *one* future restore operation. Thus, if CRead value is less than two, keeping two copies can avoid at least half of the restore operations for such narrow width ($0 < CW \leq 32B$) blocks.

5.3.4 Action on Read and Write Operations

We now discuss the action taken by SHIELD on any write and read operation.

Write Operations: On any write operation, the cache controller compresses the data and computes *CW*. For $0 < CW \leq 32$, two copies of data are stored, and for the other *CW* values, only one data copy is stored. Fig. 5.2 summarizes the action taken on a write operation and benefit/cost incurred for different *CW* values. Table 5.1 shows the four-bit encoding used for different compression states for BDI algorithm. Out of 16 combinations from four bits, 13 combinations are utilized and the remaining 3 are not utilized. For each block, this four-bit encoding is stored at the time of a write operation using a memory technology that does not suffer from RDE (e.g., SRAM). Note that due to our optimizations to BDI algorithm (refer Section 5.3.3), the sizes of compressed blocks are different in our work (shown in Table 5.1) than those in the original BDI algorithm [14].

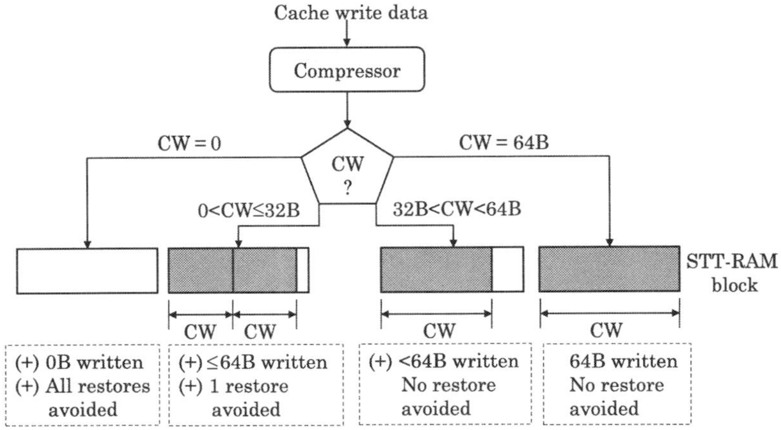

Figure 5.2
Overview of SHIELD and action taken on a cache write operation.

Table 5.1: Encoding and compressed sizes for different BDI compression states, all sizes are in bytes (Enc. = encoding, Uncomp. = uncompressed).

Enc.	State	Copies	Size	Enc.	State	Copies	Size
0000	Zeros	1	0	1100	$B_4\Delta_1$	1	19
0001	Repeat	1	8	1101	$B_4\Delta_1$	2	19 × 2
0011	Repeat	2	8 × 2	0100	$B_4\Delta_2$	1	34
0010	$B_8\Delta_1$	1	15	1110	$B_2\Delta_1$	1	33
0110	$B_8\Delta_1$	2	15 × 2	1000	$B_8\Delta_4$	1	36
0101	$B_8\Delta_2$	1	22	1111	Uncomp.	1	64
0111	$B_8\Delta_2$	2	22 × 2				

Read Operations: On any read operation, first the encoding (E) is consulted. If $E = 0000$, then the zero data are reconstructed without accessing the block which avoids RDE and need of restoration. If $E = 0011$, 0110, 1101, or 0111, it implies that $0 < CW \leq 32B$ and two copies of data are stored. Hence, one copy is read, no restore is performed, and the encoding is changed to 0001, 0010, 1100, and 0101, respectively, which indicates that now only one (error free) copy of data is stored in the block. If $E = 0001$, 0010, 1100, 0101, 0100, 1110, 1000, or 1111, then data value is read, a restore operation is issued, and the encoding remains unchanged.

5.3.5 Overhead Assessment

Latency Overhead: The latency overhead of BDI compression and decompression is two and one cycles, respectively [14]. We account for these values in performance simulation. Since STT-RAM write latency is high, a small value of BDI compression latency is easily hidden.

Table 5.2: STT-RAM L2 cache parameters.

	2 MB	4 MB	8 MB	16 MB
Hit latency (ns)	4.063	3.737	4.058	4.350
Miss latency (ns)	1.976	1.567	1.805	1.814
Write latency (ns)	4.920	4.970	5.003	5.145
Hit energy (nJ)	0.264	0.304	0.333	0.391
Miss energy (nJ)	0.107	0.105	0.112	0.113
Write energy (nJ)	0.366	0.389	0.427	0.490
Leakage power (W)	0.019	0.044	0.072	0.138

Energy Overhead: We now compute the energy overhead of compression and decompression. From Wu *et al.* [32], we note that a one-byte (eight-bit) adder consumes nearly 15 fJ energy. Assuming that a subtractor also consumes same amount of energy, we first compute the overhead of compression by counting the number of subtractions in BDI algorithm. Each of $B_8\Delta_1$, $B_8\Delta_2$, and $B_8\Delta_4$ perform seven eight-byte subtractions. Both $B_4\Delta_1$ and $B_4\Delta_2$ perform 15 four-byte subtractions and $B_2\Delta_1$ performs 31 two-byte subtractions. Repeated value detection requires seven eight-byte subtractions. Thus, there are 406-byte ($=3 \times 7 \times 8 + 2 \times 15 \times 4 + 1 \times 31 \times 2 + 7 \times 8$) subtractions, which consume 6.09 pJ ($=406 \times 15$ fJ). To account for other overheads and zero detection circuit, we increase the value and assume the overhead of compression as 8 pJ. As for decompression, it requires a maximum of 31 two-byte additions, which consume 0.93 pJ ($=31 \times 2 \times 15$ fJ). Hence, we assume the decompression overhead as 1 pJ. We account for these overheads in energy computations.

To see the energy overhead of encoding bits, we note that a two-bit counter consumes 0.1pJ in each access [33], thus a four-bit counter would consume 0.20pJ. Since the encoding is written at the time of a cache write, we compare this energy with the write energy of STT-RAM cache. From Table 5.2, write energy of 4MB L2 cache is 0.389nJ, and thus the dynamic energy of encoding bits is 0.05% of the cache energy. Hence, we ignore its contribution in energy.

5.4 Salient Features of SHIELD and Qualitative Comparison

Some researchers have proposed techniques for addressing RDEs in STT-RAM caches. The technique of Wang *et al.* [10] seeks to avoid restore operations for blocks which are expected to see a write operation in the near future. To perform L2 restore operations and updating L1 or L2 metadata bits, their technique requires observing the state of the block in other (i.e., L2 or L1) cache. Also, their technique postpones restore operations and hence, in their cache design, a dirty L2 block may have RDE. Since this RDE-affected L2 block cannot be written back to memory, their technique writes back the corresponding block in L1 cache directly to memory. Due to these, their technique complicates cache management and incurs overhead, which increases with rising number of cores. Also, their technique would require significant modifications to work with different cache coherence schemes; however, they do not present these details. By comparison, SHIELD performs restore operations immediately and hence always keeps the L2 cache RDE free. Also, SHIELD works based on data width only and hence can be integrated with any cache coherence scheme.

Jiang *et al.* [11] propose selectively performing short latency read (requires restore) and LCLL read (long latency read with no requirement of restores), depending on whether a particular bank is idle. However, short latency reads still require restores and LCLL reads increase read latency. A key benefit of SHIELD is that it avoids read, write, and restore operations for all zero data and restore operations for narrow data. Thus, SHIELD addresses both write latency/energy overhead and RDE issue by using data compression. By comparison, previous techniques [10,11] reduce some restore operations only and do not reduce read operations or address the overhead of write operations. Also, the technique of Jiang *et al.* [11] is proposed for main memory where the memory controller buffers read/write requests. Due to differences in cache and main memory architecture, their technique may not be applicable or effective for caches.

To reduce the restore energy, some researchers propose restoring only cells with "1" value [10], since only these cells may be disturbed during reads. These techniques are orthogonal to SHIELD and hence can be easily integrated with it.

Note that SHIELD does not use an additional buffer for postponing refresh operations and performs any required restore operation immediately after a read operation. Thus, the data in the cache are always kept error free. Also, SHIELD actually reduces restore operations, whereas techniques that use a buffer for postponing restore operations do not reduce restore operations but merely reschedule them to avoid interfering with normal accesses.

5.5 Experimentation Platform

5.5.1 Simulator Parameters

We use Gem5 simulator to perform simulations using detailed timing model. L1 data/instruction caches have 32KB size with two-way associativity. The L1 caches are private to core and the L2 cache (LLC) is shared between cores. The size of L2 cache in single and dual-core configurations is 4MB and 8MB, respectively. All caches are write back and use LRU replacement policy. The L2 cache parameters for 16-way STT-RAM L2 are obtained using DESTINY [34,35] for 32 nm feature size. We assume that the cache is optimized for write EDP (energy-delay-product) and uses sequential tag data access. The values obtained are shown in Table 5.2.

Table 5.3: Workloads used in the chapter.

Single-core workloads and their acronyms
As(astar), Bw(bwaves), Bz(bzip2), Cd(cactusADM) Ca(calculix), Dl(dealII), Ga(gamess), Gc(gcc) Gm (gemsFDTD), Gk(gobmk), Gr(gromacs), H2(h264ref) Hm(hmmer), Lb(lbm), Ls(leslie3d), Lq(libquantum) Mc(mcf), Mi(milc), Nd(namd), Om(omnetpp) Pe(perlbench), Po(povray), Sj(sjeng), So(soplex) Sp (sphinx), To(tonto), Wr(wrf), Xa(xalancbmk) Ze(zeusmp), *Co(comd)*, *Lu(lulesh)*, *Xb(xsbench)* Dual-core workloads (using acronyms shown above)
BwLu, PeLq, XaTo, CaMc, GkLb, GmWr, PoSp, NdGr, GcMi, BzZe, LsSo, SjXb, OmH2, CoCd, AsDl, HmGa

5.5.2 Workloads

We use all the 29 benchmarks from SPEC2006 suite with *reference* inputs and 3 benchmarks from HPC field (shown as italics in Table 5.3) as single-core workloads. Using these, we randomly create 16 dual-core multiprogrammed workloads such that a benchmark is used exactly once. Table 5.3 shows the workloads.

5.5.3 Simulation Completion Strategy

In multiprogrammed workloads, different programs have different IPCs and thus different rate of progress. Hence, they may execute the same number of instructions in different amount of time. Due to this, a strategy is required for ensuring that in each run with a workload (with baseline or a technique) the number of instructions simulated should be nearly equal.

One strategy for this is to simulate each workload for a fixed number of cycles; however, in this strategy, the number of instructions simulated may be different for baseline and a technique, depending on the performance impact of a technique.

We use another strategy where we simulate every workload till each application in the workload has executed nInst instructions. The value of nInst is 150 M for single-core workloads and 100 M for dual-core workloads. This keeps the simulation turnaround time manageable, since we perform detailed timing simulation with a full-system simulator and simulate a large number of workloads, techniques (baseline, SHIELD, HCRR, LCLL), and their parameter configurations/variants. The early-completing benchmarks continue to run, but their IPC (instruction per cycle) is recorded only for the first nInst instructions following the well-established simulation methodology [30,36]. The remaining metrics are computed for the whole execution, since they are system wide metrics (whereas IPC is a per core metric).

5.5.4 Comparison with Related Schemes

Our baseline is an ideal scheme that assumes a cache free of RDE. We compare our technique with two other schemes for mitigating RDE.

High current restore required read: This technique uses normal current ($20\mu A$) to read the STT-RAM cell and performs a restore (i.e., write) operation after every read operation to correct the RDE. This technique is also referred to as refresh (or restore) after read [10] and "disruptive reading and restoring" [27] technique.

Low current long latency read: To avoid RDEs, low current ($\sim 10\mu A$) SAs can be used which have larger sensing duration [11,22]. These SAs also require extra sensing stages to minimize sensing margin degradation due to variations in STT-RAM cells. This approach, referred to as LCLL, does not create RDE but incurs $3\times$ the sensing latency of traditional SAs [11,22].

5.5.5 Evaluation Metrics

We use the following evaluation metrics:

1. L2 cache energy (leakage + dynamic)
2. Weighted speedup (referred to as relative performance), defined as $\Sigma_n(\text{IPC}_n(\text{technique})/\text{IPC}_n(\text{baseline}))/N$, where N = number of cores.
3. Average CRead value for baseline (Section 5.3.2).
4. ΔBWPKI ($=\text{BWPKI}_{\text{technique}}$-$\text{BWPKI}_{\text{baseline}}$), where BWPKI refers to bytes written to cache per kilo instruction. BWPKI shows the write traffic to cache, and since compression changes the size of each block, we have used BWPKI instead of writes per kilo instruction. Thus, ΔBWPKI shows the increase in write traffic due to RDE under any technique (SHIELD or HCRR) compared to an ideal RDE-free cache. Since this metric for LCLL is nearly zero, we do not show this value for LCLL in Figs. 5.3(d) and 5.4(d).
   For SHIELD, we also show
5. *CW of data on each cache write*. We classify the *CW* values in four ranges (refer Table 5.1): $CW = 0$, $0 < CW \leq 32B$ (i.e., {8B,15B,19B,22B}), $32B < CW < 64B$ (i.e., {33B,34B,36B}), $CW = 64B$ (i.e., uncompressed).
6. Percentage of restore operations avoided (RstAvd). Let $Reads_{CW=0}$ be the reads to blocks with $CW = 0$ data and $Reads_{0<CW\leq 32\_2copy}$ be the reads to blocks with $0 < CW \leq 32B$ data having two copies. Then, $RstAvd = \frac{(Reads_{CW=0} + Reads_{0<CW\leq 32\_2copy}) \times 100}{TotalReads}$
   Note that for BDI algorithm, $Reads_{CW=0}$ corresponds to $E = 0000$ and $Read_{s0<CW\leq 32\_2copy}$ corresponds to $E = 0011$, 0110, 1101, or 0111.

5.6 Results and Analysis

5.6.1 Main Results

Figs. 5.3 and 5.4 show the results and Table 5.4 summarizes the results. We now analyze the results.

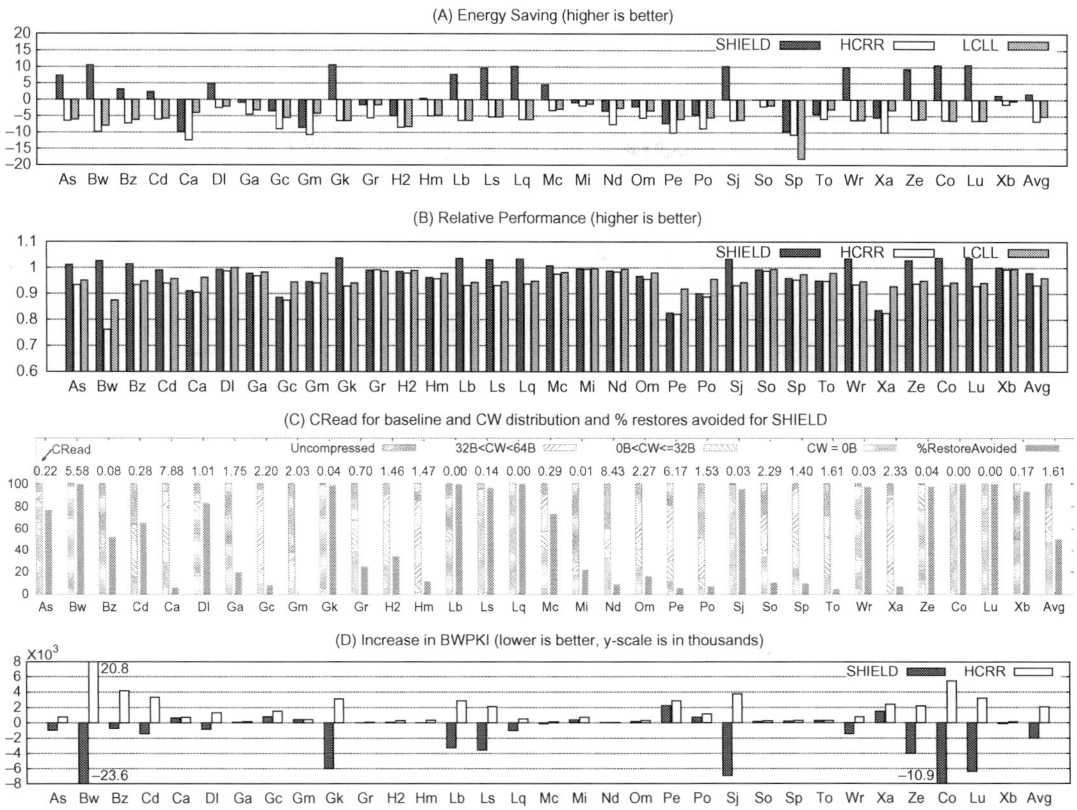

Figure 5.3
Results for single-core system.

SHIELD: Compared to the RDE-free cache (baseline), both HCRR and LCLL incur *energy loss*, whereas SHIELD provides *energy saving*. Also, SHIELD provides higher performance than HCRR and LCLL. The energy and performance improvement of SHIELD for an application depends on the fraction of restores avoided. This, in turn, depends on the compressibility of application's data blocks and its read/write access pattern, specifically its CRead value. Out of a total of 48 workloads, compared to baseline, SHIELD saves energy for 28 workloads and improves performance for 18 workloads.

Results on percentage of restores avoided: For single- and dual-core configurations, SHIELD reduces 50.7% and 48.5% of restore operations, respectively. For many applications, more than 95% of restores are avoided, e.g., Bw (99.3%), Gk (98.8%), Lb (99.3%), Ls (96.3%), Lq (99.9%), Sj (95.4%), Wr (97.0%), Ze (97.4%), Co (99.9%), Lu (99.9%), BwLu (99.7%), GkLb (99.4%), SjXb (96.3%), etc. This happens because these applications have highly compressible data and/or their CRead values are small.

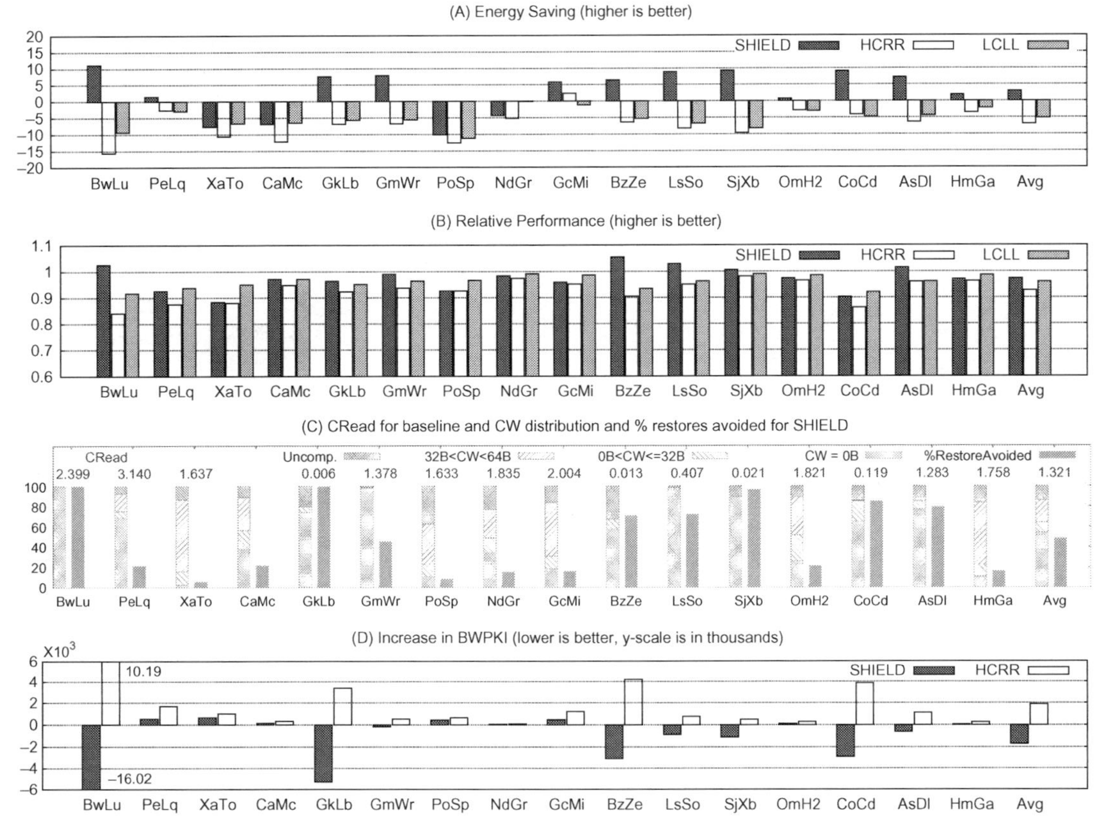

Figure 5.4
Results for dual-core system.

Table 5.4: Summary of results (rel. perf. = relative performance).

	Single Core			Dual Core		
	SHIELD	HCRR	LCLL	SHIELD	HCRR	LCLL
Energy Saving	1.73%	−6.57%	−5.09%	3.03%	−6.96%	−5.23%
Rel. Perf.	0.98 ×	0.93 ×	0.96 ×	0.97 ×	0.93 ×	0.96 ×
ΔBWPKI	−1980	2085	−4	−1725	1877	−5

Data compressibility and CRead results: For some applications, more than 95% of blocks are all zero ($CW = 0$), e.g., Bw (97.8%), Gk (98.2%), Lq (99.9%), Sj (98.9%), Wr (97.9%), Co (99.9%), Lu (99.9%), BwLu (98.7%), GmWr (98.2%), etc. Similarly, low CRead values for Gk (0.04), Lb (0.0003), Lq (0.0003), Sj (0.03), Wr (0.02), Ze (0.03), Co (0.002), Lu (0.00006), GkLb (0.006), BzZe (0.01), SjXb (0.02), etc., indicate that only few restore operations are required for these applications since SHIELD avoids restores for narrow data ($0 < CW \leq 32B$) and these applications are read-unintensive. For these applications, large performance improvement and energy saving are achieved and write traffic is significantly reduced.

For some applications CRead values are small (e.g., Mi(0.006), etc.), but only few blocks are compressible. Opposite is true for applications where CRead values are high (e.g., Bw [5.58], BwLu [2.40], PeLq [3.14], etc.), but most blocks are compressible. From the perspective of energy saving with SHIELD, these effect partially cancel each other and hence these applications show only small energy gain/loss.

For some applications, CRead values are high, e.g., Gc (2.20392), Gm (2.0307), Ca (7.88), Nd (8.42531), Om (2.27), Pe (6.16507), So (2.286), Xa (2.32), etc. The restore requirement for these applications is also high and thus restore operations lead to performance and energy loss for these applications. On using SHIELD, only 4 out of 48 workloads show relative performance less than 0.90 ×, viz., Gc (0.89), Pe (0.83), Xa (0.84), and XaTo (0.88 ×). Similarly, in some applications, most blocks are either incompressible or have *CW* greater than 32B, e.g., Gc, Gm, Hm, Mi, Pe, Po, To, Xa, XaTo, PoSp, HmGa, etc. Hence, compared to RDE-free baseline, SHIELD incurs energy loss or very small energy saving due to restore operations.

ΔBWPKI results: SHIELD brings large reduction in cache write traffic, which is due to the use of compression and avoiding many restore operations. Especially for applications with compressible data, SHIELD brings large reduction in BWPKI, e.g., Bw (23606), Gk (6011), Sj (6945), Co (10942), Lu (6397), BwLu (16015), GkLb (5281), etc. Clearly, in addition to RDE, SHIELD also addresses the write overhead in STT-RAM caches.

Overhead of (de)compression: For single- and dual-core systems, the compression and decompression energy (refer Section 5.3.5) together account for 0.14% and 0.15% of the total L2 energy. Clearly, the energy overhead of SHIELD is negligible.

We conclude that SHIELD can bridge the large performance gap between an ideal RDE-free cache and an RDE-affected cache that uses restore-after-read (HCRR) scheme. Also, SHIELD completely removes the energy overhead of restore operations, and hence it has lower energy consumption and lower LLC write traffic than an ideal RDE-free cache.

HCRR: With HCRR, each read operation also leads to a write operation and thus HCRR causes significant increase in write traffic to STT-RAM cache. The increase in BWPKI is especially high for read-intensive applications, e.g., Bw (20844), Bz (4181), Sj (3766), Co (5478), BwLu (10194), BzZe (4218), CoCd (3858), etc. Since STT-RAM writes have high energy/latency overhead, HCRR incurs large energy and performance loss and is also likely to cause bandwidth issues. In fact, the worst case performance with HCRR can be very poor, e.g., Bw (0.76 ×), Gc (0.87 ×), Pe (0.82 ×), Po (0.89 ×), Xa (0.83 ×), BwLu (0.84 ×), PeLq (0.88 ×), XaTo (0.88 ×), CoCd (0.86 ×), etc. Also, for some applications, HCRR incurs more than 10% energy loss, e.g., Ca (−12.4%), Gm (−10.7%), Pe (−10.1%),

Sp (−10.6%), XaTo (−10.6%), CaMc (−12.1%), PoSp (−12.6%), etc. With an increasing number of cores, the LLC access intensity increases and hence restore operations will cause port obstruction and make the LLC unavailable for serving accesses. Clearly, due to its large penalty, use of HCRR in performance-critical systems is challenging.

LCLL: LCLL degrades performance by slowing down the read operations. Since read operations happen on critical access path, the large read latency in LCLL may not be easily hidden. In worst case, LCLL can cause large energy loss, e.g., Sp (−18.1%), PoSp (−11.2%), etc. On average, LCLL incurs energy loss for both single- and dual-core configurations, whereas SHIELD saves energy. Compared to RDE-free cache (baseline), LCLL has negligible impact on BWPKI, whereas SHIELD brings large reduction in BWPKI (as shown earlier). Thus, SHIELD reduces the write traffic to cache more effectively than LCLL.

Although LCLL provides better performance and energy efficiency compared to HCRR, a crucial limitation of LCLL is that use of small read current in LCLL can lead to *decision failure* [27], such that it may not be possible to distinguish between two states of a bit cell during read operation. Also, since process variation affects device parameters [37], to guarantee RDE-free read operations, the value of read current needs to be set even lower than that assumed here [8]; this, however, would further increase the read latency and cause performance loss. Given this, SHIELD presents as a better technique for mitigating RDE than LCLL.

5.6.2 Parameter Sensitivity Results

We, henceforth, focus exclusively on SHIELD. Also, we omit per workload results for brevity and only present average results in Table 5.5. For comparison purpose, the results with default parameters are also shown.

1. *Effect of Replication*: As shown in Section 5.3.3, SHIELD replicates the narrow ($0 < CW \leq 32B$) data to reduce restore operations. We now present two variants of SHIELD which explore the tradeoffs associated with it.

Table 5.5: Parameter sensitivity results (Rstavd = % restores avoided).

	Single Core				Dual Core			
	Energy Saving	Rel. Perf.	ΔBWPKI	RstAvd	Energy Saving	Rel. Perf.	ΔBWPKI	RstAvd
Default	1.73%	0.981 ×	−1980	50.7%	3.03%	0.973 ×	−1725	48.4%
SHIELD1	1.84%	0.978 ×	−2003	45.3%	3.16%	0.967 ×	−1726	43.9%
SHIELD3	0.79%	0.980 ×	−1938	50.4%	2.70%	0.972 ×	−1692	48.5%
Half-size L2	2.84%	0.980 ×	−1983	50.3%	3.77%	0.983 ×	−1799	48.1%
Double-size L2	1.1%	0.977 ×	−1970	51.0%	3.39%	0.992 ×	−1710	49.0%

SHIELD1: SHIELD1 does not replicate a data item with width $0 < CW \leq 32B$, but otherwise works same as SHIELD. For BDI algorithm, four states, viz., repeated value, $B_8\Delta_1$, $B_8\Delta_2$, and $B_4\Delta_1$ have *CW* at most 32B.

SHIELD1 reduces the number of bits written originally at the cost of incurring one extra restore operation later. Due to this, SHIELD1 saves higher energy but achieves lower performance than SHIELD (Table 5.5), and thus a designer can choose an appropriate variant to optimize a metric of interest. It is noteworthy that the higher performance of SHIELD can reduce execution time which leads to lower leakage power in other processor components, and hence SHIELD may be preferable over SHIELD1. Also, SHIELD avoids larger percentage of restore operations than SHIELD1. For single- and dual-core systems, $0 < CW \leq 32B$ data account for 14.3% and 10.0% of data blocks (refer Figs. 5.3(c) and 5.4(c)), and thus due to the relatively smaller contribution of $0 < CW \leq 32B$ data, the difference between performance/energy of SHIELD and SHIELD1 is relatively small.

SHIELD3: On any write operation, SHIELD3 makes three copies of data with $0 < CW < 22B$, but otherwise works same as SHIELD (i.e., two copies of data with $0 < CW \leq 32B$ and one copy for other data). For BDI algorithm, three states, viz., repeated value, $B_8\Delta_1$, and $B_4\Delta_1$, have *CW* at most 22B.

SHIELD3 avoids the need of restore on two read operations at the cost of increasing the bits written to cache initially. However, SHIELD3 reduces energy saving without providing corresponding performance improvement. This is because average CRead values for single- and dual-core configurations are 1.61 and 1.32, respectively (refer Figs. 5.3(c) and 5.4(c)). Thus, most L2 cache lines see less than two consecutive reads, and hence keeping three copies does not generally avoid an extra restore operation compared to keeping two copies. Further, SHIELD3 may have higher complexity than SHIELD due to keeping two and three copies of data of different widths.

2. *Effect of Cache Size*: To evaluate the impact of cache size, we experiment with half and double of default L2 cache size (4MB for single core and 8MB for dual core).

In general, restore operations incur larger overhead at lower cache sizes due to increased cache contention, and hence the scope of energy/performance gains with lower sized caches is also higher. Hence, SHIELD brings larger gains at smaller cache sizes (refer Table 5.5). Overall, SHIELD performs consistently well at different cache sizes: maintaining performance close to baseline, saving energy, avoiding nearly half of the restores, and bringing large reduction in write traffic.

5.7 Conclusion and Future Work

In this chapter, we presented a technique to mitigate RDEs in STT-RAM caches. Our technique uses compression and selective duplication of compressed data to address both RDE and write overhead issue in STT-RAM. Experimental results performed with single- and dual-core system configurations have shown that our technique is effective in improving performance and energy efficiency in the presence of RDE. Also, it outperforms two other techniques for mitigating RDE.

We now list some possible directions for future work:

1. Our future work will focus on synergistic integration of SHIELD with device-level techniques [15] to reduce RDEs and approximate computing techniques to tolerate RDEs [38].
2. By reducing the read accesses to an STT-RAM cache, the RDEs can be reduced. For achieving this, cache bypassing technique [39] or read buffers can be used.
3. We also plan to explore SRAM-STTRAM hybrid caches [40,41], where frequently accessed blocks are migrated to SRAM. This will reduce pressure on STT-RAM ways and mitigate the impact of RDE on performance.
4. In this chapter, we explored STT-RAM for designing CPU caches. CPUs primarily focus on optimizing latency in serial applications and hence they use large caches and small register file. By comparison, GPUs focus on optimizing throughput, and hence they use small caches and a very large register file to support their massively multithreaded architecture. Due to its high density, STT-RAM is suitable for designing GPU register file; however, the RDE may make this challenging. Some recent techniques seek to address this issue [25]. We plan to propose novel techniques for managing RDE in STT-RAM-based GPU register file for improving performance and energy efficiency.
5. Several other emerging memory technologies provide attractive properties. For example, SOT-RAM (spin-orbit torque RAM) which does not suffer from read disturbance issue [42]. We also plan to explore use of domain wall memory for designing caches [43].

References

[1] J.S. Vetter, S. Mittal, Opportunities for nonvolatile memory systems in extreme-scale high performance computing, Comput. Sci. Eng. 17 (2) (2015) 73–82.

[2] B.K. Kaushik, S. Verma, A.A. Kulkarni, S. Prajapati, Next Generation Spin Torque Memories. Springer, Singapore, 2017, ISBN: 978-981-10-2719-2 . Available from: https://doi.org/10.1007/978-981-10-2720-8.

[3] K.C. Chun, H. Zhao, J.D. Harms, T.-H. Kim, J.-P. Wang, C.H. Kim, A scaling roadmap and performance evaluation of in-plane and perpendicular MTJ based STT-MRAMs for high-density cache memory, IEEE J. Solid-State Circuits 48 (2) (2013) 598–610.

[4] S. Mittal, J.S. Vetter, D. Li, A survey of architectural approaches for managing embedded DRAM and non-volatile on-chip caches, IEEE Trans. Parallel Distrib. Syst. 26 (6) (2014) 1524–1537.

[5] R. Takemura, T. Kawahara, K. Ono, K. Miura, H. Matsuoka, and H. Ohno, "Highly-scalable disruptive reading scheme for Gb-scale SPRAM and beyond," in *IEEE International Memory Workshop (IMW)*, 2010, pp. 1–2 Seoul, South Korea, 16-19 May, 2010.

[6] S. Mittal, A survey of soft-error mitigation techniques for non-volatile memories, Comput. 6 (8) (2017).

[7] W. Kang, Y. Cheng, Y. Zhang, D. Ravelosona, and W. Zhao, "Readability challenges in deeply scaled STT-MRAM," in *Non-Volatile Memory Technology Symposium (NVMTS)*, Jeju, South Korea, 27–29 October, 2014, IEEE, pp. 1–4.

[8] Y. Zhang, W. Wen, Y. Chen, The prospect of STT-RAM scaling from readability perspective, IEEE Trans. Magn. 48 (11) (2012) 3035–3038.

[9] Y. Ran, W. Kang, Y. Zhang, J.-O. Klein, and W. Zhao, "Read disturbance issue for nanoscale STT-MRAM," in *IEEE Non-Volatile Memory System and Applications Symposium (NVMSA)*, Hong Kong, China, 19–21 Aug, 2015, pp. 1–6.

[10] R. Wang, L. Jiang, Y. Zhang, L. Wang, and J. Yang, "Selective restore: an energy efficient read disturbance mitigation scheme for future STT-MRAM," in *Des. Autom. Conf.*, 2015, p. 21.

[11] L. Jiang, W. Wen, D. Wang, and L. Duan, "Improving read performance of STT-MRAM based main memories through smash read and flexible read," Macao, China, 25–28 Jan, 2016. Asia and South Pacific Design Automation Conference (ASP-DAC).

[12] S. Mittal, J. Vetter, A survey of architectural approaches for data compression in cache and main memory systems, IEEE Trans. Parallel Distrib. Syst. 27 (5) (2016) 1524–1536.

[13] A. Chakraborty, H. Homayoun, A. Khajeh, N. Dutt, A. Eltawil, and F. Kurdahi, "E < MC2: less energy through multi-copy cache," in International Conference on Compilers, Architectures and Synthesis for Embedded Systems, Scottsdale, Arizona, USA, — October 28, 2010, pp. 237–246.

[14] G. Pekhimenko, V. Seshadri, O. Mutlu, P.B. Gibbons, M.A. Kozuch, T.C. Mowry, Base-delta-immediate compression: practical data compression for on-chip caches," Minneapolis, Minnesota, USA, 19-23 September, 2012. International Conference on Parallel Architectures and Compilation Techniques (PACT), 2012, pp. 377–388.

[15] X. Fong, Y. Kim, S.H. Choday, K. Roy, Failure mitigation techniques for 1T-1MTJ spin-transfer torque MRAM bit-cells, IEEE Trans. VLSI Syst. 22 (2) (2014) 384–395.

[16] S. Mittal, J. Vetter, A survey of software techniques for using non-volatile memories for storage and main memory systems, IEEE Trans. Parallel Distrib. Syst. 27 (5) (2016) 1537–1550.

[17] S. Agarwal and H.K. Kapoor, "Towards a better lifetime for non-volatile caches in chip multiprocessors," in *International Conference on VLSI Design (VLSID)*. IEEE, Hyderabad, India, 7–11 January 2017, pp. 29–34.

[18] R. Wang, S. Mittal, Y. Zhang, and J. Yang, "Decongest: Accelerating super-dense PCM under write disturbance by hot page remapping," *IEEE Computer Architecture Letters (CAL)*, 16 (2) (2017) 107–110.

[19] S. Mittal, J. Vetter, A survey of techniques for modeling and improving reliability of computing systems, IEEE Trans. Parallel Distrib. Syst. 27 (4) (2016) 1226–1238.

[20] S. Mittal, J. Vetter, Reliability tradeoffs in design of volatile and non-volatile caches, J. Circuits, Syst., Comput. 25 (11) (2016) 1650139.

[21] S.M. Seyedzadeh, R. Maddah, A. Jones, and R. Melhem, "Leveraging ECC to mitigate read disturbance, false reads and write faults in STT-RAM," in *Int. Conf. Dependable Syst. and Netw. Workshops*, 2016, pp. 215–226.

[22] W. Kang, W. Zhao, J.-O. Klein, Y. Zhang, C. Chappert, D. Ravelosona, High reliability sensing circuit for deep submicron spin transfer torque magnetic random access memory, Electron. Lett. 49 (20) (2013) 1283–1285.

[23] J. Kong, A novel technique for technology-scalable STT-RAM based L1 instruction cache, IEICE Electr. Express 13 (11) (2016) 20. 160 220–20 160 220.

[24] S. Mittal, J. Vetter, L. Jiang, Addressing read-disturbance issue in STT-RAM by data compression and selective duplication, IEEE Comput. Architect. Lett. 16 (2) (2017).

[25] H. Zhang, X. Chen, N. Xiao, F. Liu, and Z. Chen, "Red-shield: Shielding read disturbance for STT-RAM based register files on GPUs," in *Great Lakes Symposium on VLSI*, Boston, MA, USA, 18–20 May, 2016, pp. 389–392.

[26] W. Kang, Z. Li, J.-O. Klein, Y. Chen, Y. Zhang, D. Ravelosona, et al., Variation-tolerant and disturbance-free sensing circuit for deep nanometer stt-mram, IEEE Trans. Nanotechnol. 13 (6) (2014) 1088–1092.

[27] R. Bishnoi, M. Ebrahimi, F. Oboril, and M.B. Tahoori, "Read disturb fault detection in STT-MRAM," in *IEEE International Test Conference (ITC)*, Seattle, Washington, USA, 21–23 October, 2014, pp. 1–7.

[28] A. Raychowdhury, "Pulsed READ in spin transfer torque (STT) memory bitcell for lower READ disturb," in *International Symposium on Nanoscale Architectures (NANOARCH)*, New York City, USA. 15–17 July 2013, pp. 34–35.

[29] T. Na, J.P. Kim, S.H. Kang, S.-O. Jung, Read disturbance reduction technique for offset-canceling dual-stage sensing circuits in deep submicrometer stt-ram, IEEE Trans. Circuits Syst. II Express Briefs 63 (6) (2016) 578–582.

[30] S. Mittal and J. Vetter, "Reducing soft-error vulnerability of caches using data compression," in *ACM Great Lakes Symposium on VLSI (GLSVLSI)*, Boston, Massachusetts, USA, 18-20 May, 2016.

[31] S. Mittal, A survey of architectural techniques for improving cache power efficiency, Elsevier Sustainable Computing: Informatics and Systems 4 (1) (2014) 33–43.

[32] X. Wu, F. Wang, and Y. Xie, "Analysis of subthreshold FinFET circuits for ultra-low power design," in *IEEE International SOC Conference*, Seoul, South Korea, 26–27 October, 2006, pp. 91–92.

[33] S. Mittal, J.S. Vetter, EqualWrites: Reducing intra-set write variations for enhancing lifetime of non-volatile caches, IEEE Trans. VLSI Syst. 24 (1) (2016) 103–114.

[34] M. Poremba, S. Mittal, D. Li, J.S. Vetter, and Y. Xie, "DESTINY: A tool for modeling emerging 3D NVM and eDRAM caches," in *Design Automation and Test in Europe*, Grenoble, France. 9–13 March, 2015.

[35] S. Mittal, M. Poremba, J. Vetter, Y. Xie, Exploring design space of 3D NVM and eDRAM caches using DESTINY tool, Oak Ridge National Laboratory, USA, Tech. Rep. ORNL/TM-2014/636, 2014.

[36] S. Mittal, J. Vetter, A technique for improving lifetime of non-volatile caches using write-minimization, J. Low Power Electron. Appl. 6 (1) (2016). Available from: http://www.mdpi.com/2079-9268/6/1/1.

[37] S. Mittal, A Survey of architectural techniques for managing process variation, ACM Comput. Surveys 48 (4) (2016) 54. 1–54:29.

[38] S. Mittal, A survey of techniques for approximate computing, ACM Comput. Surveys 48 (4) (2016) 62. 1–62:33.

[39] S. Mittal, A survey of cache bypassing techniques, J. Low Power Electron. Appl. 6 (2) (2016) 5:1–5:30. Available from: https://doi.org/10.3390/jlpea6020005.

[40] S. Mittal, J. Vetter, AYUSH: A technique for extending lifetime of SRAM-NVM hybrid caches, Comput. Architect. Lett. 14 (2) (2015) 115–118.

[41] S. Agarwal and H.K. Kapoor, "Restricting writes for energy-efficient hybrid cache in multi-core architectures," in *IFIP/IEEE International Conference on Very Large Scale Integration (VLSI-SoC)*, Tallinn, Estonia, 26–28 September, 2016, pp. 1–6.

[42] S. Mittal, R. Bishnoi, F. Oboril, H. Wang, M. Tahoori, A. Jog, et al., "Architecting SOT-RAM based GPU register file," in *IEEE Computer Society Annual Symposium on VLSI (ISVLSI)*, Bochum, Germany, 3–5 July, 2017.

[43] S. Mittal, A survey of techniques for architecting processor components using domain wall memory, ACM J. Emerg. Technol. Comput. Syst. 13 (2) (2016) 29.

CHAPTER 6

Multi-Functionality of Spintronic Materials

Kuldeep C. Verma[1], R.K. Kotnala[2] and Navdeep Goyal[1,*]

[1]Department of Physics, Panjab University, Chandigarh, India

[2]CSIR-National Physical Laboratory, New Delhi, India

6.1 Introduction—What Is Spintronics?

Spintronics: the electron's spin degree of freedom is expected to provide additional functionalities in future electronic devices [1–7]. This is the ability to inject spins into a semiconductor by electrical means along with the possibility to convert a spin accumulation into an electrical signal is a prerequisite for spin transport and spin manipulation. For the magnetic storage and sensing of spintronic applications, the tunneling magnetoresistance (TMR), giant magnetoresistance, and electrical spin injection must be included [4]. However, the nanostructures of the spintronic materials have much broader impact to control the spin and magnetism. Brataas et al. [5] explain how currents can generate torques that affect the magnetic orientation and the reciprocal effect in a wide variety of magnetic materials and structures. For example, it is imagined that a computer that powers up exactly as one can left it the night before with the cursor flashing on the last word that one typed—a computer that needs no long, frustrating wait while it searches and loads the suitable software from the hard disk. This is for a computer in which the distinction between active and passive memory has been removed and the data is processed in the same part of memory in which it is stored. This is a consumer dream that spintronics may one day offer to us. Spintronics is an information-transfer phenomenon in which one part of a spin device, the encoder, imposes information—in the form of spin polarization—onto the electrical carriers, which then transfer it to a different part of the device where readout takes the form of a change in the device characteristic as shown in Fig. 6.1 [6]. The electrons injected into the electrode on the left are randomly spin-oriented. Their spin orientation is subsequently defined in the encoding region. These spin-polarized electrons then move toward the other electrode where the decoding takes place. The decoding occurs through recombination with a positive carrier (hole) with consequent emission of light of either right- or left-handed circular polarization depending on the encoder orientation. The spin coded information partially decays in transit between encoder and decoder that is determined by the rate of spin-memory loss due to spin-flip processes [6].

* Corresponding author: dkuldeep.physics@gmail.com; kcv0309@gmail.com

Nanoelectronics. DOI: https://doi.org/10.1016/B978-0-12-813353-8.00004-X

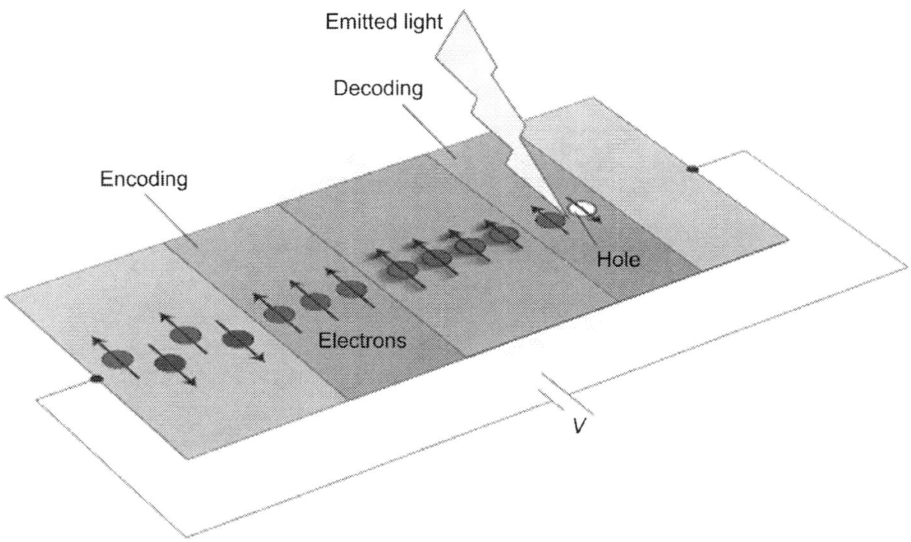

Figure 6.1
Representation of spintronic device. Source: *From J.F. Gregg, Spintronics: a growing science, Nat. Mater. 6 (2007) 798–799.*

Recently there have been dramatic advances made toward spintronics to process information using the spin of the electron, as well as the emergence of several low-power, nonvolatile, and information-storage technologies [7]. This is based on exploring magnetism with electric fields. It could be understood that the use of the latter necessitates the use of insulators because free charge carriers would otherwise screen the applied electric fields. This may be realized with the use of multiferroics (such as antiferromagnetic ferroelectrics) in contact with a ferromagnet as a pathway to control the ferromagnetic state. Ferromagnetic layers in contact with an antiferromagnet exhibit an interfacial magnetic coupling, termed as exchange bias coupling. If such an antiferromagnet is a magnetoelectric (ME) multiferroic, then in principle the interface coupling can be controlled and manipulated by the application of an electric field. Indeed, theoretical calculations have predicted the possibility of significant changes in the interfacial magnetization and spin polarization in a ferromagnet in response with ferroelectric polarization state across the interface [8]. Thus, the interplay between the nonvolatile polarization states of the ferroelectric and its influence on the spin polarization could lead to unusual tunneling characteristics. This type of the spintronic approach has created a possibility towards understanding both multiferroic and diluted magnetic semiconductors (DMSs).

6.1.1 Spintronics Based on Multiferroics

Multiferroics have the magnetic and ferroelectric properties manifested together; the two types are often coupled to each other. The currents limit the computer's processing speed and produce so much heat that energy consumption and the need for cooling become serious problems. However, in a multiferroic memory element, a voltage pulse could be used to control the ferroelectric state and, through an internal ME coupling, activate the ferromagnetic state that represents the memory bit [9]. Voltage pulses can be transmitted more quickly than electric currents and consume less power. Significantly, multiferroics could lead to a new generation of memory devices that can be electrically written and magnetically read [10]. Gajek et al. [11] reported a significant breakthrough in the quest for the ultimate memory device. A high-density four-state multiferroic memory has been demonstrated for Mn-doped $BaTiO_3$ (BTO) nanorods as shown in Fig. 6.2 [12]. The polarization (P) switching by an external electric field does not influence the magnetization (M) of the nanorods owing to a negligible degree of the P–M cross-coupling and vice versa. On the basis of these, it is possible to implement a four-state nonvolatile multiferroic memory, namely, (+P, +M), (+P, −M), (−P, +M), and (−P, −M) with the reliability in the P and M switching demonstrated by the switching cycle-dependent piezoelectric and magnetic responses. The ME coupling enables the control of the ferroelectric polarization by a magnetic field and, conversely, the manipulation of magnetization by an electric field [13]. As ferroelectric polarization and magnetization are used to encode binary information in FeRAMs (ferroelectric random access memories) and MRAMs (magnetic random access memories), respectively, the coexistence of magnetization and polarization in a multiferroic allows the realization of four-state logic in a single device [11,14]. The multiferroic tunnel

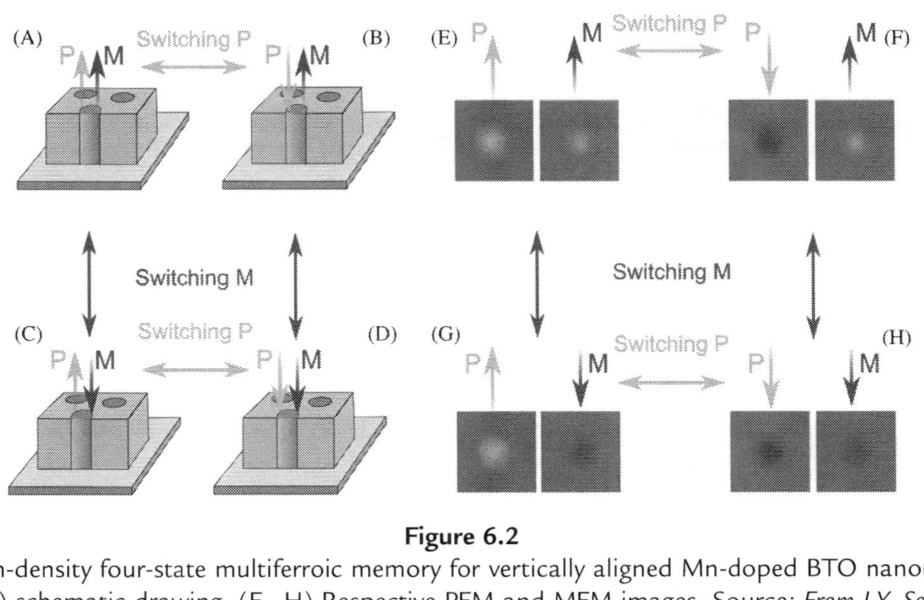

Figure 6.2
High-density four-state multiferroic memory for vertically aligned Mn-doped BTO nanorods (A–D) schematic drawing. (E–H) Respective PFM and MFM images. Source: *From J.Y. Son, J.H. Lee, S. Song, Y.H. Shin, H.M. Jang, Four-states multiferroic memory embodied using Mn-doped $BaTiO_3$ nanorods, ACS Nano 7 (2013) 5522–5529.*

junction can exhibit four states, characterized by the TMR, which is determined by the spin polarization of the tunneling electrons and by the tunnel electroresistance [15].

To achieve spintronic approaches in multiferroics, one could use a spin current to create a "spin torque" which could be subsequently transferred to a magnetic element, such as a domain wall as shown in Fig. 6.3 [1]. Garcia et al. [16] presented an exciting new development in this direction by demonstrating low-power control of magnetoresistance (MR) across a ferroelectric tunnel junction with ferromagnetic electrodes. For this, the use of multiferroics (such as antiferromagnetic ferroelectrics) in contact with a ferromagnet is a pathway to control the ferromagnetic state. Ferromagnetic layers in contact with an antiferromagnet exhibit an interfacial magnetic coupling, termed as exchange bias coupling. If such an antiferromagnet is an ME multiferroic, then in principle the interface coupling can be controlled and manipulated by the application of an electric field [17]. Generally, the multiferroicity has been achieved due to cation off-centering in ferroelectrics (empty d orbitals), and the formation of magnetic moments (partially filled d orbitals). In the magnetic perovskite-structure oxides, multiferroicity is most commonly achieved by making use of the stereochemical activity of the lone pair on the large (A-site) cation to provide the ferroelectricity, while keeping the small (B-site) cation magnetic. This is the mechanism for ferroelectricity in the perovskites (ABO_3) such as $BiFeO_3$ (BFO), BTO, $PbTiO_3$, $CaTiO_3$, etc. A second route to multiferroism is provided by "geometrically driven" ferroelectricity, which is compatible with the coexistence of magnetism; the antiferromagnetic ferroelectrics $YMnO_3$ and $BaNiF_4$ fall into this class.

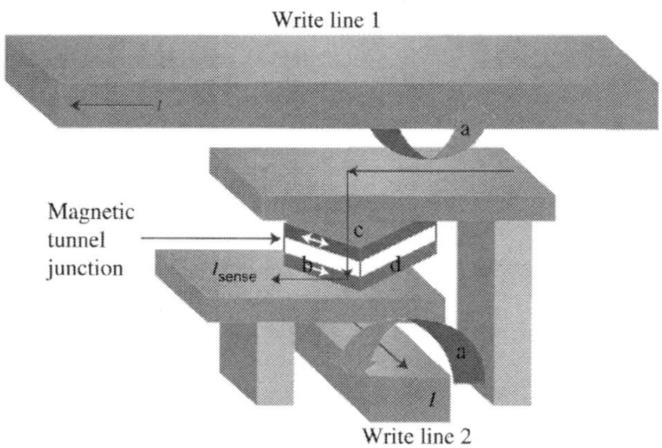

Figure 6.3
For a storage device, electric control of magnetism. In magnetic random-access memory, an electric current I (Isense is sensing current) is used to apply a magnetic torque (arrow: a) that switches the magnetization (arrow: b) in one layer (marked as: c) with respect to a layer with a fixed magnetization (marked as: d). Source: *From R. Ramesh, A new spin on spintronics, Nat. Mater. 9 (2010) 380–381.*

6.1.2 Spintronics Based on DMSs

In the efforts to reduce the power dissipation and variability with continuous scaling of Si technology, spintronics has emerged as an appealing solution, which utilizes both charge and spin of the electron for information processing [18,19]. Recently, DMS materials have generated potential in spintronics because the DMS has the collective magnetic ordering that is mediated by the semiconductor charge carriers, as well as electron scattering at localized magnetic impurities and electron—electron interactions [20,21]. The ferromagnetic ordering in these DMS systems could lead to spin-polarized charge transport. Spin-polarized transport will occur naturally in any material for which there is an imbalance of the spin populations at the Fermi level. This imbalance usually occurs in ferromagnetic metals because the density of states available to spin-up and spin-down electrons is often nearly identical, but the states are shifted in energy with respect to each other as shown in Fig. 6.4A [19]. This shift results in an unequal filling of the bands, which is the source of the net magnetic moment for the DMS, but it can also cause the spin-up and spin-down carriers at the Fermi level to be unequal in number, character, and mobility. This inequality can produce a net spin polarization in a transport measurement. The basic action in a spin-polarized deviceis assumed to be electrons traveling from a ferromagnetic metal, through a normal metal, and into a second ferromagnetic metal as shown in Fig. 6.4B and C [19]. When the magnetizations of the two ferromagnetic metals are in an aligned state, the resistance is low, whereas the resistance is high in the antialigned state.

For practical spintronic applications, ferromagnetic DMSs with Curie temperatures, T_C, greatly exceeding room temperature will be required. Theoretical predictions of high-T_C ferromagnetism in DMSs of GaN and ZnO stimulated efforts to develop these and related wide-band gap materials. For spintronics applications, the relatively long room-temperature

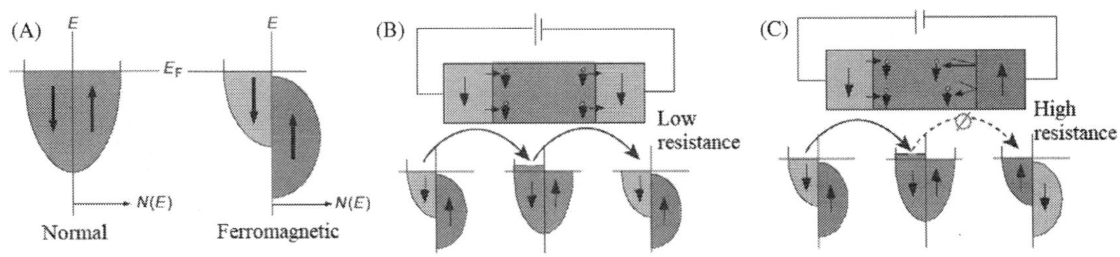

Figure 6.4

Schematic representations: (A) Density of electronic states for the electrons available in a normal metal and in a ferromagnetic metal whose majority spin states are completely filled. (B and C) Spin-polarized transport from a ferromagnetic metal, through a normal metal, and into a second ferromagnetic metal for aligned and antialigned magnetic moments. E, Electron energy; E_F, Fermi level; $N(E)$, density of states; Ø, disallowed channel. Source: *From G.A. Prinz, Magnetoelectronics, Science 282 (1998) 1660—1663.*

spin-coherence time of *n*-type ZnO is advantageous. Additionally, It is the potential for generation of both p- and n-type ZnO with low resistivity that made bipolar spintronics which has a realistic possibility, to induce both hole and electron-mediated ferromagnetism for such DMS ZnO.

6.2 Methods of Synthesis of the Spintronic Materials

6.2.1 Synthesis of Multiferroics

6.2.1.1 Sol-gel method

If the sol-gel method relates with synthesis in the solution (or liquid) state, one can ensure complete homogeneity with lower synthesis temperatures [22,23]. It has the ability to use molecular precursors to control the morphology of a ceramic product. For example, the sol-gel synthesis applies for the synthesis of Fe-doped $PbTiO_3$ nanoparticles [24]. Starting materials are lead acetate, $C_{16}H_{36}O_4Ti$, and ferric chloride. The $C_{16}H_{36}O_4Ti$ is added in a solution of ethyl alcohol and acetic acid (3:1). Lead acetate solution in distilled water is mixed with the above solution and is constant stirring for 3 h. Ferric chloride is added to the resultant mixture. This precursor solution M was added to a PVA solution in the molar ratio of M:PVA, 5:2. The gel was dried at 250°C and heated at 700°C/3 h to crystallize the sample.

6.2.1.2 Chemical combustion

The chemical combustion synthesis, given a choice of molecular precursor, can influence a material structure even after the solution or gel has been dried and heated. For this, a strong exothermic reaction between an oxidant such as nitrate and fuel such as citrate or glycine produces large volumes of gas that result in an open foam-like structure. For example, the combustion has been used for $Bi_{1-x}Pb_xFeO_3$ multiferroic nanostructures using the Bi nitrate, Pb nitrate, Fe nitrate, polyethylene glycol (PEG), and urea in a desired stoichiometric ratio [25]. To prepare this precursor for combustion, Bi nitrate, Pb nitrate, and Fe nitrate are added in PEG and constantly stirred while raising the room temperature to 50°C. Urea is added and the temperature is then raised to 70°C till the combustion took place. The brown powder so obtained is subjected to annealing at 500°C/5 h.

6.2.1.3 Hydrothermal method

The main advantages of hydrothermal processing are the production of particles that are monodispersed with total control over their shape and size in addition to their chemical homogeneity with the highest dispersibility [26,27]. Moreover, the use of capping agents, surfactants, and other organic molecules contribute greatly to the surface modification of these nanocrystals. For example, the hydrothermal synthesis of pure and Fe-doped BTO nanostructures was used as starting materials such as the barium acetate, $C_{16}H_{36}O_4Ti$, ferric chloride, and NaOH as a mineralizer [28]. Firstly, barium acetate of 0.01 M is dissolved in distilled water and 0.01 M of $C_{16}H_{36}O_4Ti$ is dissolved in ethanol to form a uniform solution by stirring, respectively. Subsequently, the latter solution is slowly dropped into the barium acetate solution under constant-rate stirring with pH = 13 adjusted by adding NaOH. The final reaction mixtures are poured into a Teflon vesseltill 80% of its total volume, and then hydrothermal treatment is performed at 180°C/48 h. The autoclave cooldown to room temperature naturally, and the obtained products are washed with distilled water and ethanol and dried at 80°C/8 h.

6.2.1.4 Metallo-organic decomposition synthesis

This MOD method is similar to a sol-gel process. For example, the multiferroic nanocomposites of MFe_2O_4/BTO [M = Mn, Co, Ni, Zn] thin films were synthesized by a metallo-organic decomposition (MOD) method [29]. The precursor solutions are prepared from $(C_7H_{15}COO)_2Ba$, $C_{16}H_{36}O_4Ti$, $(C_7H_{15}COO)_3Fe$, $(C_7H_{15}COO)_2Mn$, $(C_7H_{15}COO)_2Co$, $(C_7H_{15}COO)_2Ni$, $(C_7H_{15}COO)_2Zn$ in xylene. The stoichiometric ratio of the multiferroic composites is taken as MFO:BTO1:1 in each sample. The final solution is refluxed at 110°C/10 h to get a homogeneous mixing. The PEG at ~3% (in volume) is added dropwise to the MFO/BTO solution to get final coating solution. The solution was coated on Pt/Ti/SiO_2/Si and Si substrates using spin-coating technique (4500 rpm for 60 s) and annealed at 600°C/3 h. The film thickness after three coatings is ~600 nm.

6.2.1.5 Spark plasma sintering

Spark plasma sintering (SPS) has become a routine method for the densification of thermoelectric materials [30,31]. SPS inherited the same graphite die/punch assembly as used in hot-pressing. The design of direct current heating has increased the ramp rate and shortened the holding time, which is vital to maintain the fine microstructure of the starting powder to the sintered material. For a high-resistance powder, most of the current passes through the conductive die and there is little direct current Joule heating of the sample. For example, the polycrystalline samples of $Bi_{1-x}Tb_xFeO_3$ were prepared by SPS method [32]. These as-prepared Tb-doped BFO crystalline ceramic was placed in a cylindrical graphite die with a 20 mm inner diameter. The consolidation was carried out under vacuum for 3 min at a maximum temperature of 1023–1085 K and a uniaxial stress of 50 MPa using an SPS furnace. After the SPS consolidation, the power and pressure were switched off and the samples were allowed to cool in the SPS chamber to room temperature naturally. The SPS processed pellets were subjected to an additional heat treatment in air at 900 K/10 h.

6.2.1.6 Conventional solid-state reaction

The solid-state reaction method for the multiferroic synthesis is milling the appropriate metal oxides or carbonates and then calcinating at high temperature to allow interdiffusion of the cations. For example, multiferroic BTO–$CoFe_2O_4$ composites were fabricated using either the simplified or conventional solid-state reaction [33]. In the simplified solid-state reaction route, all four of the basic chemicals $BaCO_3$, TiO_2, Fe_2O_3, CoO were directly mixed together at suitable stoichiometric proportions in a ball mill for 6 h. The powders were calcined at 850°C/5 h to get solid-state reaction of the BTO–$CoFe_2O_4$ composite. The as-obtained powders were pressed into disks and sintered at 1000°C/5 h. However, for the conventional solid-state reaction, firstly, the individual BTO and $CoFe_2O_4$ powders were synthesized separately using a standard ceramic method. The BTO and $CoFe_2O_4$ powders were both calcined at 850°C/5 h. Secondly, these as-prepared BTO and $CoFe_2O_4$ powders were mixed together in suitable proportions in a ball mill for 6 h. The mixed powders were pressed into disks and sintered at 1000°C/5 h.

6.2.1.7 Pulsed laser deposition

Pulsed laser deposition is a synthesis technique where a high-power pulsed laser beam is focused onto a target of the desired composition. Material is then vaporized from the target and deposited as a thin film on a substrate facing the target [34]. For example, the BFO targets with a diameter of 3 cm were prepared through a conventional solid-state reaction method and annealed at 830°C/2 h. [35]. For the BFO growth, the oxygen pressure ($P_O = 1 \times 10^{-4}$, 0.01, 0.1, 0.5, and 1 Pa) was systematically varied while the deposition temperature ($T = 450°C$) was fixed. A pulsed Nd:YAG (yttrium aluminum garnet) laser (532 nm wavelength, 5 ns duration) operated with an energy of 60 mJ/pulse and repetition rate of 10 Hz was used as the ablation source. The distance between the target and the *c*-sapphire substrate was kept at 3 cm. The deposition time was set to ~30 min. Finally, the films were annealed at 600°C in air atmosphere.

6.2.1.8 Electrospray method

Electrospray is an effective method for the fabrication of nanostructured materials, such as nanospheres and nanofibers. Recently, it has been widely used for the fabrication of hollow nanomaterials [36]. Typically, the first step of the process is to pump a liquid into a stainless steel capillary tube. A strong static electric field is then applied to compete with the surface tension of the liquid, distorting the liquid into a sharp cone. The electric field helps to break the sharp cone into pieces of charged liquid droplets. The nanoparticles, which are finally deposited on a collection substrate, are generated after the evaporation of the solvent [37]. Du et al. [38] reported the synthesis of hollow BFO nanospheres, the diameters of the hollow nanospheres range between 50 and 200 nm, with a shell thickness of ~20 nm.

6.2.1.9 Sol-gel precipitation

This sol-gel precipitation is the coprecipitation which is used by Liu et al. [39] for transition metal (TM) doped BTO samples. In a typical synthesis, 1 mL of $TiCl_4$ was put into 5 mL of absolute ethanol. After 10 min of stirring, 5 mL of water was added to the above-mentioned yellow solution. The mixture was heated slowly, simultaneously stirred to form a transparent solution. Separately, 10 mmol of $BaCl_2$ was dissolved in 30 mL of distilled water, which was subsequently added to the above-mentioned solution. The TM doping was achieved by the addition of their chlorides. Then 50 mL of 12 M NaOH aqueous solution was slowly added dropwise and the temperature of the final solution was raised to 90°C in 30 min and then kept constant for 5 h under agitation. After the system temperature was cooled to room temperature, the resulting solid product was centrifuged, washed with distilled water and ethanol, and the final product was dried.

6.2.1.10 RF sputtering

A synthesis technique where a solid target is bombarded with energetic ions (e.g., $Ar+$) generated in a magnetron causing target atoms to be ejected ("sputtering"). The ejected atoms then deposit, as a thin film, on a substrate. For example, the bilayer thin films of BFO/BTO were prepared on $Pt/TiO_2/SiO_2/Si(100)$ substrate by RF-magnetron sputtering [40]. Targets of BFO and BTO were synthesized by conventional solid-state reaction method. BTO layer was first deposited on $Pt/Ti/SiO_2/Si(100)$ substrate maintained at a temperature of 650°C followed by subsequent deposition of BFO layer on $BTO/Pt/Ti/SiO_2/Si(100)$ at the same substrate temperature. The deposition was carried out at 50 mTorr with a mixture of Ar and O_2 in the ratio of 4:1 with a base pressure of 3×10^{-6} Torr.

6.2.2 Synthesis of DMSs

6.2.2.1 Thermal evaporation method

This is a synthesis technique in which the material to be deposited is heated until evaporation in a vacuum ($<10^{-4}$ Pa) and eventually deposits as a thin film by condensing on a substrate, e.g., the Co/Mn-doped ZnO nanowires deposited on Si substrate [41]. First, high-purity Zn, $MnCl_2$, $CoCl_2$, and graphite powders were mixed with mass ratio 10:1:1:1. The compounds were laid in an alumina boat which was placed at the center of a horizontal tube furnace. A silicon substrate plated by a 20 nm Au layer was placed above the boat. Then, the tube furnace was heated to 500°C with a constant argon flow of 50 sccm. After 180 min, a thin layer was deposited on the silicon wafer.

6.2.2.2 Chemical vapor deposition

The chemical vapor deposition (CVD) is the thin films deposition by chemical precursors in the vapor phase. For example, the $Zn_{1-x}Mn_xSe$ nanoribbons were synthesized by a CVD method [42]. The mixed powder of ZnSe and $MnCl_2$ with mole ratio of 40:1 was put in a ceramic boat as precursors and placed into the central region of the quartz tube. The cleaned silicon pieces coated with Au of ~10 nm thickness were placed downstream, ~10 cm from the precursors. The tube was purged with mixed gas of 90% Ar and 10% hydrogen for 1 h before heating, and then the mixed powder was heated to 1050°C at a rate of 85°C/min. The growth furnace was held for 90 min at 1050°C.

6.2.2.3 Sol-gel spin-coating technique

The $Zn_{0.95}Mn_{0.05}O$ films were synthesized by a sol-gel spin-coating technique [43]. A required amount of Zn acetate and Mn acetate was added to a solution containing 2-propanol and diethanolamine. The solution was used for spin coating on the glass substrates. The spinning rate and period were optimized to be 2000 rpm/35 s, respectively. After coating, the films were dried at 300°C/15 min to evaporate the solvent and finally it was heat treated at 500°C to obtain crystalline films.

6.2.2.4 Spray pyrolysis technique

It is a one-step combustion process where the precursor is in liquid form, with significantly higher combustion enthalpy, usually in an organic solvent. It is one of the most exploited techniques for the production of complex and functional nanoparticles. The samples of $Zn_{1-x}Mg_xO$ were deposited on glass substrates keept at 350°C by spray pyrolysis technique; the distance between the spray nozzle and the substrate was 20 cm [44]. A homogenous solution was prepared by dissolving Zn acetate (0.05 M) and Mg sulfate in distilled water in the desired stoichiometric ratio. In addition, 5 mL ethanol is used as solvents and the prepared solution was sprayed onto the heated substrates with the spray rate of 1.5 mL/min.

6.3 Spintronics Based on BTO Multiferroic Systems

6.3.1 Perovskite (ABO_3) Multiferroics

The term "perovskite" originally refers to $CaTiO_3$, which has ABO_3 as the general formula. For an ideal cubic-symmetry structure, the A-cation is at the cubic corners in 12-fold cuboctahedral coordination while the B-cation is at the body center positions surrounded by oxygen octahedra in sixfold coordination. Depending on the relative size of the atoms, the perovskite structure is prone to slight bucklings and distortions that can produce several lower-symmetry variants, notably tetragonal and orthorhombic phases. This is accompanied with the off-centering of the B atom would lead to an electric dipole that induces ferroelectricity in BTO and BFO. The perovskite structure can host magnetic atoms at both A and B positions and the corresponding spins can interact via different mechanism such as the long-range magnetic orders. These orders coexist with, are affected by, and even promote structural changes that can be related with ferroelectricity. This caused multiferroicity in the perovskite structure [45,46]. However, the origin of ferroelectricity leads to the non-centrosymmetric distortions of BO_6 octahedra inversion symmetry and a net macroscopic electric polarization [47].

The ABO_3 cubic crystal structure of BTO is shown in Fig. 6.5A [12,48]. The A-cation is 12-fold coordinated with oxygen and, generally, the interaction with the surrounding O ions is ionic, in that the A-cation valence electrons are transferred to the O ion 2p band without significant hybridization. On the other hand, the B-cation is sixfold coordinated to oxygen, and the B-cation d states will generally show a significant covalent interaction with the O 2p states. The O 2p and B-cation d states will form bonding and antibonding states, with the former having predominant O 2p character, and the latter

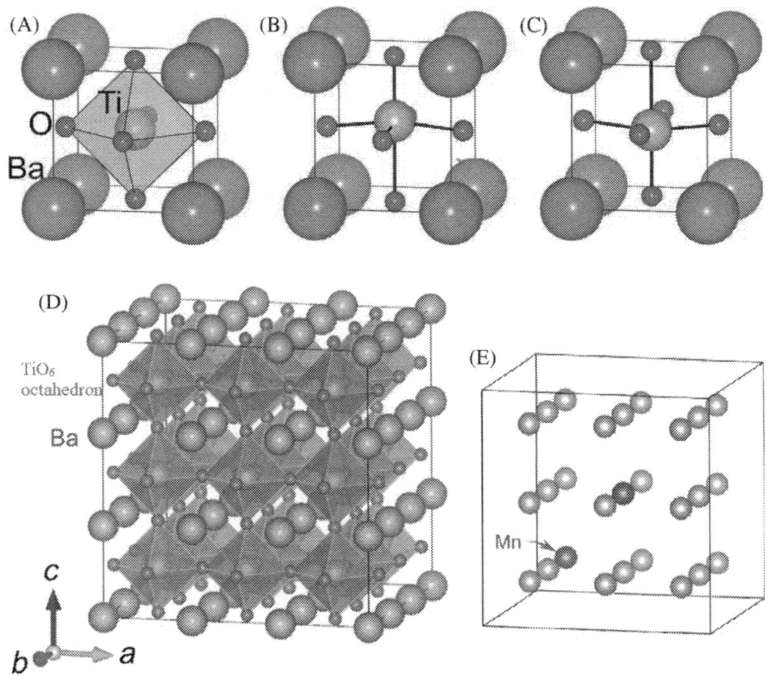

Figure 6.5

Crystal structure of BTO. (A) High-temperature paraelectric cubic phase. (B and C) Room-temperature ferroelectric tetragonal phase, showing up and down polarization variants. The lattice distortion by elongation of the *c*-axis as well as a shift of the B-cation with respect to octahedron O ion. (D) Optimized $3a \times 3b \times 3c$ supercell structure of the ferroelectric *P4mm* BTO with lattice constants $a = b = 3.997$ Å and $c = 4.0413$ Å. (E) The substitution pattern of Mn-doped BTO. Source: *From J.Y. Son, J.H. Lee, S. Song, Y.H. Shin, H.M. Jang, Four-states multiferroic memory embodied using Mn-doped* $BaTiO_3$ *nanorods, ACS Nano 7 (2013) 5522–5529; P.M. Rorvik, T. Grande, M.A. Einarsrud, One-dimensional nanostructures of ferroelectric perovskites, Adv. Mater. 23 (2011) 4007–4034.*

having predominant B-cation d character. Due to the octahedral crystal field, the free ion d band is split into a t_{2g} triplet and an e_g doublet, for which the p–d interaction is π bonding and σ bonding, respectively.

6.3.2 Single-Phase Multiferroic BTO Systems

6.3.2.1 Structure and phase transition of doped BTO

Recently, multiferroics have been widely investigated in spintronics, magnetic field sensors, MRAM, etc. due to their possible ferroelectric and ferromagnetic coupling parameters [49,50]. Unfortunately, single-phase multiferroics are extremely rare and BTO is one of them. BTO is well known for room-temperature multiferroic behaviors [12]. The three crystallographic phases of BTO are: rhombohedral <190 K, orthorhombic $190\ K < T < 278$ K, and tetragonal $278\ K < T < 395$ K. At higher temperatures, BTO is a paraelectric. For tetragonal ($a = b = 3.99$ Å and $c = 4.03$ Å) BTO with P4*mm* space group, the occurrence of the spontaneous electric polarization is related with displacement of the Ti^{4+} ion along the *c*-axis (Fig. 6.5B and C) [12,48]. It is sensitive to charge hybridization of the 3d states of the Ti cation with the 2p states of the surrounding O anions. Using density functional theory (DFT), it has been revealed that the off-center displacement connected to a substantial charge rearrangement changes the occupation of the d_z^2, d_{xz}, and d_{yz} orbitals of Ti [51]. The p orbitals of the oxygen ions located on the *z*-axis yield a destabilization of the centrosymmetric configuration. With a wide survey of literature on BTO, the researchers have decided that a low doping at the Ti site can introduce itinerant electrons, leading to the presence of inhomogeneous state for multiferroicity [28,52]. In perovskite BTO, Ba^{2+} (A-site cation) induces required distortion for ferroelectricity, while magnetism can be achieved by the doping along B-site cation [53]. For this, the TM = Cr, Mn, Fe, Co, Ni, Cu ions are the dopants for BTO lattice [50]. The replacement of the Ti^{4+} with small dopant level of the TM ions would compact ferroelectricity [54].

Ab initio DFT calculations were performed to understand the Mn (7.4 at.%) doping effect on BTO [12]. For this purpose, the $3a \times 3b \times 3c$ supercell contains two Mn ions distributed among BTO formula cells (Fig. 6.5D and E). The computed off-center ferroelectric displacement of the substituted Mn ion in the MnO_6 octahedron unit is smaller than that of the pure Ti ion in the TiO_6 octahedron unit by 0.05 Å, with a concomitant decrease in the ferroelectric polarization. According to the Kohn–Sham (K–S) energy, the antiferromagnetic state corresponds to the spin ground state and is more stable than the paramagnetic state by 115.96 meV/Mn ion. However, the difference in the computed K–S energy between the ordered ferromagnetic and antiferromagnetic states is negligibly small, <0.002 meV/Mn ion. This result is somewhat different from the previous result done at a higher Mn-doping level, where the antiferromagnetic state is the most stable phase with K–S energy difference, $E_{FM} - E_{AF} > 0$ of ~3.5 meV/Mn ion [55].

6.3.2.1.1 X-ray diffraction of Ce-, La-substituted $BaFe_{0.01}Ti_{0.99}O_3$ nanostructures

Fig. 6.6 shows the Rietveld refinement (using Full-Prof program) of X-ray diffraction (XRD) pattern ($2\theta = 20° - 80°$) of the $Ba(Fe_{0.67}Ce_{0.33})_{0.01}Ti_{0.99}O_3$ (BFTO:Ce) and $Ba(Fe_{0.67}La_{0.33})_{0.01}Ti_{0.99}O_3$ (BFTO:La) nanostructures using tetragonal phase model (space group P4*mm*) [56]. The position of atoms are Ba (0 0 0), Ti/Fe/Ce/La (0.5 0.5 *z*), O_1 (0.5 0.5 z), and O_2 (0.5 0 *z*). All diffraction lines of BFTO:Ce and BFTO:La show polycrystalline with tetragonal BTO phase. The Rietveld refined parameters are a (Å) = 4.001(3) and 3.998(2), c (Å) = 4.087(2) and 4.026(1), c/a = 1.021 and 1.007, R_p = 5.80 and 7.77, R_{wp} = 7.76 and 11.10, and χ^2 = 1.60 and 0.967, respectively, for BFTO:Ce and BFTO:La. For bulk BTO, c/a is 1.01. However, the observed value of a, c, and c/a in Ce-, La-, and Fe-doped BTO is enhanced than nanostructural BTO [57]. The effect on sample densities of BTO with Fe, Ce, and La doping is calculated using X-ray density (theoretical), $d_{XRD} = ZM/Na^2c$, where Z is the number of molecules per unit cell ($Z = 5$ for tetragonal BTO), M is the molecular weight, a and c are the lattice constants, and

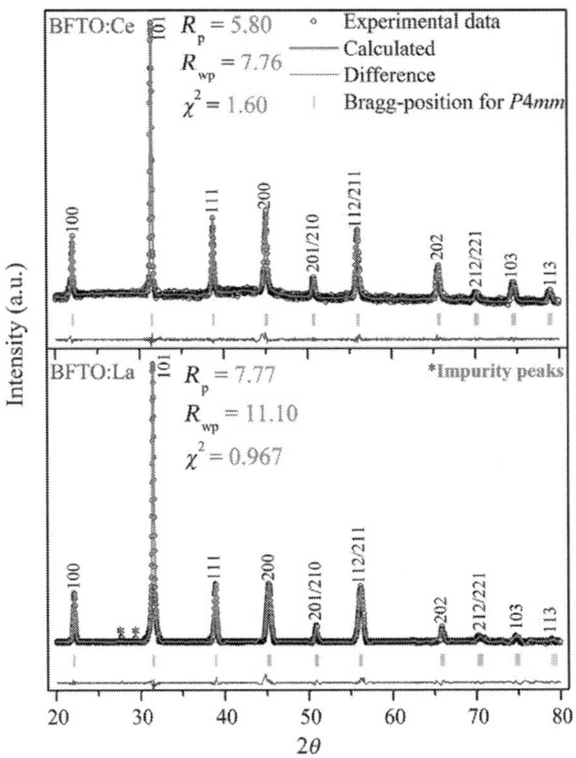

Figure 6.6

Rietveld refinement of the XRD pattern of BFTO:Ce and BFTO:La nanostructures. Source: *From K. C. Verma, R.K. Kotnala, Lattice defects induce multiferroic responses in Ce, La-substituted $BaFe_{0.01}Ti_{0.99}O_3$ nanostructures, J. Am. Ceram. Soc. 99 (5) (2016) 1601–1608.*

N is the Avogadro's number. The experimental (apparent) density is calculated using method of geometrical shape by assuming the cylindrical shape of the pellet, $d_a = m/V = m/\pi r^2 t$, where m, V, r, t are mass, volume, radius, and thickness of the pellet sample, respectively. The calculated values of XRD density, $d_x = 6.057$ and 6.018 g/cm^3, and apparent density, $d_a = 5.993$ and 5.869 g/cm^3, respectively, for BFTO:Ce and BFTO:La sample. For pure BTO, $d_x = 6.017$ g/cm^3. The porosity, p, in the samples ($p = 1 - d_a/d_{XRD}$) is 1.057% and 2.48%, respectively, measured for BFTO:Ce and BFTO:La.

6.3.2.2 Induction of multiferroicity of the BTO with doping

For perovskite structure of BTO, the magnetism is induced due to partially filled d orbitals (ordering of spins of electrons in incomplete ionic shells) and ferroelectricity by off-centered displacement, d^0 (relative shifts of negative and positive ions that induce surface charges), which are essential for multiferroicity in single-phase multiferroic [56].

6.3.2.2.1 TM impurity in BTO

In single-phase BTO, BTO with doping of TM ions, the Fe doping is widely studied due to its magnetic order that originated from the unpaired spin of Fe ions [57]. Apostolova et al. [58] suggested superexchange interactions between Fe^{3+} ions due to different occupational sites associated with oxygen vacancies (V_O). It has been recently reported that the ME voltage is induced by shape/size and lattice distortion effects in $BaFe_{0.01}Ti_{0.99}O_3$ nanowires [29]. With wide survey on multiferroic BTO, we eventually decided that a low doping at the Ti site can introduce itinerant electrons, leading to the presence of inhomogeneous state to influence multiferroic parameters [50]. Since BTO has the Ba^{2+} (A-site cation) required distortion for ferroelectricity, magnetism is achieved by the doping along B-site cation. The spontaneous ferroelectric phase stabilization is induced by the hybridization between the d orbitals of the central cation and the 2p orbitals of the O. Ju et al. [59] suggested the incorporation of Cr^{3+} into BTO which suppresses the ferroelectric tetragonal phase at room temperature with long-range antiferromagnetic interactions. Yan et al. [49] invented Co-doped BTO for nonvolatile memory cells. Das et al. [60] suggested magnetism due to carrier-mediated exchange interactions for Ni-doped BTO. It is also reported that the weak ferromagnetism and ferroelectricity disappearance have been observed in Mn-doped BTO [61].

6.3.2.2.2 Low doping level of Fe impurity ions in BTO influence multiferroicity

It is reported that the multiferroic Fe-doped BTO has room-temperature ferromagnetism (RTFM) with low Fe concentration (0–4 mol% of Fe) [62]. The ferromagnetism of these Fe-substituted BTO multiferroic systems is intrinsic in which the electrons trapped in the V_O via the F-center will overlap the d shells of both neighboring Fe. At a low Fe concentration, the $3d^5$ orbitals have unoccupied minority spin orbitals, resulting in weak ferromagnetism. However, the higher Fe concentration leads to antiferromagnetic superexchange interaction between Fe ions via O. The ME coupling in TM-doped perovskite is a product of piezoelectric and piezomagnetic phases in which elastic strain field puts in relationships with the electric polarization in the piezoelectric phase with the magnetization arising in the piezomagnetic component. Fig. 6.7A and B reports the Fe-concentration-dependent magnetic phase transition temperature T_C, magnetization and polarization with doping concentration $x =$ Fe, of $Ba(Ti_{1-x}Fe_x)O_3$. From this observation, magnetization (M) and the T_C increase with Fe doping; it is obtained that the polarization P is decreased and consequently the ferroelectric T_C is also decreased. This is because $BaFeO_3$ is a pure magnetic material and does not have any ferroelectric polarization. It means the ferroelectric–paraelectric phase transition temperature T_C is strongly dependent on the Fe-dopant level in BTO [58]. For pure BTO, the phase transition is observed at 125°C that is shifted to a lower temperature with Fe doping.

6.3.2.2.3 Tetragonal distortion by splitting/shifting of (200) XRD peak of BTO with TM ions

Table 6.1 provides the XRD, transmission electron microscopy (TEM), and field emission scanning electron microscope (FESEM) results of TM (1 mol%) doped BTO nanoparticles [50]. This data clearly determines that the TM ions crystallized to BTO tetragonal phase. The elongation of the lattice constant, a, revealed the competition between the long-range Coulombic attractive interactions and the short-range repulsive interactions in ionic nanoparticles. It creates an effective negative pressure that leads to an expansion in unit cell. The elongation of c-axis must create tensile strain. The small increase in tetragonal distortion leads to a higher deformation of the TiO_6 octahedra with TM in BTO lattice. These variations of the lattice constants (a and c) as given in Table 6.1 indicate distortion/strain in the BTO lattice. Fig. 6.7C shows the tetragonal splitting of (200) diffraction peak at $2\theta = 44.3°-45.7°$. The diffraction peak of pure BTO has been shifted toward a lower angle with TM doping that introduced tensile strain in BTO. The splitting of BTO (200) peak into (002/200) confirms the tetragonal phase. It is due to electrostatic repulsions between 3d electrons of Ti^{4+} ions and 2p electrons of O^{2-} ions; the structure becomes distorted and attains a stable tetragonal structure. These distortions definitely lead to defects/vacancies formation in the BTO unit cell [63].

6.3.2.2.4 Rare earth ions impurity in multiferroic BTO

The ionic modification of Fe 3d shell with rare earth (RE) ions may have given significance for further multiferroic advancement. Tokunaga et al. [64] give symmetric exchange striction between RE 4f moments and Fe 3d spins, which is antiferromagnetic at 2.65 K and far above room temperature, respectively. The ferromagnetic moment arises due to canted antiferromagnetic Fe spins from Dzyaloshinskii–Moriya (DM) interaction that required polarization reversal to change the antiferromagnetic phase either by RE 4f moment or Fe spins. The V_O plays an important role in exchange interaction among TM 3d and RE 4f transition due to formation of multiferroic ordering [56]. Since for the multiferroic system

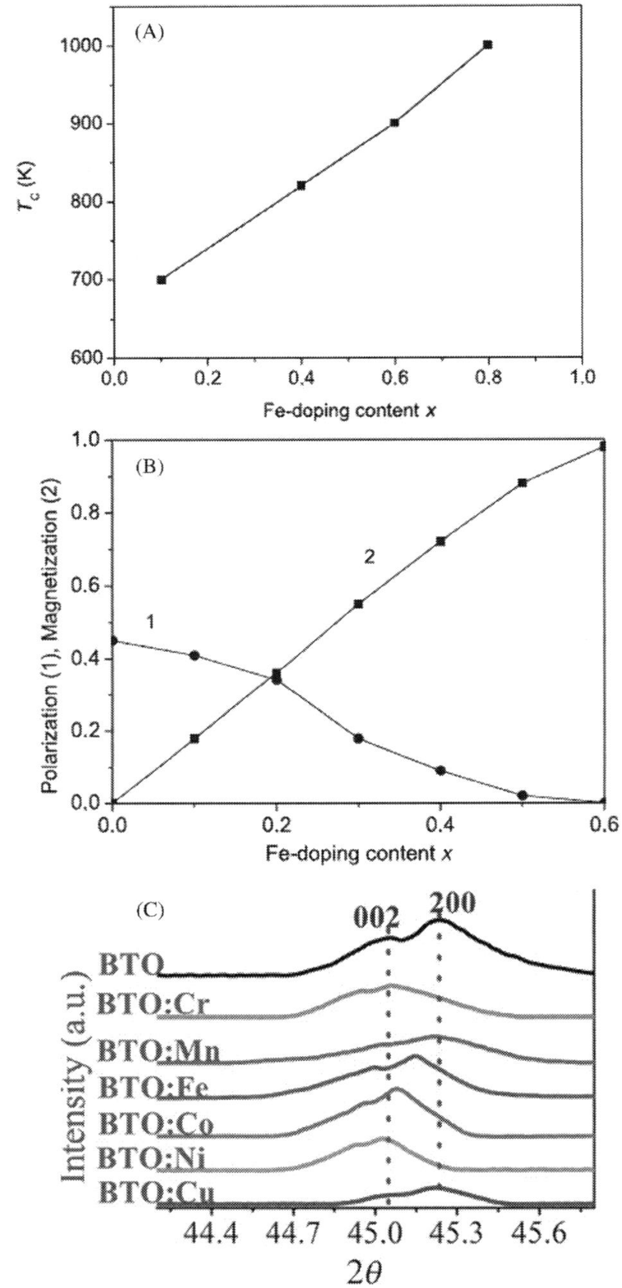

Figure 6.7

With Fe doping (A) magnetic phase transition temperature, T_C, (B) polarization (1) and the magnetization (2) (at $T = 100$ K) of $Ba(Ti_{1-x}Fe_x)O_3$. (C) Splitting of the (200) peak of pure TM: $BaTiO_3$ nanoparticles. Source: *From K.C. Verma, R.K. Kotnala, Multiferroic approach for Cr,Mn,Fe,Co,Ni, Cu substituted $BaTiO_3$ nanoparticles, Mater. Res. Exp. 3 (2016) 055006–055013; I.N. Apostolova, A.T. Apostolov, S.G. Bahoosh, J.M. Wesselinowa, Origin of ferromagnetism in transition metal doped $BaTiO_3$, J. Appl. Phys. 113 (2013) 203904.*

Table 6.1: Rietveld refined lattice parameters and particles size (*D*) from TEM/FESEM of $BaTM_{0.01}Ti_{0.99}O_3$.

Sample	*a* (Å)	*c* (Å)	*c*/*a*	*V* (Å^3)	D_{TEM} (nm)	D_{SEM} (nm)
BTO	3.991(1)	4.0293(3)	1.00959	64.179	–	–
BTO:Cr	3.992(1)	4.0292(2)	1.00932	64.2095	20 ± 3	23 ± 1
BTO:Mn	3.9948(3)	4.0223(1)	1.00688	64.1895	13 ± 1	15 ± 1
BTO:Fe	4.0138(3)	4.0503(1)	1.00909	65.2527	33 ± 5	36 ± 3
BTO:Co	3.9961(2)	4.0271(2)	1.00776	64.3080	35 ± 3	41 ± 7
BTO:Ni	4.0011(3)	4.0261(3)	1.00625	64.4530	17 ± 1	19 ± 3
BTO:Cu	3.997(2)	4.0173(2)	1.00508	64.181	47 ± 7	51 ± 8

TM = Cr (BTO:Cr), Mn(BTO:Mn), Fe (BTO:Fe), Co(BTO:Co),Ni (BTO:Ni),Cu (BTO:Cu)] nanoparticles. Source: *From K.C. Verma, R.K. Kotnala, Multiferroic approach for Cr,Mn,Fe,Co,Ni,Cu substituted $BaTiO_3$ nanoparticles, Mater. Res. Exp. 3 (2016) 055006–055013.*

of Ce- and La-substituted $BaFe_{0.01}Ti_{0.99}O_3$, the multiferroicity is enhanced because the unpaired electrons of Ce^{3+} ([Xe] $4f^1$ $5d^1$ $6s^2$) and La^{3+} ([Xe] $5d^1$ $6s^2$) may influence lone pair of electrons along A-site (d^0) formation. The large electronegativity differences between dopant Ce^{3+} and La^{3+} and Ba^{2+}, Fe^{3+}, and Ti^{4+} ions must increase the bond length between atoms inside the unit cell, the stretching between atoms responsible for polarization enhancement.

6.3.2.3 Multiferroic nanostructures

It is well known that at least one of the dimensions of the materials below 100 nm could be considered as nanomaterials [65]. On the basis of their dimensionality, the multiferroic nanomaterials are categorized in the form of nanostructures, i.e., nanoparticles, nanowires, nanosponges, nanowires networks, hollow nanospheres, nanosheets, nanorods, nanoflowers (schematically shown in Fig. 6.8). The ferroelectric and multiferroic nanostructural categorization is described in the following sections.

6.3.2.3.1 Zero-dimensional nanostructures

Any system having all three dimensions well below 100 nm (critical length) is considered as zero-dimensional (0D), most common examples are quantum dots, spheres, and cubes [66]. Park et al. [67] reported a decrease in T_C and an increase in tetragonal crystal distortion of BTO nanoparticles with increasing diameter (33–140 nm). The minimum critical diameter required for a ferroelectric phase transition in the BTO nanoparticles is 37 nm.

6.3.2.3.2 One-dimensional nanostructures

The one-dimensional (1D) nanostructure has only two dimensions (area/diameter) below the critical length. The 1D multiferroic has potential in the field effect transistors (FET), sensors, flexibility in electronics components, etc. [66]. Weber et al. [68] fabricated Si

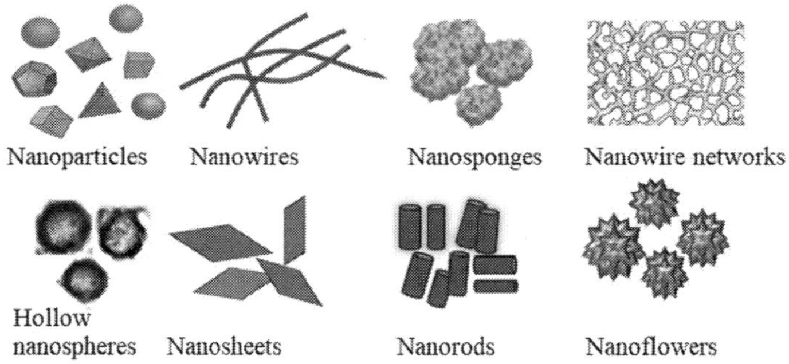

Figure 6.8
Schematic nanostructural classification.

nanowires with dimensions: only one atom tall and four atoms wide. In addition to 1D ferroelectric BTO, $PbZr_{1-x}Ti_xO_3$, $SrTiO_3$ are most extensively investigated due to their excellent ferroelectric characteristics [68].

6.3.2.3.3 Two-dimensional nanostructures

The two-dimensional (2D) nanostructures have only one dimension less than the critical length, with thin films (< 100 nm) grown on any of the substrates. The formation of periodic arrays of isolated ferroelectric nanoislands or dots has already been shown to enhance the performance of FeRAM, by providing a high-density ferroelectric single-domain per unit area [69].

6.3.2.3.4 Three-dimensional nanostructures

The three-dimensional (3D) nanostructures do not follow the critical length along all the three directions but its basic building components may possess 0D, i.e., nanocomposites, nanocrystalline materials, etc. The 3D has the ability to order and align ferroelectric or piezoelectric nanostructures, especially nanotubes and nanowires, into high-density free-standing arrays. For example, an uniaxial compressive force, the 3D multiferroic nanowires generate a piezoelectric field, which in turn produces an electrical output.

6.3.2.3.5 Grain-size-dependent ME coupling of BTO nanoparticles

In the reported literature, the ME coupling of BTO nanoparticles of different grain sizes (16–26 nm) has been studied [70]. These BTO nanoparticles are synthesized with modified Pechini method which observes variation in average grains size by varying citric acid (CA) to ethylene glycol ratio. Fig. 6.9A shows XRD patterns of BTO nanoparticles heated at 750°C with varying CA concentration 10%–50%. In the XRD pattern, the BTO crystallizes into cubic phase. Grain size of each sample was estimated by using Scherrer formula ($x = \frac{0.9\lambda}{\beta\cos\theta}$, where λ is the wavelength and θ is the angle of diffraction) which found particles size in 16–26 nm. Fig. 6.9B shows linear ME coefficient versus grain size of

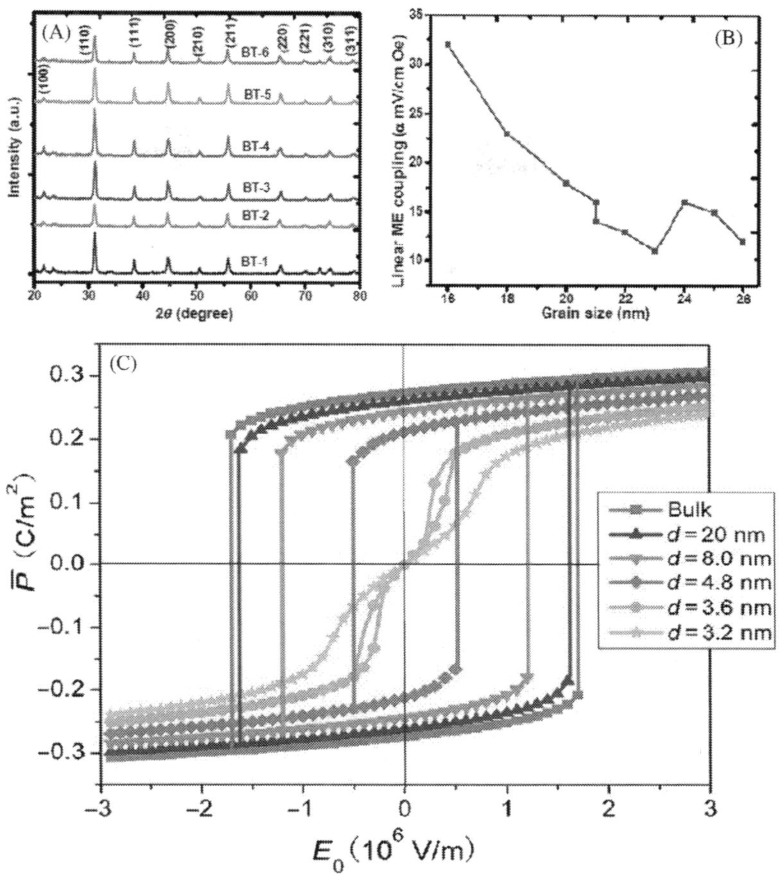

Figure 6.9

(A) XRD patterns of BTO nanoparticles obtained by varying CA percentage. (B) Linear ME coefficient versus grain size of BTO. (C) Ferroelectric hysteresis measured for BTO nanowires at different nanosize ($T = 293$ K). Source: *From T. Woldu, B. Raneesh, M.V. Ramana Reddy, N. Kalarikkal, Grain size dependent magnetoelectric coupling of $BaTiO_3$ nanoparticles, RSC Adv. 6 (2016) 7886–7892; J. Hong, D. Fang, Size-dependent ferroelectric behaviors of $BaTiO_3$ nanowires, Appl. Phys. Lett. 92 (2008) 012906-012913.*

same BTO samples of grains size 16–26 nm [70]. It is observed that the linear ME coupling coefficient decreases with increasing grain size. When the grain size gets smaller and smaller, lattice size diminishes, then the surface strain introduces coordinated distortion and in turn disorder permeate throughout the entire particle, as opposed to being confined only at the surface. Consequently, different frustrated spin structures are produced.

6.3.2.3.6 Physical significance of BTO multiferroic nanostructures

The 1D nanostructure <100 nm exhibits new physical phenomena due to finite size effects that occur during downscaling [71]. In comparison to bulk counterparts, the nanostructures cover relatively high specific surface areas with greater amount of surface atoms and higher levels of surface energy. These characteristics give low-dimension ferroelectrics with specific modifications in T_C, dielectric constants, coercive fields, spontaneous polarization levels, and piezoelectric response [66]. Moreover, decrease in size and dimensionality can facilitate the formation of single-domain structures, which can dramatically enhance ferroelectricity. Hong et al. [72] using Landau–Ginsburg–Devonshire theory investigate the size-dependent ferroelectric properties of BTO nanowires, with the polarization oriented along the radial direction (Fig. 6.9C). It is observed that the T_C, mean polarization, and area enclosed by hysteresis loop decrease with the reducing diameter of the nanowire. This size effect of BTO nanowires becomes apparent only when the diameter >20 nm. It is also directed that the BTO nanowire displays double hysteresis loops when the diameter reduces to the critical value, 3.6 nm. In this case, the value of T_C is 286 K which is slightly lower than the adopted room temperature. The coercive field decreases much faster than the remnant polarization. Moreover, when the diameter reduces to 4.8 nm, E_c diminishes at $\sim$70%, but P_r diminishes only at $\sim$20%. This result is very useful for small BTO nanowires for nonvolatile memory devices since ferroelectric memories just require that the material has a high P_r for better "writing" and "reading" and a low E_c to operate at low voltage.

6.3.2.4 Raman measurement of BTO: lattice structure, defects/vacancies evaluation

Fig. 6.10 shows the Raman scattering of Ce and La-doped $BaFe_{0.01}Ti_{0.99}$ nanostructures at room temperature [56]. The observed sets of Raman bands around 186, 243, 303, 513, and 713 cm^{-1}, respectively, due to A_1(TO + LO), [A_1(TO), E(LO)], [B_1; E(TO + LO)], [A_1(TO), E(TO)], and [A_1(LO), E(LO)] assigned to tetragonal BTO. Generally, the peak at 303 cm^{-1} indicates the asymmetry of TiO_6 octahedra within the BTO lattice which is the clear evidence for tetragonal BTO structure. The band 713 cm^{-1} is related with Ba^{2+} defects to the BTO lattice. The peaks at 513, 243, and 186 cm^{-1} are attributed by TO mode of A_1 symmetry. It gives difference among tetragonal and orthorhombic BTO phases. In comparison to pure BTO, the observed Raman active modes are slightly shifted toward lower frequency and this shifting with Ce, La, and Fe must be caused by tensile strain.

6.3.2.5 Magnetism in BTO with doping

For a single-phase BTO multiferroic, it is rarely that the magnetic, dielectric, and ferroelectric behavior coexisted and influenced with each other. It is well known that the perovskite ABO_3 is subjected to indirect superexchange magnetic interactions. The mechanism for magnetism production in Fe-doped BTO is recently reported and shown in Fig. 6.11 [73]. For this one could have need an empty d shells in which the empty d states of TM ions may establish strong covalency with surrounding O. This may favorable to shift TM ion from the center of O_6 octahedra toward one (or three) O, to form a strong covalent bond with this particular O(s) at the expense of weakening the bonds with other O. It is noted that Fe^{3+} ions substituting for Ti^{4+} will introduce V_O into BTO. Such V_O may act as

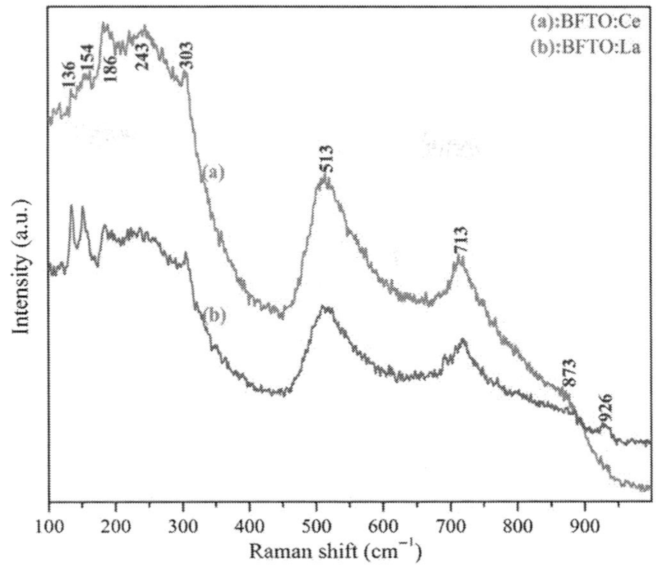

Figure 6.10
Raman spectra measured at room temperature for Ce-, La-, and Fe-doped BTO nanostructures. Source: *From K.C. Verma, R.K. Kotnala, Lattice defects induce multiferroic responses in Ce, La-substituted* $BaFe_{0.01}Ti_{0.99}O_3$ *nanostructures, J. Am. Ceram. Soc. 99 (5) (2016) 1601–1608.*

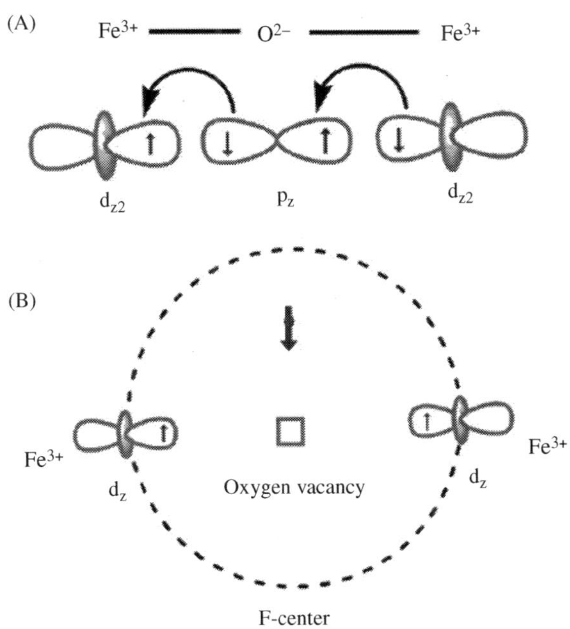

Figure 6.11
Mechanism showing (A) superexchange interactions and (B) F-center exchange.

a kind of medium through which superexchange interaction between neighboring Fe^{3+} ions occur. An electron trapped in the V_O constitutes an F-center, where electron occupies an orbital which overlaps the d shell of both iron neighbors. This exchange interaction leads to ferromagnetic coupling.

6.3.2.5.1 Magnetic ordering near ferroelectric transition in BTO:$Fe_{113\ ppm}$ system

The intrinsic ME multiferroic are quite rare because the usual requirements for the existence of ferromagnetism, such as the electrons of the semioccupied d orbitals, tend to be unpropitious to the off-center cations usually required by ferroelectricity. Fig. 6.12A

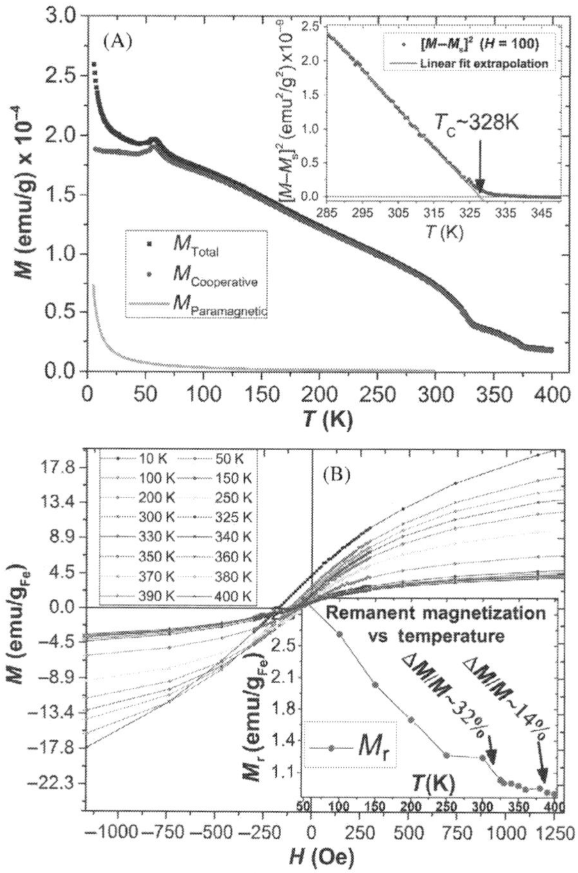

Figure 6.12

BTO with 113 ppm of Fe (A) temperature-dependent magnetization $M(T)$ showing the ordered and paramagnetic behaviors. Inset: $[M - M_s]^2$ versus T around the 325 K is the magnetic anomaly. (B) Magnetic-field-dependent magnetization (at different temperatures). Temperature-dependent remanent magnetization (inset of (B)). Source: *From C.O. Amorim, F. Figueiras, J.S. Amaral, P. Mirzadeh Vaghefi, P.B. Tavares, M.R. Correia et al., Peculiar magnetoelectric coupling in BaTiO$_3$:Fe$_{113}$ ppm nanoscopic segregations, ACS Appl. Mater. Interf. 7 (2015) 24741–24747.*

shows the paramagnetic contribution to the magnetization, obtained by using equation $M = \frac{C}{T} + AT^{3/2} + B$, where the second term accounts for the curvature of the temperature-dependent ordered part, and the remaining contribution due to cooperative phenomena [74]. From the Curie constant, product is obtained, $NJ(J+1)$, which can be written as $NS(S+1)$, assuming the orbital quenching of Fe ions. Taking these considerations into account, it is found that if the Fe ions are in the Fe^{3+} valence ($J = S = 5/2$), it is ~39 ppm of diluted paramagnetic ions, or if they are in the Fe^{2+} valence ($S = 2$), it ~56 ppm of dilute paramagnetic ions, leaving 74 or 57 ppm respectively in ordered magnetic phase. Using spontaneous magnetization models, representing the magnetization (Fig. 6.12A, inset) as $[M - M_S]^2$ versus T enables us to determine the T_C values of the magnetic transitions. Fig. 6.12B shows the low-field detail of the magnetic field dependence measured at $T = 10-400$ K. The saturation magnetization (measured at $T = 5$ K and $B = 5$ T) agrees well with the Fe concentration obtained. It is also observed that there are temperature-dependent remanent magnetizations and coercive fields which are ~170 Oe at 10 K (Fig. 6.12B, inset).

6.3.2.6 Ferroelectricity in BTO with doping

The ferroelectricity in BTO arises from the off-centering of the Ti ions with respect to a centrosymmetric cubic perovskite crystal. It is well known that ferroelectricity is suppressed at the nanoscale due to depolarization fields that arising from the bound charges at the nanostructural surface [75]. It is reported that the ferroelectric polarization of BTO is disappeared when the size of nanoparticles is <40 nm. The detailed mechanism for ferroelectric polarization in doped BTO multiferroic is given in the following sections.

6.3.2.6.1 Ferroelectricity induced by lone-pair electrons

For a multiferroic system such as Fe-doped BTO, the lone-pair electrons of Ba^{2+} cause off-centering that drive Fe^{3+} and Ti^{4+} ions to induce ferroelectricity (second-order Jahn–Teller ferroelectrics) [76,77]. This is an ideal combination of cations and anions to form (dis) connected polyhedral, i.e., tetrahedra in the spinel or octahedra in the perovskite structure, so that the metal centers undergo cooperative off-centering displacements within their coordination surroundings. It is the tendency of the cations to displace in a polar fashion, i.e., d^0 TM or lone-pair active cations. In BTO, the change in hybridization of the Ti 3d states with the O 2p states from the Ti^{4+} displacements stabilizes its off-centering in its octahedral coordination, leading to a net electric polarization.

6.3.2.6.2 Ferroelectricity due to charge ordering

It is difficult to find a material which is both ferroelectric and magnetic [78]. Most ferroelectrics are TM oxides, in which transition ions have empty d shells. These positively charged ions like to form "molecules" with one (or several) of the neighboring negative O ions. This collective shift of cations and anions inside a periodic crystal induces bulk electric polarization. The mechanism of the covalent bonding (electronic pairing) in such molecules is the virtual hopping of electrons from the filled O shell to the empty d shell of a TM ion. When charges order in a nonsymmetric fashion, they induce electric polarization. It would be interesting to explore multiferroics in which ferroelectricity appears as a combined effect of charge and magnetic ordering that together break inversion symmetry.

6.3.2.6.3 Multiferroicity due to DM interaction

For multiferroicity, the magnetic spiral states might be found which has spin rotation due to credible microscopic mechanism to induce ferroelectricity [79]. It involves the antisymmetric DM interaction. This interaction is a relativistic correction to the usual superexchange and its strength is proportional to the spin–orbit coupling constant. The DM interaction favors noncollinear spin ordering. For example, it gives rise to weak ferromagnetism in antiferromagnetic layers of La_2CuO_4 (Fig. 6.13). It also transforms the collinear Neel state in ferroelectric BFO into a magnetic spiral. Ferroelectricity induced by spiral magnetic ordering is the inverse effect, resulting from exchange striction, i.e., lattice relaxation in a magnetically ordered state. The exchange between spins of TM ions is

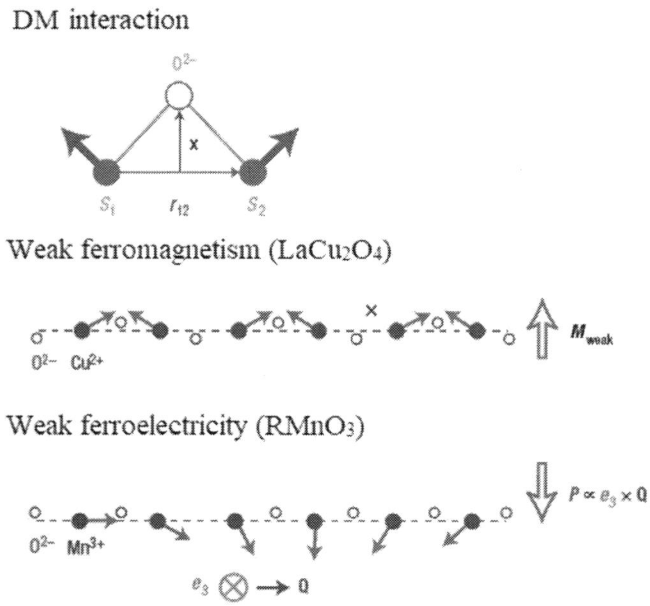

Figure 6.13

Antisymmetric DM interaction effects. Oxygen ion (open circle) and TM ions (filled circles). Source: *From S.W. Cheong, M. Mostovoy, Multiferroics: a magnetic twist for ferroelectricity, Nat. Mater. 6 (2007) 13–20.*

usually mediated by ligands, e.g., O ions, forming bonds between pairs of TM. In particular, when the spin rotation axis e_3 is not oriented along a crystal axis, the orientation of P depends on strengths of ME couplings along different crystallographic directions. As shown in Fig. 6.13, the interaction $H_{DM} = D_{12} \cdot [\mathbf{S}_1 \times \mathbf{S}_2]$. The Dzyaloshinskii vector D_{12} is proportional to spin–orbit coupling constant λ and depends on the position of the O ion between two magnetic TM ions, $D_{12} \propto \lambda \times \hat{r}_{12}$.

6.3.2.7 ME response due to an anomaly in phase transition temperatures

This anomaly in the phase transition of ferroelectric, T_{FE}, around magnetic transition, T_N, and vice versa, which has intrinsic multiferroic origin, must be correlated with inverse DM-type interactions [79]. It occurs in complex magnetic structures such as noncollinear canted antiferromagnets, where the canted spin polarizes the off-center orbital through electron–lattice interaction. This is the framework of the Landau–Devonshire theory of phase transition in ME-ordered systems. Since ferroelectromagnets are an interesting group of compounds that complement purely (anti-)ferroelectric or (anti-)ferromagnetic materials, they display simultaneous electric and magnetic order [80]. Fig. 6.14A shows the M–H hysteresis of $0.7BiFeO_3$–$0.3BaTiO_3$ (BFO–BTO) nanoparticles measured at room temperature, indicating a strong ferromagnetism [81]. The ME coupling due to intrinsic effect of BFO–BTO nanoparticles is also indicated by an anomaly in phase transition temperatures of electric and magnetic behavior. Fig. 6.14 shows the temperature-dependent inverse dc magnetization at $H = 500$ Oe in $T = 298$–725 K for BFO–BTO nanoparticles. The value of T_N is 641 K. It observed an anomaly in the temperature dependable relative permittivity, $\varepsilon(T)$, by observing peaks near 641 K correspond to T_N in $\chi^{-1}(T)$ curve.

6.3.2.8 Magnetocapacitance

The concept of magnetodielectric/magnetocapacitance (MD/MC) was first observed on $DyMnO_3$ due to electric polarization induced by a cycloidal spiral ordering [82]. This effect is achieved through a combination of MR and the Maxwell–Wagner effect. Fig. 6.15 shows the magnetic field affected dielectric behavior of $NiFe_2O_4/BaTiO_3$ (NFO/BTO) nanostructures (synthesized by chemical combustion (C:NFO/BTO), sol-gel (S:NFO/BTO), and hydrothermal (H:NFO/BTO) method) measured at room temperature [83]. It shows frequency-dependent relative permittivity (ε_r) of NFO/BTO composites within 100 Hz to 10 MHz under the influence of dc magnetic field, $H_{dc} = 0$, 1 kOe. The magnetic field was applied along the pellet thickness and the dielectric constant measured along the thickness direction (longitudinal mode). All the NFO/BTO composites showing variation in the dielectric constant at low-frequency region which follows the Maxwell–Wagner type interfacial polarization [84]. It is clearly understood that the capacitance decreases with H_{dc} indicates a negative MD effect. The MR related to spin-polarized tunneling across grain boundaries or interfaces may induce a negative MC. The MC is the change in dielectric

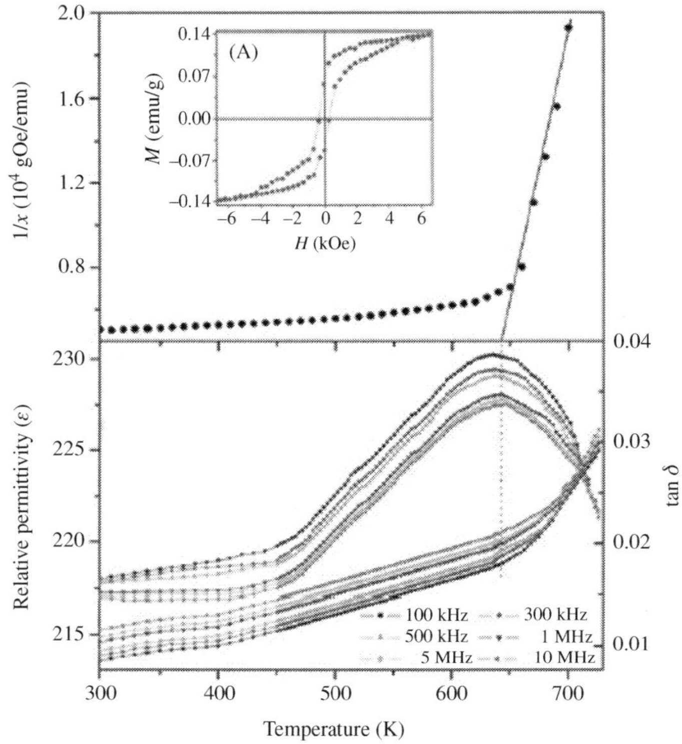

Figure 6.14
Temperature-dependent ε and tan δ without applied magnetic field, and magnetic susceptibility (χ) measured at 500 Oe. Inset shows $M-H$ curve measured at room temperature of BFO–BTO multiferroic. Source: *From K.C. Verma, R.K. Kotnala, Multiferroic magnetoelectric coupling and relaxor ferroelectric behavior in 0.7BiFeO$_3$–0.3BaTiO$_3$ nanocrystals, Solid State Commun. 151 (2011) 920–923.*

constant (ε), $[\varepsilon(H) - \varepsilon(0)]/\varepsilon(0) = \Delta\varepsilon/\varepsilon(0)$, where $\varepsilon(H)$ and $\varepsilon(0)$ denote the dielectric constants at applied dc bias magnetic field H and zero field [85]. The inset of Fig. 15 shows the variation of MC $[\Delta\varepsilon/\varepsilon(0)]$ in the frequency region 100 Hz to 10 MHz. The calculated values of MC are 2.42, 1.45, and 4.58% at 1 kHz, and at 1 MHz, 0.58, 0.87, and 0.91%, respectively, for C:NFO/BTO, S:NFO/BTO, and H:NFO/BTO. The observed MC behavior is contributed to the magnetostrictive effect that induces a change in size of the magnetic nanoparticles under an applied magnetic field, and leads to local stresses (or strains) and a consequent change in polarization of the ferroelectric phase due to the piezoelectric effect.

6.3.3 Multiferroic Composites

Recently, the ME coupling was proposed in composite multiferroics consisting of ferromagnetic grains embedded in a ferroelectric matrix [86]. In comparison with a single-phase

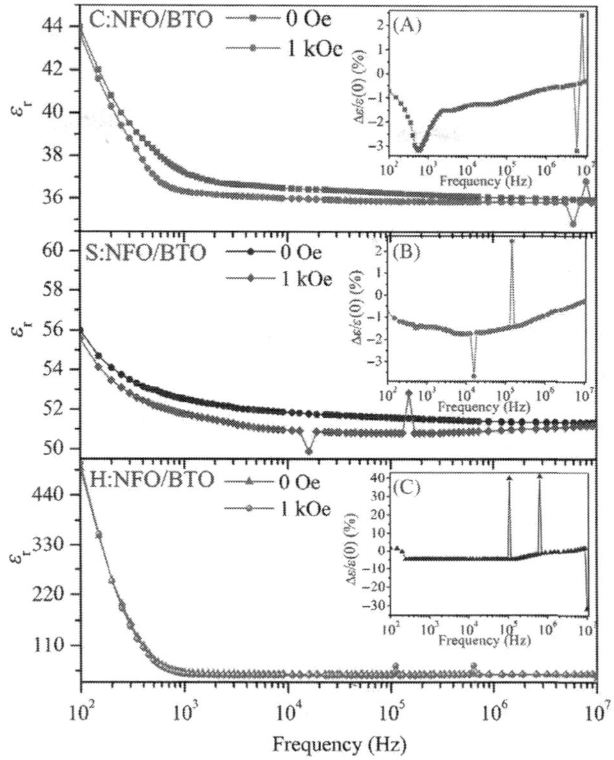

Figure 6.15
Frequency-dependent relative permittivity (ε_r) of NFO/BTO (at $H_{dc} = 0$, 1 kOe). Inset shows respective value of MD/MC $\{[\varepsilon(H) - \varepsilon(0)]/\varepsilon(0) = \Delta\varepsilon/\varepsilon(0)\}$ (in %). Source: *From K.C. Verma, R.K. Kotnala, Nanostructural and lattice contributions to multiferroism in $NiFe_2O_4/BaTiO_3$, Mater. Chem. Phys. 174 (2016) 120–128.*

multiferroic, the composite/heterostructures integrating individual magnetic and ferroelectric materials are more likely to be utilized in ME devices, because of their strong room-temperature ME effects and flexibility for materials choices and device designs [87]. The composite multiferroics are typically heterostructures consisting of a magnetostrictive and a piezoelectric phase. A typical configuration in the multiferroic heterostructures due to ferroelectric and ferromagnetic (magnetic) phases is shown in Fig. 6.16. For example, a 0–3 composite describes a particulate-based composite where 0D particles are dispersed within a 3D matrix. Particulate-based composites (0–3), pillars within a 3D matrix (1–3), and layered heterostructures (2–2) represent the most common geometries of ME heterostructures [88]. In these multiferroic composites, the ME coupling is mechanical in nature and is based on the fact that piezoelectric and magnetostrictive materials undergo a shape change in an applied electric or magnetic field, respectively. Therefore, the ME

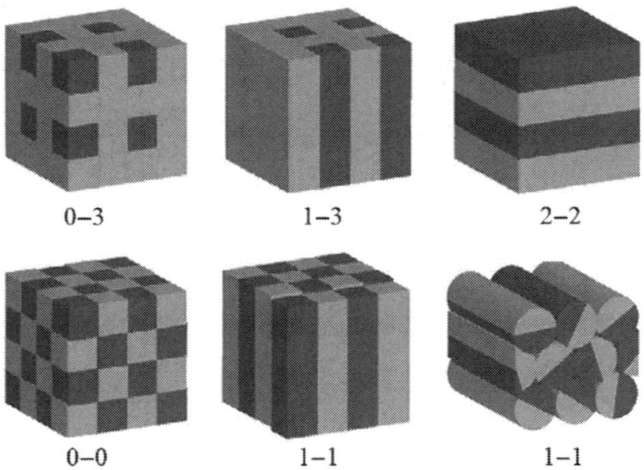

Figure 6.16
Multiferroic composite connectivities to the development of nanostructured construction. Source: *From J.S. Andrew, J.D. Starr, M.A.K. Budi, Prospects for nanostructured multiferroic composite materials, Scr. Mater. 74 (2014) 38–43.*

effect in these heterostructures is described by the direct ME effect, which is the product of the magnetostrictive (magnetic/mechanical) and the piezoelectric (mechanical/electrical) effects [89]. The less studied 0–0 type structures are mainly composed of ferroelectric–ferromagnetic epitaxial nanodot arrays which have advantages to allow the atomic level mechanical coupling and concurrently avoid the substrate clamping in multiferroic composite films, benefiting the ME coupling enhancement.

6.3.3.1 $MFe_2O_4/BaTiO_3$ (M = Mn, Co, Ni, Zn) nanocomposites

The multiferroic nanocomposite may be selected from perovskites ($BaTiO_3$, $PbTiO_3$, and $BiFeO_3$), i.e., large polarization and piezoelectric coefficients, and ferrite, MFe_2O_4 (M = TM) have high magnetostriction, high resistivity, T_N above room temperature and large magnetization [29]. In the tetragonal BTO (*P*4*mm* space group) at room temperature, Ti atoms are displaced along the *c*-axis. This means that the impurities and strain in tetragonal BTO can dramatically change the ferroelectric and piezoelectric behaviors. BTO has a piezoelectric coefficient, d_{33}, ~191 pC/N and the values of the direct ME coefficients, α_{ME}, for bulk composites are typically below 0.1 VA^{-1} (80 mV/cm/Oe). Thus, the spinel ferrites, MFe_2O_4, have been widely used as a magnetic component in multiferroic composites [90]. Among these spinel ferrites, $CoFe_2O_4$ has the largest saturation magnetostriction coefficient (1.1×10^{-4}) and thus acts as a suitable material for multiferroic heterostructures with BTO. However, $NiFe_2O_4$ has higher dc resistivity which supports ME coupling with BTO. For nanostructural $ZnFe_2O_4$, the redistribution of Fe^{3+} in tetrahedral and Zn^{2+} in octahedral interstices leads to the formation of two (A) and (B) magnetic sublattices in which the magnetization increases due to the stronger inter sublattice (A–B) superexchange interactions. It should be remembered that the exchange bias has been observed at the interface between BTO and the spinel phase, which imply a minimum interfacial diffusion/reaction between the two components [91]. It would lead to strain-mediated coupling between ferroelectric BTO and magnetic MFe_2O_4 to modify magnetic anisotropy of multiferroics.

6.3.3.2 Multiferroic $NiFe_2O_4/BaTiO_3$ nanostructures

Recently, the nanostructures of multiferroic $NiFe_2O_4/BaTiO_3$ (NFO/BTO) composite had been rarely studied [83]. The nanostructural shape/size plays an important role to modify multiferroicity. The nanostructured NFO involves mixed spinel structure $\{(Ni_{1-x}Fe_x)[Ni_xFe_{2-x}]O_4$, where x is the degree of inversion$\}$ together with the inverse one [92]. Since NFO has a cubic lattice ($a = 8.34$ Å), it has higher lattice mismatch with (001)BTO that corresponds to a growth template [a(NFO) = 2a(BTO)] [93]. Therefore, the nanocomposite of NFO with BTO may highly influence ME coupling of the NFO/BTO composite [84]. The NFO/BTO nanostructures have been synthesized by three different methods [94–96]. Fig. 6.17 shows the TEM images of NFO/BTO nanostructures [83]. The C:NFO/BTO is prepared by the chemical combustion of very small size nanoparticles of diameter, $D \sim 7 \pm 2$ nm. The mechanism for small size nanoparticles is described due to their low fabrication temperature (i.e., 500°C), nitrates as oxidizing agent, PEG, and urea as reducing agent during combustion [94]. The S:NFO/BTO shows nanorods type formation with $D \sim 14 \pm 0.5$ nm and length, $L \sim 81 \pm 0.5$ nm. The action of PVA during heating process is responsible for nanosize product and oriented growth. However, H:NFO/BTO sample prepared by the hydrothermal synthesis shows nanowires type product with $D \sim 3 \pm 0.5$ nm and $L > 40$ nm.

6.3.4 Multiferroic Thin Films

In comparison with bulk multiferroics, the thin films of BTO/spinel provide more degrees of freedom which tune the ferroelectric/piezoelectric and magnetostrictive phases at the

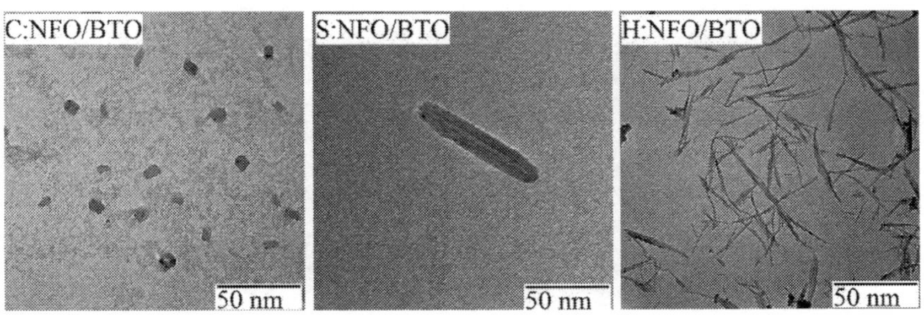

Figure 6.17
TEM images corresponding to NFO/BTO samples synthesized by different methods. Source: *From K.C. Verma, R.K. Kotnala, Nanostructural and lattice contributions to multiferroism in $NiFe_2O_4/BaTiO_3$, Mater. Chem. Phys. 174 (2016) 120–128.*

nanoscale, representing the novel ME coupling mechanisms [97]. Furthermore, the bulk ME composite is a two-phase multiferroic material (magnetic and ferroelectric) in which the constituent phases are usually formed by cosintering or adhesive bonding that inevitably results into losses at the interfaces. In composite ME thin films, the different phases are combined at the atomic level and thus reduce the interface losses significantly [98]. Multilayered epitaxial thin films (2−2 configuration) were proposed first as a structure with strong ME coupling due to the high quality of crystallography and intimate coherent interface. But the clamping effect of the substrate onto the ferroelectric phase reduces the order of magnitude of the ME coupling coefficient [99]. With the breakthrough in self-assembly vertical nanostructures (1−3 configuration) from immiscible perovskite and spinel systems, such as $CoFe_2O_4$ nanopillars embedded in a BTO or BFO matrix, significant enhancement of ME coupling and electric-field-induced magnetization switching were found. This is due to the large heteroepitaxial interface that reduced clamping effect [100]. \displaystyle

6.3.4.1 Nanostructural $MFe_2O_4/BaTiO_3$ (M = Mn, Co, Ni, Zn) thin films

Fig. 6.18 shows the atomic force microscopy (AFM) images of $MFe_2O_4/BaTiO_3$ thin films [29]. A dense, crack-free surface morphology without any pinholes is observed. The

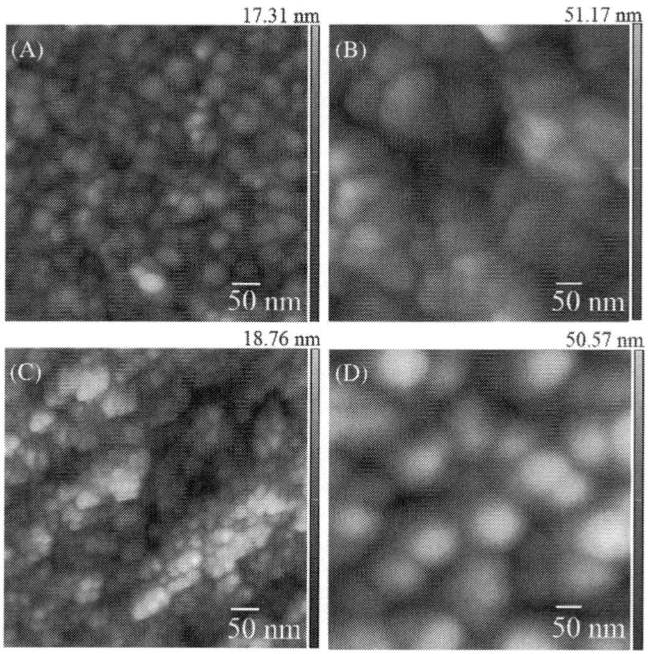

Figure 6.18
AFM images of the $MFe_2O_4/BaTiO_3$ nanocomposite: (A) MnFO/BTO, (B) CFO/BTO, (C) NFO/BTO, and (D) ZFO/BTO. Source: *From K.C. Verma, D. Singh, S. Kumar, R.K. Kotnala, Multiferroic effects in $MFe_2O_4/BaTiO_3$ (M = Mn, Co, Ni, Zn) nanocomposites, J. Alloys Compd. 709 (2017) 344−355.*

respective z-axis color scale bars of 17.31, 51.17, 18.76, and 50.57 nm indicates relative vertical or the depth profile. The observed value of average grain size is 25, 102, 24, and 133 nm, and the root mean square (rms) surface roughness is 1.1, 1.5, 1.3, and 1.5 nm, respectively, for MnFO/BTO, CFO/BTO, NFO/BTO, and ZFO/BTO films.

6.3.4.2 ME coupling due to magnetic control of ferroelectric polarization

Fig. 6.19 shows a typical ferroelectric polarization loop of the $0.67BaTiO_3-0.33BiFeO_3$ (BTO67/BFO33) film at $T = 100$ K [101]. Only minor changes of the hysteresis loop could be detected when sweeping the magnetic field from 0 to +8 T, and back to 0 and −8 T, and to 0 T. Since the electric field control of magnetism or magnetically controlled ferroelectric order in the solid state is significant both fundamentally and practically, it remains a challenge. So far the only known intrinsic room-temperature ME material is BFO. However, the origin of ferroelectricity in single-phase multiferroics is largely unrelated to the magnetic order, and the ME coupling so far observed at room temperature in these materials is still too weak to be useful for device design [102]. The results in Fig. 6.19 indicate that ME coupling results from the interaction between different orderings of the two phases in the composite. This is an interface coupled electrical and magnetic ordering via elastic interaction between the piezoelectric and piezomagnetic phases.

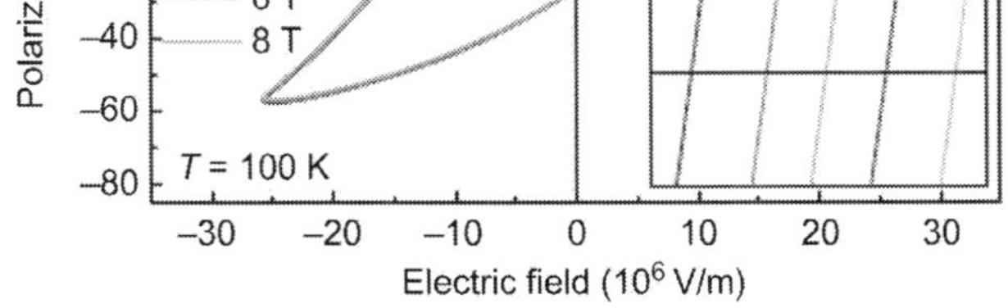

Figure 6.19

Ferroelectric polarization hysteresis at $T = 100$ K of the BTO67/BFO33 film under the influence of magnetic field up to 8 T. Source: *From M. Lorenz, V. Lazenka, P. Schwinkendorf, F. Bern, M. Ziese, H. Modarresi et al., Multiferroic $BaTiO_3-BiFeO_3$ composite thin films and multilayers: strain engineering and magnetoelectric coupling, J. Phys. D Appl. Phys. 47 (2014) 135303–135310.*

6.3.4.3 Dynamic ME coupling measurement for $MFe_2O_4/BaTiO_3$ thin films

The ME coupling coefficient, α_{ME}, of the MFO/BTO (M = Mn, Co, Ni, Zn) nanocomposites is considered as the local sublattice interactions between spin moments and the disordered electric dipole moment, which was measured using Ref. [29]: $\alpha_{ME} = \frac{\partial E}{\partial H} = \frac{1}{t}\frac{\partial V}{\partial H} = \frac{1}{t}\frac{V_{out}}{H_{ac}}$, where E is the induced electric field, H is the applied magnetic field, V_{out} is the ME voltage developed across the film, H_{ac} is the magnitude of the ac magnetic field, and t is the thickness of the film. The value of α_{ME} is measured using dynamic method involving simultaneous application of an ac as well as a dc magnetic field of MFO/BTO thin films as shown in Fig. 6.20 [29]. The thin films were biased with ac magnetic field (H_{ac}) of 0.008 × 10^5 A/m at 1093 Hz, and dc magnetic field (H_{dc}) was applied in the longitudinal mode. As shown in Fig. 6.20, the maximum value of α_{ME} is 0.027, 0.12, 0.11, and 0.081 V/A, respectively, measured for MnFO/BTO, CFO/BTO,

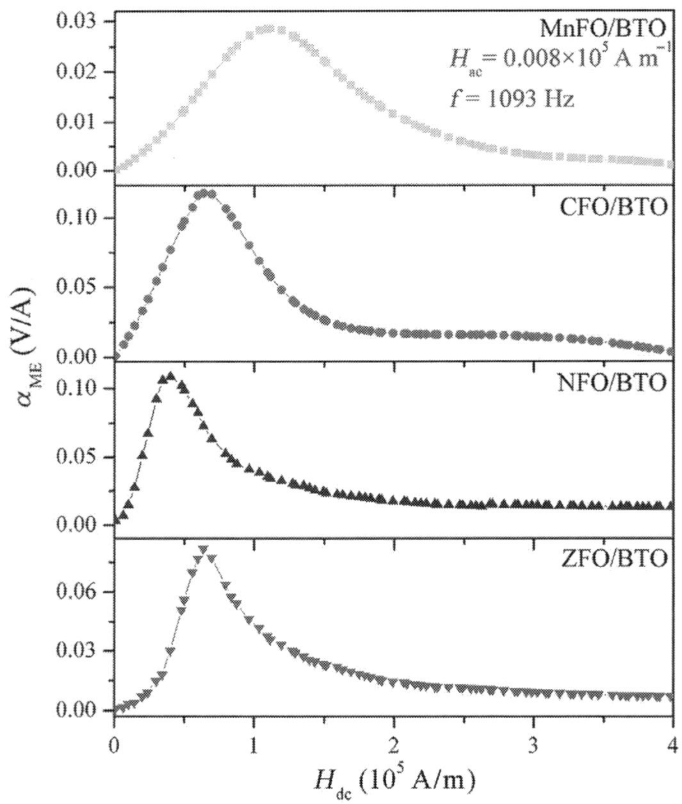

Figure 6.20

Longitudinal ME voltage coefficient (α_{ME}) (H_{dc} = 0–4 × 10^5 A/m) of the multiferroic MFe_2O_4/$BaTiO_3$ thin films. Source: *From K.C. Verma, D. Singh, S. Kumar, R.K. Kotnala, Multiferroic effects in $MFe_2O_4/BaTiO_3$ (M = Mn, Co, Ni, Zn) nanocomposites, J. Alloys Compd. 709 (2017) 344–355.*

NFO/BTO, and ZFO/BTO thin films. The ME voltage is abruptly increased at the initial stage due to an elastic enhancement. The observed ME coefficient, α_{ME} (maximum value $\sim$ 0.12 V/A (95.54 mV/cm/Oe)) has indicated an improvement over single-phase BTO [70,103]. This is because the ME interactions result from magnetostriction (proportional to dM^2/dH^2) of the magnetic sublattice, and the α_{ME} is linearly proportional to the piezomagnetic coefficient, which is highly dependent upon the H_{dc}. The piezomagnetic coefficient is close to zero at $H_{dc} = 0$ for most of the multiferroic materials that lead to a zero or very low value of α_{ME} of the ME sensors without applying H_{dc}. Therefore, the dc bias magnetic field, H_{dc}, is required to achieve maximum piezomagnetic coefficient and sensitivity for ME sensors [104].

6.4 Spintronics Based on Diluted Magnetic Semiconductor, DMS ZnO

In few decades, Ohno suggested that magnetic semiconductors, i.e., DMS, contain a small amount of magnetic impurities which preserve their magnetic characters at room temperature, i.e., T_C above 300 K, in order to be useful for technological applications [20,105]. The DMS ZnO has large direct band gap (3.37 eV) and high bulk exciton binding energy (60 meV). Kittilstved et al. [106] predicted that the ferromagnetism of ZnO DMS has strong electronic coupling between the magnetic ions and charge carriers at the Fermi level. Dietl et al. [20] reported RTFM of DMS compounds, such as 3d ions substituted ZnO. However, the nanoparticles of pure ZnO have also shown RTFM [107,108]. Gao et al. [108] suggested that oxygen vacancies, V_O, located at the surface of ZnO nanoparticles are responsible for RTFM. The oxygen atoms at the surface escape from the bondage of the chemical bond because of heating, then the unpaired electrons show the abnormal spin phenomenon to cause magnetism. Garcia et al. [109] found that the ZnO nanoparticles had absorbed certain organic molecules to modify the electronic structure for RTFM even in the absence of magnetic ions. Xu et al. [110] suggested singly charged V_O that depends upon size of nanoparticles and heating condition and located mainly near the ZnO nanoparticles surface to modulate ferromagnetism.

6.4.1 TM Ions Impurity in DMS ZnO

During the last decade, the magnetism due to TM-doped ZnO is extensively studied. Sato et al. [111] were used Korringa–Kohn–Rostoker method based on the local density approximation (LDA) to discuss ferromagnetic ordering that is more stable in Fe-doped ZnO phase if no additional carrier dopants are presented. However, Gopal et al. [112] have studied the nature of magnetic interactions in TM-doped ZnO using the LSDA + U (spin-polarized LDA with Hubbard-U parameter) method which indicate that antiferromagnetic ordering is favorable. In agreement, Karmakar et al. [113] found antiferromagnetism in Fe-doped ZnO without any native defects. However, Spaldin [114] had found that the robust ferromagnetic ordering is not possible when Zn sites are substituted with Co or Mn, unless additional hole carriers are incorporated. These additional hole carriers might be ferromagnetic clusters or precipitated secondary phases. To the survey of several theoretical works, it was predicted that a slight ($\sim$5%) TM doping into ZnO would lead to ferromagnetic coupling even at room temperature. Venkatesan et al. [115] used spin-split donor impurity-band model and observed RTFM in ZnO with 5% of Sc, Ti, V, Fe, Co, or Ni, but not Cr, Mn, or Cu ions. In this case, the magnetic moments are associated with two-electron defects as well as unpaired electrons of the 3d ions. For Cr, Mn, Cu, or Zn, no moment appreciably greater than the experimental uncertainty ($<0.1\ \mu_B$) was observed at room temperature. It is recently reported the effect of TM = Co, Fe (=0,1,3,5%) into ZnO nanorods-induced RTFM [116] that was described by Coey et al. [117].

6.4.2 RE Ions Impurity in DMS ZnO

The Rare Earth, RE-ion-doped ZnO has ferromagnetism that is induced by p–f hybridization due to defect carriers [118]. Compared with 3d TM ions, 4f RE ions have larger magnetic moments. Furthermore, the electrons may mediate the ferromagnetic coupling between the RE ions due to the coupling between f electrons and host s electrons [119]. Aravindh et al. [120] suggested that the additional charge or V_O increases the ferromagnetic coupling energy by enhancing the s–f coupling significantly, as Fermi level is shifted further into the conduction band, close to the RE f states. It was suggested that the electrons and holes with different concentrations mediate the ferromagnetism in the RE-doped ZnO. The exchange interaction by simultaneous doping from TM and RE ions is 4f–5d–3d which is antiferromagnetic when the 5d band is less than half full and the 3d band is more than half full. The RE such as Ce ions in Ni- and Cu-doped ZnO temporarily generates charge carriers which lead to lattice defects and the formation of new energy levels in the band gap [119].

6.4.3 Defects-Assisted Ferromagnetism Due to TM and RE Ions in ZnO

Some of the models proposed to explain the mechanism of ferromagnetism in DMS is Ruderman–Kittel–Kasuya–Yosida model (RKKY), double exchange, superexchange, and bound magnetic polaron (BMP). The RKKY model is based on the second-order perturbation theory. This theory describes the indirect exchange interaction between the nuclear spin and the conduction electrons leading to antiferromagnetism. The Kramers–Anderson superexchange model explains the coupling between two nearest-neighbor cations. In this model, the net antiferromagnetic coupling can only be observed because of the nearest nonmagnetic ion interaction. The BMP model deals with the ferromagnetism resulted by the formation of polarons of the shallow donors (V_O) with their surrounding magnetic ions. The overlap of these polarons yields the long-range ferromagnetism in DMSs [121]. The BMP model considered for deep acceptors (V_{Zn}) which have degenerate energy levels in the valence band has a troublesome mechanism. This is since the BMP model ascribed that the bound electrons (holes) in the defect states can couple with TM ions and cause the ferromagnetic regions to overlap, giving rise to long-range ferromagnetic ordering. When donors or acceptors are present, the sp–d interaction often leads to the formation of BMPs.

6.4.3.1 BMP in Co-substituted ZnO

Song et al. [122] give a possible explanation for the intrinsic magnetism based on the BMP model for Co-doped ZnO. For the lattice structure of Co:ZnO, Co^{2+} ions incorporate into the wurtzite lattice at Zn^{2+} ion sites. Based on the change in the lattice constant and nanostructural dimensions, the lattice defects are formed, i.e., vacancies related to O and Zn. These defects may exist due to lattice deformations. An electron associated with a particular defect is confined in a hydrogenic orbital of radius $r_H = \varepsilon_r(\frac{m}{m^*})a_o$ and $\gamma = \varepsilon_r\left(\frac{m}{m^*}\right)$, where ε_r is the dielectric constant, m is the electron mass, m^* is the effective mass of the donor electrons, and a_0 is the Bohr radius (0.53 Å). For (4 at.%) Co in ZnO, $m/m^* = 0.28$, $\varepsilon_r = 25$ [122]. Hence high ε_r results in large r_H and γ, i.e., $r_H = 3.7$ Å, $\gamma = 7$. Taking into account a sufficiently large orbital radius, say $\gamma = 7$, overlap between a hydrogenic electron and the cations within its orbit leads to ferromagnetic superexchange between them (Fig. 6.21).

\BigFig. 6.21 schematically shows the BMP percolation with respect to the presence of defects in the form of Zn and O vacancies and distribution of Co ions in the ZnO lattice. It explains that defect-mediated BMP formation within its hydrogenic radius helps to promote long-range ferromagnetic ordering with the increase of BMP concentration. There exist two cases: (1) percolation of two nearby BMPs and (2) an isolated BMP. Whereas isolated Co ions can have rare ferromagnetic interaction without the presence of defects, since they would either align antiferromagnetically in case of short-range interaction or can hardly

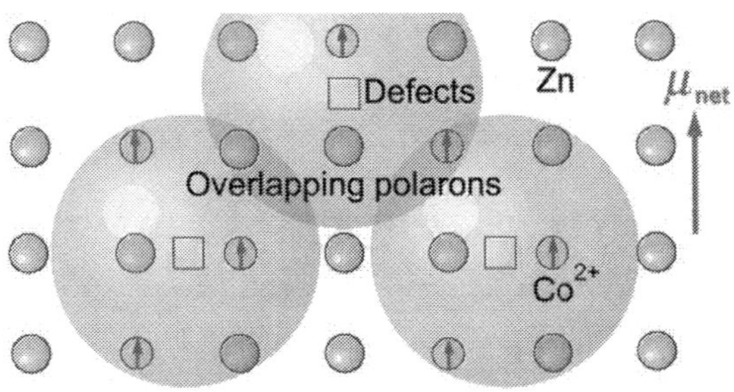

Figure 6.21
Schematic representation of the BMP. Three-polaron subsystem is representing supercoupling between polarons.

interact through the itinerant carriers due to their dilute concentration as compared with their metallic ferromagnetic phases. The doped cations provide an extra random potential, which extends the localized region as its concentration increases. Overlap between a hydrogenic electron and the cations within its orbit leads to ferromagnetic exchange coupling between them, provided the hydrogenic orbital radius is sufficiently large.

6.4.4 First-Principle Calculations for RE and TM Ions in the Wurtzite ZnO Structure

In the survey of TM-substituted DMS ZnO, it has been found that TM ions occupy the Zn atomic sites in the wurtzite structure. Chang et al. [123] revealed that the Co-dopant most likely substitutes for Zn atoms and that free charge carriers are absent for a wide range of Co concentrations. Due to this, the interstitial Co impurities and charged V_O are energetically less favorable than substitutional Co atoms. The DFT predicts occupancy of Fe ions into Zn which has antiferromagnetic ordering in ZnO [124]. Aravindh et al. [120] found that an isolated Gd atom incorporated in a Zn site exhibits a magnetic moment of $\sim 6.87\ \mu_B$, for all Gd positions. Bantounas et al. [125] also showed the occupancy of RE ions in the wurtzite ZnO structure using DFT calculation within generalized gradient approximation (GGA). They suggest that V_{Zn-Gd} complexes are highly favorable. In light of the above discussion, it is understood that both TM and RE ions replace Zn atoms in the wurtzite structure.

First-principle calculations using DFT have been carried out using ab initio calculations that were performed for Ce-, Fe-, and Co-substituted ZnO nanoparticles shown in Fig. 6.22 [126]. The DFT has been performed on $Zn_{0.94}Fe_{0.03}Ce_{0.03}O$ (ZFCeO) and $Zn_{0.94}Co_{0.03}Ce_{0.03}O$ (ZCCeO) samples within a ZnO supercell consisting of 72 atoms. The GGA has been corrected by Hubbard-U terms. Herein, we use typical values of $U = 4.5$ eV on Zn, Fe, Co, and 5.0 eV for Ce. The Ce atom is placed on Zn sites that are the nearest neighbor (Fig. 6.22A and C) and non-nearest neighbor (Fig. 6.22B and D) to the TM atoms in the ZnO supercell. For ZFCeO, the binding energy of the relaxor wurtzite structure corresponds to the nearest neighboring Ce-substituted Zn to Fe (Fig. 6.22A with $l_{Ce-Fe} = 2.80$ Å), $E_a = -2932.691869\ H_a$ and non-nearest neighboring Ce(Zn) to Fe (Fig. 6.22B with $l_{Ce-Fe} = 6.72$ Å), $E_b = -2932.676185\ H_a$. The difference $\Delta E = E_a - E_b = -426.783$ meV. However, for ZCCeO, the binding energy corresponds to the nearest neighboring Ce(Zn) to Co (Fig. 6.22C with $l_{Ce-Co} = 2.66$ Å), $E_c = -2971.987503\ H_a$ and non-nearest neighboring Ce(Zn) to Co (Fig. 6.22D with $l_{Ce-Co} = 6.74$ Å), $E_d = -2971.963521\ H_a$. The difference, $\Delta E = E_c - E_d = -652.583$ meV. The enhancement in the difference ΔE indicates that the Ce atom is more stable when replaced by Zn that is the nearest neighbor of TM: Fe or Co. Moreover, the calculated value of ΔE for ZCCeO is higher than ZFCeO. This indicates that the Ce(Zn) atom is more stable with Co than Fe.

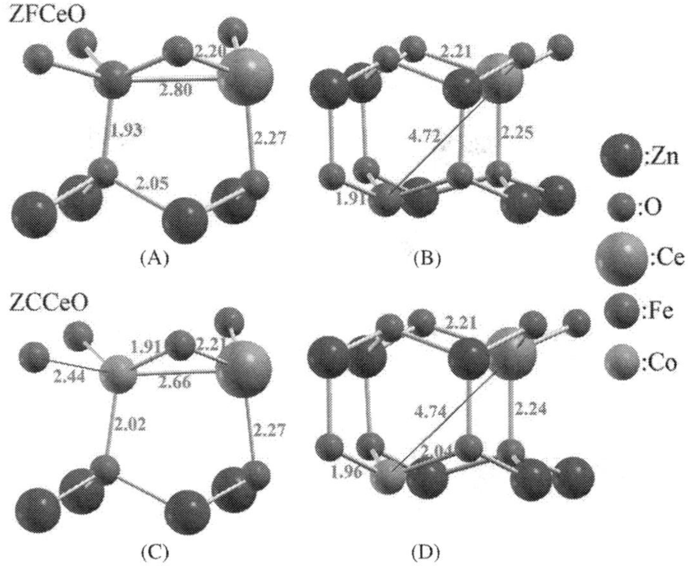

Figure 6.22
Atomic representation of the ZnO supercell used for Ce doping into Fe:ZnO and Co:ZnO carried out by first principles calculation based on DFT. The Ce atom is placed on the Zn site that is the nearest neighbor to a TM (A) and (C), and the non-nearest neighbor to a TM atom (B) and (D), respectively. The bond distance between atoms is measured in Å. Energy difference between configuration (A) and (B), ΔE, EA − EB = −426.783 MeV, and (C) and (D), E, EC − ED = −652.583 MeV. Source: *From K.C. Verma, R.K. Kotnala, Defects-assisted ferromagnetism due to bound magnetic polarons in Ce into Fe, Co:ZnO nanoparticles and first-principle calculations, Phys. Chem. Chem. Phys. 18 (2016) 5647–5657.*

6.4.5 Influence of Dopant Concentration (TM and RE ions) on Ferromagnetism of ZnO

The ferromagnetism appears in degenerate or nondegenerate semiconductors, and even in insulators, at concentrations of 3d dopants that lie far below the percolation threshold x_p associated with nearest-neighbor cation coupling. The average moment per dopant cation m_c approaches (or even exceeds) the spin-only moment at low dopant concentrations x, and it falls progressively as x increases toward x_p. The results of the concentration-dependent ferromagnetic moments in ZnO systems have been given by Coey et al. [117]. Fig. 6.23A shows the change in magnetic moment with TM ions impurity in ZnO. It indicates that the lower doping concentration of TM ions has higher magnetic moment.

6.4.6 Realizing Wurtzite Structure of ZnO With Dopant Ions

\BigIn wurtzite ZnO, four O ions form a tetrahedron around every Zn ion and vice versa. In these tetrahedra, when a Zn vacancy is formed, four Zn–O bonds are broken. The dangling

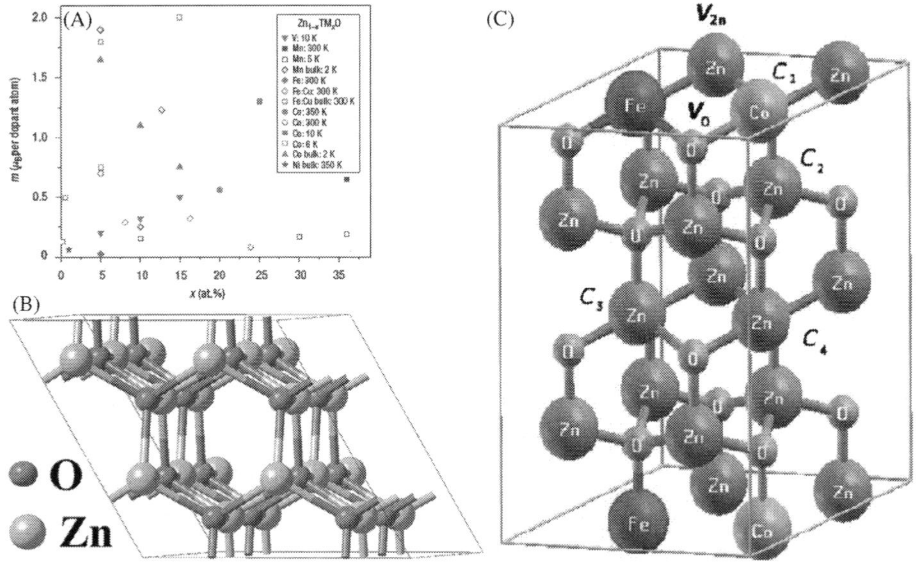

Figure 6.23
(A) Magnetic moment per TM cation in ZnO. (B) Hexagonal wurtzite ZnO structure. (C) Part of the hexagonal ZnO wurtzite structure with Fe/Co ions substitution.
Source: *From J.M.D. Coey, M. Venkatesan, C.B. Fitzgerald, Donor impurity band exchange in dilute ferromagnetic oxides, Nat. Mater. 4 (2005) 173–179; P. Nath, A. Chakraborti, D. Sanyal, Ab initio calculation of magnetic properties of p block element doped ZnO, RSC Adv. 4 (2014) 45598–45602; D. Karmakar, T.V. Chandrasekhar Rao, J.V. Yakhmi, A. Yaresko, V.N. Antonov, R.M. Kadam et al. Electronic structure and magnetic properties of (Fe, Co)-co doped ZnO: theory and experiment, Phys. Rev. B 81 (2010) 184421–184513.*

O 2p bonds that remain are partially filled by six electrons and can accommodate two more. This given Zn vacancies, V_{Zn}, acts as a double acceptor. However, it can also trap holes in polaronic states [127]. This V_{Zn} in ZnO is a deep polaronic acceptor that can bind a localized hole on each of its four nearest-neighbor O ions. Fig. 6.23B shows the hexagonal wurtzite structure of ZnO with no excess or deficiency of electrons is present. It means the pristine ZnO does not show any magnetic ordering in its ground state [128]. Incorporation of impurities leads to magnetism in ZnO. It is interesting to note that all types of impurities do not give rise to magnetism in ZnO systems instead of the absence of oxygen vacancies, V_O.

For example, the Fe/Co-doped ZnO has substituted two Zn atoms by one Fe and one Co atom, under three different situations, i.e., (1) without any defect/vacancies, (2) with V_O, and (3) with V_{Zn} [129]. The percentage of V_O and V_{Zn} is equal to the TM-dopant concentrations in the supercell (Fig. 6.23C). The V_O are located at the first-nearest-neighbor in equivalent position from the dopant TM atoms while the V_{Zn} positions are chosen to be at the same plane as the TM dopants. It is found that the ferrimagnetic state is energetically most stable for the nearest-neighbor position of the co-dopants and the presence of unusual charge state of the co-dopants. However, the exact exchange behavior depends on the distance between the two TM ions and the position of the defects with respect to them. It indicates a competitive mechanism of the parallel-spin double exchange versus antiparallel-spin superexchange within the system. It revealed that the Fe and Co co-dopant in ZnO has defect states that mediate the long-range magnetic order of the ferrimagnetic and ferromagnetic states.

6.4.6.1 XRD studies of ZnO nanoparticles with La and Fe doping

The structural information in DMS system of pure ZnO, $Zn_{0.95}Fe_{0.05}O$ (ZFO5), and $Zn_{0.92}Fe_{0.05}La_{0.03}O$ (ZFLaO53) is obtained by the XRD patterns using the Rietveld method. For this, refinements of the standard wurtzite hexagonal structure (space group $P6_3mc$) were carried out with each atom residing on the 2b Wyckoff position: Zn atoms at (1/3 2/3 0) and the O atoms at (1/3 2/3 z) coordinates. In the refinement process the Fe and La occupancies were varied for the two Zn and O sites to locate their exact site. Typical refined patterns of pure, Fe- and La-doped ZnO nanostructure is shown in Fig. 6.24 [107]. All the diffraction peaks indexed to the hexagonal wurtzite ZnO. The related Rietveld fitting parameters (such as χ^2, R_p, R_{wp}, R_{Bragg}, and R_F) are considerable low, suggesting a good wurtzite structure (Table 6.2). The refined lattice parameters, a (Å) = 3.255(2), 3.253 (1), and 3.249(1), c (Å) = 5.203(1), 5.205(3), and 5.207(2), V (Å^3) = 47.739(1), 47.699 (3), and 47.600(3), and lattice distortion, c/a = 1.598, 1.600, and 1.603, respectively, for ZnO, ZFO5, and ZFLaO53. The enhancement in lattice distortion with doping indicated defects/ vacancies formation in the wurtzite lattice [130].

6.4.6.2 Calculation for lattice constants and bond length of La-, Gd-, Co-doped ZnO

This calculation has been performed from XRD pattern reported for DMS systems of $Zn_{0.95}Co_{0.05}O$ (ZCO5), $Zn_{0.92}Co_{0.05}La_{0.03}O$ (ZCLO53), and $Zn_{0.92}Co_{0.05}Gd_{0.03}O$ (ZCGO53) nanoparticles [130]. The lattice parameters for hexagonal wurtzite ZnO are calculated:

$$\frac{1}{d^2} = \frac{4}{3}\frac{(h^2 + hk + k^2)}{a^2} + \frac{l^2}{c^2} \tag{6.1}$$

where a, c, h, k, l, and d have their usual meaning. The lattice constants a and c using Eq. (6.1) are calculated from most intense crystalline peaks of (100) and (002) planes. The value of bond length is calculated [131]:

$$l_{Zn-O} = \sqrt{\frac{a^2}{3} + \left(\frac{1}{2} - u\right)^2 * c^2} \tag{6.2}$$

where

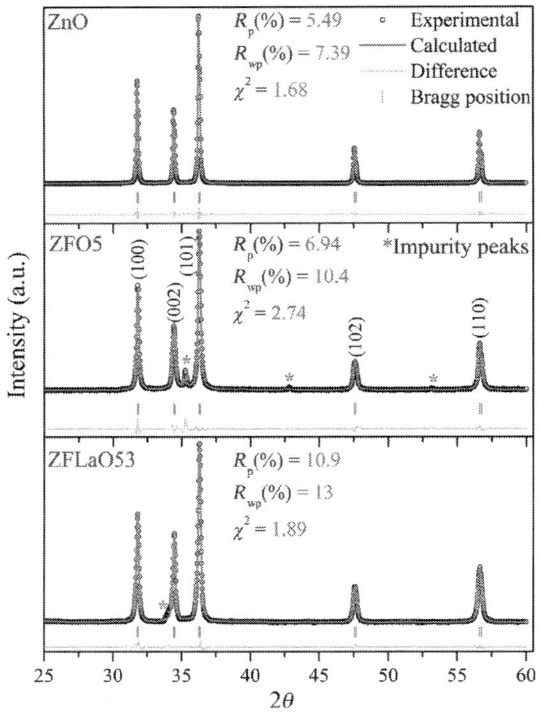

Figure 6.24

XRD pattern of pure ZnO, ZFO5, and ZFLaO53 nanostructures. Source: *From K.C. Verma, R.K. Kotnala, Oxygen vacancy induced by La and Fe into ZnO nanoparticles to modify ferromagnetic ordering, J. Solid State Chem. 237 (2016) 211–218.*

Table 6.2: Rietveld refinements of ZnO, ZFO5, and ZFLaO53 nanoparticles. Space group = $P6_3mc$.

Parameters	ZnO	ZFO5	ZFLaO53
a (Å)	3.255(2)	3.253(1)	3.249(1)
c (Å)	5.203(1)	5.205(3)	5.207(2)
V (Å^3)	47.739(1)	47.699(3)	47.600(3)
R_p (%)	5.49	6.94	10.9
R_{wp} (%)	7.39	10.4	13
χ^2	1.68	2.74	1.89
R_{Bragg} (%)	0.623	1.46	1.14
R_F (%)	0.349	0.899	0.783

Atomic positions: Zn = (1/3 2/3 0), Fe/La = (1/3 2/3 0), O = (1/3 2/3 z), where z = 0.38324 for ZnO, 0.38305 for ZFO5, and 0.39978 for ZFLaO53.

Source: *From K.C. Verma, R.K. Kotnala, Oxygen vacancy induced by La and Fe into ZnO nanoparticles to modify ferromagnetic ordering, J. Solid State Chem. 237 (2016) 211–218.*

$$u = \frac{a^2}{3c^2} + 0.25 \tag{6.3}$$

The value of lattice constant, a (Å) = 3.251, 3.253, and 3.254, c (Å) = 5.204, 5.217, and 5.211, c/a = 1.6007, 1.6037, and 1.6014, l_{Zn-O} = 1.9779, 1.9804, and 1.9801, and unit cell volume, V (Å^3) = 47.631, 47.809, and 47.783, respectively, for ZCO5, ZCLO53, and ZCGO53. The volume per unit cell for the hexagonal system, $V = 0.866 \times a^2 \times c$.

6.4.7 Nanostructural Formation in Pure and Doped DMS ZnO

The main role of nanostructure is to create oxygen vacancies that must help in electric transition of long-range magnetic interactions among dopant and host DMS ZnO. The ferromagnetic behavior due to nanostructural growth has been observed in Fe-, Co-, and Ce-doped SnO_2 that also shown superparamagnetism and spin glasses in addition to antiferromagnetic ordering [131]. The long-range ferromagnetic exchange is mediated by percolation of BMPs that formed by point defects such as V_O. The defect formation must be related with both structural deformations and nanostructural size. Zheng et al. [132] suggested a superparamagnetic state with small nanoparticles that exhibit thermally induced fluctuations in the direction of magnetization. The thermal fluctuation of the random spin orientation is frozen, which induces magnetization of the spin-glass surface layer, producing a long-range magnetic order. The spin-glass behavior on the surface arises from the larger surface-to-volume ratio in nanostructures, producing an anisotropic field that frustrates the inner spins. This frustration would result in quantum tunneling at low temperatures.

6.4.7.1 Nanostructural growth of ZnO with Fe, Co, Ce substitution

Fig. 6.25A and B shows the TEM images of ZFCeO and ZCCeO nanoparticles [126]. The pure samples of $Zn_{0.97}Fe_{0.03}O$ (ZFO) (D = 14 nm and L = 50 nm) and $Zn_{0.97}Co_{0.03}O$ (ZCO) (D = 63 nm and L = 131 nm) are nanorods [116]. This means the dopant Ce ions result to influence the shape of ZnO nanorods into nanoparticles. The average size, D, of nanoparticles is 97 $\pm$ 4 nm and 106 $\pm$ 3 nm measured for ZFCeO and ZCCeO, respectively. The variation in the value of D is explained on the basis of ionic radii differences between Zn and the dopant ions. The insets in the TEM images (Fig. 6.25A and B) are high-resolution TEM images to show the crystalline formation and lattice spacing after doping into the ZnO lattice. It is noted that the distorted lattice has an enhanced interplanar spacing d (corresponding to (101) planes) of ~0.247 and 0.250 nm measured for ZFCeO and ZCCeO, respectively. For pure ZnO, the value of d is ~0.237 nm. The nanoparticles of Ce-doped ZFO and ZCO are formed due to simulations of ZnO aggregation $[Zn_x(OH)_y]^{2x-y}$ [133]. Since for pure ZnO nanorods, the chemical behavior of Zn^{2+} ions involves [116]:

$$Zn(CH_3COO)_2 \leftrightarrow Zn^{2+} + 2CH_3COO^- \tag{6.4}$$

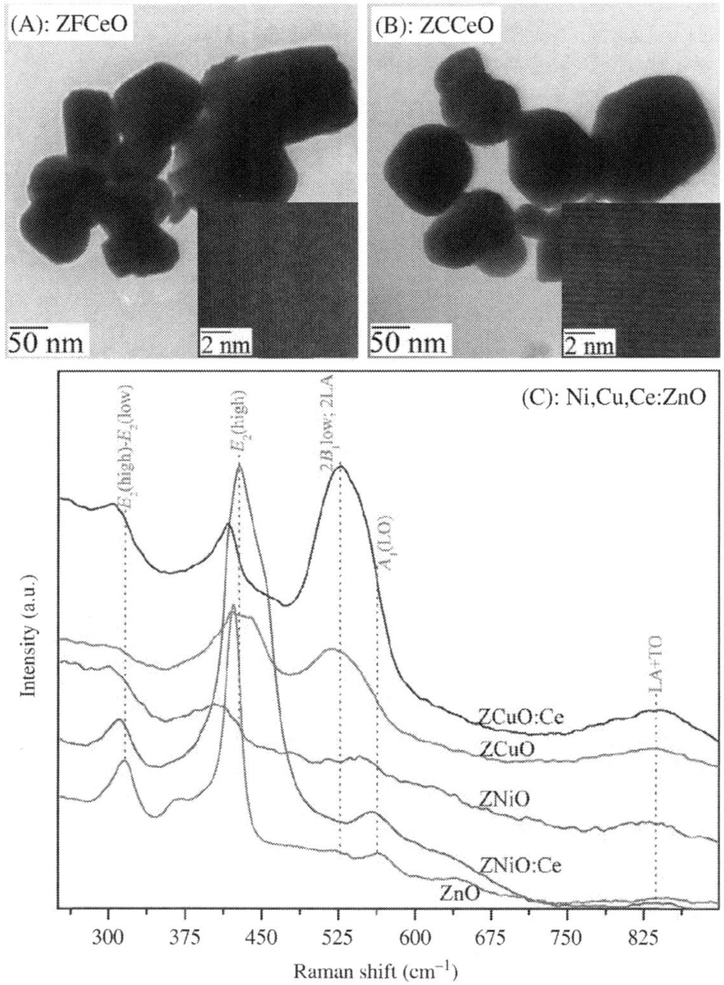

Figure 6.25

TEM images of (A) ZFCeO, (B) ZCCeO nanoparticles. Respective high-resolution TEM image (inset).(C) Raman spectra for pure, Ni-, Cu-, Ce-substituted ZnO nanoparticles at room temperature. Source: *From K.C. Verma, R.K. Kotnala, Understanding lattice defects to influence ferromagnetic order of ZnO nanoparticles by Ni, Cu, Ce ions, J. Solid State Chem. 246 (2017) 150–159; K.C. Verma, R.K. Kotnala, Defects-assisted ferromagnetism due to bound magnetic polarons in Ce into Fe, Co: ZnO nanoparticles and first-principle calculations, Phys. Chem. Chem. Phys. 18 (2016) 5647–5657.*

$$2H_2O + 2e^- \rightarrow H_2 + 2OH^- \tag{6.5}$$

$$Zn^{2+} + 2OH^- \rightarrow Zn(OH)_2 \tag{6.6}$$

$$Zn(OH)_2 + 2OH^- \leftrightarrow [Zn(OH)_4]^{2-} \rightarrow ZnO + H_2O + 2OH^- \tag{6.7}$$

The aggregation surface mainly consists of Zn^{2+} and OH^- ions, whereas the core comprises Zn^{2+} and O^{2-} ions, only. The dissociation of water and Zn^{2+}/OH^- ions from the aggregates stabilize into (1) ripening reactions leading to the formation of O^{2-} ions and (2) nucleation and growth of a ZnO domain in the wurtzite structure. It is well known that ZnO is a polar crystal, having positively Zn^{2+}-terminated (0001) and negatively O^{2-}-terminated (000−1) polar surfaces, which induced a net dipole moment along the *c*-axis [134]. The surface energies of the polar {0001} plane are higher than the nonpolar {01−10} and {2−1−10}. Therefore, preferential growth along the *c*-axis is energetically favorable. However, doping with Co and Fe involves the reactions[126]:

$$Fe^{3+} + e^- \leftrightarrow Fe^{2+} \tag{6.8}$$

$$Fe^{2+} + 2H_2O \leftrightarrow Fe(OH)_2 + 2H^+ \tag{6.9}$$

$$Fe^{3+} + 3H_2O \leftrightarrow Fe(OH)_3 + 3H^+ \tag{6.10}$$

$$Fe(OH)_2 + OH^- \leftrightarrow FeOOH + H_2O + e^- \tag{6.11}$$

$$2Fe(OH)_3 + Fe(OH)_2 \rightarrow Fe_3O_4 + 4H_2O \tag{6.12}$$

$$Co^{2+} + 2OH^- \rightarrow Co(OH)_2 \tag{6.13}$$

$$Co(OH)_2 + OH^- \leftrightarrow CoOOH + H_2O + e^- \tag{6.14}$$

$$CoOOH + OH^- \leftrightarrow CoO_2 + H_2O + e^- \tag{6.15}$$

The reaction mechanisms (6.8)−(6.12) are given by Zhang et al. [135]. The $Fe(OH)_2$ and $Fe(OH)_3$ products easily form nanorods. It is also suggested the aggregation among Co^{2+} and OH^- ions that involve Eqs. (6.13)−(6.15) [126]. Due to this, the dimensional ratio of the ZFO nanorods is enhanced over ZCO [116]. The incorporation of Ce ions involves large ionic aggregation type products with ZFO and ZCO due to comparatively higher ionic environment of Ce than Fe or Co and follows [126]:

$$Ce^{3+} + OH^- + \frac{1}{2}H_2O_2 \rightarrow Ce(OH)_2^{3+} \tag{6.16}$$

$$Ce(OH)_2^{3+} + 2OH^- \rightarrow Ce(OH)_4 \rightarrow CeO_2 + 2H_2O \tag{6.17}$$

It is well known that Ce can exist in two 3 + and 4 + oxidation states via V_O and therefore the resulting aggregation of ZFCeO and ZCCeO is energetically favored.

6.4.8 Raman Spectra for Ni-, Cu-, Ce-Substituted ZnO Nanoparticles

The $Zn_{0.95}Ni_{0.05}O$ (ZNiO), $Zn_{0.91}Ni_{0.05}Ce_{0.04}O$ (ZNiO:Ce), $Zn_{0.95}Cu_{0.05}O$ (ZCuO), and $Zn_{0.91}Cu_{0.05}Ce_{0.04}O$ (ZCuO:Ce) nanoparticles were synthesized by a sol-gel process [119]. The room-temperature Raman modes of pure, Ni-, Cu-, Ce-substituted ZnO nanoparticles are shown in Fig. 6.25C. The presence of E_2 mode in all samples indicates that the doping does not change the wurtzite phase. The intensity enhancement of E_2 mode and weak or negligibly absent E_1(LO) mode indicates the Raman scattering of good crystalline quality [136]. It is also observed that the Ni and Cu doping into ZnO gradually decreases the intensity of E_2(high) mode than pure ZnO [116]. This E_2(high) is strengthened again with Ce co-doping. It implies that the Raman bands influenced with size/shape of nanoparticles. This would lead to defects and local lattice distortions in the ZnO lattice [137]. The observed normal modes at 314, 433, 533, 567, and 839 cm^{-1} are corresponding to E_2(high) − E_2(low), E_2(high), ($2B_1$ low; 2LA), A_1(LO), and LA + TO, respectively. The existence of E_2(high) − E_2(low) modes is the clear indication of V_O in ZnO lattice. The peak position of E_2(high) mode at 422 cm^{-1} is observed for pure ZnO nanorods which shows variations with Ni, Cu, Ce doping [116]. It ascribed with change in the level of V_O [138].

6.4.9 Photoluminescence Spectra Evaluated Defects in Co:ZnO Nanoparticles

The $Zn_{1-x}Co_xO[x = 0.002$ (ZCO02), 0.004 (ZCO04), 0.006 (ZCO06), and 0.008 (ZCO08)] nanoparticles were synthesized by a sol-gel process [139]. The low concentrated Co ions in ZnO are responsible for free charge carriers and V_O to induce long-range ferromagnetic ordering. Fig. 6.26 shows the multiple peaks in photoluminescence (PL) spectra for pure and low Co concentrated ZnO nanoparticles at room temperature (excitation wavelength 325 nm (He–Cd laser)). The UV emission peaks are observed at 399, 396, 395, 394, and 389 nm, respectively, for pure ZnO, ZCO02, ZCO04, ZCO06, and ZCO08. This corresponds to the near-band-edge (NBE) emission of ZnO [140]. The variation in NBE emission with Co-doping is the blue shifting behavior. The visible region 400–750 nm shows multiple peaks that are attributed to the deep-level defects which are generally originated from V_O or Zn interstitials (Zn_i) [141]. The peaks observed at 409–449 nm are attributed to a violet emission that formed due to radiative defects such V_O, Zn_i, or V_{Zn} of interface traps existing at the grain boundaries of ZnO nanoparticles. The blue emission in 454–479 nm is probably caused by two defect level, either transition from Zn_i to the valance band or transition from bottom of the conduction band to the interstitial O (O_i). It has been observed that the peak intensity of violet and blue emission is highly influenced with Co-doping. It indicates that the Zn or O vacancies are formed with the replacement of Zn atoms by Co in the wurtzite structure. Inamdar et al. [142] suggested the defects such as Zn_i and V_O that influenced Co ions in ZnO. The peaks in 499–528 nm is green emission that is associated with intrinsic defects such as V_O, Zn_i, O_i or antisite O [143]. The weak emission peaks (592 and 601 nm) orange and (644–748 nm) red are also observed. Overall, the Co-dopants result in the formation of defects due to oxygen or Zn vacancies in the ZnO samples. Yi et al. [144] studied Li-doped ZnO and suggested that the V_O vacancies are more favorable than Zn vacancies in the doped ZnO.

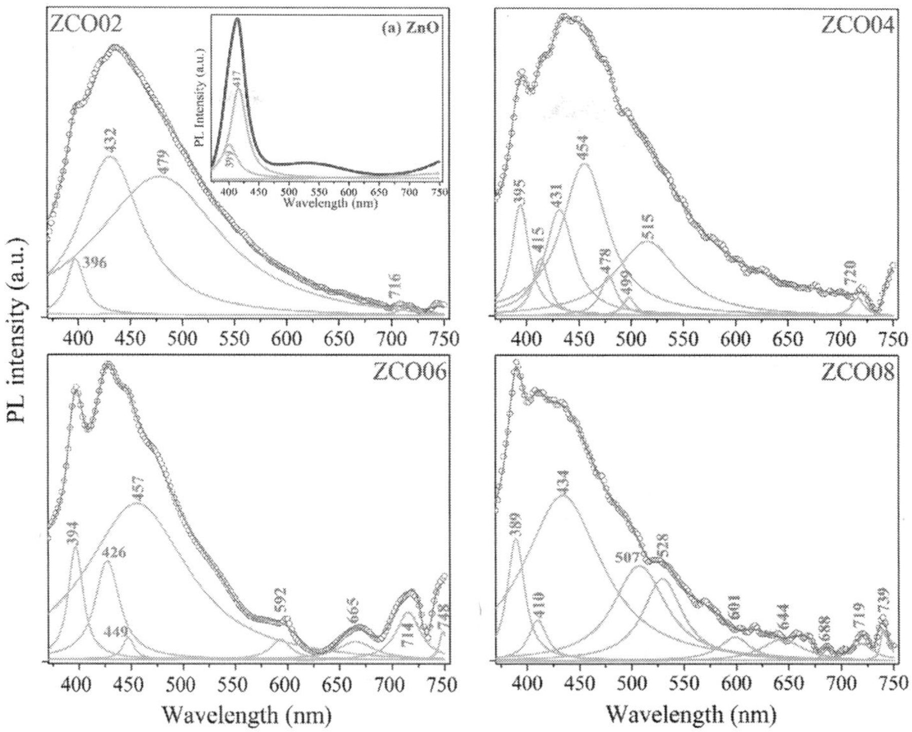

Figure 6.26
PL spectra of pure (inset) and low Co concentrated ZnO nanoparticles. Source: *From K.C. Verma, R. Bhatia, S. Kumar, R.K. Kotnala, Vacancies driven magnetic ordering in ZnO nanoparticles due to low concentrated Co ions, Mater. Res. Exp. 3 (2016) 076103–076112.*

6.4.10 Magnetism in DMS ZnO

The DMSs have drawn much attention in the recently reported spintronic devices such as nonvolatile memory, spin light-emitting diode, spin-based quantum computers, spin FET, logic devices, etc. [145]. For the basic requirement for application point of view, the T_C of DMS material is 300 K or beyond. Sato et al. [146] used DFT calculation which shows ferromagnetic ordering with T_C above room temperature for n-type ZnO with V, Cr, Co, Fe, and Ni doping. It suggests neutral V_O in ZnO which is nonmagnetic, but V_{Zn} does lead to magnetism ($\sim 1.30\ \mu_B$). However, the formation energy of neutral cation V_{Zn} in ZnO is very high. The charged state of V_{Zn} can be easily formed in n-type samples and under oxygen-rich condition. The cation vacancies introduce local magnetic moments as well as holes to the host DMS, and ferromagnetism in such a DMS could be mediated by holes. To enhance ferromagnetism in ZnO, one way is to stabilize cation vacancies and introduce holes at the same time. This can be achieved by doping impurities such as TM ions [9,112,113,115,123,126,129,147]. For example, in the low doping level of Fe, the well-separated Fe^{3+} ions can act as efficient electron as well hole traps, turning into Fe^{2+} or Fe^{4+} states, respectively, due to V_O. Coey et al. [148] reported intrinsic ferromagnetism in Fe-doped SnO_2 with a T_C of ~610 K. Walsh et al. [149] suggest carrier-mediated magnetism for Co-doped ZnO with weak antiferromagnetic ordering through Co t_{2d} states. However, the Fe-doped ZnO involves a double exchange mechanism $Fe^{2+}-Fe^{3+}$ via V_O proposed by Zener [150]. It has O 2p orbitals that could show a spin polarization induced by Fe.

6.4.10.1 RTFM in Co, Fe, ZnO nanorods

The $Zn_{1-x}Co_xO$ (ZC) [$x=0$ (Z0), 0.01 (ZC1), 0.03 (ZC3), and 0.05 (ZC5)] and $Zn_{1-x}Fe_xO$ (ZF) [$x=0.01$ (ZF1), 0.03 (ZF3), and 0.05 (ZF5)] nanorods were prepared by a sol-gel process [116]. The XRD pattern revealed the hexagonal wurtzite structure. The TEM images show the value of D (nm) = 9–13, 66–73, 57–65, 43–47, 43–51, 12–15, and 11–13, and L (nm) = 21–25, 197–207, 129–135, 115–120, 110–119, 49–53, and 27–32, respectively, for Z0, ZC1, ZC3, ZC5, ZF1, ZF3, and ZF5 nanorods. Fig. 6.27 shows

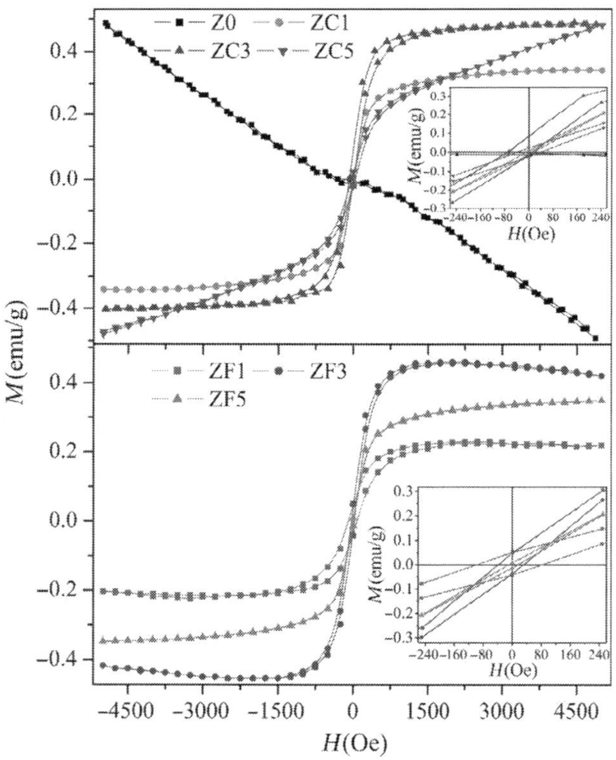

Figure 6.27

$M-H$ hysteresis measured at room temperature for $Zn_{1-x}Co_xO$ and $Zn_{1-x}Fe_xO$ nanorods. Source: *From J. Kaur, R.K. Kotnala, V. Gupta, K.C. Verma, Anionic polymerization in Co and Fe doped ZnO: nanorods, magnetism and photoactivity, Curr. Appl. Phys. 14 (2014) 749–756.*

RTFM by magnetization versus magnetic field ($M-H$) hysteresis curves for Co- and Fe-doped ZnO nanorods [116]. The $M-H$ curves show that pure ZnO exhibits diamagnetic behavior, whereas Co- and Fe-substituted ZnO display superferromagnetic behavior which is commonly observed in magnetic nanosystems. The value of $M_s = 0.344$, 0.483, 0.354, 0.233, 0.459, and 0.328 emu/g, respectively, observed for ZC1, ZC3, ZC5, ZF1, ZF3, and ZF5 nanorods. These variations in the observed M_s value depend upon factors such as nanostructural shape/size, the dopant stoichiometric ratio, and lattice parameters. With 5% of Fe and Co-doping in ZnO, the value of maximum magnetization is not saturated up to a magnetizing field of 5 kOe. This is the effect of the randomly distributed localized spins due to nonuniform grain structure, random distribution of V_O, and strong anisotropy. Liu et al. [151] suggested that the nonsaturated magnetic hysteresis due to superparamagnetism which is generally found in magnetic nanostructures. Bryan et al. [152] suggested unsaturated magnetic hysteresis at low applied magnetic field, which is related with the possibility of different dopant sites that may have different spin states to form antiferromagnetic superexchange.

Xiao et al. [153] theoretically investigated Fe-doped ZnO nanoparticles, and have shown that Fe stabilizes polar instable surfaces, whereas nonpolar surfaces, namely, $(10\bar{1}0)$ and $(11\bar{2}0)$ remain intact. Polar surfaces can be stabilized indirectly through $Fe^{2+}-Fe^{3+}$ pair-assisted charge transfer, which reduces polarity and therefore solubility in polar solvents that is responsible for RTFM. Gilliland et al. [154] studied the electronic structure of Co- and Fe-doped ZnO through DFT calculations. It is reported that the occupied α-spin Co 3d states are much closer to the valence band maximum, which strongly mixes the Co $3d_z{}^2$ A_1 singlet with the A_1 O $2p_z$, resulting in crystal field splitting. However, all β-spin Fe 3d states are inside the conduction band and α-spin bands at the top of the valence band have strong Fe 3d character, in which the O 2p contribution is dominant. The band gap is found to increase at the rate of 23 and 4 meV/% for Co and Fe, respectively. Therefore, a hopping mechanism is easily formed with Fe doping at room temperature. However, the Co-doped ZnO have effective moments per Co as 4.8 μ_B and the total magnetization is reduced, demonstrating that Co–O–Co pairs are antiferromagnetic [155].

6.4.10.2 Origin of RTFM in Ni:ZnO nanostructure

The Ni-doped ZnO nanostructures were synthesized by a wet chemical method [145]. The Rietveld refinement of XRD patterns indicate that up to 12.5% of Ni doping into ZnO have no secondary phase formation and show the decrease in Zn/O occupancy ratio, associated with the defect induced due to Ni doping. The SEM images demonstrate modification of rod-like morphology in pure ZnO to flakes like with Ni doping. The $M-H$ plots (Fig. 6.28A and B), measured at different concentration of Ni and temperatures, clearly show the typical weak ferromagnetic saturation and this saturation magnetization is found to increase from 0.02 to 0.17 emu/g with increasing Ni concentration. It can be noted that

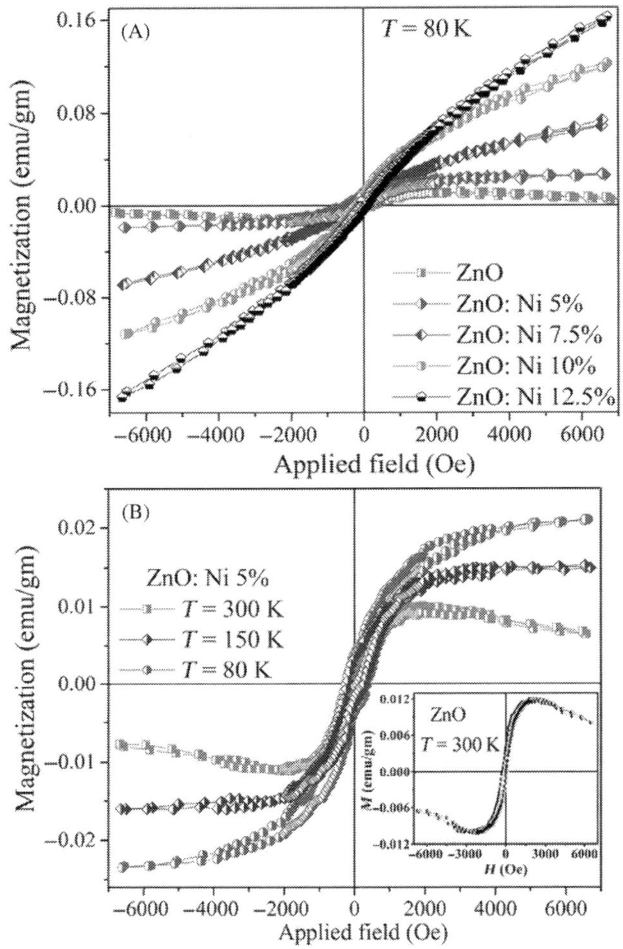

Figure 6.28

(A) $M-H$ measurement at 80 K for pure and Ni-doped ZnO nanostructure. (B) $M-H$ measurement of Ni 5% doped ZnO at 80, 150, and 300 K. Inset of (B) is the room-temperature $M-H$ measurement of ZnO. Source: *From A.K. Rana, Y. Kumar, P. Rajput, S.N. Jha, D. Bhattacharyya, P.M. Shirage, Search for origin of room temperature ferromagnetism properties in Ni-doped ZnO nanostructure, ACS Appl. Mater. Interf. 9 (2017) 7691–7700.*

the saturation tendency is not increasing with increasing Ni concentration; however, a decrease in the temperature does not alter the important ferromagnetic characteristic but significantly changes the value of M_s. The origin of RTFM in Ni-doped ZnO is still controversial. Since TM = Ni, defects produced by TMs, especially single ionized V_O, play a significant role in ferromagnetism. The reported research work has been suggested that the origin of RTFM in TM-doped ZnO is the intrinsic exchange interactions within oxygen vacancies, i.e., defects, which is assisted by BMPs [145]. But the Ni^{2+} ions cannot just produce intrinsic ferromagnetism unless and until there are formation of point defects such as singly ionized V_O-assisted BMPs for RTFM.

6.4.10.3 Lattice defects influenced ferromagnetic ordering of ZnO by Cu and Ce ions

The ZCuO and ZCuO:Ce nanoparticles were synthesized by a sol-gel process [119]. Rietveld refinement of XRD pattern results into wurtzite ZnO structure and the distortion/strain/stress formation due to variation in a and c constants, V and l_{Zn-O} creates lattice defects. The average size (from TEM images) of nanoparticles, D, is 57 and 159 nm, respectively, for ZCuO and ZCuO:Ce. Fig. 6.29A and B shows $M-H$ hysteresis at room temperature (~ 300 K) and 10 K for Cu- and Ce-substituted ZnO nanoparticles. At room temperature, the values of $M_s = 0.053$ and 0.132 emu/g, and $M_r = 0.0124$ and 0.0134 emu/g with $H_c = 393$ and 68 Oe, respectively, for ZCuO and ZCuO:Ce. The RTFM in TM-substituted ZnO has a great influence on the intrinsic defects such as V_O [156]. The

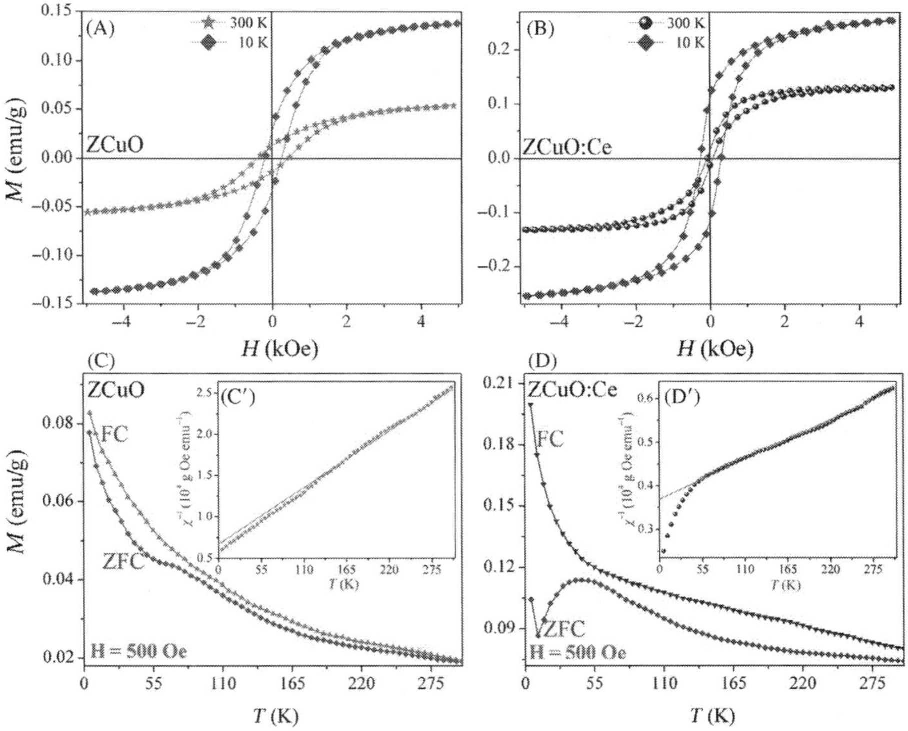

Figure 6.29

$M-H$ hysteresis measured at room temperature (~ 300 K) and 10 K (A) ZCuO and (B) ZCuO:Ce. (C and D) Magnetization (M) as a function of temperature (T) following FC and ZFC at $H = 500$ Oe. Inset shows their $\chi^{-1}-T$ and the Curie–Weiss fittings. Source: *From K.C. Verma, R.K. Kotnala, Understanding lattice defects to influence ferromagnetic order of ZnO nanoparticles by Ni, Cu, Ce ions, J. Solid State Chem. 246 (2017) 150–159.*

long-range interaction is necessary to obtain high-T_C ferromagnetism of DMS TM:ZnO systems and it can be mediated by defect states. This is since the BMP model ascribed that the bound electrons (holes) in the defect states can couple with TM ions and cause the ferromagnetic regions to overlap, giving rise to long-range ferromagnetic ordering [117]. The RTFM in Cu-doped ZnO is described due to the presence of Cu in high-spin configuration in a tetrahedral crystal field [157]. The crystal field splitting of 3d state into t_{2g} and e_g orbitals, with four electron in e_g state and one unpaired electron in t_{2g} state, results in the high-spin state of Cu^{2+} (1/2 moments per ion). It is reported that the RTFM in Cu-doped ZnO is intrinsic and probably results from the magnetic exchange interaction between V_O and Cu^{2+} ions which aligns all the Cu^{2+} ions around the V_O, forming BMPs [158]. The enhancement in RTFM with Ce doping is explained by size effect of nanoparticles and double exchange mechanism of Ce^{3+} and Ce^{4+} ions via V_O. The Ce-substituted Cu:ZnO could influence the RTFM due to the $Ce^{3+}-O^{2-}$ transition with Cu ions, which is expected to be antiferromagnetic [159]. The $M-H$ hysteresis is also measured at low temperature ($\sim$10 K) (Fig. 6.29A and B). The value of $M_s = 0.136$ and 0.251 emu/g, $M_r = 0.0273$ and 0.113 emu/g, and $H_c = 176$ and 250 Oe, respectively, for ZCuO and ZCuO:Ce.

The origin of observed RTFM of Cu,Ce:ZnO nanoparticles is also evaluated by the temperature-dependent magnetization [$M(T)$] with field cooling (FC) at 500 Oe and zero field cooling (ZFC) measurement (Fig. 6.29C and D). The $M-T$ measurements show that ZFC/FC curves did not coincide with each other or slightly coincide at around 300 K. This splitting of the ZFC/FC curves usually appears to be a coexistent system of the antiferromagnetic and ferromagnetic phases. This splitting in FC/ZFC curves for ZCuO is smaller than Ce co-doping. It means the antiferromagnetic behavior becomes stronger with Ce ions into Cu:ZnO. Gopal et al. [112] theoretically investigated weak antiferromagnetic ordering in TM:ZnO. The steep increase of magnetization in FC curve with decreasing temperatures below 50 K is the characteristic of all DMS materials and is probably related with defect structure and possible fraction of weakly coupled/uncoupled dopant ions [160]. For Ce-doped samples, the shape of $M(T)$ curves is almost concave suggesting predominance of BMP interaction [161]. The dip in the ZFC curve indicates the existence of the blocking temperature, $T_B = 45$ K for ZCuO:Ce. This T_B here may involve processes such as superparamagnetic relaxation, glass transition, Neel temperature, T_N [162]. The upward curvature observed in the FC curve in the $M(T)$ measurements (Fig. 6.29C′ and D′) of Cu,Ce:ZnO nanoparticles involve a Curie–Weiss-like behavior. However, the $\chi^{-1}(T)$ (inversion of susceptibility vs temperature) of FC curve demonstrates a notable deviation from the Curie–Weiss law at low temperature due to relation:

$$\chi = \frac{C}{T - \theta} \tag{6.18}$$

where C is the material specific Curie constant, T is the absolute temperature, and θ is the Weiss constant (in Kelvin). The estimated value of θ is found to be negative in both the samples which indicate the antiferromagnetic interactions [163].

6.4.10.4 The ac susceptibility SQUID measurement of Co,Fe,Ce:ZnO nanoparticles

The ac magnetic susceptibility measurements have been determined by the antiferromagnetic to ferromagnetic ordering in ZFCeO and ZCCeO nanoparticles [126]. The temperature-dependent real (χ') and imaginary (χ'') components of the ac magnetic susceptibility of Ce-doped ZFO and ZCO nanoparticles at frequencies 100, 300, 500, and 1000 Hz are shown in Fig. 6.30. The applied oscillating field is $H_{ac} = 2.5$ Oe without any dc field bias in $T = 10-100$ K. The frequency-independent peak maxima of χ' corresponds to the magnetic phase transition into an antiferromagnetic state with $T_N = 56$ and 34 K

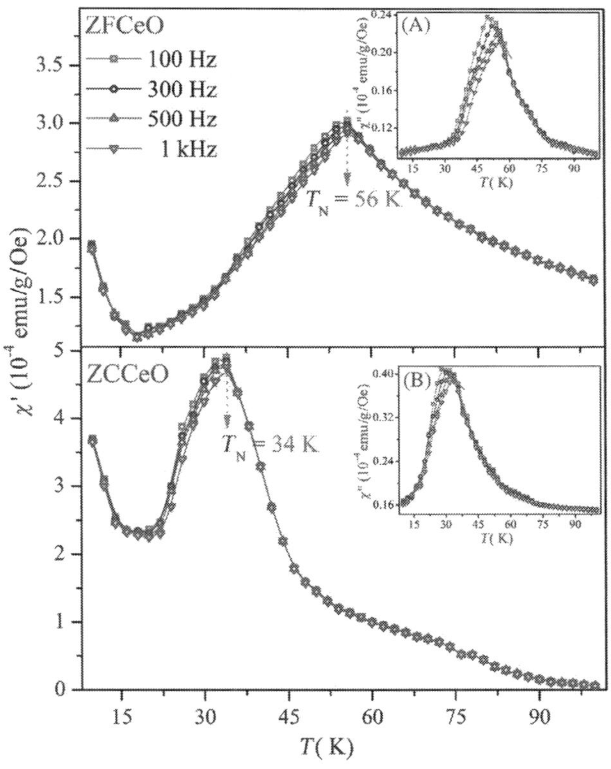

Figure 6.30

Temperature-dependent ac magnetic susceptibility of ZFCeO and ZCCeO nanoparticles at $T = 10-100$ K. Source: *From K.C. Verma, R.K. Kotnala, Defects-assisted ferromagnetism due to bound magnetic polarons in Ce into Fe, Co:ZnO nanoparticles and first-principle calculations, Phys. Chem. Chem. Phys. 18 (2016) 5647–5657.*

observed for ZFCeO and ZCCeO, respectively, compared with reports [25]. A slight dispersion in the $\chi'(T)$ curve after the peak maxima is also observed which indicates certain spin-glass, ferromagnetic clusters and cluster-glass magnetic type states [164]. However, the peak in $\chi''(T)$ increases with decreasing frequency. This is qualitatively different from the behavior of most spin glasses in which we expect an increase in the peak magnitude with increasing frequency [165]. This also induced certain peak shifts with frequency. Thus, the obtained ac magnetization data supports the antiferromagnetic DMS materials with certain cluster-glass formation over spin glass.

6.4.10.5 Vacancies induce ferromagnetism of pure and doped ZnO

The RTFM has been widely investigated in ZnO doped with TM ions. Correspondingly, the models based on the exchanging coupling of localized spins or the segregation of the magnetic metal clusters/precipitates has adapted to explain the RTFM [166]. It was widely accepted that the oxygen-related defects such as V_O are responsible for RTFM of pure and doped ZnO. Ozgur et al. [167] suggested that the deviation from an ideal wurtzite crystal is probably due to the point defects such as Zn antisites or V_O. Wang et al. [168] have shown that the V_{Zn} always prefers to reside on the surface of the material. The magnetic moment of V_{Zn} defects lyingon the surface becomes larger compared to the V_{Zn} that lies deep inside the bulk. It is based on DFT that the Zn vacancy introduces spin polarization on the top of the valence band and the observed magnetism in ZnO arises indeed from V_{Zn}.

6.4.10.5.1 XPS spectra for Zn 2p, Co 2p and O 1s for Co:ZnO nanoparticles

Fig. 6.31A–C shows the X-ray photoelectron spectroscopy (XPS) spectra for low Co concentrated ZnO nanoparticles measured at room temperature [139]. The peaks corresponding to the binding energies of Zn $2p_{3/2}$ ($\sim$1021.72 eV) and Zn $2p_{1/2}$ ($\sim$1044.93 eV), and a spin–orbital splitting (difference, $\Delta E \sim$ 23.21 eV), confirm that the Zn ions present in +2 oxidation state. Since smaller nanoparticles have larger surface-to-volume ratio, this result into formation of many defect sites such as V_O on their nanosurfaces. Fig. 6.31B also shows the XPS spectra of Co 2p regions characterized by a doublet that arise due to spin–orbit coupling ($2p_{3/2}$ and $2p_{1/2}$). The value of binding energy, Co $2p_{3/2} \sim$ 780.019 eV, $2p_{1/2} \sim$ 795.51 eV, $\Delta E \sim$ 15.51 eV, and shake-up satellite peak (S) $\sim$785.48 eV are observed. The binding energies of Co $2p_{3/2}$ and $2p_{1/2}$ of low Co concentrated ZnO nanoparticles indicate that the Co ions exist either in +3 or +2 oxidation states [169].

In order to evaluate the presence of V_O, the O 1s spectra are given in Fig. 6.31C, which deconvoluted into three peaks (O_a, O_b, O_c) [139]. The peak located on the low binding energy side (O_a) around 528.79 eV is attributed to O^{2-} ions on the wurtzite structure of a

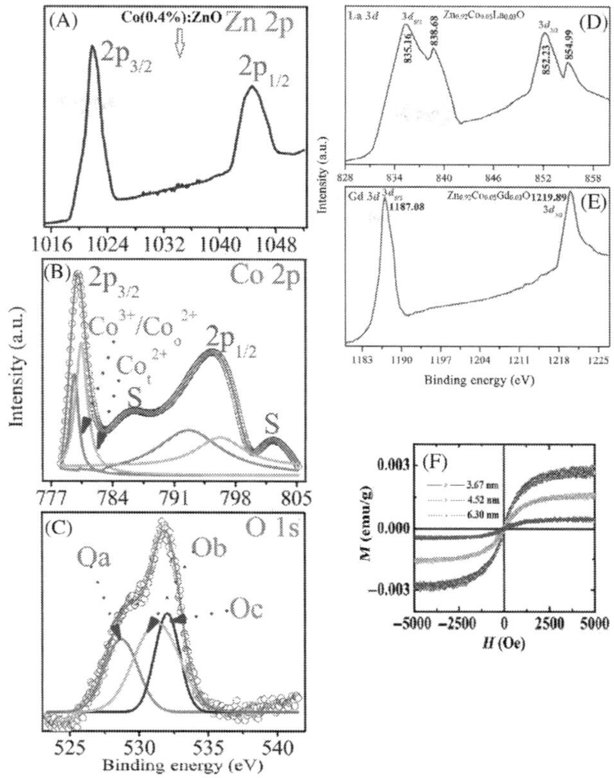

Figure 6.31

(A–C) XPS spectra for low Co concentrated (0.4%) ZnO nanoparticles of Zn 2p, Co 2p, and O 1s. (D) ZCLO53 for La 3d, (E) ZCGO53 for Gd 3d nanostructures. (F) Room-temperature *M–H* loops of ZnO of different nanosizes. Source: *From X. Xu, C. Xu, J. Dai, J. Hu, F. Li, S. Zhang, Size dependence of defect-induced room temperature ferromagnetism in undoped ZnO nanoparticles, J. Phys. Chem. C 116 (2012) 8813–8818; K.C. Verma, R.K. Kotnala, Defects due to lattice distortion and nano-size intermediate ferromagnetism in La, Gd substituted Zn0.95Co0.05O, Curr. Appl. Phys. 16 (2016) 175–182; K.C. Verma, R. Bhatia, S. Kumar, R.K. Kotnala, Vacancies driven magnetic ordering in ZnO nanoparticles due to low concentrated Co ions, Mater. Res. Exp. 3 (2016) 076103–076112.*

hexagonal Zn^{2+} ion array, which is surrounded by Zn atoms with the full supplement of nearest-neighbor O^{2-} ions. This O_a peak of the O 1s spectrum is associated with Zn–O bonds and the intensity of this component is a measure of the amount of O atoms in fully oxidized stoichiometric surrounding. The O_b peak at ~531.19 eV is associated with O^{2-} ions in the O-deficient regions within the ZnO matrix which indicate the formation of the defects. These observed V_O or defects are formed during heating process. The O_c peak at ~532.16 eV is usually attributed to chemisorbed oxygen on the surface of the ZnO, such as adsorbed $-CO_3$, H_2O, or adsorbed O_2.

6.4.10.5.2 Valence states of RE La, Gd ions influence ferromagnetism of ZnO nanoparticles

The DMSs ZCLO53 and ZCGO53 nanoaggregation were synthesized by sol-gel process [130]. The average value of *D* of nanoaggregation is 142 and 86 nm, respectively, for ZCLO53 and ZCGO53. The long-range ferromagnetic ordering occurs due to strong antiferromagnetic interactions performed by La and Gd doping into Co:ZnO. Fig. 6.31D shows La 3d states of ZCLO53 sample at room temperature. It is well known that the La 3d states in the XPS spectrum are split not only due to a spin–orbit interaction into two lines $3d_{5/2}$ and $3d_{3/2}$, but additionally, each line is split due to a transfer of an electron from oxygen ligands to the La 4f (initially empty) [170]. The La 3d line has a main line M (ascribed to $3d^9 4f^0$) and a satellite line *S* (ascribed to a charge transfer $3d^9 4f^1 L$, here L represents the hole in a ligand site) at higher binding energy. The corresponding binding energy of main peak, *M*, is 835.16 and 852.23 eV, respectively, to $3d_{5/2}$ and $3d_{3/2}$. The respective satellite peak, *S*, is observed at 838.68 and 854.99 eV. Energy difference between the $3d_{3/2}$ and $3d_{5/2}$ levels is 17.07 eV. The binding energies and the multiplet splitting agree well with reported values for La^{3+} [171]. Fig. 6.31E represents the Gd 3d spectrum of ZCGO53. The two strong peaks at 1187.08 and 1219.89 eV corresponding to a spin–orbit splitting of 32.81 eV correspond, respectively, to the $3d_{5/2}$ and $3d_{3/2}$ energy levels of Gd. These observations indicate that Gd exists in the +3 oxidation states.

The superexchange interactions between La^{3+} and Gd^{3+} ions with O^{2-}, respectively, induce magnetic interactions. Trivalent Gd ions have larger localized magnetic moments because of their $4f^7$ electron configuration, compared to the TM ions. Therefore, it is expected that the Gd-doped oxides could exhibit a large ferromagnetic moment and the substitution of trivalent Gd ions at the divalent Zn site. This can also introduce vacancies in the DMS system, which can be favorable to sustain ferromagnetic coupling. This has been attributed to the F-center exchange mechanism where electrons are trapped in V_O (F-center) created by La^{3+} and Gd^{3+} ions.

6.4.10.6 RTFM of DMS ZnO influenced with nanostructural formation

It is increasingly evident that d^0 RTFM is often observed only in thin film and nanostructures. It exhibits notable surface effects, which is suggested to contact with superficial defects [110]. Therefore, nanoparticles with the high surface–volume ratio are expected to be an excellent model for studying the defect-related ferromagnetism because most of the defects must exist near the surface of nanoparticles. This implies that the size of nanoparticles should strongly affect the ZnO ferromagnetism. The *M–H* curves of the ZnO nanoparticles of sizes 3.67, 4.52, and 6.30 nm were measured at room temperature and shown in Fig. 6.31F. It can be seen that all samples exhibit clear hysteresis loops, suggesting unambiguously reproducible RTFM. More importantly, it can also been found that M_s decreases monotonically with increasing size of the nanoparticles. This size dependence of RTFM supports further the grain boundary effect and the surface effect of magnetic behaviors in ZnO [172,173].

6.5 Conclusion

To increase the density of memory devices, the detection of spin current in multiferroic and DMS nanostructures are the major aspects of spintronics. The nanostructural formation of both multiferroic BTO and DMS ZnO depends upon the synthesis route, dopant material type, stoichiometric ratio, and temperature conditions. In perovskites, the multiferroicity appears as the superposition of two phenomena formally independent ($T_N \neq T_{FE}$), magnetism generally emerges in structures that already present some distortions (including ferroelectric). The ferromagnetism in Fe-doped BTO and ZnO systems is explained due to *F*-centre exchange (FCE) and the BMP model. The survey from multiferroic studies has demonstrated that the ME coupling may enhance effectively from the multilayered heterostructures of BTO, BFO, and ferrite nanocomposites. The high-temperature ferromagnetism in DMS is analyzed in terms of an indirect exchange via shallow donors, treated in two sublattices that require hybridization and charge transfer from a donor-derived impurity band to unoccupied 3d states at the Fermi level. The DFT calculations suggest that the Fe, Co co-doped ZnO system with and without defects, irrespective of the system size, tends to stabilize in a ferrimagnetic (antiferromagnetic) configuration, for a nearest-neighbor position of the TM dopants. The magnetization is enhanced with TM doping into ZnO which is the joint effects of the TM ions as well as the intrinsic exchange interactions arising from defect such as V_O-assisted BMPs. The ZFC/FC magnetization measurement at $H = 500$ Oe and $T = 5$-300 K shows antiferromagnetic–ferromagnetic transitions in TM-doped ZnO. The RE ions co-doping into TM:ZnO is responsible for high T_C, i.e. long-range ferromagnetic ordering.

Acknowledgment

This work was financially supported by the SERB Department of Science and Technology (DST), Government of India under Fast Track Scheme for Young Scientist (SR/FTP/PS-180/2013).

References

[1] R. Ramesh, A new spin on spintronics, Nat. Mater. 9 (2010) 380–381.

[2] S.A. Wolf, D.D. Awschalom, R.A. Buhrman, J.M. Daughton, S. von Molnar, M.L. Roukes, et al., Spintronics: a spin-based electronics vision for the future, Science 294 (2001) 1488–1495.

[3] D. Vasilyev, F.O. Schumann, F. Giebels, H. Gollisch, J. Kirschner, R. Feder, Spin-entanglement between two freely propagating electrons: experiment and theory, Phys. Rev. B 95 (2017) 115134–115137.

[4] J. Sinova, I. Zutic, New moves of the spintronics tango, Nat. Mater. 11 (2012) 368–371.

[5] A. Brataas, A.D. Kent, H. Ohno, Current-induced torques in magnetic materials, Nat. Mater. 11 (2012) 372–381.

[6] J.F. Gregg, Spintronics: a growing science, Nat. Mater. 6 (2007) 798−799.

[7] D.D. Awschalom, M.E. Flatte, Challenges for semiconductor spintronics, Nat. Phys. 3 (2007) 153−159.

[8] M.Y. Zhuravlev, R.F. Sabirianov, S.S. Jaswal, E.Y. Tsymbal, Giant electroresistance in ferroelectric tunnel junctions, Phys. Rev. Lett. 94 (2005) 246802−246804.

[9] D. Szaller, V. Kocsis, S. Bordacs, T. Feher, T. Room, U. Nagel, et al., Magnetic resonances of multiferroic $TbFe_3(BO_3)_4$, Phys. Rev. B 95 (2017) 024427.

[10] J.F. Scott, Data storage: multiferroic memories, Nat. Mater. 6 (2007) 256−257.

[11] M. Gajek, M. Bibes, S. Fusil, K. Bouzehouane, J. Fontcuberta, A. Barthelemy, et al., Tunnel junctions with multiferroic barriers, Nat. Mater. 6 (2007) 296−302.

[12] J.Y. Son, J.H. Lee, S. Song, Y.H. Shin, H.M. Jang, Four-states multiferroic memory embodied using Mn-doped $BaTiO_3$ nanorods, ACS Nano 7 (2013) 5522−5529.

[13] M. Bibes, A. Barthelemy, Multiferroics towards a magnetoelectric memory, Nat. Mater. 7 (2008) 725−726.

[14] D. Pantel, S. Goetze, D. Hesse, M. Alexe, Reversible electrical switching of spin polarization in multiferroic tunnel junctions, Nat. Mater. 11 (2012) 289−293.

[15] E.Y. Tsymbal, H. Kohlstedt, Tunneling across a ferroelectric, Science 313 (2006) 181−183.

[16] V. Garcia, M. Bibes, L. Bocher, S. Valencia, F. Kronast, A. Crassous, et al., Ferroelectric control of spin polarization, Science 327 (2010) 1106−1110.

[17] V. Laukhin, V. Skumryev, X. Marti, D. Hrabovsky, F. Sanchez, M.V. Garcıa-Cuenca, et al., Electric-field control of exchange bias in multiferroic epitaxial heterostructures, Phys. Rev. Lett. 97 (2006) 227201−227204.

[18] W. Liang, B.D. Yuhas, P. Yang, Magnetotransport in Co-doped ZnO nanowires, Nano Lett. 9 (2) (2009) 892−896.

[19] G.A. Prinz, Magnetoelectronics, Science 282 (1998) 1660−1663.

[20] T. Dietl, H. Ohno, F. Matsukura, J. Clbert, D. Ferrand, Zener model description of ferromagnetism in zinc-blende magnetic semiconductors, Science 287 (2000) 1019−1022.

[21] Y. Ohno, D.K. Young, B. Beschoten, F. Matsukura, H. Ohno, D.D. Awschalom, Electrical spin injection in a ferromagnetic semiconductor heterostructures, Nature 402 (1999) 790−792.

[22] A.E. Danks, S.R. Hall, Z. Schnepp, The evolution of "sol-gel" chemistry as a technique for materials synthesis, Mater. Horiz 3 (2016) 91−112.

[23] Anju, A. Agarwal, P. Aghamkar, B. Lal, Structural and multiferroic properties of barium substituted bismuth ferrite nanocrystallites prepared by sol-gel method, J. Magn. Magn. Mater. 426 (2017) 800−805.

[24] K.C. Verma, R.K. Kotnala, N.S. Negi, Improved dielectric and ferromagnetic properties in Fe-doped $PbTiO_3$ nanoparticles at room temperature, Appl. Phys. Lett. 92 (2008) 152902−152903.

[25] K.C. Verma, R.K. Kotnala, Tailoring the multiferroic behavior in $BiFeO_3$ nanostructures by Pb doping, RSC Adv. 6 (2016) 57727−57738.

[26] K. Byrappa, T. Adschiri, Hydrothermal technology for nanotechnology, Prog. Cryst. Growth Charact. Mater. 53 (2007) 117−166.

[27] Y. Qiao, Y. Zhou, S. Wang, L. Yuan, Y. Du, D. Lu, et al., Composition dependent magnetic and ferroelectric properties of hydrothermally synthesized $GdFe_{1-x}Cr_xO_3$ ($0.1 \leq x \leq 0.9$) perovskites, Dalt. Trans. 46 (2017) 5930−5937.

[28] K.C. Verma, V. Gupta, J. Kaur, R.K. Kotnala, Raman spectra, photoluminescence, magnetism and magnetoelectric coupling in pure and Fe doped $BaTiO_3$ nanostructures, J. Alloys Compd. 578 (2013) 5−11.

[29] K.C. Verma, D. Singh, S. Kumar, R.K. Kotnala, Multiferroic effects in $MFe_2O_4/BaTiO_3$ (M = Mn, Co, Ni, Zn) nanocomposites, J. Alloys Compd. 709 (2017) 344−355.

[30] B. Du, F. Gucci, H. Porwal, S. Grasso, A. Mahajan, M.J. Reece, Flash spark plasma sintering of magnesium silicide stannide with improved thermoelectric properties, J. Mater. Chem. C 5 (2017) 1514−1521.

[31] J. Gonzalez-Julian, K. Jahnert, K. Speer, L. Liu, J. Rathel, M. Knapp, et al., Effect of internal current flow during the sintering of zirconium diboride by field assisted sintering technology, J. Am. Ceram. Soc. 99 (2016) 35−42.

[32] V. Koval, I. Skorvanek, J. Durisin, G. Viola, A. Kovalcikova, P. Svec Jr., et al., Terbium-induced phase transitions and weak ferromagnetism in multiferroic bismuth ferrite ceramics, J. Mater. Chem. C 5 (2017) 2669−2685.

[33] Y. Shen, J. Sun, L. Li, Y. Yao, C. Zhou, R. Su, et al., The enhanced magnetodielectric interaction of $(1-x)BaTiO_3 - xCoFe_2O_4$ multiferroic composites, J. Mater. Chem. C 2 (2014) 2545−2551.

[34] K. Jiang, J.J. Zhu, J.D. Wu, J. Sun, Z.G. Hu, J.H. Chu, Influences of oxygen pressure on optical properties and interband electronic transitions in multiferroic bismuth ferrite nanocrystalline films grown by pulsed laser deposition, ACS Appl. Mater. Intf 3 (2011) 4844−4852.

[35] W.W. Li, J.J. Zhu, J.D. Wu, J. Gan, Z.G. Hu, M. Zhu, et al., Temperature dependence of electronic transitions and optical properties in multiferroic $BiFeO_3$ nanocrystalline film determined from transmittance spectra, Appl. Phys. Lett. 97 (2010) 121102−121103.

[36] A.J. Rulison, R.C. Flagan, Synthesis of yttria powders by electrospray pyrolysis, J. Am. Ceram. Soc. 77 (1994) 3244−3250.

[37] B. Almeria, A. Gomez, Electrospray synthesis of monodisperse polymer particles in a broad (60 nm−2 μm) diameter range: guiding principles and formulation recipes, J. Colloid Interf. Sci. 417 (2014) 121−130.

[38] Y. Du, Z.X. Cheng, S.X. Dou, D.J. Attard, X.L. Wang, Fabrication, magnetic, and ferroelectric properties of multiferroic $BiFeO_3$ hollow nanoparticles, J. Appl. Phys. 109 (2011) 109.

[39] H. Liu, B. Cao, C. O'Connor, Intrinsic magnetism in $BaTiO_3$ with magnetic transition element dopants (Co, Cr, Fe) synthesized by sol-precipitation method, J. Appl. Phys. 109 (2011) 07B516. 07B523.

[40] R. Gupta, S. Chaudhary, R.K. Kotnala, Interfacial charge induced magnetoelectric coupling at $BiFeO_3$/$BaTiO_3$ bilayer interface, ACS Appl. Mater. Interf. 7 (2015) 8472−8479.

[41] H. Li, Y. Huang, Q. Zhang, Y. Qiao, Y. Gu, J. Liu, et al., Facile synthesis of highly uniform Mn/Co-codoped ZnO nanowires: optical, electrical, and magnetic properties, Nanoscale 3 (2011) 654−660.

[42] L. Hou, W. Zhou, B. Zou, Y. Zhang, J. Han, X. Yang, et al., Spin−exciton interaction and related microphotoluminescence spectra of ZnSe:Mn DMS nanoribbon, Nanotechnology 28 (2017) 105202−105211.

[43] N. Midya, S.K. Neogi, Md. A. Ahmed, A. Banerjee, P. Kumar, D. Kanjilald, et al., Correlation between magnetic and microstructural properties of low energy ion irradiated and un-irradiated $Zn_{0.95}Mn_{0.05}O$ films, RSC Adv. 7 (2017) 771−781.

[44] M. Rouchdi, E. Salmani, B. Fares, N. Hassanain, A. Mzerd, Synthesis and characteristics of Mg doped ZnO thin films: experimental and ab-initio study, Results Phys. 7 (2017) 620−627.

[45] L. Weston, X.Y. Cui, S.P. Ringer, C. Stampf, Multiferroic crossover in perovskite oxides, Phys. Rev. B 93 (2016) 165210−165211.

[46] D.I. Khomskii, Multiferroics: different ways to combine magnetism and ferroelectricity, J. Magn. Magn. Mater. 306 (2006) 1−8.

[47] R.E. Cohen, Origin of ferroelectricity in perovskite oxides, Nature (London) 358 (1992) 136.

[48] P.M. Rorvik, T. Grande, M.A. Einarsrud, One-dimensional nanostructures of ferroelectric perovskites, Adv. Mater. 23 (2011) 4007−4034.

[49] Z. Yan, Y. Guo, G. Zhang, J.M. Liu, High-performance programmable memory devices based on Co-doped $BaTiO_3$, Adv. Mater. 23 (2011) 1351−1355.

[50] K.C. Verma, R.K. Kotnala, Multiferroic approach for Cr,Mn,Fe,Co,Ni,Cu substituted $BaTiO_3$ nanoparticles, Mater. Res. Exp. 3 (2016) 055006−055013.

[51] C. Schmitz-Antoniak, D. Schmitz, P. Borisov, F.M.F. de Groot, S. Stienen, A. Warland, et al., Electric in-plane polarization in multiferroic $CoFe_2O_4$/$BaTiO_3$ nanocomposite tuned by magnetic fields, Nat. Commun. 4 (2013) 2051−2058.

[52] T. Kolodiazhnyi, M. Tachibana, H. Kawaji, J. Hwang, E. Takayama-Muromachi, Persistence of ferroelectricity in $BaTiO_3$ through the insulator–metal transition, Phys. Rev. Lett. 104 (2010) 147602–147604.

[53] J.A. McLeod, Z.V. Pchelkina, L.D. Finkelstein, E.Z. Kurmaev, R.G. Wilks, A. Moewes, et al., Electronic structure of $BiMO_3$ multiferroics and related oxides, Phys. Rev. B 81 (2010) 144103–144110.

[54] N.A. Hill, Why are there so few magnetic ferroelectrics? J. Phys. Chem. B 104 (2000) 6694–6709.

[55] H. Nakayama, H. Katayama-Yoshida, Theoretical prediction of magnetic properties of $BaTi_{1-x}M_xO_3$ (M = Sc,V,Cr, Mn,Fe,Co,Ni,Cu), Jpn. J. Appl. Phys. 40 (2001) L1355–L1358.

[56] K.C. Verma, R.K. Kotnala, Lattice defects induce multiferroic responses in Ce, La-substituted $BaFe_{0.01}Ti_{0.99}O_3$ nanostructures, J. Am. Ceram. Soc. 99 (5) (2016) 1601–1608.

[57] B. Xu, K.B. Yin, J. Lin, Y.D. Xia, X.G. Wan, J. Yin, et al., Room-temperature ferromagnetism and ferroelectricity in Fe-doped $BaTiO_3$, Phys. Rev. B 79 (2009) 134109–134115.

[58] I.N. Apostolova, A.T. Apostolov, S.G. Bahoosh, J.M. Wesselinowa, Origin of ferromagnetism in transition metal doped $BaTiO_3$, J. Appl. Phys. 113 (2013) 203904.

[59] L. Ju, T. Sabergharesou, K.G. Stamplecoskie, M. Hegde, T. Wang, N.A. Combe, et al., Interplay between size, composition, and phase transition of nanocrystalline Cr^{3+}-doped $BaTiO_3$ as a path to multiferroism in perovskite-type oxides, J. Am. Chem. Soc. 134 (2012) 1136–1146.

[60] S.K. Das, R.N. Mishra, B.K. Roul, Magnetic and ferroelectric properties of Ni doped $BaTiO_3$, Solid State Commun. 191 (2014) 19–24.

[61] D. Bin-Feng, Z. Sheng-Qiang, The coexistence of ferroelectricity and ferromagnetism in Mn-doped $BaTiO_3$ thin films, Chin. Phys. B 20 (12) (2011) 127701–127707.

[62] K.C. Verma, J. Kaur, N.S. Negi, R.K. Kotnala, Multiferroic and magnetoelectric properties of nanostructured $BaFe_{0.01}Ti_{0.99}O_3$ thin films obtained under polyethylene glycol conditions, Solid State Commun. 178 (2014) 11–15.

[63] R. Kumari, A. Sahaib, N. Goswami, Effect of nitrogen doping on structural and optical properties of ZnO nanoparticles, Prog. Nat. Sci. Mater. Int. 25 (2015) 300–309.

[64] Y. Tokunaga, Y. Taguchi, T. Arima, Y. Tokura, Magnetic biasing of a ferroelectric hysteresis loop in a multiferroic orthoferrite, Phys. Rev. Lett. 112 (2014) 037203–037205.

[65] A. Kumar, R.S. Katiyar, R. Guo, A.S. Bhalla, Magnetoelectric characterization of multiferroic nanostructure materials, Ferroelectrics 473 (2014) 137–153.

[66] J. Varghese, R.W. Whatmore, J.D. Holmes, Ferroelectric nanoparticles, wires and tubes: synthesis, characterisation and applications, J. Mater. Chem C1 (2013) 2618–2638.

[67] Y. Park, W.J. Lee, H.G. Kim, Particle-size-induced diffuse phase transition in the fine-particle barium titanate porcelains, J. Phys. Condens. Matter 9 (1997) 9445–9456.

[68] B. Weber, S. Mahapatra, H. Ryu, S. Lee, A. Fuhrer, T.C.G. Reusch, et al., Ohm's law survives to the atomic scale, Science 335 (2012) 64–67.

[69] W. Lee, H. Han, A. Lotnyk, M.A. Schubert, S. Senz, M. Alexe, et al., Individually addressable epitaxial ferroelectric nanocapacitor arrays with near Tb $inch^{-2}$ density, Nat. Nanotechnol. 3 (2008) 402–407.

[70] T. Woldu, B. Raneesh, M.V. Ramana Reddy, N. Kalarikkal, Grain size dependent magnetoelectric coupling of $BaTiO_3$ nanoparticles, RSC Adv. 6 (2016) 7886–7892.

[71] L. Liang, X. Kang, Y. Sang, H. Liu, One-dimensional ferroelectric nanostructures: synthesis, properties, and applications, Adv. Sci 3 (2016) 1500358–1500421.

[72] J. Hong, D. Fang, Size-dependent ferroelectric behaviors of $BaTiO_3$ nanowires, Appl. Phys. Lett. 92 (2008) 012906–012913.

[73] J. Kaur, R.K. Kotnala, K.C. Verma, Multiferroic properties of $Ba(Fe_xTi_{1-x})O_3$ nanorods, Mater. Lett. 65 (2011) 3160–3163.

[74] C.O. Amorim, F. Figueiras, J.S. Amaral, P. Mirzadeh Vaghefi, P.B. Tavares, M.R. Correia, et al., Peculiar magnetoelectric coupling in $BaTiO_3$:Fe_{113} ppm nanoscopic segregations, ACS Appl. Mater. Interf 7 (2015) 24741–24747.

[75] D.P. Dutta, M. Roy, N. Maiti, A.K. Tyagi, Phase evolution in sonochemically synthesized Fe^{3+} doped $BaTiO_3$ nanocrystallites: structural, magnetic and ferroelectric characterisation, Phys. Chem. Chem. Phys. 18 (2016) 9758–9769.

[76] M. Kunz, I.D. Brown, Out-of-center distortions around octahedrally coordinated d^0 transition metals, J. Solid State Chem. 115 (1995) 395–406.

[77] P.S. Halasyamani, K.R. Poeppelmeier, Noncentrosymmetric oxides, Chem. Mater. 10 (1998) 2753–2769.

[78] S.W. Cheong, M. Mostovoy, Multiferroics: a magnetic twist for ferroelectricity, Nat. Mater. 6 (2007) 13–20.

[79] A.M. Kadomtseval, A.K. Zvezdin, Y.F. Popov, A.P. Pyatakov, G.P. Vorobev, Space-time parity violation and magnetoelectric interactions in antiferromagnets, JETP Lett. 79 (2004) 571–581.

[80] M. Fiebig, T. Lottermoser, D. Frohlich, A.V. Goltsev, R.V. Pisarev, Observation of coupled magnetic and electric domains, Nature 419 (2002) 818–820.

[81] K.C. Verma, R.K. Kotnala, Multiferroic magnetoelectric coupling and relaxor ferroelectric behavior in $0.7BiFeO_3–0.3BaTiO_3$ nanocrystals, Solid State Commun. 151 (2011) 920–923.

[82] S.N. Dong, et al., Magnetodielectric effect and tunable dielectric properties of $LaMn_{1-x}Fe_xO_3$, J. Am. Ceram. Soc. 93 (2010) 3814–3818.

[83] K.C. Verma, R.K. Kotnala, Nanostructural and lattice contributions to multiferroism in $NiFe_2O_4/BaTiO_3$, Mater. Chem. Phys. 174 (2016) 120–128.

[84] K.W. Wagner, Electricity magnetism, Ann. Phys. 40 (1913) 817.

[85] K. Raidongia, A. Nag, A. Sundaresan, C.N.R. Rao, Multiferroic and magnetoelectric properties of core-shell $CoFe_2O_4$@$BaTiO_3$ nanocomposites, Appl. Phys. Lett. 97 (2010) 062904–062913.

[86] O.G. Udalov, N.M. Chtchelkatchev, I.S. Beloborodov, Coupling of ferroelectricity and ferromagnetism through Coulomb blockade in composite multiferroics, Phys. Rev. B 89 (2014) 174203–174205.

[87] J.M. Hu, L.Q. Chen, C.W. Nan, Multiferroic heterostructures integrating ferroelectric and magnetic materials, Adv. Mater. 28 (2016) 15–39.

[88] J.S. Andrew, J.D. Starr, M.A.K. Budi, Prospects for nanostructured multiferroic composite materials, Scr. Mater. 74 (2014) 38–43.

[89] J. Ma, J. Hu, Z. Li, C.W. Nan, Recent progress in multiferroic magnetoelectric composites: from bulk to thin films, Adv. Mater. 23 (2011) 1062–1087.

[90] S.Z. Lu, X. Qi, RF magnetron co-sputtering growth and characterisation of multiferroic composite films of $Ni_{0.5}Zn_{0.5}Fe_2O_4$ + $BiFeO_3$, J. Mater. Chem. C 4 (2016) 8679–8686.

[91] W.J. Lin, X. Qi, G. Chern, J.C.A. Huang, Epitaxial growth and exchange coupling of spinel ferrimagnet $Ni_{0.3}Zn_{0.7}Fe_2O_4$ on multiferroic $BiFeO_3$, Thin Solid Films 519 (2011) 8326–8329.

[92] A. Shan, X. Wu, J. Lu, C. Chen, R. Wang, Phase formations and magnetic properties of single crystal nickel ferrite ($NiFe_2O_4$) with different morphologies, CrystEngComm 17 (2015) 1603–1608.

[93] V. Tileli, M. Duchamp, A.K. Axelsson, M. Valant, R.E. Dunin-Borkowski, N.M. Alford, On stoichiometry and intermixing at the spinel/perovskite interface in $CoFe_2O_4/BaTiO_3$ thin films, Nanoscale 7 (2015) 218–224.

[94] K.C. Verma, S. Singh, S.K. Tripathi, R.K. Kotnala, Multiferroic $Ni_{0.6}Zn_{0.4}Fe_2O_4–BaTiO_3$ nanostructures: magnetoelectric coupling, dielectric, and fluorescence, J. Appl. Phys. 116 (2014) 124103–124106.

[95] K.C. Verma, R.K. Kotnala, Magneto-electric/dielectric and fluorescence effects in multiferroic $xBaTiO_3$-$(1-x)ZnFe_2O_4$ nanostructures, RSC Adv. 4 (2014) 60234–60242.

[96] S. Singh, M. Singh, R.K. Kotnala, K.C. Verma, Hydrothermal synthesis of $NiFe_2O_4$, $Ni_{0.6}Zn_{0.4}Fe_2O_4$ and $Ni_{0.6}Zn_{0.4}Fe_2O_4/SrFe_2O_4$: nanostructure, magnetic and dielectric properties, Ind. J. Pure Appl. Phys. 52 (2014) 550–555.

[97] L. Zhang, J. Zhai, W. Mo, X. Yao, Electric and magnetic properties of $(x)CoFe_2O_4–(1-x)BaTiO_3$ thick film prepared by electrophoretic deposition technique, Solid State Sci. 13 (2011) 321.

[98] Y. Wang, J. Hu, Y. Lin, C.W. Nan, Multiferroic magnetoelectric composite nanostructures, NPG Asia Mater. 2 (2010) 61–68.

[99] M.I. Bichurin, V.M. Petrov, G. Srinivasan, Theory of low-frequency magnetoelectric coupling in magnetostrictive–piezoelectric bilayers, Phys. Rev. B 68 (2003) 054402–054413.

[100] H. Zheng, J. Wang, S.E. Lofland, Z. Ma, L. Mohaddes-Ardabili, T. Zhao, et al., Multiferroic $BaTiO_3-CoFe_2O_4$ nanostructures, Science 303 (2004) 661–663.

[101] M. Lorenz, V. Lazenka, P. Schwinkendorf, F. Bern, M. Ziese, H. Modarresi, et al., Multiferroic $BaTiO_3-BiFeO_3$ composite thin films and multilayers: strain engineering and magnetoelectric coupling, J. Phys. D Appl. Phys. 47 (2014) 135303–135310.

[102] J. Wang, J.B. Neaton, H. Zheng, V. Nagarajan, S.B. Ogale, B. Liu, et al., Epitaxial $BiFeO_3$ multiferroic thin film heterostructures, Science 299 (2003) 1719–1722.

[103] V.V. Shvartsman, F. Alawneh, P. Borisov, D. Kozodaev, D.C. Lupascu, Converse magnetoelectric effect in $CoFe_2O_4-BaTiO_3$ composites with a core-shell structure, Smart Mater. Struct. 20 (2011) 075006.

[104] T. Nan, Y. Hui, M. Rinaldi, N.X. Sun, Self-biased 215 MHz magnetoelectric NEMS resonator for ultra-sensitive DC magnetic field detection, Sci. Rep 3 (2013) 1985–1986.

[105] H. Ohno, Making nonmagnetic semiconductors ferromagnetic, Science 281 (1998) 951–955.

[106] K.R. Kittilstved, W.K. Liu, D.R. Gamelin, Electronic structure origins of polarity-dependent high-T_C ferromagnetism in oxide-diluted magnetic semiconductors, Nat. Mater. 5 (2006) 291–297.

[107] K.C. Verma, R.K. Kotnala, Oxygen vacancy induced by La and Fe into ZnO nanoparticles to modify ferromagnetic ordering, J. Solid State Chem. 237 (2016) 211–218.

[108] D. Gao, Z. Zhang, J. Fu, Y. Xu, J. Qi, D. Xue, Room temperature ferromagnetism of pure ZnO nanoparticles, J. Appl. Phys. 105 (2009) 113928–113934.

[109] M.A. Garcia, J.M. Merino, E. Fernandez Pinel, A. Quesada, J. de la Venta, M.L. Ruız Gonzalez, et al., Magnetic properties of ZnO nanoparticles, Nano Lett. 7 (6) (2007) 1490–1494.

[110] X. Xu, C. Xu, J. Dai, J. Hu, F. Li, S. Zhang, Size dependence of defect-induced room temperature ferromagnetism in undoped ZnO nanoparticles, J. Phys. Chem. C 116 (2012) 8813–8818.

[111] K. Sato, H. Katayama-Yoshida, Ferromagnetism in a transition metal atom doped ZnO, Phys. E 10 (2001) 251–255.

[112] P. Gopal, N.A. Spaldin, Magnetic interactions in transition-metal-doped ZnO: An ab initio study, Phys. Rev. B 74 (2006) 094418–094419.

[113] D. Karmakar, S.K. Mandal, R.M. Kadam, P.L. Paulose, A.K. Rajarajan, T.K. Nath, et al., Ferromagnetism in Fe-doped ZnO nanocrystals: experiment and theory, Phys. Rev. B 75 (2007) 144404–144414.

[114] N.A. Spaldin, Search for ferromagnetism in transition-metal-doped piezoelectric ZnO, Phys. Rev. B 69 (2004) 125201–125207.

[115] M. Venkatesan, C.B. Fitzgerald, J.G. Lunney, J.M.D. Coey, Anisotropic ferromagnetism in substituted zinc oxide, Phys. Rev. Lett. 93 (2004) 177206–177214.

[116] J. Kaur, R.K. Kotnala, V. Gupta, K.C. Verma, Anionic polymerization in Co and Fe doped ZnO: nanorods, magnetism and photoactivity, Curr. Appl. Phys. 14 (2014) 749–756.

[117] J.M.D. Coey, M. Venkatesan, C.B. Fitzgerald, Donor impurity band exchange in dilute ferromagnetic oxides, Nat. Mater. 4 (2005) 173–179.

[118] K.C. Zhang, Y.F. Li, Y. Liu, F. Chi, Density-functional study on the robust ferromagnetism in rare-earth element Yb-doped SnO_2, J. Magn. Magn. Mater. 360 (2014) 165–168.

[119] K.C. Verma, R.K. Kotnala, Understanding lattice defects to influence ferromagnetic order of ZnO nanoparticles by Ni, Cu, Ce ions, J. Solid State Chem. 246 (2017) 150–159.

[120] S.A. Aravindh, U. Schwingenschloegl, I.S. Roqan, Ferromagnetism in Gd doped ZnO nanowires: a first principles study, J. Appl. Phys. 116 (2014) 233906–233915.

[121] A. Kaminski, S.D. Sarma, Polaron percolation in diluted magnetic semiconductors, Phys. Rev. Lett. 88 (2002) 247202.

[122] C. Song, K.W. Geng, F. Zeng, X.B. Wang, Y.X. Shen, F. Pan, et al., Giant magnetic moment in an anomalous ferromagnetic insulator: Co-doped ZnO, Phys. Rev. B 73 (2006) 024405–024406.

[123] G.S. Chang, E.Z. Kurmaev, D.W. Boukhvalov, L.D. Finkelstein, S. Colis, T.M. Pedersen, et al., Effect of Co and O defects on the magnetism in Co-doped ZnO: experiment and theory, Phys. Rev. B 75 (2007) 195215–195217.

[124] J. Xiao, T. Frauenheim, T. Heine, A. Kuc, Temperature-mediated magnetism in Fe-doped ZnO semiconductors, J. Phys. Chem. C 117 (2013) 5338–5342.

[125] I. Bantounas, V. Singaravelu, I.S. Roqan, U. Schwingenschlogl, Structural and magnetic properties of Gd-doped ZnO, J. Mater. Chem. C 2 (2014) 10331–10336.

[126] K.C. Verma, R.K. Kotnala, Defects-assisted ferromagnetism due to bound magnetic polarons in Ce into Fe, Co:ZnO nanoparticles and first-principle calculations, Phys. Chem. Chem. Phys. 18 (2016) 5647–5657.

[127] Y.K. Frodason, K.M. Johansen, T.S. Bjorheim, B.G. Svensson, A. Alkauskas, Zn vacancy as a polaronic hole trap in ZnO, Phys. Rev. B 95 (2017) 094105–094108.

[128] P. Nath, A. Chakraborti, D. Sanyal, Ab initio calculation of magnetic properties of p block element doped ZnO, RSC Adv. 4 (2014) 45598–45602.

[129] D. Karmakar, T.V. Chandrasekhar Rao, J.V. Yakhmi, A. Yaresko, V.N. Antonov, R.M. Kadam, et al., Electronic structure and magnetic properties of (Fe, Co)-co doped ZnO: theory and experiment, Phys. Rev. B 81 (2010) 184421–184513.

[130] K.C. Verma, R.K. Kotnala, Defects due to lattice distortion and nano-size intermediate ferromagnetism in La, Gd substituted $Zn_{0.95}Co_{0.05}O$, Curr. Appl. Phys. 16 (2016) 175–182.

[131] K.C. Verma, R.K. Kotnala, Realizing ferromagnetic ordering in SnO_2 and ZnO nanostructures with Fe, Co, Ce ions, Phys. Chem. Chem. Phys. 18 (2016) 17565–17574.

[132] W. Zheng, P. Kumar, A. Washington, Z. Wang, N.S. Dalal, G.F. Strouse, et al., Quantum phase transition from superparamagnetic to quantum superparamagnetic state in ultrasmall $Cd_{1-x}Cr(II)_xSe$ quantum dots? J. Am. Chem. Soc. 134 (2012) 2172–2179.

[133] A. Kawska, P. Duchstein, O. Hochrein, D. Zahn, Atomistic mechanisms of ZnO aggregation from ethanolic solution: ion association, proton transfer, and self-organization, Nano Lett. 8 (8) (2008) 2336–2340.

[134] F. Ahmed, S. Kumar, N. Arshi, M.S. Anwar, B.H. Koo, Morphological evolution between nanorods to nanosheets and room temperature ferromagnetism of Fe-doped ZnO nanostructures, CrystEngComm 14 (2012) 4016–4026.

[135] M. Zhang, K. Chen, X. Chen, X. Peng, X. Sun, D. Xue, Crystallization of FeOOH via iron salts: an anionchemoaffinity controlled hydrolysis toward high performance inorganic pseudocapacitor materials, CrystEngComm 17 (2015) 1917–1922.

[136] D.E. Motaung, G.H. Mhlongo, S.S. Nkosi, G.F. Malgas, B.W. Mwakikunga, E. Coetsee, et al., Shape-selective dependence of room temperature ferromagnetism induced by hierarchical ZnO nanostructures, ACS Appl. Mater. Interf 6 (2014) 8981–8995.

[137] X. Wang, J. Xu, B. Zhang, H. Yu, J. Wang, X. Zhang, J. Yu, Q. Li, Signature of intrinsic high-temperature ferromagnetism in cobalt-doped zinc oxide nanocrystals, Adv. Mater. 18 (2006) 2476–2480.

[138] W.W. Li, W.L. Yu, Y.J. Jiang, C.B. Jing, J.Y. Zhu, M. Zhu, et al., Structure, optical, and room-temperature ferromagnetic properties of pure and transition-metal-(Cr, Mn, and Ni)-doped ZnO nanocrystalline films grown by the sol-gel method, J. Phys. Chem. C 114 (2010) 11951–11957.

[139] K.C. Verma, R. Bhatia, S. Kumar, R.K. Kotnala, Vacancies driven magnetic ordering in ZnO nanoparticles due to low concentrated Co ions, Mater. Res. Exp. 3 (2016) 076103–076112.

[140] S. Kuriakose, B. Satpati, S. Mohapatra, Enhanced photocatalytic activity of Co doped ZnO nanodisks and nanorods prepared by a facile wet chemical method, Phys. Chem. Chem. Phys. 16 (2014) 12741–12749.

[141] S. Kumar, S. Basu, B. Rana, A. Barman, S. Chatterjee, S.N. Jha, et al., Structural, optical and magnetic properties of sol-gel derived ZnO:Co diluted magnetic semiconductor nanocrystals: an EXAFS study, J. Mater. Chem. C 2 (2014) 481–495.

[142] D.Y. Inamdar, A.D. Lad, A.K. Pathak, I. Dubenko, N. Ali, S. Mahamuni, Ferromagnetism in ZnO nanocrystals: doping and surface chemistry, J. Phys. Chem. C 114 (2010) 1451–1459.

[143] P. Klason, T.M. Borseth, Q.X. Zhao, B.G. Svensson, A.Y. Kuznetsov, P.J. Bergman, et al., Temperature dependence and decay times of zinc and oxygen vacancy related photoluminescence bands in zinc oxide, Solid State Commun. 145 (2008) 321–326.

[144] J.B. Yi, C.C. Lim, G.Z. Xing, H.M. Fan, L.H. Van, S.L. Huang, et al., Ferromagnetism in dilute magnetic semiconductors through defect engineering: Li-doped ZnO, Phys. Rev. Lett. 104 (2010) 137201.

[145] A.K. Rana, Y. Kumar, P. Rajput, S.N. Jha, D. Bhattacharyya, P.M. Shirage, Search for origin of room temperature ferromagnetism properties in Ni-doped ZnO nanostructure, ACS Appl. Mater. Interf 9 (2017) 7691–7700.

[146] K. Sato, H. Katayama-Yoshida, Material design of GaN-based ferromagnetic diluted magnetic semiconductors, Jpn. J. Appl. Phys. 40 (2001) L485–L487.

[147] M. Ram, K.C. Verma, N.S. Negi, Room temperature ferromagnetism and photoluminescence of nanocrystalline $Zn_{1-x}Co_xO$ thin films prepared by sol-gel MOD, Phil. Mag. Lett. 94 (7) (2014) 404–414.

[148] J.M.D. Coey, A.P. Douvalis, C.B. Fitzgerald, M. Venkatesan, Ferromagnetism in Fe-doped SnO_2 thin films, App. Phys. Lett. 84 (2004) 1332–1334.

[149] A. Walsh, J.L.F. Da Silva, S.H. Wei, Theoretical description of carrier mediated magnetism in cobalt doped ZnO, Phys. Rev. Lett. 100 (2008) 256401–256404.

[150] C. Zener, Interaction between the d-shells in the transition metals. II. Ferromagnetic comyountls of manganese with perovskite structure, Phys. Rev 82 (1951) 403–405.

[151] X. Liu, J. Liu, S. Zhang, Z. Nan, Q. Shi, Structural, magnetic, and thermodynamic evolutions of Zn-doped Fe_3O_4 nanoparticles synthesized using a one-step solvothermal method, J. Phys. Chem. C 120 (2016) 1328–1341.

[152] J.D. Bryan, S.M. Heald, S.A. Chambers, D.R. Gamelin, Strong room-temperature ferromagnetism in Co^{2+}-doped TiO_2 made from colloidal nanocrystals, J. Am. Chem. Soc. 126 (2004) 11640–11647.

[153] J. Xiao, A. Kuc, T. Frauenheim, T. Heine, Stabilization mechanism of ZnO nanoparticles by Fe doping, Phys. Rev. Lett. 112 (2014) 106102–106104.

[154] S.J. Gilliland, J.A. Sans, J.F. Sanchez-Royo, G. Almonacid, B. Garcıa-Domene, A. Segura, et al., Role of p–d and s–d interactions in the electronic structure and band gap of $Zn_{1-x}M_xO$ (M = Cr, Mn, Fe, Co, Ni, and Cu): photoelectron and optical spectroscopy and first-principles band structure calculations, Phys. Rev. B 86 (2012) 155203–155213.

[155] A. Ney, K. Ollefs, S. Ye, T. Kammermeier, V. Ney, T.C. Kaspar, et al., Absence of intrinsic ferromagnetic interactions of isolated and paired co dopant atoms in $Zn_{1-x}Co_xO$ with high structural perfection, Phys. Rev. Lett. 100 (2008) 1572014.

[156] B. Pal, D. Sarkar, P.K. Giri, Structural, optical, and magnetic properties of Ni doped ZnO nanoparticles: correlation of magnetic moment with defect density, Appl. Surf. Sci. 356 (2015) 804–811.

[157] D. Chakrabortia, J. Narayan, J.T. Prater, Room temperature ferromagnetism in $Zn_{1-x}Cu_xO$ thin films, Appl. Phys. Lett. 90 (2007) 062504–062513.

[158] S. Yılmaz, E. McGlynn, E. Bacaksız, S. Ozcan, D. Byrne, M.O. Henry, et al., Effects of Cu diffusion-doping on structural, optical, and magnetic properties of ZnO nanorod arrays grown by vapor phase transport method, J. Appl. Phys. 111 (2012) 013903–013907.

[159] B. Yao, P. Wang, S. Wang, M. Zhang, Ce doping influence on the magnetic phase transition in In_2S_3:Ce nanoparticles, CrystEngComm 16 (2014) 2584–2588.

[160] A. Singhal, S.N. Achary, J. Manjanna, S. Chatterjee, P. Ayyub, A.K. Tyagi, Chemical synthesis and structural and magnetic properties of dispersible cobalt- and nickel-doped ZnO nanocrystals, J. Phys. Chem. C 114 (2010) 3422–3430.

[161] D.Y. Inamdar, A.K. Pathak, I. Dubenko, N. Ali, S. Mahamuni, Room temperature ferromagnetism and photoluminescence of Fe doped ZnO nanocrystals, J. Phys. Chem. C 115 (2011) 23671–23676.

[162] J.C. Denardin, A.L. Brandl, M. Knobel, P. Panissod, A.B. Pakhomov, H. Liu, et al., Thermoremanence and zero-field-cooled/field-cooled magnetization study of $Co_x(SiO_2)_{1-x}$ granular films, Phys. Rev. B 65 (2002) 064422–064428.

[163] S. Basu, D.Y. Inamdar, S. Mahamuni, A. Chakrabarti, C. Kamal, G.R. Kumar, et al., Local structure investigation of cobalt and manganese doped ZnO nanocrystals and its correlation with magnetic properties, J. Phys. Chem. C 118 (2014) 9154–9164.

[164] S.S. Rathore, S. Vitta, Effect of divalent Ba cation substitution with Sr on coupled "multiglass" state in the magnetoelectric multiferroic compound $Ba_3NbFe_3Si_2O_{14}$, Sci. Rep 5 (2015) 9751–9811.

[165] E. Fertman, S. Dolya, V. Desnenko, A. Beznosov, M. Kajnakova, A. Feher, Cluster glass magnetism in the phase-separated $Nd_{2/3}Ca_{1/3}MnO_3$ perovskite, J. Magn. Magn. Mater. 324 (2012) 3213–3217.

[166] J. Wang, S. Hou, H. Chen, L. Xiang, Defects-induced room temperature ferromagnetism in ZnO nanorods grown from ε-$Zn(OH)_2$, J. Phys. Chem. C 118 (2014) 19469–19476.

[167] U. Ozguur, Y.I. Alivov, C. Liu, A. Teke, M. Reshchikov, S. Dogan, et al., A comprehensive review of ZnO materials and devices, J. Appl. Phys. 98 (2005) 041301.

[168] Q. Wang, Q. Sun, G. Chen, Y. Kawazoe, P. Jena, Vacancy-induced magnetism in ZnO thin films and nanowires, Phys. Rev. B 77 (2006) 205411–205416.

[169] J. Marco, J. Gancedo, M. Gracia, J. Gautier, F. Berry, Characterization of the nickel cobaltite, $NiCo_2O_4$, prepared by several methods: an XRD, XANES, EXAFS, and XPS study, J. Solid State Chem. 153 (2000) 74–81.

[170] D.A. Pawlak, M. Ito, M. Oku, K. Shimamura, T. Fukuda, Interpretation of XPS O (1s) in mixed oxides proved on mixed perovskite crystals, J. Phys. Chem. B 106 (2002) 504–507.

[171] M.S.J. Marshall, D.T. Newell, D.J. Payne, R.G. Egdell, M.R. Castell, Atomic and electronic surface structures of dopants in oxides: STM and XPS of Nb- and La-doped $SrTiO_3(001)$, Phys. Rev. B 83 (2011) 035410–035416.

[172] B.B. Straumal, A. Mazilkin, S.G. Protasova, A.A. Myatiev, P.B. Straumal, G. Schutz, et al., Magnetization study of nanograined pure and Mn-doped ZnO films: formation of a ferromagnetic grain-boundary foam, Phys. Rev. B 79 (2009) 205206.

[173] T. Li, C.S. Ong, T.S. Herng, J.B. Yi, N.N. Bao, J.M. Xue, et al., Surface ferromagnetism in hydrogenated-ZnO film, Appl. Phys. Lett. 98 (2011) 152505.

PART III

Optics and Photonics

CHAPTER 7

Photonics Integrated Circuits

Shibnath Pathak

Research and Development, Lumentum Holdings Inc., San Jose, CA, United States

7.1 Introduction to Photonics

In 1962, the first semiconductor laser was demonstrated along with a rapid development of optical fiber, which opened a new research area called *photonics*. The name *photonics* came with analogy to electronics: electronics deal with electrons while photonics deal with photons. According to the Merriam-Webster Dictionary, the definition for photons is "a branch of physics that deals with the properties and applications of photons especially as a medium for transmitting information." The definition states the main application which drives the photonics development: *transmitting information*.

The journey of integrated photonics began at the same time as integrated electronics. In 1968, the first complementary metal oxide semiconductor (CMOS) integrated circuit was demonstrated. After that, electronics integrated circuit developed rapidly fulfilling the Moore's law: the number of transistors in a dense integrated circuit doubles approximately every two years. One year later, in 1969, Stewart Miller published a paper entitled "Integrated Optics: An Introduction." In this paper, Miller proposed that the photolithographic technique can be used to realize complex circuit. This paper is considered as the official beginning of integrated photonics. But, over the decade, integrated photonics remained under development in the lab. The reason behind this slow development is that integrated photonics has several drawbacks compared to electronics. Where an electronics circuit based on a single material platform required silicon, whereas photonics circuit required multiple materials to realize all the functionalities, which dilutes the research focus. Another reason could be the market demand, which started changing over the last three decades due to the exponential increase of the data traffic sent over global communication networks.

7.2 Material Platform

Over the years, photonics has developed in several material platforms. The most famous material platforms are silica-on-silicon, III-V semiconductor materials, lithium niobate

Nanoelectronics. DOI: https://doi.org/10.1016/B978-0-12-813353-8.00008-7

($LiNbO_3$), silicon nitride-on-insulator (Si_3N_4), and silicon-on-insulator. The basic design of photonics material has a core and cladding material, where the refractive index of the core material is higher compared to the refractive index of the cladding material, and where both these core and cladding material should be transparent in the operation wavelength region. The relative difference in refractive index of the core and cladding is known as index contrast. Details of the waveguide design are discussed in Section 7.3.

7.2.1 Silica-on-Silicon

Silica-on-silicon is one of the most robust mature photonics platform, also known as planer lightwave circuit (PLC). Today, silica is the backbone of most of the photonics chips used in telecommunication and data communication. Optical waveguide in this platform has a germanium (Ge)-doped silica core and an undoped silica cladding on top of a silicon substrate. The refractive index of the core is 0.8%–3.5% higher compared to the refractive index of the cladding. The advantages of this platform are mature fabrication technology, low loss waveguide, and better coupling with optical fiber. But due to the low index contrast between the core and the cladding, the bend radius of the waveguide for low loss propagation is large, which makes the optical components less suitable for applications require high-density integration like on-chip interconnects.

7.2.2 III-V Semiconductor Materials

III-V semiconductor materials are important photonics materials due to their direct bandgap, which makes it suitable for active photonics components. Over the last few decades, these materials were used for making integrated laser, modulator, and photodetectors. We discuss these components in Section. III-V semiconductor materials are also used for passive photonics components, but due to high cost, low-yield fabrication technology, and lower integration density, the use of these materials are limited only for active photonics components.

7.2.3 Lithium Niobate

Lithium niobate ($LiNbO_3$) is commercially used for optical modulator. Other than modulation, $LiNbO_3$ is not possible to be used for laser or photodetector. Although some passive photonics component like AWG (see Section 7.4.2) is demonstrated in this waveguide platform, $LiNbO_3$ is limited to optical modulation due to large bend radius, high cost, and less reliable fabrication process. Often, $LiNbO_3$ modulators are integrated with silica passive components for optical communication applications.

7.2.4 Silicon Nitride

During the last few years, silicon nitride (Si_3N_4) gained interest for optical applications due to its interesting refractive index and CMOS compatibility, which means it is possible to use mature electronics fabrication facility for Si_3N_4 waveguide. At 1550 nm wavelength, the refractive index of the Si_3N_4 (waveguide core) is ~ 2 and silica (cladding) is 1.44. Therefore, Si_3N_4 is neither a very high index contrast platform like silicon-on-insulator (SOI) nor a low index contrast platform like silica-on-silicon. The advantages are that it does not suffer from high loss or low coupling with optical fiber like Si waveguide and it is possible to design a compact device due to small bend radius. The main disadvantage is that the platform does not have a mature fabrication process: high stress for higher than 400 nm thick Si3N4 layer, low quality, etc.

7.2.5 Silicon-on-Insulator

In SOI, a silicon waveguide layer is used as a core on top of a buried oxide which works as cladding. The refractive index of the silicon and the silicon dioxide are 3.476 and 1.444 for 1550 nm wavelength, respectively. The main advantages of this platform are CMOS compatibility, high index contrast waveguide which makes smaller bend, i.e., compact footprint. Due to the high integration density, silicon is capable of realizing multifunctionalities on a single chip. Small footprint increases the number of chips per wafer, which reduce the single chip cost significantly. The main disadvantage is that the silicon waveguides are highly sensitive to fabrication defects: sidewall roughness of the waveguide increases the loss and also induces phase error. On top of this, the silicon waveguides are strongly birefringent, have significant temperature dependence, and high coupling loss with optical fiber.

7.3 Waveguide Geometries

The simple definition of a waveguide should be a structure used for efficient electromagnetic power transfer from one point to another. In practice, various geometries are used as a waveguide, which depends on materials, operating wavelength of the light, etc. In this section, we discuss mainly the waveguide geometries used in integrated photonics, and for simplicity we use the waveguides designed for SOI platform as an example to derive the basics waveguide functionalities.

A general geometry of an optical waveguide has a higher refractive index (η_{core}) material surrounded by a lower refractive index ($\eta_{cladding}$) material, i.e., $\eta_{core} > \eta_{cladding}$. The refractive index of the waveguide determines the number of optical modes guided by the waveguide. In general, optical mode is a time-harmonics solution of Maxwell equations. These optical modes can be TE or TM polarized. An optical mode has two basic properties: effective refractive index (η_{eff}) and group index (η_{group}). Effective index can be determined from the phase velocity (ν_p) of the mode, which is the velocity of the phase at any frequency components of the mode propagate through the waveguide.

$$\eta_{eff} = \frac{c_0}{\nu_p} \tag{7.1}$$

Therefore, an effective index is a wavelength-dependent property of the waveguide. The effective index provides important information: the confinement of the mode into the waveguide core. Effective index cannot be higher than core index. The mode will be guided if the effective index lies in between the core and cladding index ($\eta_{core} > \eta_{eff} > \eta_{cladding}$). If the effective index is below the cladding index, the mode will not be guided by the waveguide ($\eta_{eff} < \eta_{cladding}$). If the effective index is close to the core index, light is tightly confined in the core ($\eta_{core} \sim \eta_{eff}$). If the effective index is close to the cladding index, light is poorly confined in the core ($\eta_{cladding} \sim \eta_{eff}$).

The group velocity (ν_g) is the velocity of the wave packets propagate through the waveguide. The group velocity determines the group index of the waveguide.

$$\eta_{group} = \frac{c_0}{\nu_g} \tag{7.2}$$

Unlike effective index, group index of the waveguide can be higher than the core index.

In photonics, waveguides of different shapes are used for different applications and material platform. Fig. 7.1 shows a list of waveguide geometries used in integrated photonics. Later, we discuss some commonly used waveguide geometries for SOI. In SOI, a 220 nm thick silicon layer is used as a core on top of a 2 μm thick thermal silicon dioxide, which is developed on top of a silicon substrate. For most of the applications, a silicon dioxide or air is used as a top and side cladding of the waveguides.

7.3.1 Slab Waveguide

Fig. 7.1(A) shows a schematic of silicon slab waveguide. A slab waveguide can propagate both TE and TM polarization light. Z-axis is the propagation direction and the geometry of the waveguide is invariant along X-axis, i.e., TE polarization will have only electric field components along X-axis and TM polarization will have only electric field components along Y-axis.

7.3.2 Ridge Waveguide

At 1550 nm wavelength, 450 nm-wide ridge waveguide work as a single-mode waveguide in SOI platform. Depending on applications, various materials are used as cladding. In this

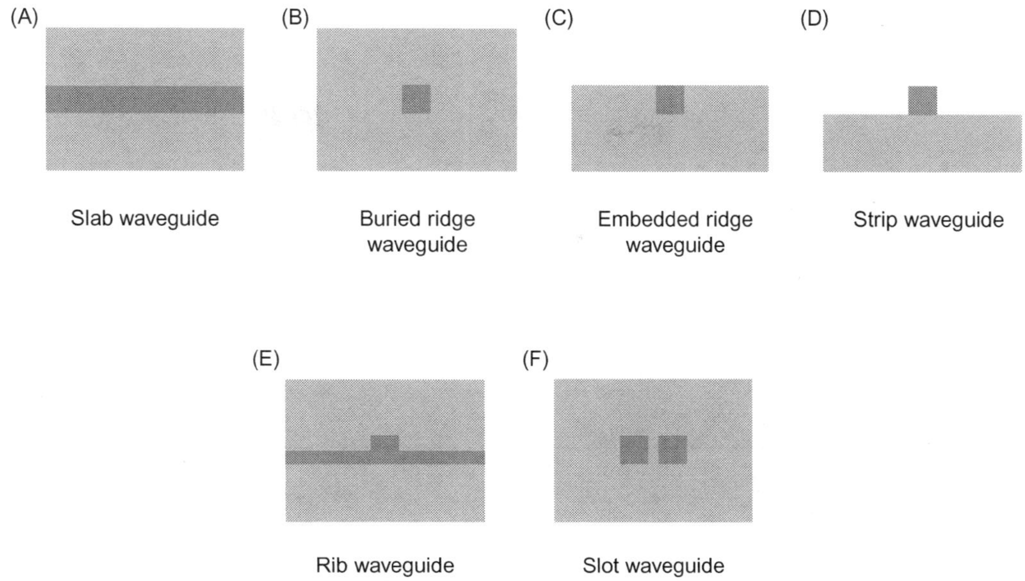

Figure 7.1
Schematic of optical waveguide geometries.

section, we discuss about ridge waveguide with silicon dioxide cladding and air cladding (see Fig. 7.2(A) and (D)). For simplicity, let us call the ridge waveguide with oxide cladding as buried waveguide and the ridge waveguide with air cladding as strip waveguide. The buried waveguide has symmetry in both x and y direction, but strip waveguide has symmetry only in x direction.

The electric field along X, Y, and Z axis is denoted as Ex, Ey, and Ez, respectively. For TE polarization light, Ex is the dominant field component and for TM polarization, Ey is the dominant field component. The other field components are weak but not zero, which is the reason often these modes are denoted as TE-like or TM-like modes. Fig. 7.2(B) and (C) shows these dominate field components for buried waveguide and Fig. 7.2(E) and (F) for strip waveguide. If we bring our attention to the TE Ex field for both waveguides, we see that the field profiles are symmetric as the waveguides are symmetric along X-axis. But the field has a discontinuity at the sidewalls and strong near the boundary. Any fabrication imperfection in these interfaces, i.e., sidewall roughness, will cause a high loss and phase error for TE polarization light. For TM polarization, Ey field of the two waveguides are strong at the top and bottom of the waveguide. Therefore, sidewall roughness will not have high impact on the loss or phase error for TM polarization. However, the effective index of this polarization is strongly sensitive to the effective width and height variation. The buried waveguide is symmetry along Y-axis, but the strip waveguide does not have the symmetry. From Fig. 7.2(F), we see that confinement near the top of the waveguide is much higher

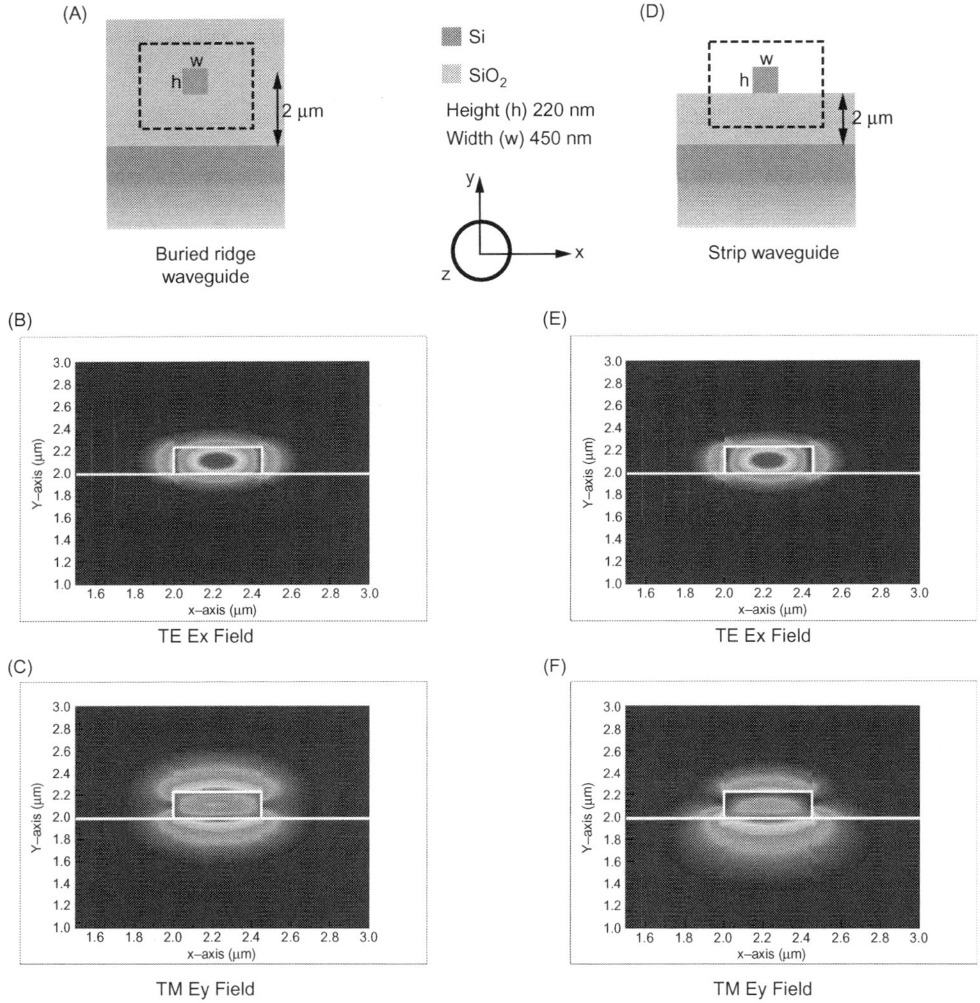

Figure 7.2

(A) Schematic of SOI buried ridge waveguide. (B) Ex field of TE polarized light for buried ridge waveguide. (C) Ey field of TM polarized light for buried ridge waveguide. (D) Schematic of SOI strip waveguide. (E) Ex field of TE polarized light for strip waveguide. (F) Ey field of TM polarized light for strip waveguide.

compared to the bottom of the waveguide as the index contrast at the silicon-air boundary is higher compared to the silicon-silica boundary.

Fig. 7.3(B) and (E) shows effective index variation with waveguide width at 1550 nm wavelength for TE and TM polarization. For 0.2 μm width, TE and TM modes are close to cutoff as the effective index is near the refractive index of the cladding (silicon dioxide/silica). The effective index of the TE mode at 0.45 μm width is higher compared to the

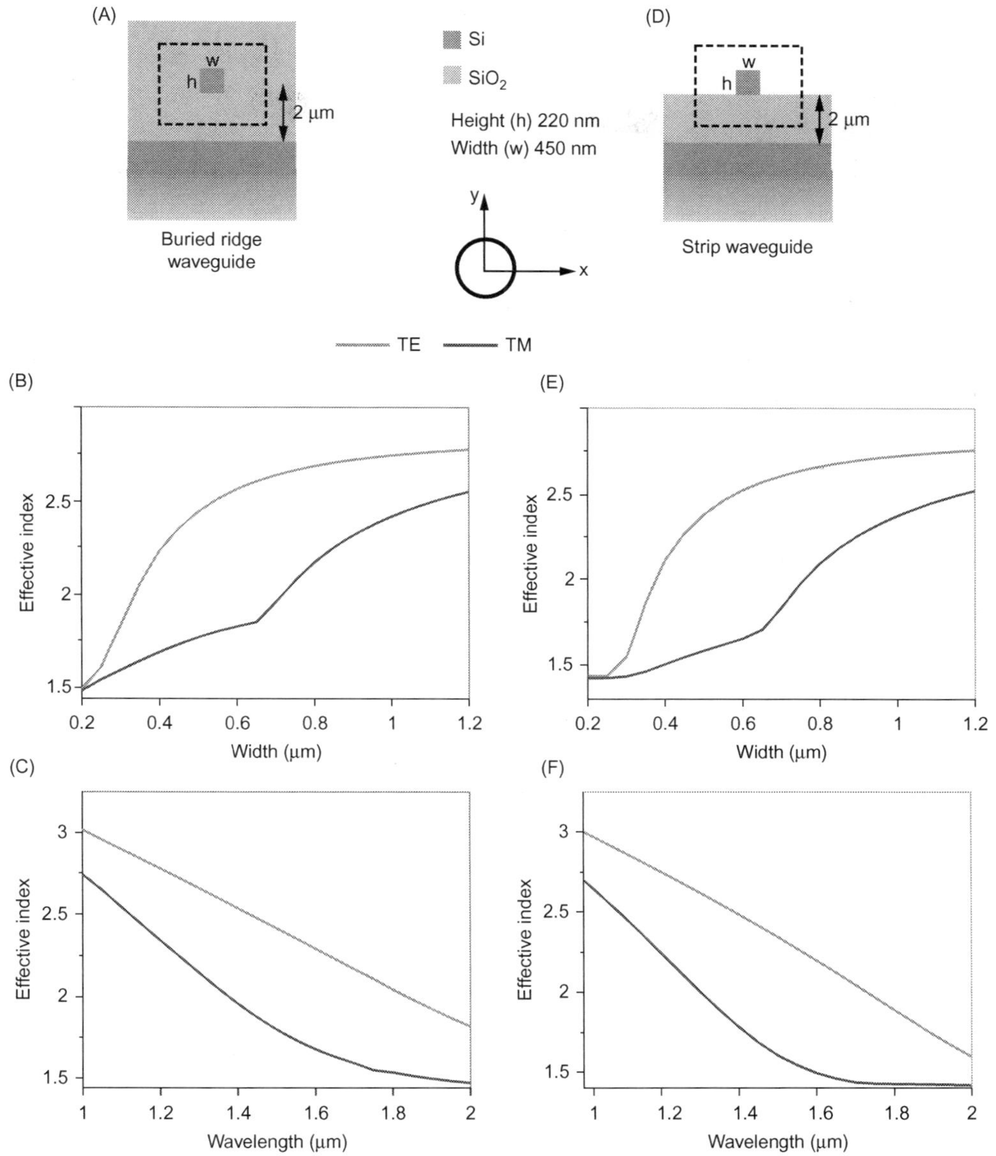

Figure 7.3

(A) Schematic of SOI buried ridge waveguide. (B) Effective index variation with waveguide width at 1550 nm wavelength for buried ridge waveguide. (C) Effective index variation with wavelength for 450 nm-wide buried ridge waveguide. (D) Schematic of SOI strip waveguide. (E) Effective index variation with waveguide width at 1550 nm wavelength. (F) Effective index variation with wavelength for 450 nm-wide strip waveguide.

refractive index of cladding for both buried and strip waveguide. The 0.45 μm-wide waveguide does not support higher order mode propagation and can be used for single-mode waveguide propagation. But the effective index of the TM mode is close to cutoff for 0.45 μm-wide strip waveguide. Therefore, 0.45 μm-wide strip waveguide will suffer from high loss. For TE mode, the effective index variation with width above 0.40 μm is small compared to the TM mode, which implies that the effective width variation due to fabrication imperfections will have higher impact on the effective index of the TM mode. The waveguide above 0.6 μm wide can propagate higher order TE modes, but has high fabrication tolerant and less impact due to sidewall roughness. In silicon photonics for long-distance propagation or phase error sensitive device, an adiabatic taper is used to transfer the light from single-mode waveguide to multimode waveguide without exciting unwanted higher order modes.

Fig. 7.3(C) and (F) shows effective index variation with wavelength for 0.45 μm-wide buried and strip waveguide, respectively. For TE mode, the effective index decreases almost linearly as we increase the wavelength from 1 to 2 μm. But for TM mode the decrement is linear up to 1.4 μm wavelength, after which the effective index decreases exponentially and saturates at the refractive index of silica. The decrement rate is faster for strip waveguide compared to the buried waveguide. From Fig. 7.3(F), it is clear that a 0.45 μm-wide strip waveguide will not guide TM mode above 1.6 μm wavelength.

7.3.3 Rib Waveguide

Fig. 7.4(A) shows the schematic of a silicon rib waveguide: 450 nm wide and 150 nm slab thickness, which is commonly used in integrated silicon photonics. Fig. 7.4(B) shows the Ex field component of the TE-like mode, which is confined in the slab region but has strong discontinuity at the boundary. As the etch is only 80 nm to fabricate this waveguide, the effect of sidewall roughness for rib waveguide is smaller compared to the ridge waveguide. Therefore, the loss and phase error of the waveguide is lower, but as the confinement is in the slab region the bend radius of the waveguide is large. Fig. 7.4(C) shows Ey field component of the TM-like mode. But it has been experimentally verified that TM-like mode in rib waveguide is inherently leaky as the effect index of the TM-like mode is in the same range of the higher order TE slab modes. These slab modes propagate in the slab at any random angle, which allow having phase match with guided TM-like mode and TM-like mode will be coupled to the slab mode. Therefore, Fig. 7.4(D) and (E) shows effective index variation with waveguide width and wavelength for TE polarization only. For fundamental TE-like mode, if the width of the waveguide decreases, the mode will be pushed to the slab region and it will no longer be guided. From Fig. 7.4(D), it is clear that the effective index variation with width variation is smaller compared to the ridge

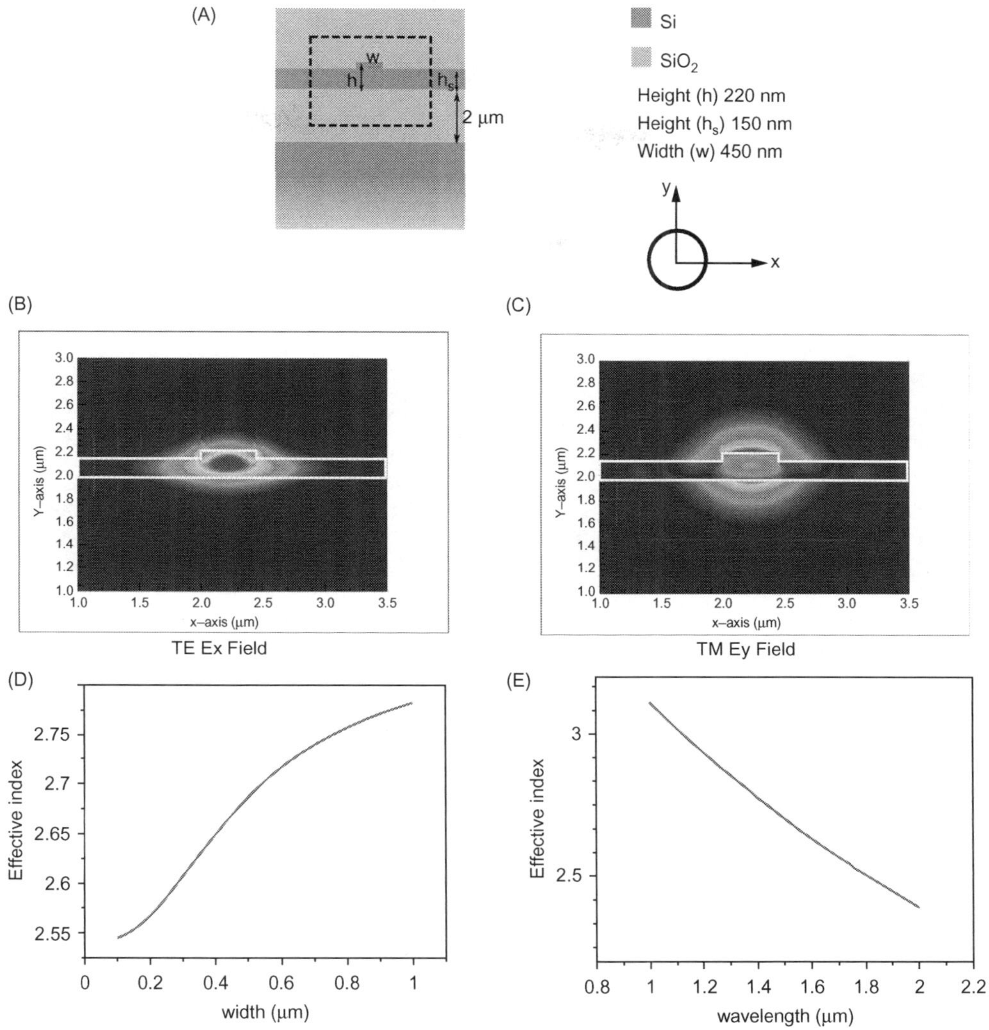

Figure 7.4
(A) Schematic of SOI rib waveguide. (B) Ex field of TE polarized light. (C) Ey field of TM polarized light. (D) Effective index variation with waveguide width at 1550 nm wavelength for TE-like mode. (E) Effective index variation with wavelength for 450 nm-wide rib waveguide.

waveguide, which makes rib waveguide less fabrication error dependent. But unlike the ridge waveguide, rib waveguide has an extra fabrication-dependent parameter: slab height.

7.3.4 Slot Waveguide

Two silicon ridge waveguides are placed close to create a slot waveguide, as shown in Fig. 7.5(A). For our simulation, we used two 250 nm-wide ridge waveguide separated by

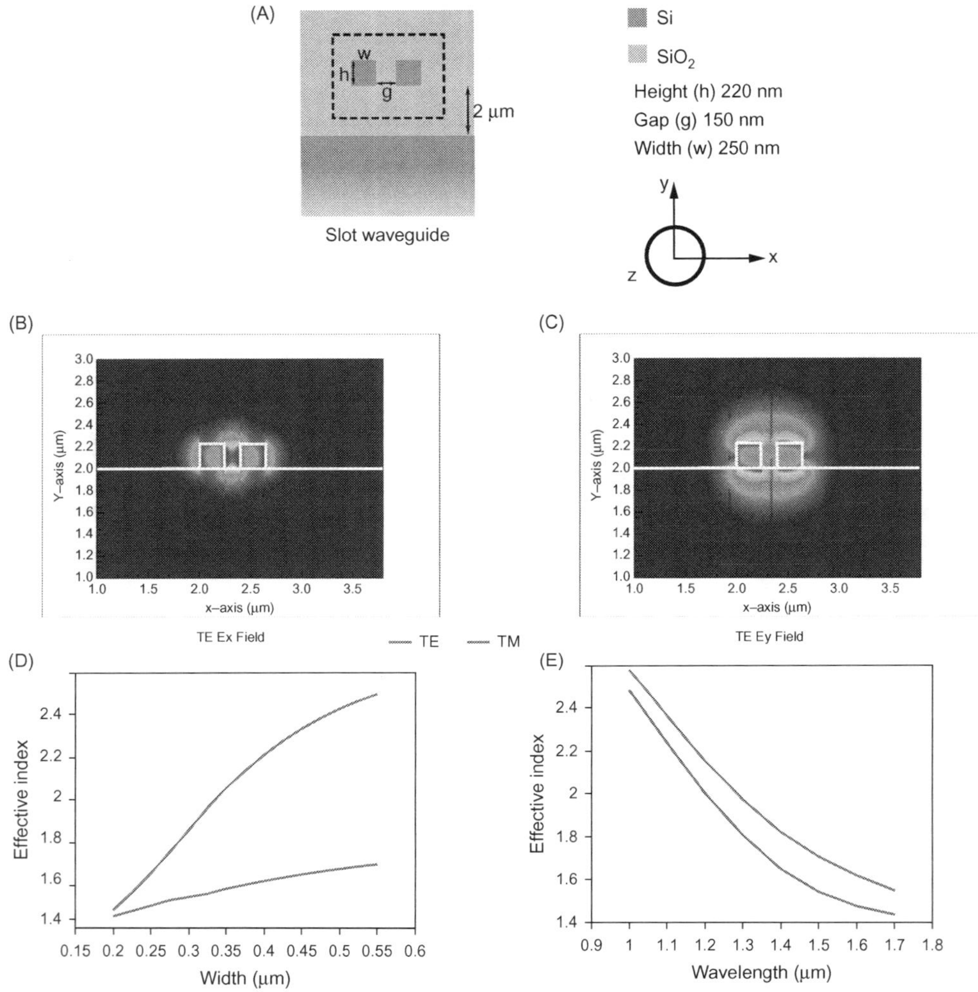

Figure 7.5

(A) Schematic of SOI slot waveguide. (B) Ex field of TE polarized light. (C) Ey field of TM polarized light. (D) Effective index variation with waveguide width at 1550 nm wavelength. (E) Effective index variation with wavelength for 250 nm wide and 150 nm gap slot waveguide.

150 nm gap with silicon dioxide cladding. From Fig. 7.5(B) and (C), we see that Ex field is highly confined in the gap for TE polarization and for TM polarization the Ey field components are highly confined at the top and bottom of the waveguide. As the Ex field is confined in the slot region, the slot waveguide is suitable for sensing application. Instead of oxide cladding, any gas or liquid can be used, which shifts the effective index of the waveguide. Using prior calibration, one can use this shift to detect unknown gas or liquid. Fig. 7.5(C) and (D) shows the effective index as function of waveguide width and wavelength for both TE and TM polarization. If the waveguide is small enough, the mode will be guided by the gap, otherwise the mode will be guided by the individual silicon ridge

waveguides and work as a directional coupler (DC). This phenomena is also true for lower wavelength and for higher wavelength the waveguide width needs to increase to guide the mode. The biggest challenge to fabricate this waveguide is to fabricate the gap with high precision. The TE-like mode has highly confined in the gap, therefore sidewall roughness will cause a high loss and phase error.

7.4 Passive Devices

Several components/devices are used to build an integrated photonics circuit. These components are divided into two category: passive and active devices. A simple definition of active and passive components can be an optical functionality obtained with or without using any electrical inputs. In this section, we discuss passive components.

7.4.1 Optical Couplers

An optical coupler is a component to couple the light from an optical fiber to the waveguide and vice versa. There are two kinds of coupling mechanism used in photonics: in-line coupling and vertical coupling. An edge couple is used for in-line coupling, where grating coupler is a commonly used component for vertical coupling. Fig. 7.6(A) and (B) shows schematic of the edge and grating coupler for in-line and vertical, respectively. In general, the single-mode fiber has a core diameter of 8–10 μm and cladding diameter of 125 μm with a mode field diameter (MFD) of ~10.5 μm. The MFD is that at which the power density is reduced to $1/e^2$ of the maximum power density. For our simulation, we use this parameter to estimate the coupling efficiency from the fiber to the waveguide.

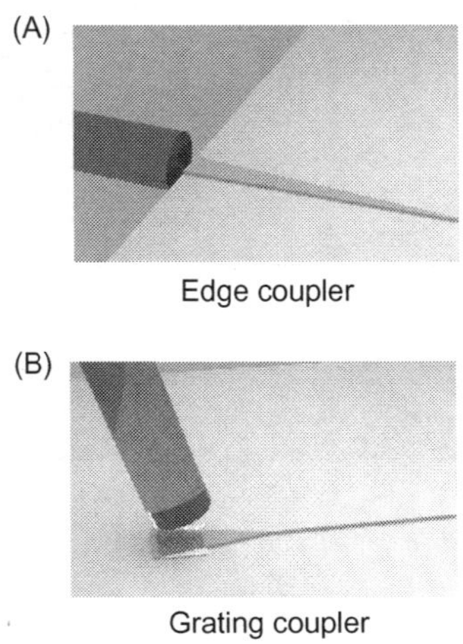

Figure 7.6
Schematic of (A) an edge coupler and (B) a grating coupler.

7.4.1.1 Edge Coupler

Edge coupler is the commonly used method to couple the light as it allows to couple both TE and TM polarizations with a wide range of wavelength bandwidth. Fig. 7.6(A) shows schematic of an edge coupler. The optical loss during coupling the light from a single-mode fiber to a single-mode waveguide is known as coupling loss. The coupling loss depends on modal shape mismatch between the fiber and waveguide. From Fig. 7.7, we see that the

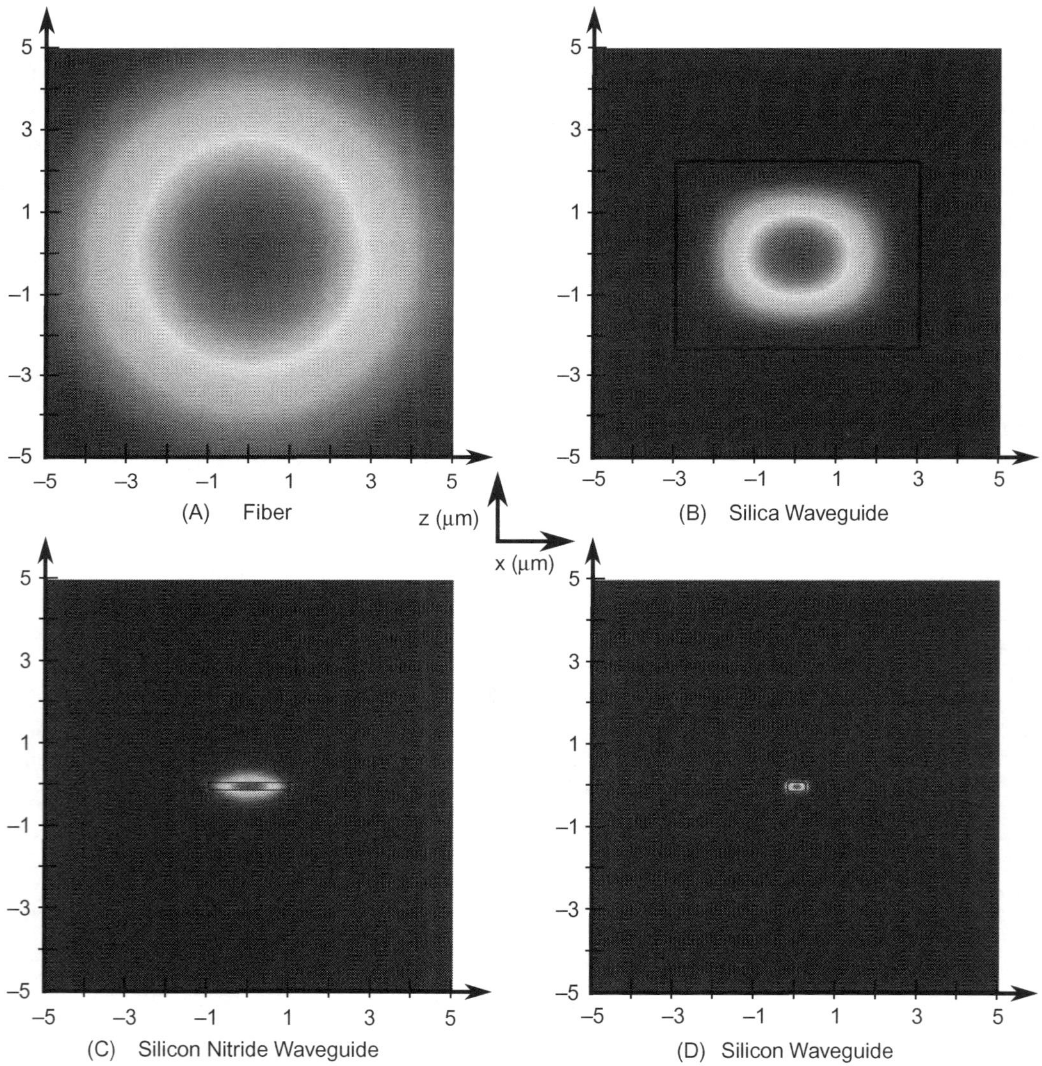

Figure 7.7

Intensity distribution of (A) single-mode fiber, (B) $6 \times 4.5\ \mu m^2$ silica (1.5% RI) waveguide, (C) $2 \times 0.2\ \mu m_2$ silicon nitride, and (D) $0.45 \times 0.22\ \mu m^2$ silicon for TE polarization light.

MFD became smaller as the index contrast of the waveguide increased, which increased the modal shape mismatch with fiber mode, and hence the coupling loss increased. The simulated coupling loss between 10.5 μm MFD fiber and single-mode silica waveguide (Fig. 7.7(B)) is 1.4 dB, silicon nitride waveguide is 10 dB, and SOI waveguide is 20 dB. Several research has been conducted to improve these coupling losses. For low index contrast waveguide like silica, the coupling loss can be improved to below 0.5 dB just by increasing the waveguide width adiabatically (positive taper waveguide).

But for silicon, the change is minimum with positive taper waveguide (Fig. 7.8), as we increase the waveguide dimension only in the x direction but not in the y direction. A gray scale mask can be used to create a vertical taper, but that will increase the complexity of the fabrication. The coupling loss improved as we adiabatically decreased the waveguide width (negative/inverse taper waveguide). The challenges for this approach are that the width is in the same range of minimum feature size available in the CMOS fabrication facility and a highly fabrication-dependent coupling loss. The silicon wafer has 2 μm oxide on top of a silicon substrate. Therefore, an expanded mode from an inverse taper waveguide will leak to the substrate mode increasing the coupling loss. To avoid the leaking, the inverse taper length needs to be short and the tip of the waveguide needs to be at the edge of the chip. But this will increase the chance to accidentally damage the tip while handling the chip. Fig. 7.9(A) shows an example of an inverted coupler [1]. These challenges are possible to overcome with a low index contrast polymer/silicon oxinitride (SiON) overlay waveguide on top of an inverted taper, as shown in Fig. 7.9(B). Fig.7.9(C) and (D) shows fabricated inverted taper with polymer and SiON overlay waveguide, respectively [2].

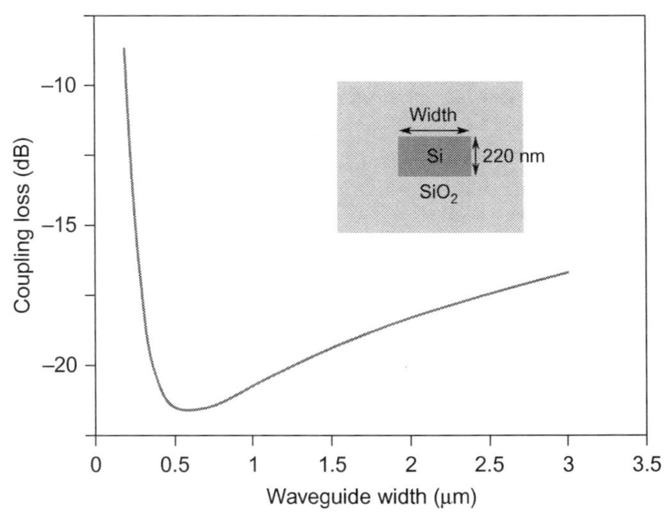

Figure 7.8
Simulated coupling loss variation with waveguide width for silicon waveguide.

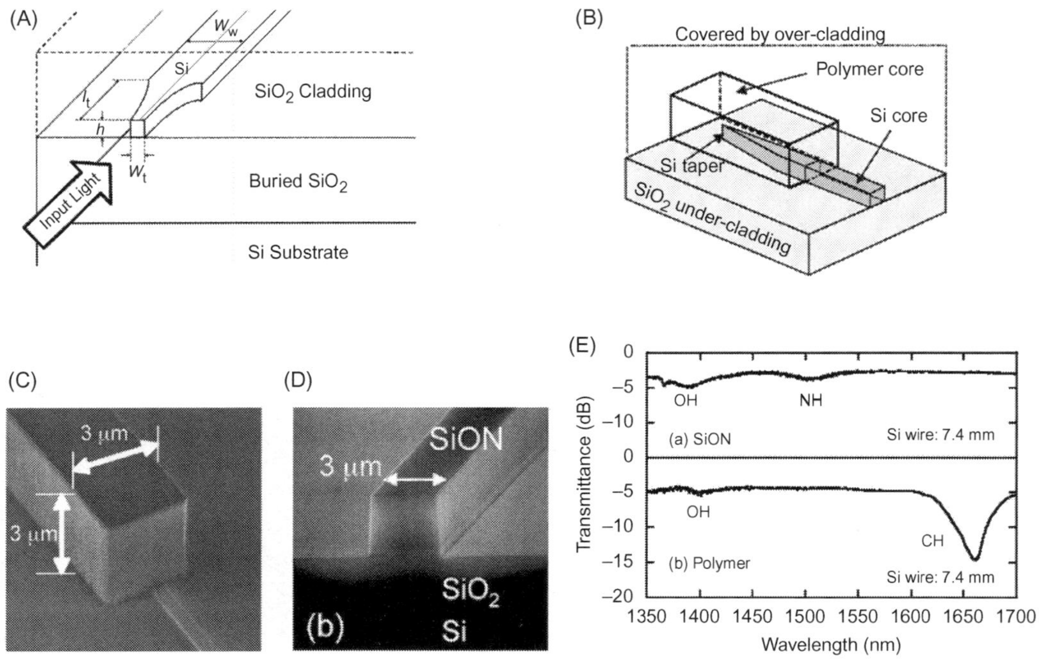

Figure 7.9

(A) Inverted taper without overlay waveguide. (B) Inverted taper with polymer overlay waveguide. (C) Fabricated inverted taper with polymer overlay waveguide. (D) Fabricated inverted taper with SiON overlay waveguide. (D) Performance comparison between inverted taper with polymer and SiON overlay waveguides. *(A) From [1] V.R. Almeida, R.R. Panepucci, M. Lipson, Nanotaper for compact mode conversion, Opt. Lett. 28 (15) (2003) 1302–1304; (B–E) [2] from T. Tsuchizawa, K. Yamada, H. Fukuda, T. Watanabe, J. Ichi Takahashi, M. Takahashi, et al., Microphotonics devices based on silicon microfabrication technology, IEEE J. Sel. Top. Quantum Electron. 11 (1) (2005) 232–240.*

A similar coupling efficiency was obtained for both polymer and SiON overlay waveguides, where $3 \times 3\ \mu m^2$ overlay waveguide is used on top of a 80 nm-wide tip and 300 μm-long silicon inverted taper: 2.5 dB coupling loss per connection for a fiber with 9 μm MFD. Fig. 7.9(D) shows transmission spectra of fiber-to-fiber coupling through a 7.4 mm and 6.7 mm-long silicon ridge waveguide with SiON and polymer overlay inverted taper, respectively. Although both SiON and polymer spot-size converter show similar optical performance, SiON has advantages over polymer spot-size converter due to CMOS compatibility, high durability, and no degradation with high intensity light. The big challenge is to fabricate a 80 nm silicon tip using 193 nm deep UV lithography, which is a stander silicon photonics fabrication tool with 120 nm minimum feature size. This can be overcome by using low index dielectric layer as a spacer in between the silicon tip and polymer overlay waveguide [3], as shown in Fig. 7.10(A). A $3 \times 1.3\ \mu m^2$ polymer waveguide is fabricated on top of a 175 nm-wide tip, with 175 μm-long silicon inverted

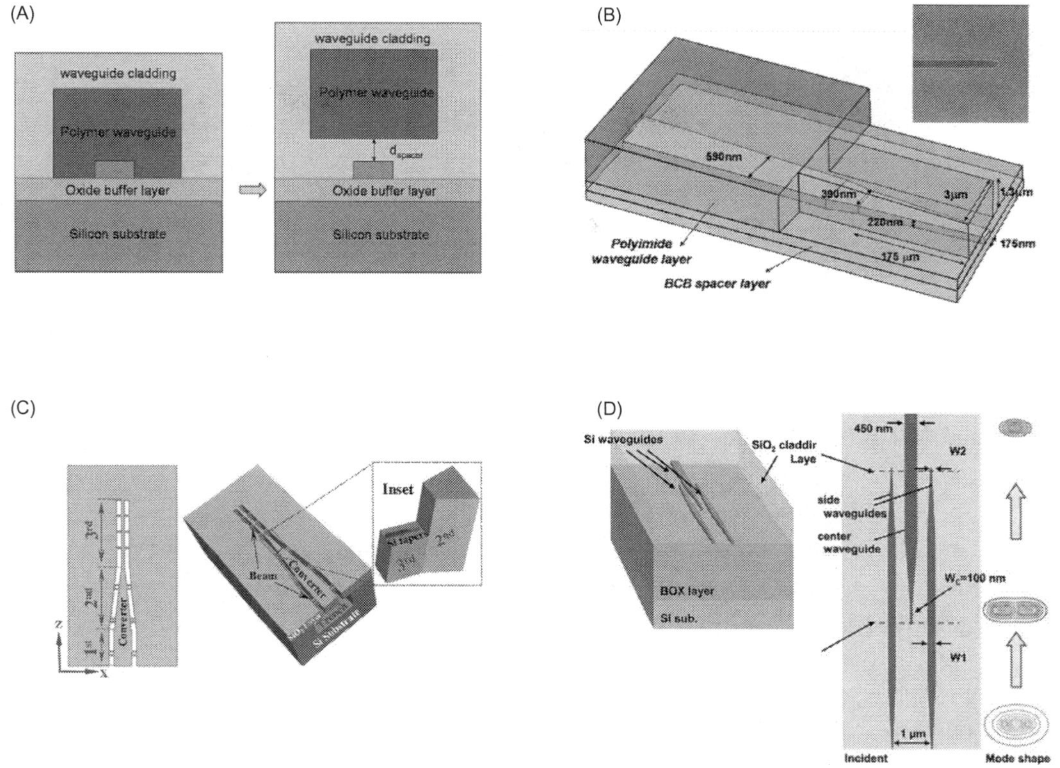

Figure 7.10

(A) Modified inverted taper design. (B) Schematic of the modified inverted taper design. (C) Suspended silicon dioxide spot-size converter. (D) Trident spot-size converter. *(A) From [3] G. Roelkens, P. Dumon, W. Bogaerts, D.V. Thourhout, R. Baets, Efficient silicon-on-insulator fiber coupler fabricated using 248-nm-deep uv lithography, IEEE Photonics Technol. Lett. 17 (12) (2005) 2613–2615; (B) [3] from G. Roelkens, P. Dumon, W. Bogaerts, D.V. Thourhout, R. Baets, Efficient silicon-on-insulator fiber coupler fabricated using 248-nm-deep uv lithography, IEEE Photonics Technol. Lett. 17 (12) (2005) 2613–2615; (C) [4] from Q. Fang, J. Song, X. Luo, M. Yu, G. Lo, Y. Liu, Mode-size converter with high coupling efficiency and broad bandwidth, Opt. Express 19 (22) (2011) 21588–21594; (D) [5] from N. Hatori, T. Shimizu, M. Okano, M. Ishizaka, T. Yamamoto, Y. Urino, et al., A hybrid integrated light source on a silicon platform using a trident spot-size converter, J. Lightwave Technol. 32 (7) (2014) 1329–1336.*

taper with 200 nm thick BCB as a spacer. The measured coupling loss is 0.5 dB per connection from a lensed fiber with 2.5 μm MFD.

Further research has been done to improve the performance and reliable packaging. Fang et al. [4] showed that a $10 \times 13\ \mu m^2$ suspended silica (SiO_2) waveguide taper laterally on top of an inverted taper improves the coupling efficiency without leakage to the Si substrate, as shown in Fig. 7.10(C). This design has three sections: first section to couple the light from 10.5 μm MFD single-mode fiber to $10 \times 13\ \mu m^2$ suspended silica waveguide, second section is the adiabatic taper to a 3.5 μm-wide silica waveguide, and the last section is the inverted Si waveguide with 100 nm tip in at the center of the 3 μm silica waveguide. The measured coupling loss of this suspended silica spot-size converter is 1 dB and 1.8 dB per connection for TE and TM polarized light, respectively. To avoid complicated fabrication steps like under etch silicon substrate, Hatori et al. [5] demonstrate a trident spot-size converter, as shown in Fig. 7.10(D). This coupler shows 0.9 dB coupling loss per connection with 3 μm MFD lensed fiber.

7.4.1.2 Grating Coupler

A wafer scale testing is not possible with an edge coupler. Therefore, there is a need to develop vertical coupling mechanism. Fig. 7.6(B) shows schematic of vertical coupling from an optical fiber to a waveguide using a grating coupler. A grating coupler consists of periodic etch structures to diffract the light in a certain direction. One of the advantages of the grating coupler is that it requires less amount of space of the chip as it does not require to route the waveguide at the edge of the chip. But the coupling efficiency is very poor for a grating coupler. In 2012, Taillaert et al. [6] demonstrated a grating coupler with 7 dB coupling loss for SOI platform. There are three main contributors behind this high coupling loss: diffraction toward the substrate, mode mismatches between the fiber mode and diffracted mode, and second order Bragg diffraction. The last loss contribution can be minimized by placing the fiber with an angle, as shown in Fig. 7.11(A).

The working principal of a grating coupler can be described as: When the light propagates through a grating coupler, it will create an array of scattering sources. Field of these sources will interfere and create a directional field. For a low loss vertical coupling, it is desired to have a constructive interference towards the top of the chip and distractive interference toward the substrate of the chip. The period of grating can be calculated from the Bragg's condition:

$$\beta - K = k_z \tag{7.3}$$

Where $\beta = 2\pi n_{eff}/\lambda$, denotes propagation constant of the waveguide, n_{eff} is the effective index of the waveguide and λ is the wavelength of the light. $K = 2\pi\Lambda$, is the grating vector of the grating structure, where Λ is the period of the grating, and is directed toward the direction of periodicity of the grating. $k_z = 2\pi Sin\theta/\lambda$, is the projected wave vector of the incident light, where θ is the angle of the fiber [7]. The directionality of the light for an efficient coupling can be improved by optimizing waveguide thickness and etch depth used to create the grating structures. Chen et al. [9] demonstrate that the coupling loss can be improved to 1.2 dB for a Si waveguide with 340 nm thickness and 200 nm etch depth to create the grating structures. Vermeulen et al. demonstrate that the coupling efficiency can be improved to 80% for a popular 220 nm thick silicon platform, using a 160 nm thick amorphous silicon layer on top of a 220 nm thick crystalline silicon on the grating area (Fig. 7.11(A)) and 220 nm etch depth to fabricate the grating structures. The directionality

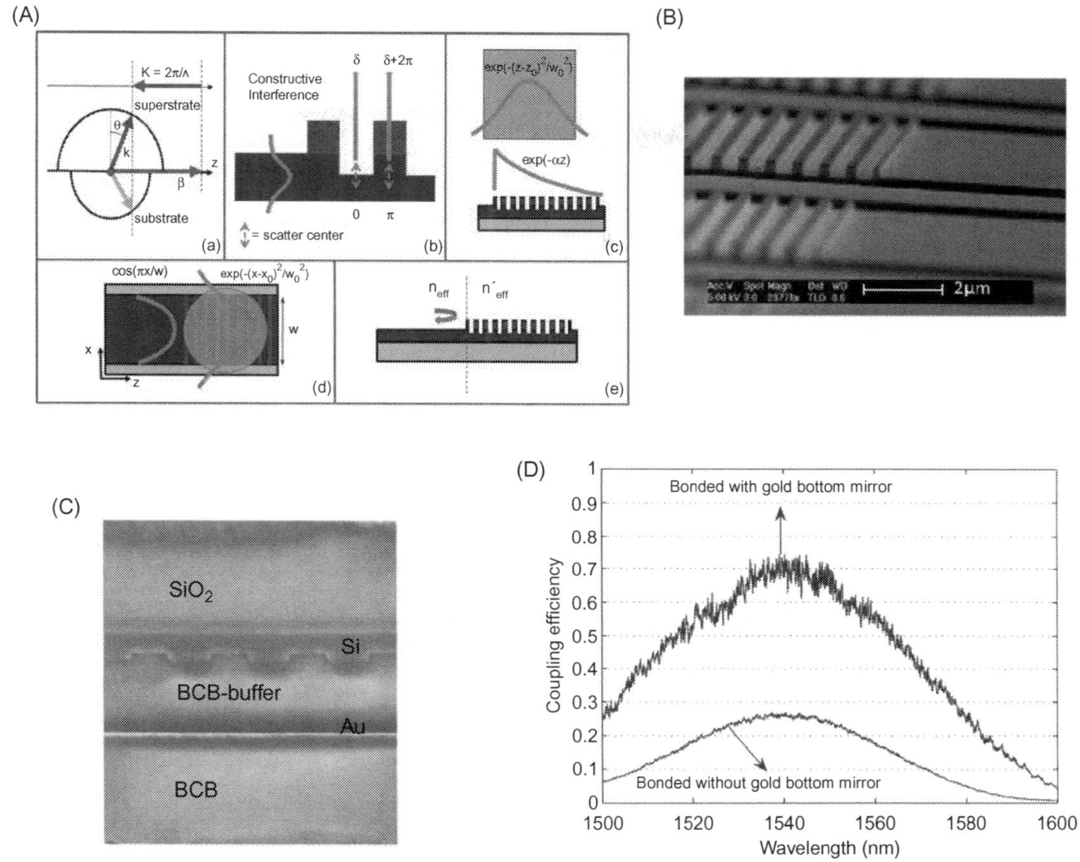

Figure 7.11

(A) Schematic illustrations of silicon grating coupler. (B) SEM view of fabricated silicon grating coupler. (C) SEM cross-sectional view of fabricated silicon grating coupler with a bottom gold mirror. (D) Measured coupling efficiency of Si grating coupler with and without bottom gold mirror. *(A) [7] From D. Vermeulen, S. Selvaraja, P. Verheyen, G. Lepage, W. Bogaerts, P. Absil, et al., High-efficiency fiber-to-chip grating couplers realized using an advanced cmos-compatible silicon-on-insulator platform, Opt. Express 18 (17) (2010) 18278–18283; (B) [7] from D. Vermeulen, S. Selvaraja, P. Verheyen, G. Lepage, W. Bogaerts, P. Absil, et al., High-efficiency fiber-to-chip grating couplers realized using an advanced cmos-compatible silicon-on-insulator platform, Opt. Express 18 (17) (2010) 18278–18283; (C) [8] from F.V. Laere, G. Roelkens, M. Ayre, J. Schrauwen, D. Taillaert, D.V. Thourhout, et al., Compact and highly efficient grating couplers between optical fiber and nanophotonic waveguides, J. Lightwave Technol. 25 (1) (2007) 151–156; (D) [8] from F.V. Laere, G. Roelkens, M. Ayre, J. Schrauwen, D. Taillaert, D.V. Thourhout, et al., Compact and highly efficient grating couplers between optical fiber and nanophotonic waveguides, J. Lightwave Technol. 25 (1) (2007) 151–156.*

of the light can be further improved by fabricating a mirror at the bottom of the grating coupler, which will redirect the light propagating toward the substrate [8], as shown in Fig. 7.11(C). Fig. 7.11(D) shows the measured coupling efficiency of a grating coupler with the presence of a mirror compared to a grating coupler without a mirror. The biggest disadvantages of a grating coupler are the narrower bandwidth and polarization selective operation.

For a polarization insensitive operation, it is not possible to use one-dimensional (1-D) grating coupler due to its strong polarization dependency. This problem can be overcome by using a two-dimensional (2-D) grating coupler, which can be considered as two orthogonally overlapped 1-D grating couplers. Fig. 7.12(B) shows schematic of a 2-D grating coupler. A 2-D grating coupler can couple any arbitrary fiber polarization light by

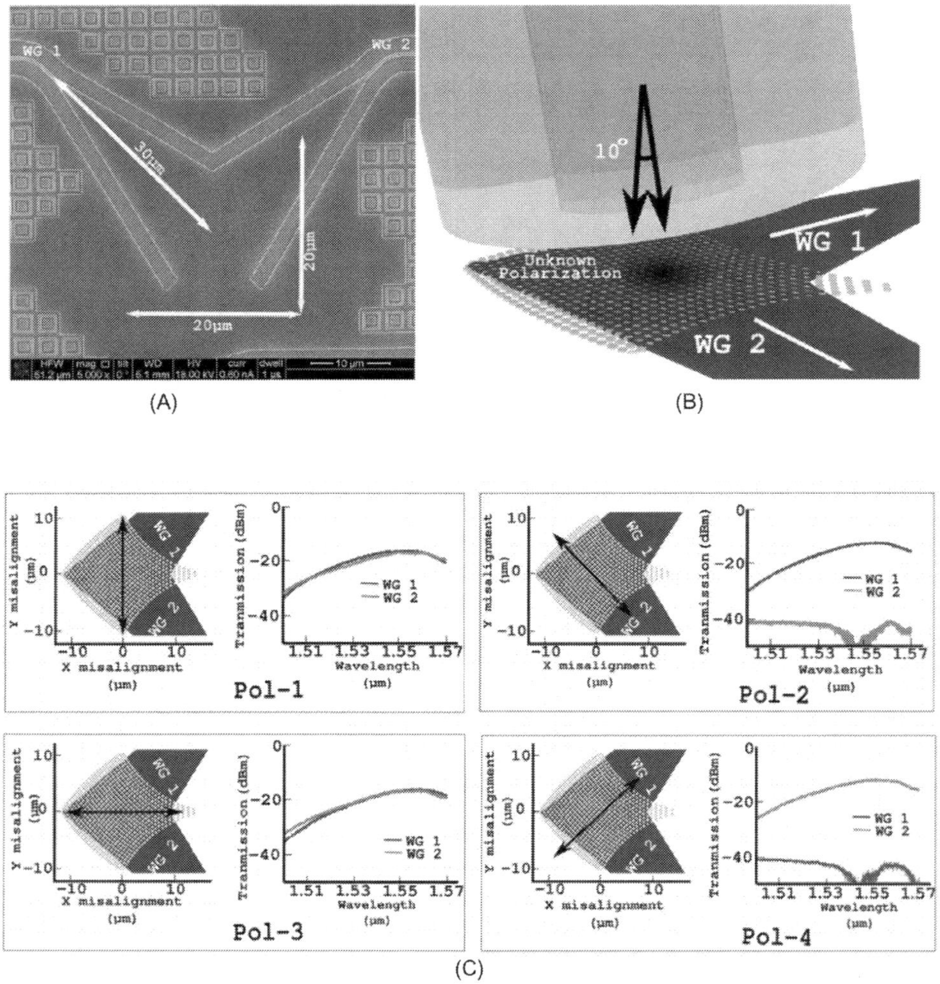

Figure 7.12

(A) Fabricated 2-D grating coupler. (B) Schematic diagram of the 2-D grating coupler. (C) Spectral response of a 2-D grating coupler for four different fiber polarizations. *(A–C) [10] From S. Pathak, M. Vanslembrouck, P. Dumon, D.V. Thourhout, W. Bogaerts, Compact soi-based polarization diversity wavelength de-multiplexer circuit using two symmetric awgs, Opt. Express 20 (26) (2012) B493–B500.*

splitting into two orthogonally TE modes. Fig. 7.12(C) shows the spectral response of the 2-D grating coupler for four different fiber polarizations. 2-D grating couplers are commonly used for polarization diversity circuit, where inputs of two identical single polarization optical circuits are connected to a 2-D grating coupler and the outputs of these two circuits are connected to the same photodetectors or combined by another 2-D grating coupler [10].

7.4.2 Arrayed Waveguide Grating

Wavelength division de/multiplexing (WDM) is the backbone of today's optical communications systems. Since the first demonstration, AWG became a key component in these WDM systems. An AWG multiplexed or demultiplexed the light into different wavelength channels. Fig. 7.13 shows schematic and operation principal of an AWG. An AWG consists of two star couplers, also known as free space region and an array of waveguides with period increment of length. The operation principle of an AWG as a demultiplexer can be described as: The light beam gets diverse in the input star coupler

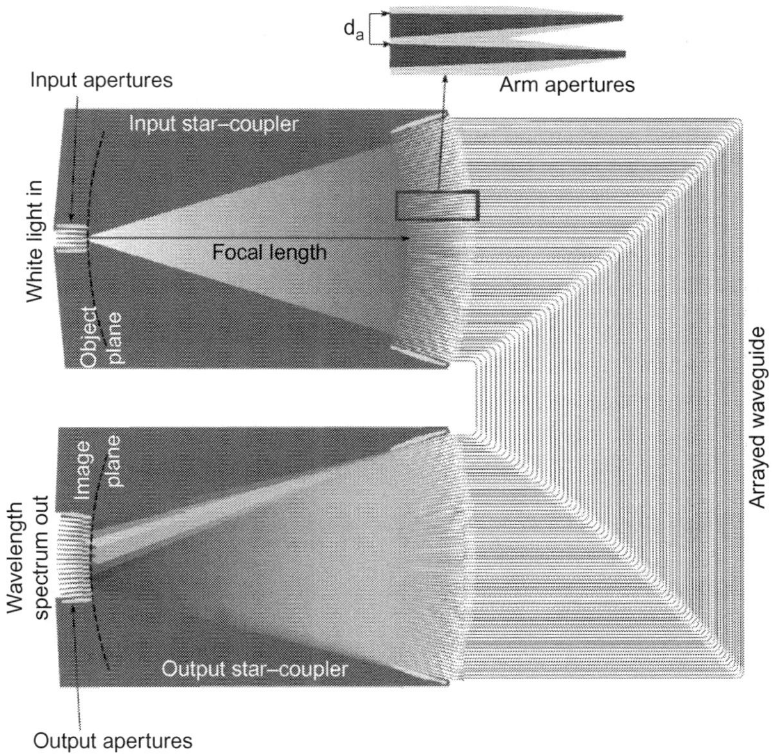

Figure 7.13
Operation principle of an AWG.

after entering from a input waveguide. This diverging light is distributed in the array of waveguides and propagate toward the second starcoupler. The optical path length difference between two conjugated waveguides is constant $\Delta L = m\lambda_c/n_{eff}$, where m is the order of the phased array, λ_c is center wavelength of the targeted spectrum, and n_{eff} is effective index of the waveguide used in the array for center wavelength. After traveling through this array of waveguides, the light will reach to the second star coupler with a wavelength-dependent phase. For a given wavelength, the phase difference between the lights propagating through two successive waveguides is constant. These lights will interfere in the second star coupler and will be focused to the desired output waveguide. As the phase difference is wavelength dependent, other wavelengths will not focus at the same position; instead, will have a lateral shift and will be coupled to other output waveguides. In this way, an AWG demultiplex the input light into multiple output channels. Due to reciprocity, an AWG can also work as a multiplexer by combining multiple wavelength signal into a output waveguide.

Over the years, a lot of materials has been used to develop AWGs. Table 7.1 shows an overview of the typical performance of AWGs fabricated in different material systems. The performance of the AWG depends on the effective index mismatch between design and fabricated waveguides in the array. The sidewall roughness of the waveguide or mask discretization will introduce a phase error, which will degrade the performance by introducing large crosstalk, high loss, spectral shape deformation, etc. As the high contrast waveguides are more effected by phase errors, the crosstalk of the silicon AWGs are higher compared to the silica AWGs. But the bend radius of silicon waveguide is small, which allows to design an AWG with compact footprint. Over the last decade, optimized design schemes and improved fabrication technology have significantly improved the performance of the Si AWGs: An array of waveguides with combination of single-mode and multimode waveguides [18] and use of 1 nm grid mask instead of 5 nm grid mask demonstrate AWG with low loss and crosstalk [19]. Fig. 7.14 shows typical measured spectrum of a Si AWG, the performance of the AWG improve as the mask grid snapping reduces from 5 nm to 1 nm.

Table 7.1: Comparison of AWGs for different materials.

Material Platform	Channel Spacing (GHz)	Channel Number	Bend Radius (um)	Crosstalk (dB)
Silica [11]	50	64	5000	−40
Silica [12]	25	400	2000	−25
InP [13]	10	100	500	−17
InP [14]	50	64	500	−25
Silicon [15]	400	8	5	−24
Si_2N_3 [16]	50	16	150	−20
Germanium [17]	200	5	100	−20

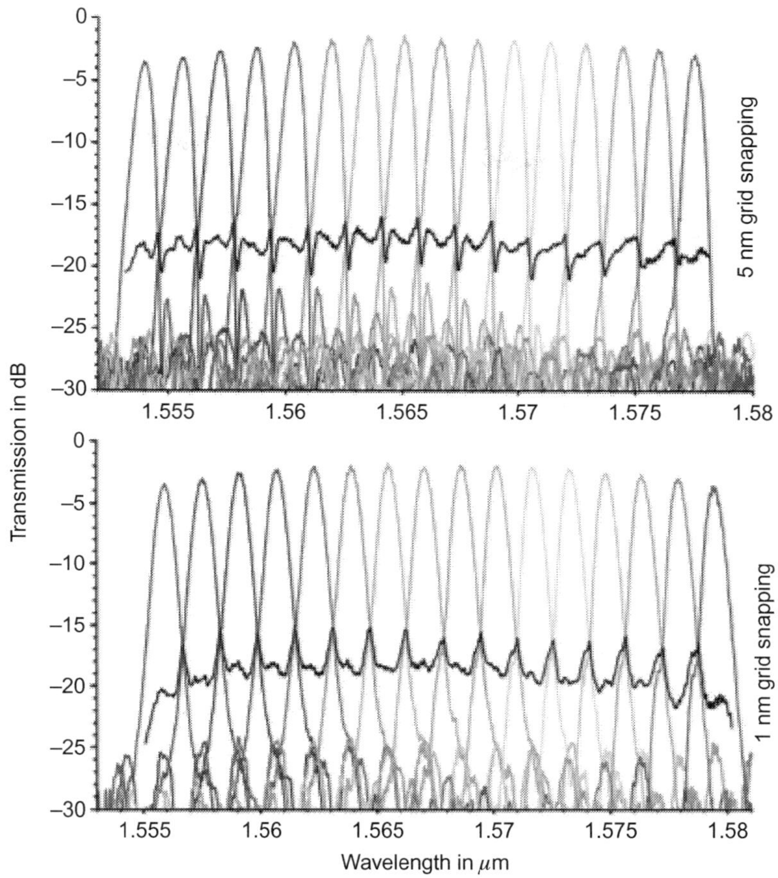

Figure 7.14

Measured spectrum of a 16 × 200 GHz Si AWG with 1 nm and 5 nm grid snapping. *From [19] S. Pathak, M. Vanslembrouck, P. Dumon, D. Van Thourhout, P. Verheyen, G. Lepage, et al., Effect of mask discretization on performance of silicon arrayed waveguide gratings, IEEE Photon. Technol. Lett. 26 (7) (2014) 718–721.*

From Fig. 7.14, we see that the passband of a conventional AWG is Gaussian-like, however, a flat passband is desirable and is necessary for many WDM applications. To create a flat passband, one of the commonly used methods is use of a multimode interferometer (MMI) as an input of the AWG [20], as shown in Fig. 7.15(A). The design of the MMI is chosen in such a way that the MMI creates a twofold field image at the object plane (junction between input and fist star couple). When this twofold field image goes through the AWG and overlap with output waveguide, a flat passband will be created. Fig. 7.15(B) shows flat spectral response of 12 × 400 GHz Si AWG with MMI input. The flat-top passband of the AWG using an MMI aperture as an input is inherently lossy due to the mode profile mismatch between the MMI input and output waveguide. The loss can be

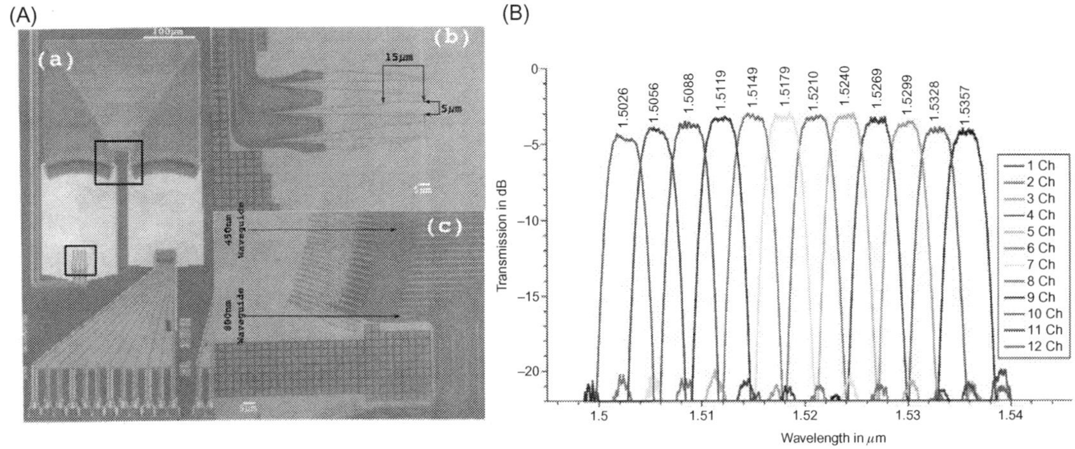

Figure 7.15

(A) Fabricated 12 × 400 GHz MMI-AWG. (B) Measured spectrum of a 12 × 400 GHz MMI-AWG. *From [20] S. Pathak, M. Vanslembrouck, P. Dumon, D.V. Thourhout, W. Bogaerts, Optimized silicon awg with flattened spectral response using an mmi aperture, J. Lightwave Technol., 31 (1) (2013) 87–93.*

improved by using a Mach-Zehnder interferometer (MZI) (discussed in detail in Section 4.3) as an input [21]. Thus, the input field at the object plane will be wavelength dependent and will have better field overlap with the output waveguide. To achieve flat spectral response, the MZI and the AWG are designed in such a way that for each wavelength channel the lateral shift with wavelength at the MZI-AWG interface is synchronized. The free spectral range (FSR) of the MZI and the channel spacing of the AWG need to be equal and both the MZI and the AWG must have the same curvature. High precision design and controlled fabrication technology is required to have a synchronized MZI and AWG, MZI-AWG is still not demonstrated in SOI platform.

7.4.3 Mach–Zehnder Interferometer

The MZI is a basic building block of integrated phonics, as it can play many roles with an engineering adaptation. MZI can work as a modulator, switch, isolator, wavelength multiplexer, demultiplexer, etc. A MZI consists of one splitter, one combiner, and two symmetric/asymmetric waveguides. The splitter splits the input light into two symmetric/asymmetric waveguides and the outputs of these two waveguides are combined by a combiner. If two symmetric waveguides are used, the MZI is known as balanced MZI. There are various components used as a splitter/combiner: DC, MMI, y-junction, and adiabatic coupler. All these splitters/combiners have advantages and disadvantages. Therefore, the requirement of an application and use of material platform will determine an appropriate splitter/combiner. Fig. 7.16(A) and (B) shows the schematic and expected spectral representation of an asymmetric 2 × 2 MZI. The extinction ratio depends on the

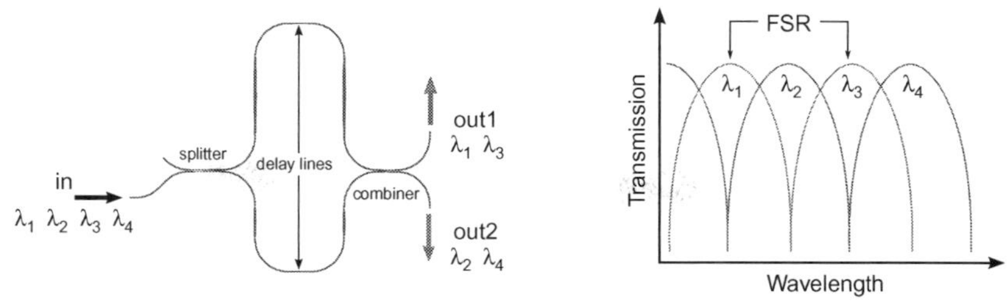

Figure 7.16
(A) Schematic of an asymmetric 2 × 2 MZI. (B) Spectral response of an asymmetric 2 × 2 MZI.

splitting ratio of the splitter/combiner and the loss mismatch between the two waveguides. For an asymmetric MZI, the optical path length difference between these two waveguides (delay length ΔL) can be written as:

$$\Delta L = m\lambda_c/\eta_{eff} \tag{7.4}$$

where m is the order, λ_c is center wavelength of the targeted spectrum, and n_{eff} is the effective index of the waveguide for center wavelength. The order m will determine the FSR, which is a wavelength separation between two conjugative peaks of the spectral response (Fig. 7.16(B)). The relation between the FSR and the order m can be written as:

$$\lambda_{FSR} = \frac{n_{eff} \cdot \lambda_c}{n_g \cdot m} \tag{7.5}$$

where n_g is group index of the waveguide.

From 7.16 (A), it is clear that a 2 × 2 MZI works as a demultiplexer when the light travels from left to right separating λ_1, λ_3 into top output and λ_2, λ_4 into bottom output. Due to reciprocity, if all these wavelengths enter from the same port on right-hand side, all these wavelengths will be combined to the same output at the left-hand side: working as a multiplexer. To enable multiple output ports, a cascaded formation is created (as shown in Fig. 7.17(A)), where two MZIs have been connected to outputs of the first MZI. The number of outputs can be increased further by increasing the number of cascading steps. The main goal of multiplexer is to combine the light into one port. Therefore, a large extinction ratio is not required, which allows cascaded MZI to be used as a multiplexer without any further modification. But for a demultiplexer application, it is required to modify the design. One of the main requirement is the large extinction ratio. The extinction ratio can be improved by using a cleanup filter at the output. To obtain a flat-top spectral response, one needs to use lattice formation of MZIs (Fig. 7.17(B)). A lattice filter has multiple coupling sections and delay lengths. The number of delay length sections of the lattice filter are denoted as stage of the lattice filter; Fig. 7.17(B) shows a three-stage lattice filter. A flat-top spectral response will be generated by choosing appropriate coupling ratio

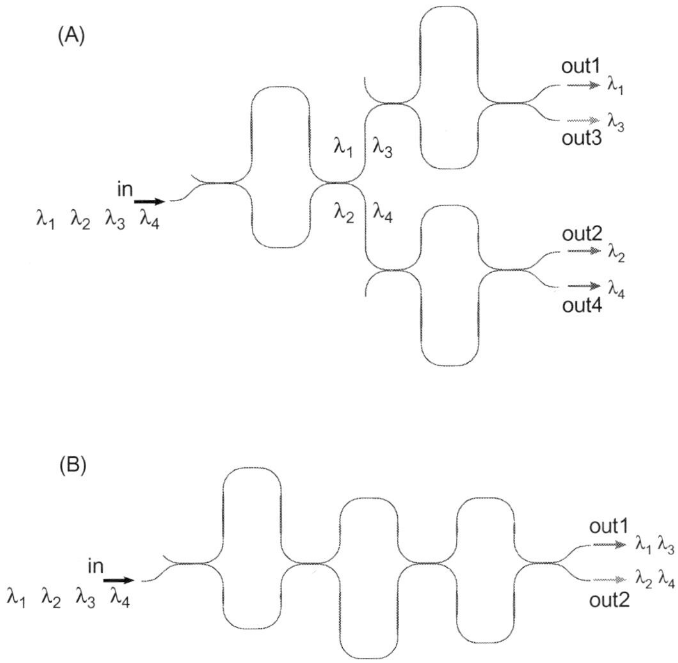

Figure 7.17
(A) Mach–Zehnder cascade. (B) Mach–Zehnder lattice.

and delay lengths combination. Fig. 7.18 shows design and spectral response of the three-stage lattice filter for silicon platform. The delay length (ΔL) will decide FSR of the spectrum. The aspect ratio (ratio between 1 dB and 10 dB bandwidth for a wavelength channel) will increase with increment of number of stages.

Seok-Hwan Jeong et al. demonstrate a four-channel, 800 GHz channel spacing lattice filter in 220 nm thick silicon platform. Fig. 7.19(A) shows configuration of the 4×800 GHz lattice filter. The lattice filter is designed with two cascaded steps, where each step used two cleanup filters. The first and second cascaded step are designed with three- and two-stage lattice filter, respectively. To achieve 800 GHz channel spacing, first cascaded step is designed for 2×800 GHz FSR and second cascaded step is designed for 4×800 GHz FSR. The measured spectral response is shown in Fig. 7.19(B) as low loss and -24dB crosstalk.

7.4.4 Ring Resonator

Ring resonator is one of the vital components for integrated photonics circuits due to its applicability in many photonics applications: optical filter, add-drop filter, optical sensing, etc. A ring resonator consists of a looped waveguide called ring and one or two bus waveguides to access the ring [24]. Schematic of the ring resonator is shown in

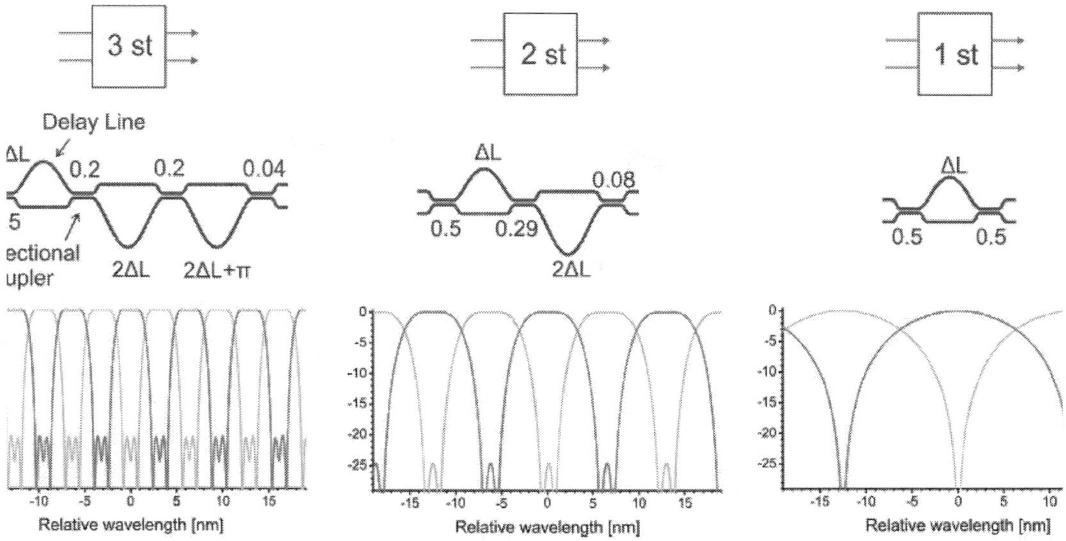

Figure 7.18
Silicon lattice filter design and simulated spectral response for three, two, and one stages. *From [22] F. Horst, W.M. Green, S. Assefa, S.M. Shank, Y.A. Vlasov, B.J. Offrein, Cascaded Mach-Zehnder wavelength filters in silicon photonics for low loss and flat pass-band wdm (de-)multiplexing, Opt. Express 21 (10) (2013) 11652–11658.*

Fig. 7.20(A) and (B). The light coupled to the ring from a bus waveguide will start rotating inside the ring. After a round trip, a certain set of periodic wavelengths (depending on path length of the round trip) will have phase change integer multiple of 2π. These wavelengths will have constructive interference and keep rotating inside the ring. These wavelengths are called resonant wavelengths. The gap between two consecutive resonant wavelengths is the FSR of the ring. These wavelengths which are resonating inside the ring will be missing in the through port and can be coupled back to a similar bus waveguide by adding it to other side of the input bus waveguide. The resonance wavelength of a ring, whose loopback path length is L, can be denoted as:

$$\lambda_{res} = \frac{n_{eff}\mathrm{L}}{m} \tag{7.6}$$

where m is the order of the ring and n_{eff} is the effective index of the waveguide. The FSR (λ_{FSR}) of the ring can be written as:

$$\lambda_{\mathrm{FSR}} = \frac{\lambda^2}{n_g\mathrm{L}} \tag{7.7}$$

where n_g is the group index of the waveguide. The maximum limit of the FSR depends on minimum path length required to form a loopback, which is the circumference of a circle

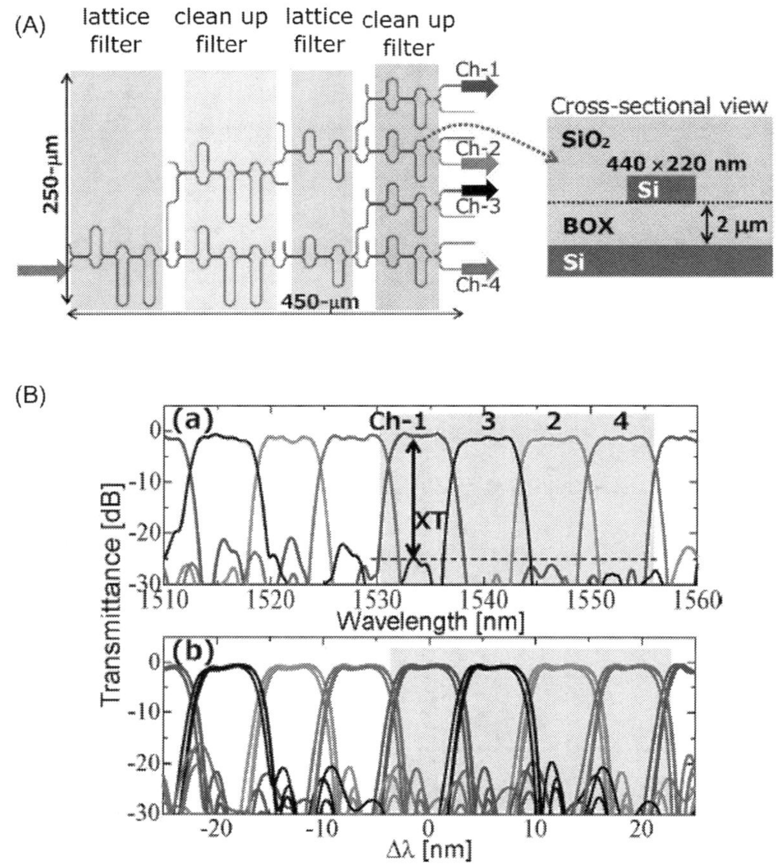

Figure 7.19

(A) Silicon four-channel 800 GHz lattice filter design. (B) Measured spectral response four-channel 800 GHz lattice filter. *From [23] S.H. Jeong, Y. Tanaka, K. Morito, Silicon photonics optical demultiplexer technology for wdm optical interconnects, in: 2015 Opto-Electronics and Communications Conference (OECC), 2015, pp. 1–3.*

$2\pi r$, where r is the minimum bend radius of the waveguide for lossless propagation. As the bend radius is smaller for high index contrast waveguide compared to the low index contrast waveguide, the ring resonator is the more suitable component for high index contrast waveguide platform like silicon.

The spectral response of a single ring has a much narrower passband, as shown in Fig. 7.21 (B), which makes it difficult to be used as a multiplexer/demultiplexer. Also, the resonance wavelengths are sensitive to the fabrication variations: the resonance wavelength changes from its designed value due to small effective index variation of the ring waveguide by means of effective width, thickness, or temperature variations. The passband can be broadened by cascading multiple rings [25,26]. The resonance wavelength shift can be overcome by using integrated heater to compensate the effective index variation. Fig. 7.21(A)

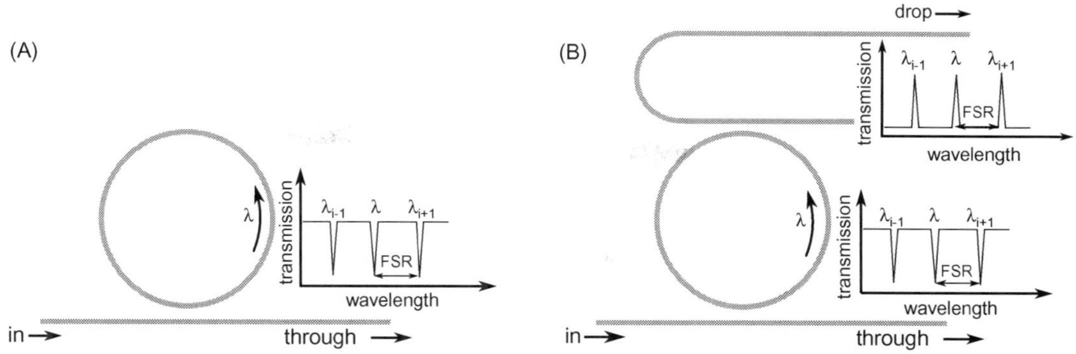

Figure 7.20
(A) Working principle of a ring resonator with single bus waveguide. (B) Working principle of a ring resonator with double bus waveguide.

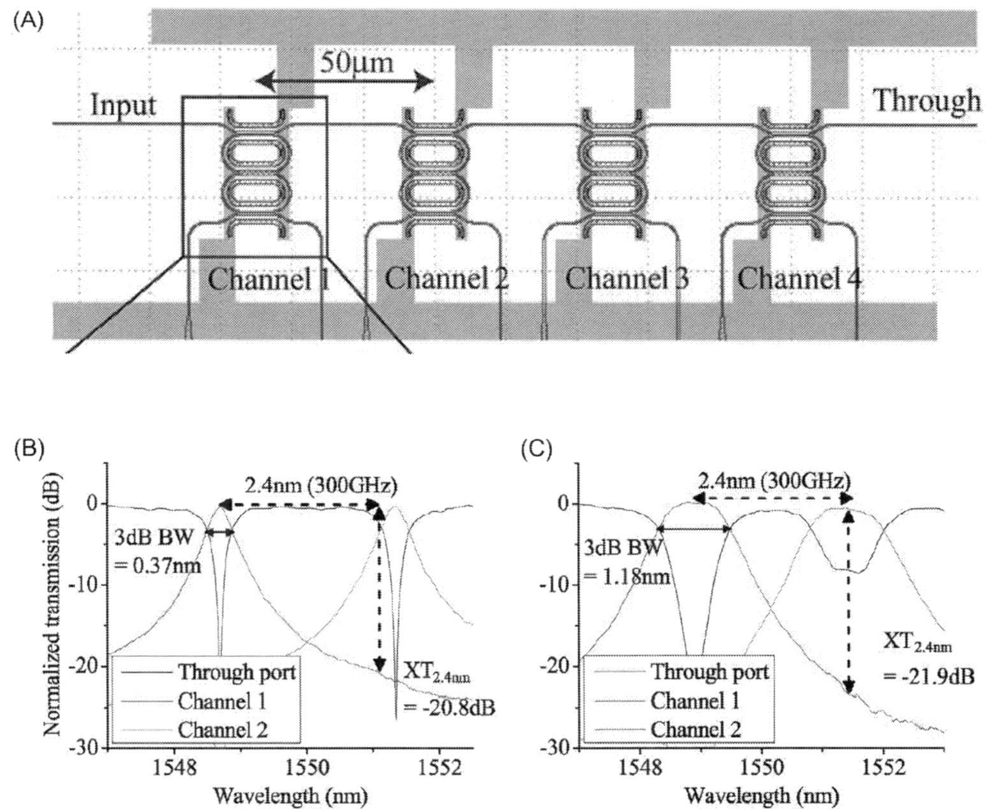

Figure 7.21
(A) Schematic view of a four-channel double ring resonator-based multiplexer/demultiplexer filter. (B) Spectral response of single ring-based filter. (C) Spectral response of double ring-based filter. *From [25] P.D. Heyn, J.D. Coster, P. Verheyen, G. Lepage, M. Pantouvaki, P. Absil, et al., Fabrication-tolerant four-channel wavelength-division-multiplexing filter based on collectively tuned si microrings, J. Lightwave Technol. 31 (16) (2013) 2785–2792.*

shows schematic view of a four-channel 300 GHz channel spacing two rings cascaded filter with integrated heater for tuning [25]. Fig. 7.21(C) shows the measured spectral response of the filter. The insertion loss and crosstalk of the worst performing wavelength channel is 1.47 dB and 17.3 dB, respectively. Comparing with single ring filter, the average 3 dB bandwidth of a double ring filter improves to 1.27 nm from 0.43 nm.

As discussed, the resonance wavelength of the ring is highly sensitive to the effective index variation of the ring waveguide. This property makes the ring a suitable component for sensing. The effective index of the waveguide changes due to molecule interaction near the surface of the waveguide, which shifts the resonance wavelength. Unlike any other sensor, the response of the ring resonator-based sensor does not decrease with decreasing effective sensing area. Another advantage of the ring resonator-based sensor is that it requires a less amount of molecule for sensing due to compactness of the ring [27,28].

7.5 Active Devices

In this section, we discuss a few basics photonics active components used in photonics integrated circuit. We cover three vital components: laser, modulator, and photodetector.

7.5.1 Laser

"Light Amplification by Stimulated Emission of Radiation" (Laser) was first demonstrated by Gordon Gould in 1957. After that, researchers around the globe developed big bulky gas laser to on-chip semiconductor laser. Our interest is in the laser principle use to generate light: laser oscillators. A laser oscillator consists of a cavity and gain medium: cavity to circulate the light and gain medium to amplify the light, shown in Fig. 7.22(A). To accomplish successful lasing, the gain in the gain medium needs to be higher than the loss in the cavity and the feedback light needs to have the phase matched (integer multiply of 2π) with the new light in the gain medium. For integrated photonics, semiconductor laser is most interesting as it can be fabricated in the same material platform. Fig. 7.22(B) shows a basic working principle of a semiconductor laser. The photon emitting in the semiconductor occurs by transitioning an electron from the conduction band to the valance band. In a steady-state condition, electrons will be in the valence band. When the energy of an external electrical/optical pump becomes higher than the bandgap energy, the electrons are starting excited from the valence band to the conduction band. As the electrons has a tendency to return to the steady state, an electron can knock from the bottom of the conduction band to the valance band, emitting photons with an energy equal to the band gap energy. In general, multiple stacks of III-V layers are used to form a laser. Then, the light couples to a waveguide using an edge coupler, which is the best possible method for low index contrast waveguide like silica. But over the years, the research has been done to

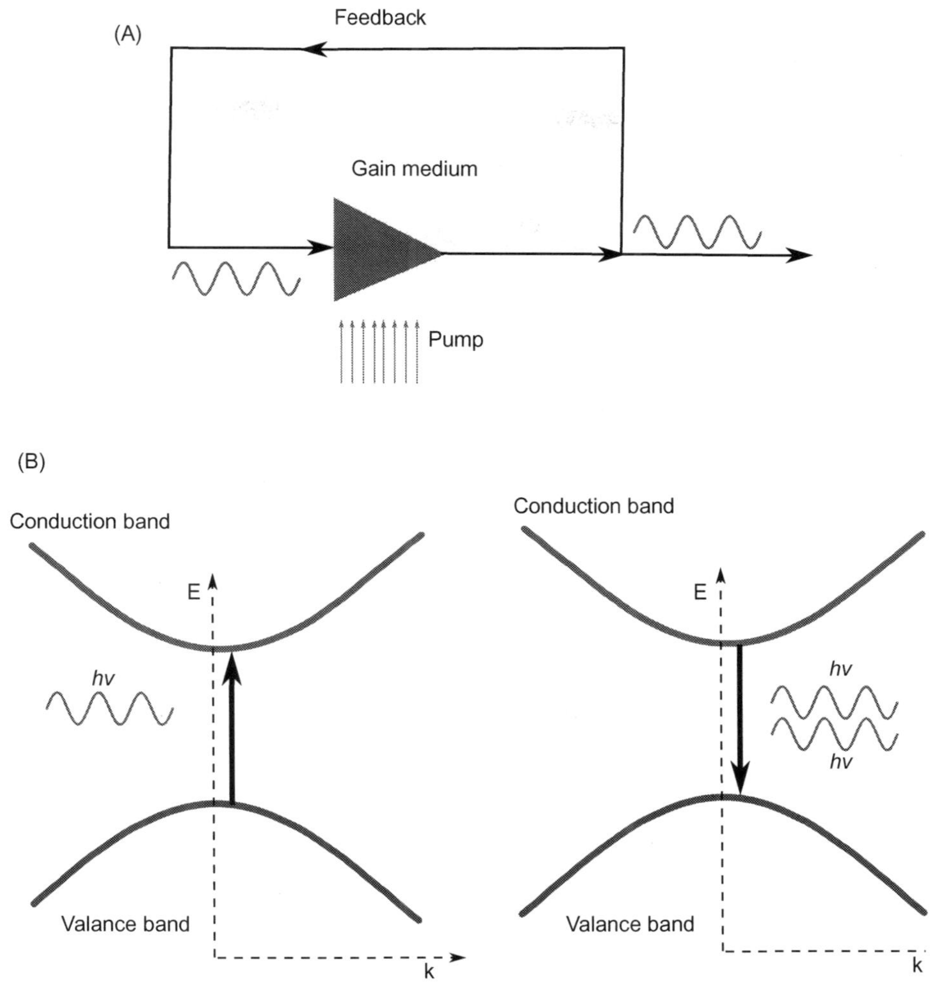

Figure 7.22
(A) Working principle of laser oscillator. (B) Working principle of semiconductor laser.

fabricate a III-V laser on top of a silicon waveguide, where the light has been coupled to the waveguide using evanescent coupling. To fabricate a III-V laser on top of the silicon, several techniques have been used. Fig. 7.23(A) and (B) shows two famous bonding techniques to fabricate the III-V laser on top of the silicon: direct bonding and BCB bonding, respectively. For both techniques, a III-V epitaxial layer is stack bonded on top of a silicon rib waveguide. On top of a rib waveguide, a thin bonding layer has been used to bond the III-V laser. Top to bottom, the III-V laser consists of a highly doped p-type ohmic contact for anode, a p-type cladding layer, a p-type separated confinement heterostructure (SCH) layer, a carrier-blocking layer, a multiple quantum well (MQW) active region, and an n-type spacer layer for cathode. The laser emitting wavelength depends on the active

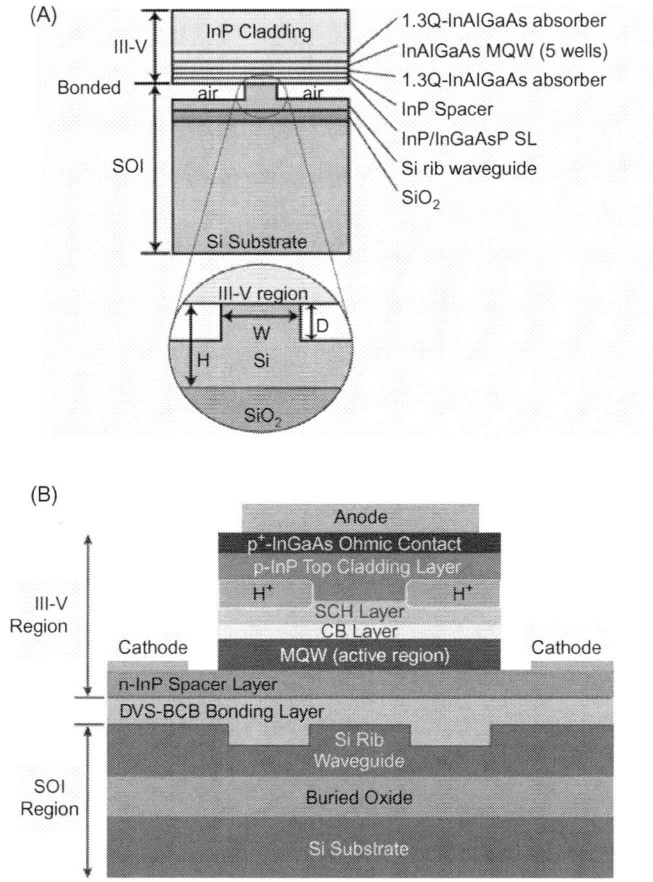

Figure 7.23

Cross-sectional diagram (A) III-V laser on silicon with direct bonding. (B) III-V laser on silicon with BCB bonding. *(A) [29] From A.W.-L. Fang, Silicon evanescent lasers, Ph.D. thesis, University of California Santa Barbara (2008); (B) [30] from S. Stankovic, Hybrid iii-v/si dfb lasers based on polymer bonding technology, Ph.D. thesis, Ghent University (2013).*

region design. In 2011, Stankovic et al. [30,31] published a III-V laser for 1310 nm wavelength. The III-V material stacks of this laser is shown in Fig. 7.24. The gap between the silicon rib waveguide and the III-V laser is small, allowing low loss evanescent coupling. To achieve a fundamental mode lasing, the waveguide width is chosen in such a way that the confinement of the light in waveguide is higher than 70%, where in the active region it is higher than 3%. Fig. 7.25 shows the optical spectra of the III-V laser with 700 μm cavity length, which consists of a Fabry-Pérot response [31]. The central peak wavelength is obtained at 1318.18 nm. The peak wavelength has a shift of 41.3 nm/W with an increase of input pump power, while the wavelength shift with surrounding temperature is 0.42 nm/°C.

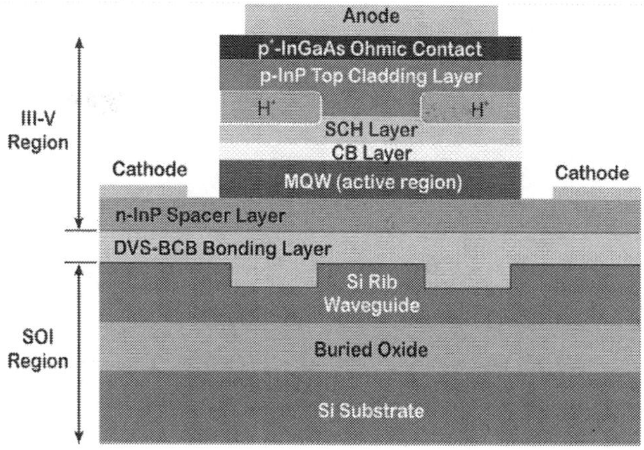

Layer	Material	Doping (cm^{-3})	Thickness (nm)	Remarks
Ohmic contact	$In_{0.53}Ga_{0.47}As$	p-type $> 1 \times 10^{19}$	100	lattice-matched to InP
Cladding	InP	p-type 1×10^{18}	1500	
SCH	$In_{0.5305}Al_{0.4055}Ga_{0.064}As$	p-type 1×10^{17}	325*	$\lambda g = 0.92$ μm lattice-matched to InP
CB layer	$In_{0.5047}Al_{0.4764}Ga_{0.0189}As$	undoped	10	$\lambda g = 0.82$ μm tensile strain: −0.12%
Barrier (×9)	$In_{0.4003}Al_{0.2695}Ga_{0.3302}As$	undoped	9	$\lambda g = 0.96$ μm tensile strain: −0.8%
Well (×8)	$In_{0.6986}Al_{0.178}Ga_{0.1234}As$	undoped	7	$\lambda PL = 1.3$ μm compress. strain: +1.2%
Spacer	InP	n-type 1×10^{18}	240*	

Figure 7.24

III-V material stack for the lasing wavelength at 1310 nm. *From [30] S. Stankovic, Hybrid iii-v/si dfb lasers based on polymer bonding technology, Ph.D. thesis, Ghent University (2013).*

For many applications, a single wavelength laser is required, which enhanced the research of distributed feedback (DFB) [32] and distributed Bragg reflector (DBR) laser [33]. To create a DFB laser, a grating structure was formed on top of the silicon rib waveguide, as shown in Fig. 7.26(A). The grating structure consists of two first-order Bragg grating separated by a $\lambda/4$ phase shifter. Fig. 7.26(B) shows the measured optical spectra of the DFB laser. The central peak wavelength measured at 1307.9 nm with 45 dB side mode suppression ratio. The peak wavelength shift with an increase of input pump power is reported, 7.2 nm/W, while the wavelength shift with surrounding temperature is reported, 0.07 nm/°C. Both this peak wavelength shift is significantly smaller compared to the Fabry-Pérot laser.

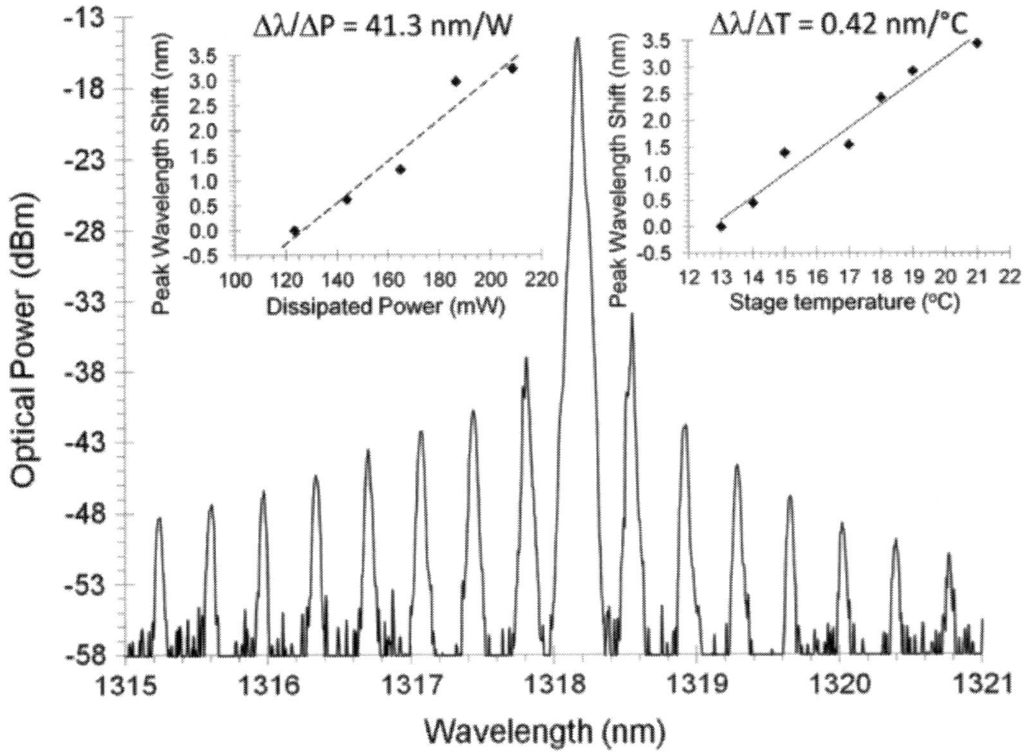

Figure 7.25

Measured optical spectra of a III-V laser with 700 μm cavity length. *From [31] S. Stankovic, R. Jones, M.N. Sysak, J.M. Heck, G. Roelkens, D.V. Thourhout, 1310-nm hybrid iii-v/si fabry-perot laser based on adhesive bonding, IEEE Photon. Technol. Lett. 23 (23) (2011) 1781–1783.*

A high power and low thermal impedance laser requirement drives the research interest for the DBR laser. A DBR laser consists of two passive Bragg reflector mirrors placed in between the gain region to obtain a laser cavity, as shown in Fig. 7.26(C). The front mirror is designed for partial reflection while the back mirror is designed to obtain a maximum possible reflection. Fang et al. [34] demonstrated III-V DBR laser, as shown in Fig. 7.26(C). The design cavity length is 600 μm. A 300 μm-long DBR is designed in silicon as a back mirror with 97% reflection, while the front mirror is 100 μm-long DBR with 44% reflection. A 80 μm-long taper region has been used on both sides of the III-V region to couple the light from III-V region to silicon waveguide. The length of the gain region is 440 μm. The spectral response of the DBR laser is shown in Fig. 7.26(D). The peak wavelength obtained is 1597.5 nm with 50 dB sidemode suppression ratio. This can be further expanded to achieve multichannel and tunable laser, as discussed in Section 6.

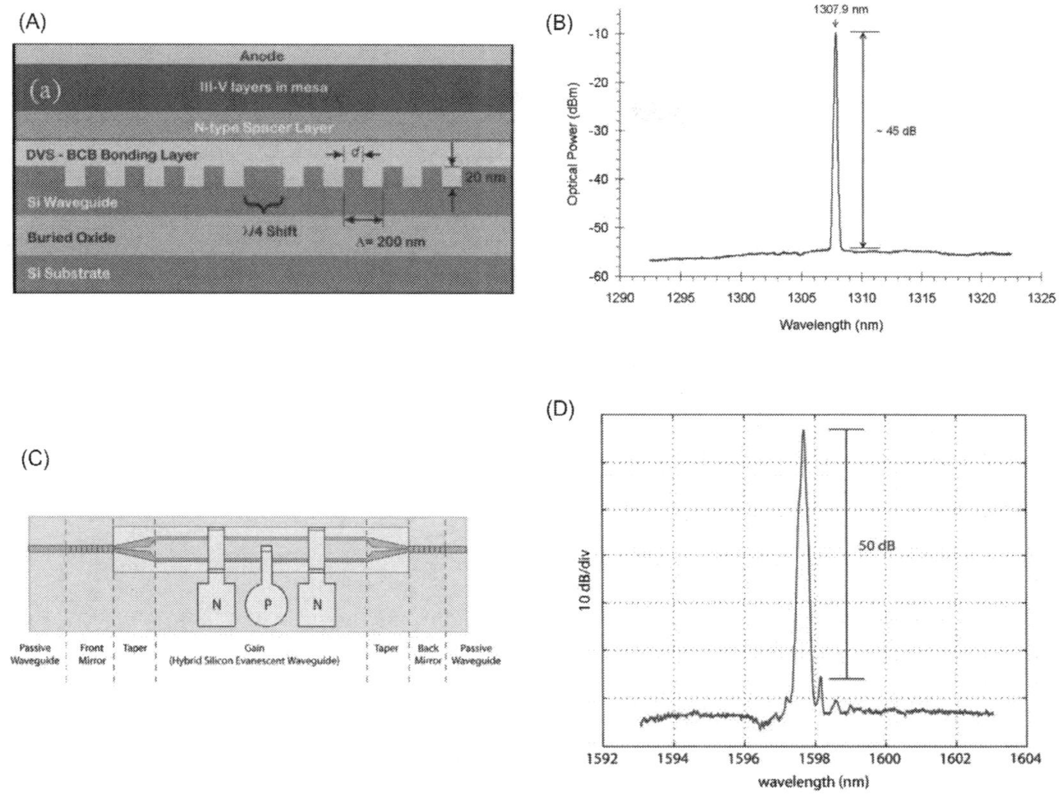

Figure 7.26

(A) Design of III-V DFB laser. (B) Measured optical spectra of III-V DFB laser. (C) Design of III-V DBR laser. (D) Measured optical spectra of III-V DBR laser. *(A) [32] From S. Stankovic, R. Jones, M.N. Sysak, J.M. Heck, G. Roelkens, D.V. Thourhout, Hybrid iii-v/si distributed-feedback laser based on adhesive bonding, IEEE Photon. Technol. Lett. 24 (23) (2012) 2155–2158; (B) [32] from S. Stankovic, R. Jones, M.N. Sysak, J.M. Heck, G. Roelkens, D.V. Thourhout, Hybrid iii-v/si distributed-feedback laser based on adhesive bonding, IEEE Photon. Technol. Lett. 24 (23) (2012) 2155–2158; (C) [33] from A.W. Fang, M.N. Sysak, B.R. Koch, R. Jones, E. Lively, Y.H. Kuo, et al., Single-wavelength silicon evanescent lasers, IEEE J. Sel. Top. Quantum Electron. 15 (3) (2009) 535–544; (D) [33] from A.W. Fang, M.N. Sysak, B.R. Koch, R. Jones, E. Lively, Y.H. Kuo, et al., Single-wavelength silicon evanescent lasers, IEEE J. Sel. Top. Quantum Electron. 15 (3) (2009) 535–544.*

7.5.2 Optical Modulator

An optical modulator can be defined as an optical device to convert the data/information from electrical domain to optical domain by manipulating the optical properties. Optical modulation technique can be classified into two categories: direct modulation (Fig. 7.27(A)) and indirect modulation (Fig. 7.27(B)). Direct modulation can be obtained by controlling the bias current of the semiconductor laser. A digital data signal consists of 1s and 0s,

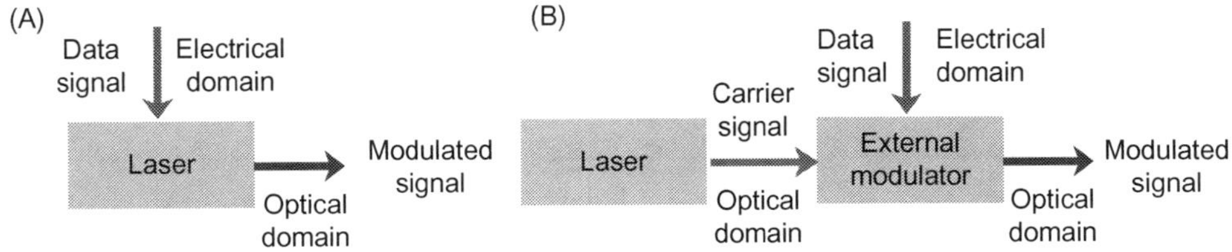

Figure 7.27
(A) Direct modulation principle. (B) Indirect/External modulation principle.

which is also referred to as an electrical on-off states or two different current/voltage states that can be used to manipulate bias current. Although a semiconductor laser is capable of high speed modulation, it is rarely used in practical applications. In practice, an external modulator is connected to the laser, where the laser is operated with constant bias current to provide a continuous wave (CW). The main disadvantages of direct modulation are the frequency shift (also called frequency chirp) of the laser with bias current change, requirement of large current swing for modulation, and reduced lifetime of the laser.

Fig. 7.27(B) illustrates the working principle of the indirect modulation. The CW unmodulated light from laser is also known as carrier signal. In the modulator, a certain parameter of the carrier signal is modified, depending on the data signal, which is required to convert from electrical to optical domain. The properties of the light modified in the modulators are amplitude, frequency, phase, polarization, etc. The data signal can be either CW or discrete. If the data signal is CW, the modulation formats are amplitude modulation (AM), frequency modulation (FM), phase modulation (PM), polarization modulation (PolM), etc. If the data signal is digital, the modulation formats are amplitude-shift keying (ASK), frequency-shift keying (FSK), phase-shift keying (PSK), polarization-shift keying (PolSK), etc.

Over the years, various kinds of optical modulators have been realized: Mach-Zehnder modulator (MZM), electro-absorption modulators, acousto-optic modulator, etc. MZM is mainly used in integrated photonic. Fig. 7.28 shows schematic of MZM. MZM design is based on balanced MZI (for details see Section 4.3), where one arm consists of a phase shifter/switch capable of introducing 2π phase shift depending on input electrical power. When the data signal has 1s, the phase shifter has zero phase change, allowing a constructive interference in the combiner as a result high amplitude signal is generated as an output. For 0s in the data signal, the phase shifter will introduce 2π phase, which will cause a destructive interference in the combiner, resulting in low amplitudes signal as an output. The extinction ratio between the low and high states depends on splitting the ratio of the splitter and combiner and loss mismatch between the arms. The phase shift in a waveguide can be obtained by changing the refractive index of the waveguide. This concept of modulation can be further expanded to an unbalanced MZI and a ring resonator. In the previous sections, we see that both these devices have wavelength dependence constructive

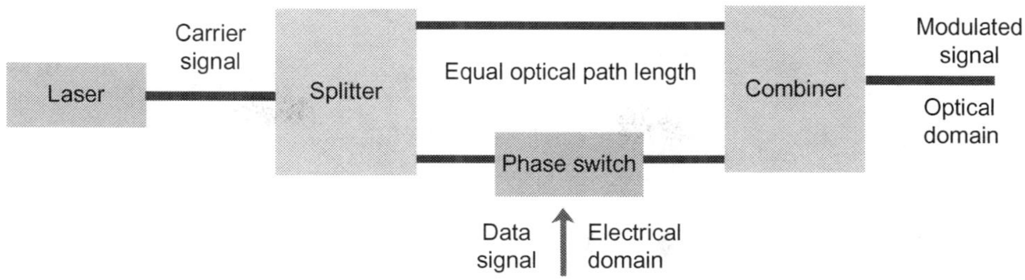

Figure 7.28
Mach–Zehnder modulator's working principle.

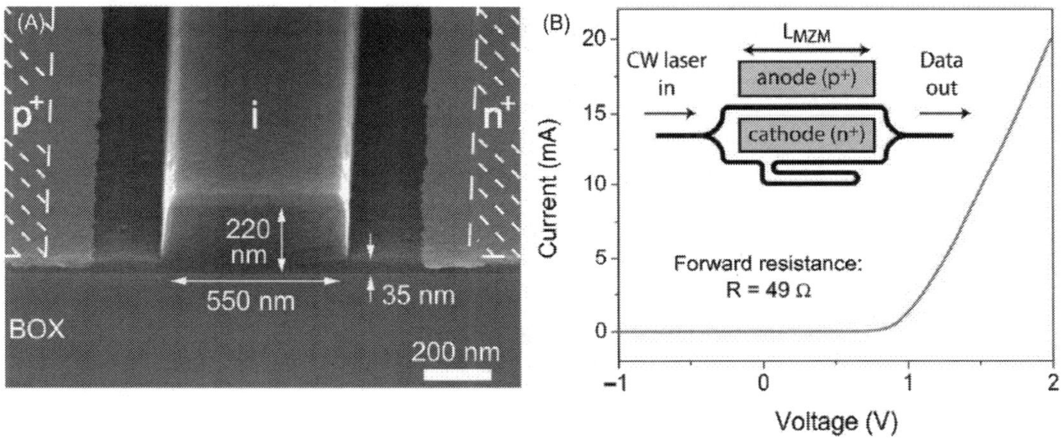

Figure 7.29
(A) SEM image of silicon PIN diode for modulator. (B) I-V plot of a modulator with modulator length (LMZM) 100 μm. *From [36] W.M.J. Green, M.J. Rooks, L. Sekaric, Y.A. Vlasov, Ultra-compact, low rf power, 10 gb/s silicon mach-zehnder modulator, Opt. Express 15 (25) (2007) 17106–17113.*

and destructive interference. The wavelengths of constructive and destructive interference can be shifted by changing the refractive index of one arm of the MZI and by changing the refractive index of the ring. The simplest method to change the refractive index of the waveguide is temperature change of the waveguide. For MZM, one can integrate heater in one arm to relatively change the temperature in that arm in order to gain a relative phase change [35]. But a high speed modulation is not possible using a thermal tuning.

During the last few decades, research has been carried out to develop a free carrier electro-optic effects-based silicon modulators for high speed optical communication link. The high speed modulation is achieved by changing carrier density, which changes the refractive index of the waveguide. Few configurations have been reported to gain a carrier density variation: carrier injection in a forward biased PIN diode [36], carrier accumulation in an metal-oxide-semiconductor (MOS) capacitor [37], and carrier depletion in a reverse biased p-n diode [38–40]. In 2007, Green et al. demonstrated a forward biased PIN diode MZM

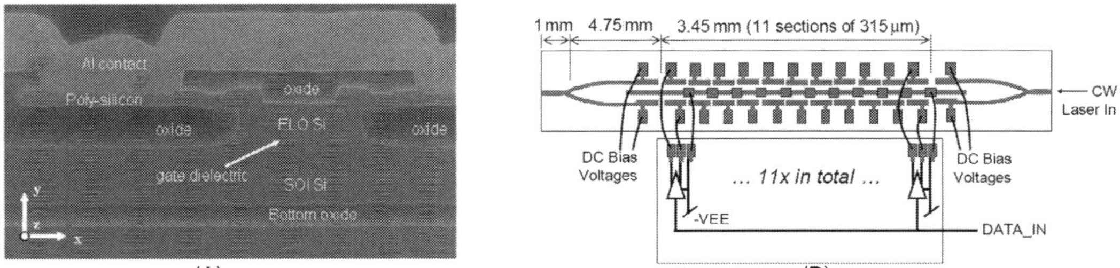

(A) (B)

Figure 7.30

(A) SEM image of MOS capacitors-based silicon modulator; (B) Schematic of MOS capacitors-based Mach-Zehnder modulator *From [37] L. Liao, D. Samara-Rubio, M. Morse, A. Liu, D. Hodge, D. Rubin, et al., High speed silicon mach-zehnder modulator, Opt. Express 13 (8) (2005) 3129–3135.*

on silicon, as shown in Fig. 7.29. A light doped silicon rib waveguide was used as an intrinsic region (i-region) to obtain refractive index change by carrier injection. The mode overlap between the waveguide and the modulator (for this modulator, it is the i-region) is large, leading to high modulation efficiency. The modulation efficiency (V_π L_π) is the product of the modulator length (L_π) and required bias voltage (V_π) to obtain a π phase shift. High efficiency corresponds to lower value of this product. The measured modulation efficiency (V_π L_π) is 0.36Vmm, with operating voltage ~1.8 V and 10 Gb/s modulation speed has been demonstrated [36]. The main disadvantage of injection-based modulators is that it suffers from the long free carrier lifetime (~1 ns) in silicon. As a result, the 3 dB electrical bandwidths are 1 GHz. To achieve high speed modulation, Xu et al. [41] demonstrated a 12.5 Gb/s carrier-injection-based silicon modulators using signal pre-emphasis. But this method increased the complexity and bias voltage of the modulator. In 2005, Intel [37] published a MOS capacitors-based MZM, as shown in Fig. 7.30(A) and (B). The modulator consists of 1μm n-type doped silicon at the bottom and 0.55 μm p-type doped silicon at the top. In between these two layers, there is a 10.5 nm thick gate dielectric. To form a waveguide, 0.65 μm silicon has been etched from top of the p-type doped silicon. The width of the waveguide is 1.6 μm. A 3 μm-wide polysilicon is used on top of the p-type doped silicon for aluminum contacts. The optical filed has higher confinement close to gate dielectric region, which allows strong interact with the free carriers accumulated within the gate dielectric layer. The measured modulation efficiency (V_π L_π) of this modulator is 3.3V-cm and 10 Gb/s modulation speed has been measured. Since free carriers accumulation of a MOS capacitor is obtained by majority carries, the modulator does not suffer from the long minority carrier lifetime in silicon. The disadvantage of the MOS modulators are that it requires a large footprint (3.5 to 15 mm long) and a high driving voltage (4–10 V), which reduces integrability of this modulator in a large photonics circuits.

One of the most popular choice of modulator in silicon platform is the p-n diode modulator based on carrier depletion. In this modulator, a reverse bias voltage is applied to obtain free carrier density change. Therefore, the speed of the modulator does not suffer from the long minority carrier lifetime in silicon. Also, the modulator does not require high voltage as the diode is operating at reverse biased. There are several designs published for carrier depletion modulators. Among them, three promising design from our point of interest are: horizontal p-n junction, vertical p-n junction, and interleave p-n junction. In a horizontal p-n junction design, the bottom part of the silicon waveguide is lightly doped with either p-type or n-type dopant and the top part is doped with opposite dopant of the bottom. In 2007, Liu et al. [42] published a horizontal p-n diode design, as shown in Fig. 7.31(A) and (B). For the modulator, a 0.6 μm-wide rib waveguide has been fabricated on a 0.5 μm thick silicon wafer with 0.22 μm etch. The measured modulation efficiency (V_π L_π) of this modulator is ~4 V-cm, which can be calculated from the voltage response of the phase shifter shown in Fig. 7.31(D). From Fig. 7.31(C), we see that the measured extinction ratio of the MZM is better than 20 dB. The measured on-chip loss (without coupling loss) of the

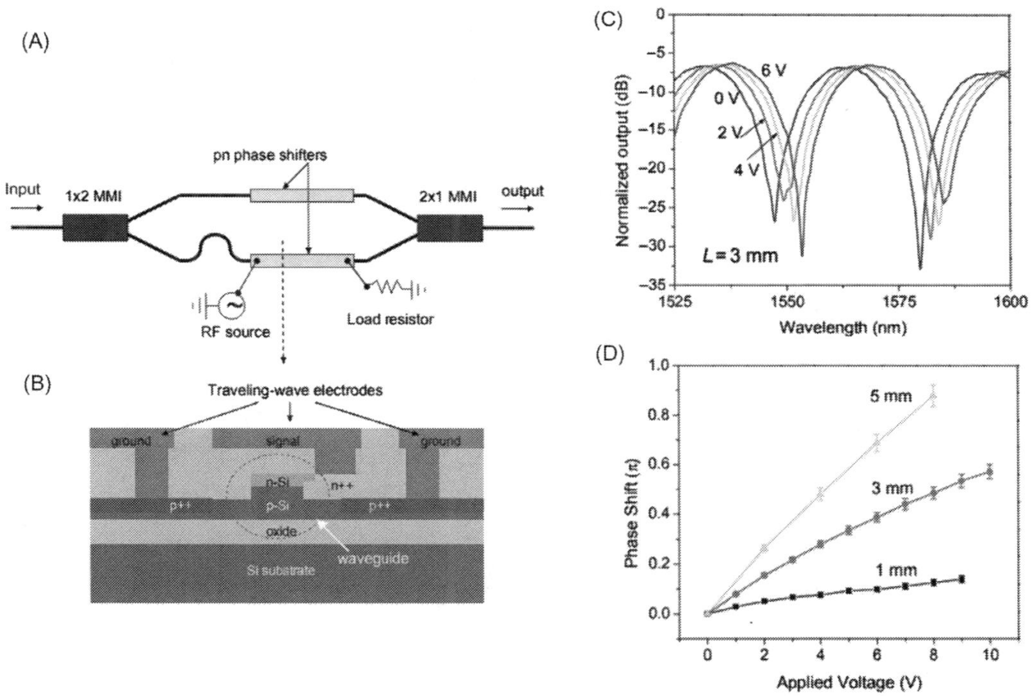

Figure 7.31

(A) Schematic of unbalanced Mach-Zehnder modulator using MMI as a splitter and combiner; (B) Cross-sectional view of a horizontal p-n junction; (C) Spectral response of the Mach-Zehnder modulator with 3 mm-long phase shifter for different biased voltage; (D) Voltage response of the phase shifter with different lengths. *From [42] A. Liu, L. Liao, D. Rubin, H. Nguyen, B. Ciftcioglu, Y. Chetrit, et al., High-speed optical modulation based on carrier depletion in a silicon waveguide, Opt. Express 15 (2) (2007) 660–668.*

MZM is 7 dB and estimated loss of the phase shifter is 5.4 dB, which is mainly attributed by doping of the waveguide. This modulator demonstrated transmitting data at 30 Gb/s speed with 3 dB electrical bandwidths greater than 20 GHz. In a vertical p-n junction, the two sides of the silicon waveguide doped with p -type and n-type form a depletion junction at the middle region of the waveguide, as shown in Fig. 7.32(A). The center part of the waveguide is lightly doped with p-type and n-type materials followed by high doped p-type and n-type slab region for ohmic contacts, respectively. Fig. 7.32(B) shows spectral response of a MZM with 2 mm long phase shifter for different biased voltages published by Park et al. in 2009 [43]. The modulation efficiency is measured from 1.83V-cm depending on the p-type doping consternation. The carrier-induced loss of the phase shifter is ~4 dB/mm. This modulator demonstrated transmitting data at 12.5 Gb/s speed with 3-dB electrical bandwidths greater than 7 GHz. If we look at the design of these two p-n junctions, we see that in both cases the depletion region is parallel to the propagation direction of the light. For an interleave p-n junction, design p-type and n-type region has been placed alternatively along the propagation direction of the waveguide, as shown in Fig. 7.33(A). Therefore, the depletion region of the interleave design is orthogonal to the propagation direction. In 2012, Xu et al. [38] reported an interleaved p-n junction-based MZM. The modulator has been fabricated on a 340 nm thick silicon. The silicon waveguide is 450 nm-wide and 80 nm high slab. Both p-type and n-type regions are 300 nm long. Fig. 7.33(B) shows the spectral response of the MZM with 0.75 mm-long phase shifter for different biased voltages. The measured modulation efficiency is varied from 1.5–2.0V-cm depending on the reverse bias voltage. The carrier-induced loss of the phase shifter is nearly 10 dB/cm. The modulator is capable of transmitting data at 44 Gb/s speed.

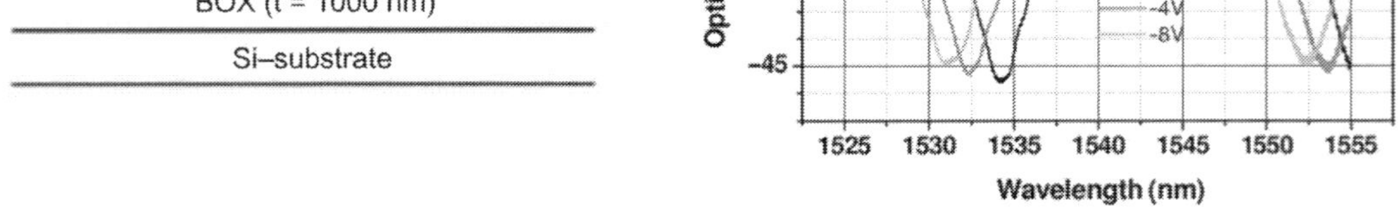

Figure 7.32

(A) Cross-sectional view of a vertical p-n junction; (B) Spectral response of the Mach-Zehnder modulator with 2 mm-long phase shifter for different biased voltage. *From [43] J.W. Park, J.-B. You, I.G. Kim, G. Kim, High-modulation efficiency silicon mach-zehnder optical modulator based on carrier depletion in a pn diode, Opt. Express 17 (18) (2009) 15520–15524.*

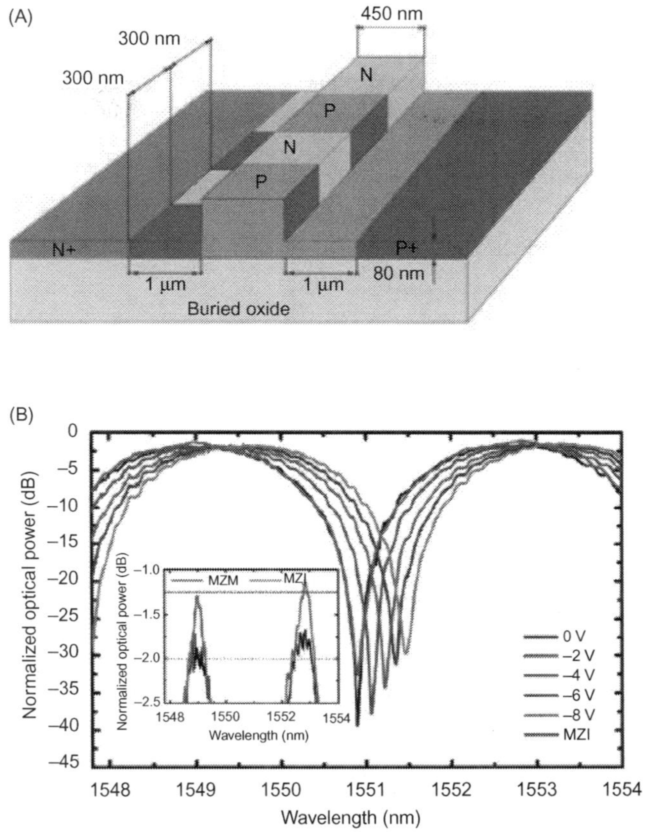

Figure 7.33

(A) Three-dimensional view of an interleaved p-n junction; (B) Spectral response of the Mach-Zehnder modulator with 0.75 mm-long phase shifter for different biased voltage. *From [38] H. Xu, X. Xiao, X. Li, Y. Hu, Z. Li, T. Chu, et al., High speed silicon mach-zehnder modulator based on interleaved pn junctions, Opt. Express 20 (14) (2012) 15093–15099.*

All these devices are fabricated in different thickness silicon platform, which makes it hard to compare the performance of these modulators with different p-n junction configuration. In this contest, Hu et al. [44] published a comparison between vertical and interleaved p-n junction-based designs in 220 nm thick silicon platform. Both of the design was fabricated with the same doping concentration: $1 \times 10^{18}/\text{cm}^3$. At 3 V reverse bias voltage, the modulator efficiency is ~1.3V-cm and ~1.9V-cm for the interleaved and vertical p-n junction-based modulator, respectively. The carrier-induced loss of the phase shifter is ~1.8 dB/mm and ~1.7 dB/mm for the interleaved and vertical p-n junction-based modulator, respectively. Under the test of high speed data transmission, both modulators show clear indication of 10 Gb/s speed. The extinction ratio for 10 Gb/s data transmission are 7.5 dB and 8.5 dB for the interleaved and vertical p-n junction-based modulator, respectively. From this work, we conclude that interleaved p-n junction-based modulator has better modulation efficiency but slightly higher optical loss and lower extinction ratio compared to the vertical p-n junction-based modulator.

7.5.3 Photodetectors

Photodetectors are essential building blocks of optical communications. A simple definition of the photodetectors is that of a device used to detect the photon. There are various kinds of photodetectors demonstrated over the years. But PIN junction-based photodiodes and metal-semiconductor-metal (MSM) photodetectors are mainly used in integrated photonics circuits. A PIN photodetector consists of a p-type doped region, n-type doped region, and an intrinsic (i)-region sandwiched between them. Schematic of PIN photodiodes are shown in Fig. 7.34 (A) and (B). Light absorbed in the i-region generates electron–hole pairs, which generate a photocurrent. The photocurrent is proportional to the intensity of the absorbed light. Such devices are known as photovoltaic devices. In integrated photonics, a photodetector mainly operates under reverse bias. It is desired that the photocurrent should be zero with an applied reverse bias voltage in the absence of photon. But in reality, under a reverse bias a small current generates without any presence of light, which is known as dark current. The dark current is the key property of the photodetectors which need to minimize for a better performances. A MSM photodetectors consist of two metal electrode on a semiconductor. MSM photodetectors are photoconductive devices, whose conductivity changes in the presence of light. Therefore, an external bias is always required for this kind of photodetectors. Schematic of MSM photodetectors are shown in Fig. 7.34(C) and (D). The main disadvantage of these photodetectors is high dark current, where the advantage is intrinsically low capacitance. Before going forward, here are the few parameters which

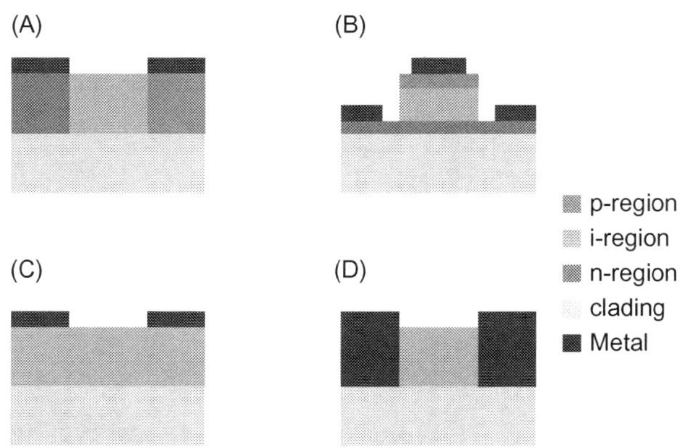

Figure 7.34
Schematic of (A) surface contact MSM photodetectors, (B) lateral contact MSM photodetectors, (C) lateral PIN photodiode, and (D) vertical PIN photodiode.

determine the quality of photodetectors: dark current, responsivity, and bandwidth. Dark current is an important parameter for an integrated photodetectors. High dark current increases the power consumption and noise. For optical communication applications, less than 1 μA dark current is desired. Responsivity (S) is the ratio between the photocurrent (I) and input optical power (P): $S = I/P$. The responsivity depends on the quantum efficiency of the photodiodes. The quantum efficiency (η) is the faction of absorbed photon that contributes to the photocurrent. Therefore, the photocurrent per watt due to a photon absorbed is equal to $i_p = \eta_e/h\nu$: e electron charge and $h\nu$ photon energy. So, the total photocurrent (I) is equal to $i_p \cdot P$. from this equation, we derive that the responsivity $S = i_p = \eta_e/h\nu$. The bandwidth of a photodetector depends on the response time. The response time of a photodetector mainly depends on carrier transit time and RC time constant. When the photon is absorbed in the i-region, it generates an electron–hole pair, which travels toward the electrodes and is collected by the electrodes. Carrier transit time is the time required for this process. The 3-dB electrical bandwidth due to the carrier transit time can be written as [45]:

$$f_t \cong \frac{3.5\overline{u}}{2\pi D} \tag{7.8}$$

$$1/\overline{u}^4 = 0.5(1/u_e^4 + 1/u_h^4) \tag{7.9}$$

where D is the carrier transit distance, v_e and v_h are the electrons and hole saturation velocities, respectively. The RC time constant can be determined by the equivalent electrical circuit of the photodiode. If the total resistance and capacitance of the equivalent electrical circuit is R and C, respectively, the 3-dB electrical bandwidth due to the RC time constant can be written as [45]:

$$f_{RC} = \frac{1}{2\pi RC} \tag{7.10}$$

The total resistance and capacitance depend on both internal and external electrical design of the photodetectors: diode series resistance, junction capacitance, parasitic capacitance due to metallization, load resistance, etc. The total 3-dB electrical bandwidth can be written as [45]:

$$f_{3dB} = \frac{f_t \cdot f_{RC}}{\sqrt{f_t^2 + f_{RC}^2}} = \frac{f_t}{\sqrt{1 + (f_t/f_{RC})^2}} = \frac{f_{RC}}{\sqrt{1 + (f_{RC}/f_t)^2}} \tag{7.11}$$

From this equation, it is clear that the bandwidth of the photodetector will be dominated by f_{RC} when f_{RC} is smaller than f_t. This implies that a smaller resistance and capacitance is required. On the other hand, a thinner i-region is required to increase f_t by minimizing the carrier transit distance D. But a thinner i-region also increases the capacitance. Therefore, a optimize design is required to have high speed photodetectors. A PIN junction-based photodiodes can also work as avalanche photodiodes (APD) by operating it with high reverse bias voltage just before breakdown. Under a strong internal electric field, APD can generate secondary carriers which work as an amplification of the signal. Therefore, the responsivity of APD is higher compared to the PIN junction photodiode, although the avalanche process may introduce some level of quantum noise.

The basic operation of a photodetector is photon absorption. This photoabsorption phenomena depends on material and wavelength due to materials direct and indirect bandgap. For instance, silicon (Si) is an appropriate photodetector material for visible wavelength from 400 nm to 1000 nm. For telecoms' wavelength range (1300 nm to 1600nm), Ge and indium gallium arsenide (InGaAs) are appropriate choices. In this section, we discuss Ge and III-V photodetector integrated in silicon platform.

In Section 5.1, we mentioned two bonding strategies used to develop III-V laser on silicon. Similar strategies were also used to develop III-V photodetector on silicon. Fig. 7.35(A) shows cross-sectional view of a BCB-bonded InGaAs PIN junction photodiode integrated

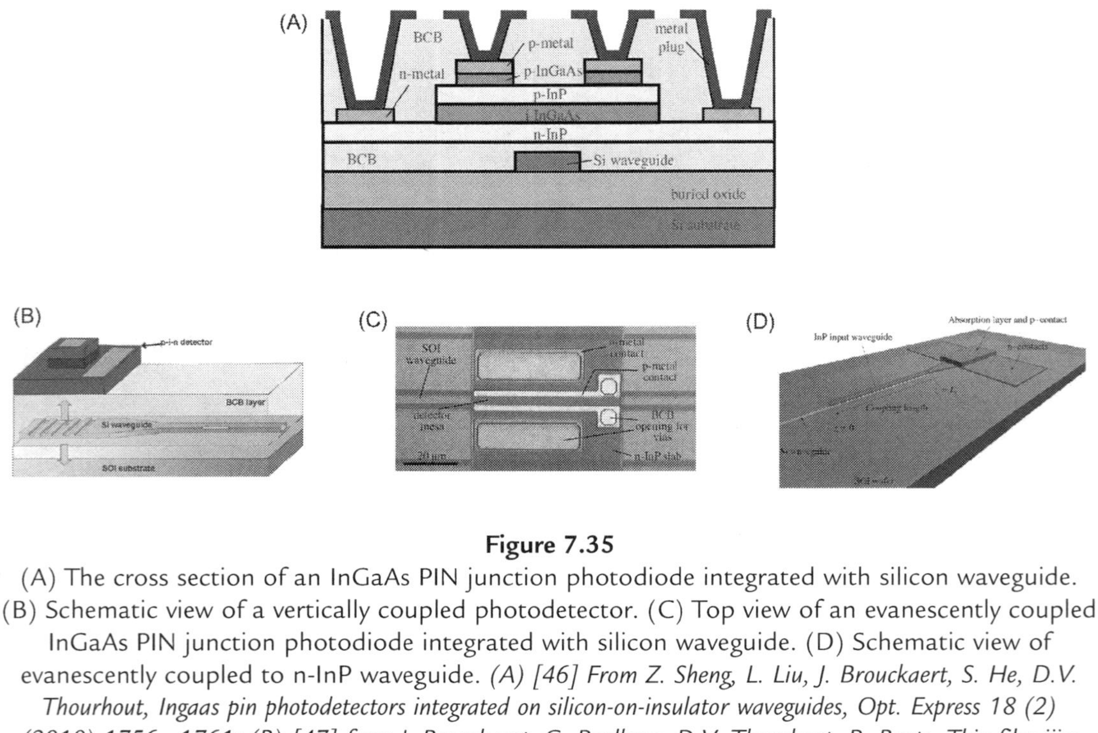

Figure 7.35

(A) The cross section of an InGaAs PIN junction photodiode integrated with silicon waveguide. (B) Schematic view of a vertically coupled photodetector. (C) Top view of an evanescently coupled InGaAs PIN junction photodiode integrated with silicon waveguide. (D) Schematic view of evanescently coupled to n-InP waveguide. *(A) [46] From Z. Sheng, L. Liu, J. Brouckaert, S. He, D.V. Thourhout, Ingaas pin photodetectors integrated on silicon-on-insulator waveguides, Opt. Express 18 (2) (2010) 1756–1761; (B) [47] from J. Brouckaert, G. Roelkens, D.V. Thourhout, R. Baets, Thin-film iii-v photodetectors integrated on silicon-on-insulator photonic ics, J. Lightwave Technol. 25 (4) (2007) 1053–1060; (C) [46] from Z. Sheng, L. Liu, J. Brouckaert, S. He, D.V. Thourhout, Ingaas pin photodetectors integrated on silicon-on-insulator waveguides, Opt. Express 18 (2) (2010) 1756–1761; (D) [48] from P.R.A. Binetti, X.J.M. Leijtens, T. de Vries, Y.S. Oei, L.D. Cioccio, J.M. Fedeli, et al., Inp/ingaas photodetector on soi photonic circuitry, IEEE Photon. J. 2 (3) (2010) 299–305.*

on top of silicon waveguide. Few coupling mechanisms are used to couple the light from silicon waveguide to the photodetector. In Section 4.1.2, we discussed the grating coupler for vertical coupling. The photodetector can be bonded directly on top of a grating coupler, as shown in Fig. 7.35(B). The advantage of this coupling method is that bonding process becomes more easier: thicker BCB can be used and coupling does not suffer from BCB surface quality. But this method suffers from all the disadvantages of a grating coupler: single polarization, lower percentage of light directed toward vertical direction, and wavelength-dependent coupling loss. In this method, light absorption depends on the thickness of the i-region. Therefore, a thick intrinsic layer is required for high responsivity. As shown in Fig. 7.35(C), the photodetctor can be bonded directly on top of silicon waveguide using thin BCB, which allows to couple the light evanescently from the silicon waveguide to the i-region of the photodetector. The light absorption depends on the length of the photodetector as well as the thickness of the BCB which will define the required length for evanescent coupling. Fig. 7.35(D) shows another approach of evanescent coupling, where the light first couple from Si waveguide to a waveguide form in n-type InP layer and then light get absorbed in the i-region. But this method required additional lithography and etch steps to fabricate a n-type InP waveguide. Fig. 7.35(A) and (C) shows the InGaAs PIN junction photodiodes developed by Sheng et al. [46]. The PIN junction consists of a lightly doped 100 nm thick p-type InP layer, 100 nm thick $In_{0.53}Ga_{0.47}$ layer as i-region and lightly doped 100 nm thick n-type InP layer. The Si waveguide is 220 nm thick and 3 μm wide. For higher evanescent coupling from Si waveguide to i-region of the photodetector, a thin BCB (200 nm) has been used for bonding. Fig. 7.36(A) and (B) shows the performance of the 4 μm-long photodetector for an input light of 1550 nm wavelength.

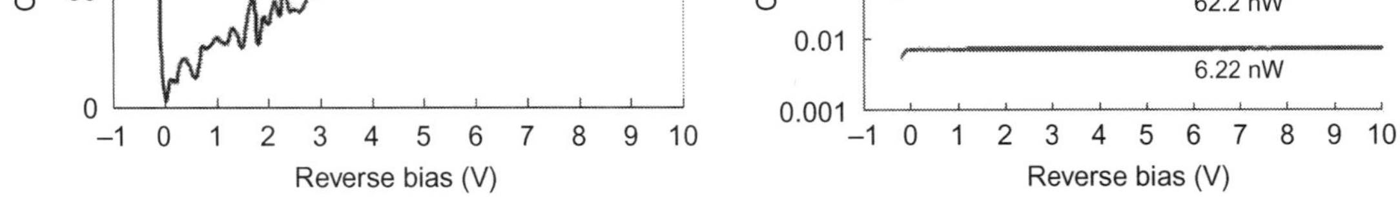

Figure 7.36
(A) Measured dark current versus reverse bias voltage; (B) Measured photocurrent versus reverse bias voltage for different input light intensity. *(A and B) From [46] Z. Sheng, L. Liu, J. Brouckaert, S. He, D.V. Thourhout, Ingaas pin photodetectors integrated on silicon-on-insulator waveguides, Opt. Express 18 (2) (2010) 1756–1761.*

The dark current of this photodiode is in the order of picoampere and the photocurrent remains stable in the range of reverse bias from 0 V to 10 V, which implies that the photodetector can operate without an external reverse bias voltage. The measured internal responsivity is around 1.1 A/W, which corresponds to a quantum efficiency of 88% for 1550 nm wavelength. As the quantum efficiency of this device is dominated by the evanescent coupling, the quantum efficiency can be further improved by using thinner BCB and longer photodetector. The 3-dB bandwidth of InGaAs PIN junction photodiodes reported by Binetti et al. [48] is 33 GHz. InGaAs PIN junction as a photodetector shows several advantages, including broad wavelength absorption range, high responsivity, low dark current, and high speed. But integration of the InGaAs PIN junction photodiodes in a silicon waveguide platform is highly expensive, unreliable due to bonding technique, and not CMOS compatible. Therefore, with Si waveguide platform, the same group IV material Ge gained more interest as a photodetectors material for telecoms wavelength.

Ge is also an indirect bandgap material like silicon. Its indirect and direct bandgap are 0.66 eV and 0.8 eV, respectively. Therefore, Ge has higher optical absorption from 1310 nm to 1550 nm wavelength range, which makes Ge-based photodetectors promising candidates for telecoms' wavelengths. However, Ge and Si has 4% lattice mismatch, which makes monolithic integration of high-quality low dislocation density Ge on Si difficult. Nevertheless, over the years, many groups developed fabrication technique to obtain high performance Ge photodetectors on silicon. One of the methods to grow crystalline Ge on Si is to use graded SiGe region as a buffer layer [49]. However, it requires 10 μm thick buffer layer for a pure crystalline Ge on Si, which makes it difficult for silicon photonics integration for both evanescent or edge coupling methods. Another method used by several groups is the two-step growth technique. First, at low temperature (~350 °C), a thin layer (~100 nm) of Ge is grown on silicon, which is used as a buffer layer. Next, at high temperature (<600 °C), few micron thick Ge is grown on top of low temperature Ge. After that, a high temperature annealing is often employed to reduce the threading dislocation density (TDD) [50]. In recent years, another method of Ge growth to become interesting is using low-energy plasma-enhanced chemical vapor deposition (LEPECVD) in the presence of hydrogen with high deposition rates [49]. The grown Ge shows a low TDD with smooth surfaces.

In 2007, Intel reported a high speed PIN junction-based Ge photodetector on Si [50]. A 0.8 μm thick Ge is grown by using the two-step growth technique on top of 1.5 μm thick Si. In the detector region, the Si waveguide is lightly doped with p-type material: boron. A 1.4 μm-wide single-mode Si rib waveguides are fabricated by 0.6 μm deep etching. Finally, the top surface of the Ge is doped with n-type material: phosphorous. Fig. 7.37(A) and (B) shows a schematic view and cross-sectional view of the Ge PIN junction photodetector on Si rib waveguide. Two different geometries photodetectors are designed: one detector (D1) is 7.4 μm wide and 50 μm long and the other detector (D2) is 4.4 μm wide and 100 μm long. The measured dark current at -2V was 169 nA and 267 nA for detector D1 and D2,

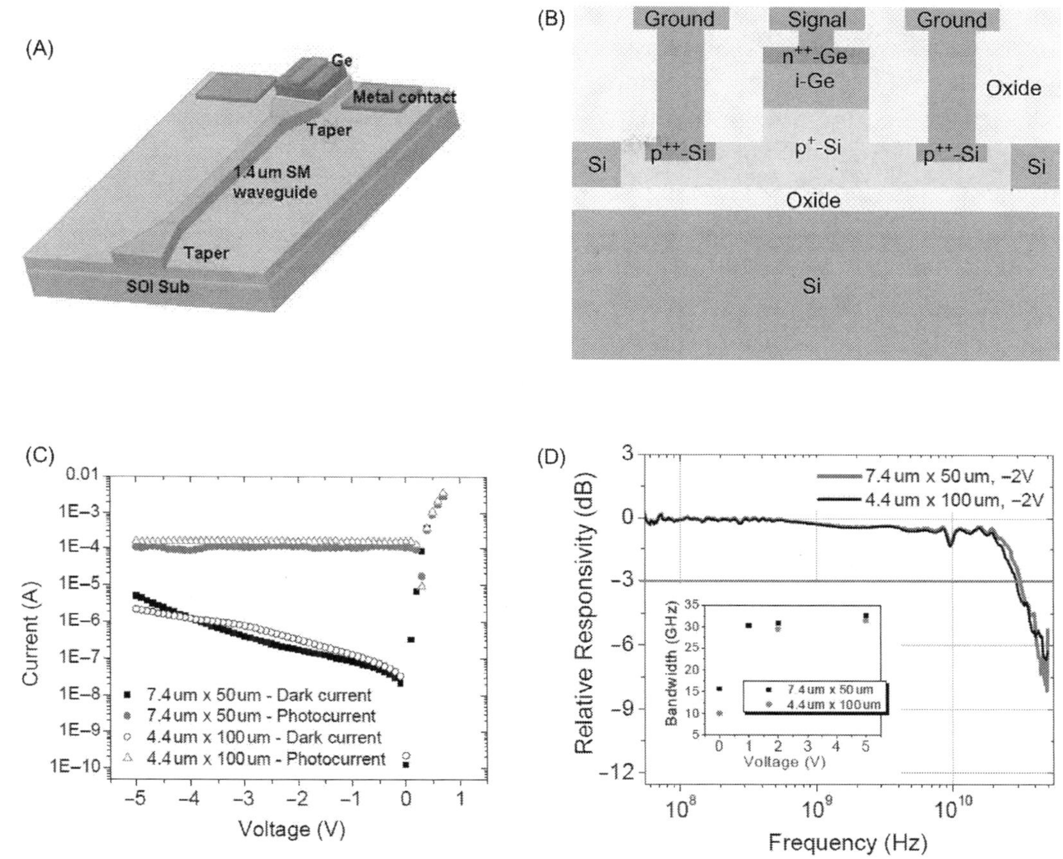

Figure 7.37

(A) Schematic view of a Ge detector integrated with Si waveguide. (B) Cross-sectional view of the Ge PIN junction photodetector on Si rib waveguide. (C) Measured dark current and photocurrent with input optical power of 126 μW at 1550 nm wavelength. (D) Measured frequency response for the Ge detectors. *From [50] T. Yin, R. Cohen, M.M. Morse, G. Sarid, Y. Chetrit, D. Rubin, et al., 31ghz ge n-i-p waveguide photodetectors on silicon-on-insulator substrate, Opt. Express 15 (21) (2007) 13965–13971.*

respectively, as shown in Fig. 7.37(C). At -2V with input optical power of 126 μW at 1550 nm wavelength, the responsivity for detectors D1 and D2 were calculated to be 0.89 A/W and 1.16 A/W, respectively. At -2V, the measured 3-dB electrical bandwidth is ~31 GHz for both detectors D1 and D2, as shown in Fig. 7.37(D). The bandwidth is limited by both carrier transit time due to thick Ge layer and RC time constant of the circuit. Therefore, an improved design is needed to overcome the following challenges of this design: low free carrier loss in the doped Si rib waveguide under the Ge layer, thinner Ge layer for small carrier transit time, and better coupling between the Si waveguide to the active region of Ge layer.

In 2016, Chen et al. published a Ge PIN junction photodetector on 220 nm thick silicon waveguide platform [51]. Fig. 7.37(A) and (B) shows a schematic view and cross-sectional view of the Ge PIN junction photodetector on Si rib waveguide. A 160 nm thick and 500 nm-wide Ge layer is grown on top of 220 nm thick silicon. An adiabatic taper has been used to couple the light from 400 nm-wide single-mode silicon waveguide to Ge-on-silicon waveguide. To improve the coupling and reduce the loss, a 120 nm thick polysilicon taper is used on top of the adiabatic taper region, as shown in Fig. 7.37(A). Under the Ge layer, each side of the silicon layer is lightly doped with p-type and n-type material to form a PIN junction. Fig. 7.37(C) shows the measured I-V performance of the photodetector with 14.2 μm-long Ge layer. The measured dark current is 2.5 nA and 3.5 nA at -1V and -2V operating bias voltage, respectively. The measured responsivity is 0.72 A/W and 0.98 A/W for 1550 nm and 1310 nm wavelength light at -1V bias voltage, respectively. From Fig. 7.37(D), we see that the device has 3-dB electrical bandwidth that is better than 50 GHz for 1550 nm wavelength light in the range from -1V to -2V bias voltage, for 1310 nm wavelength light, 3-dB electrical bandwidth is better than 45 GHz and 50 GHz when applied bias voltages is in range from -1V and -2V, respectively. At -1V, the device is capable of working as an on-off keying optical receiver with a speed of 56 Gb/s (Fig. 7.38).

7.6 Photonics Integrated Circuits

In this section, we discuss some examples of complex photonics circuit designs using multiple optical components, discussed in previous sections.

7.6.1 Laser Array

Multichannel tunable laser is attractive for many applications, which drive to expand the knowledge of III-V laser integrated with Si waveguide platform. Fig. 7.39(A) shows schematic of a four-channel laser array in silicon. III-V gain sections are integrated to the Si waveguide, where coupling the light from III-V region to silicon waveguide is obtained by using adiabatic inverse taper. Two DBR mirrors are used on both sides of the gain section. The front DBR mirror is designed for partial reflection: maximum 45% reflections, where the back DBR mirror is designed to have above 90% reflection. In between the gain region and the back reflector, a ring resonator is used for each gain sections. The loop length of those ring resonators are chosen in such a way that their resonance frequencies are separated by 250 GHz. These ring resonators will select the lasing wavelength, which is the same as their individual resonance wavelength. Fig. 7.39(B) shows measured spectral response of the laser array. The measured results shows clear indication of four narrow band lasing at four different wavelengths with 45 dB suppression of side mode [52]. Each of these ring resonators can be individually tuned by using integrated heaters, to change the resonance wavelength as well as lasing wavelength [53]. These four-ring resonators can be

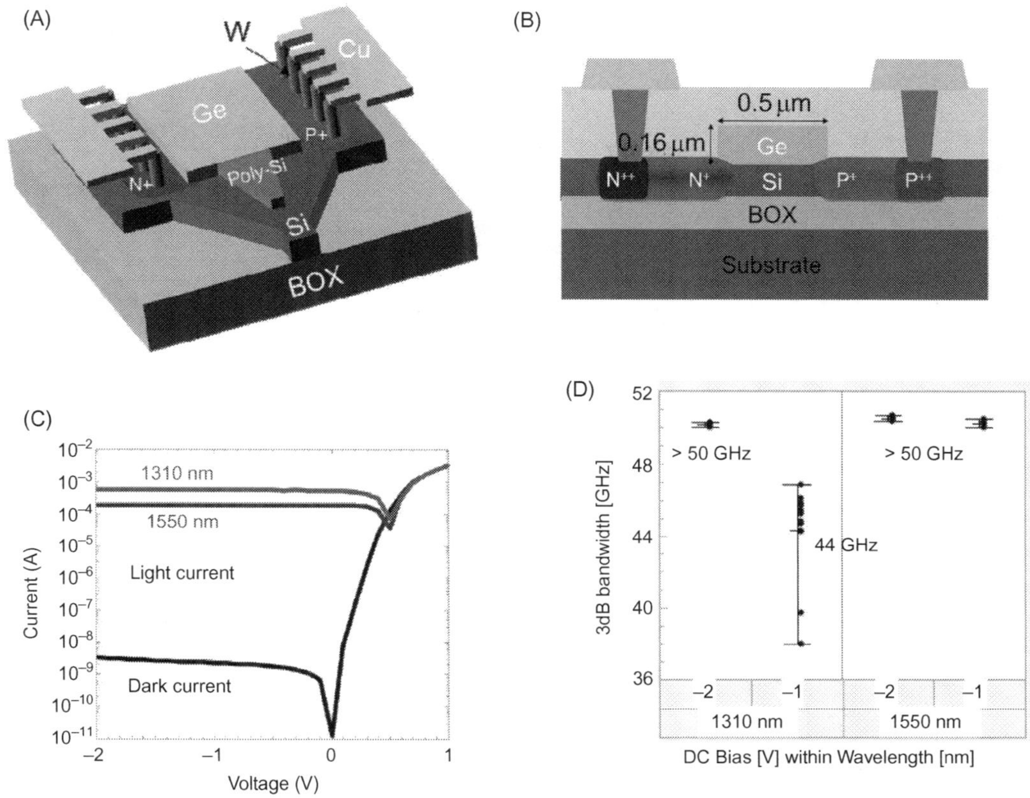

Figure 7.38

(A) Schematic view of a Ge detector integrated with Si waveguide. (B) Cross-sectional view of the Ge PIN junction photodetector on Si rib waveguide. (C) Measured dark current and photocurrent for 1310 nm and 1550 nm wavelength light. (D) Measured 3-dB electrical bandwidth for the Ge detector. *From [51] H. Chen, P. Verheyen, P.D. Heyn, G. Lepage, J.D. Coster, S. Balakrishnan, et al., -1 v bias 67 ghz bandwidth si-contacted germanium waveguide p-i-n photodetector for optical links at 56 gbps and beyond, Opt. Express 24 (5) (2016) 4622–4631.*

replaced by a four-channel AWGs, but tuning wavelength channel of an AWG is not straightforward [52].

7.6.2 Transmitter and Receiver

A few years ago, Intel published a prototype of 50 Gb/s data link for data centers. A data link consists of two-module transmitter and receiver module, as shown in Fig. 7.40. Intel use silicon photonics platform to develop these module. The transmitter chip consist of four III-V lasers on silicon, four p-n junction-based silicon modulator of 12.5 Gb/s speed, and a four-channel silicon multiplexer. Four different wavelength lasers are connected to four modulators, which are used to convert the data from electrical domain to optical domain.

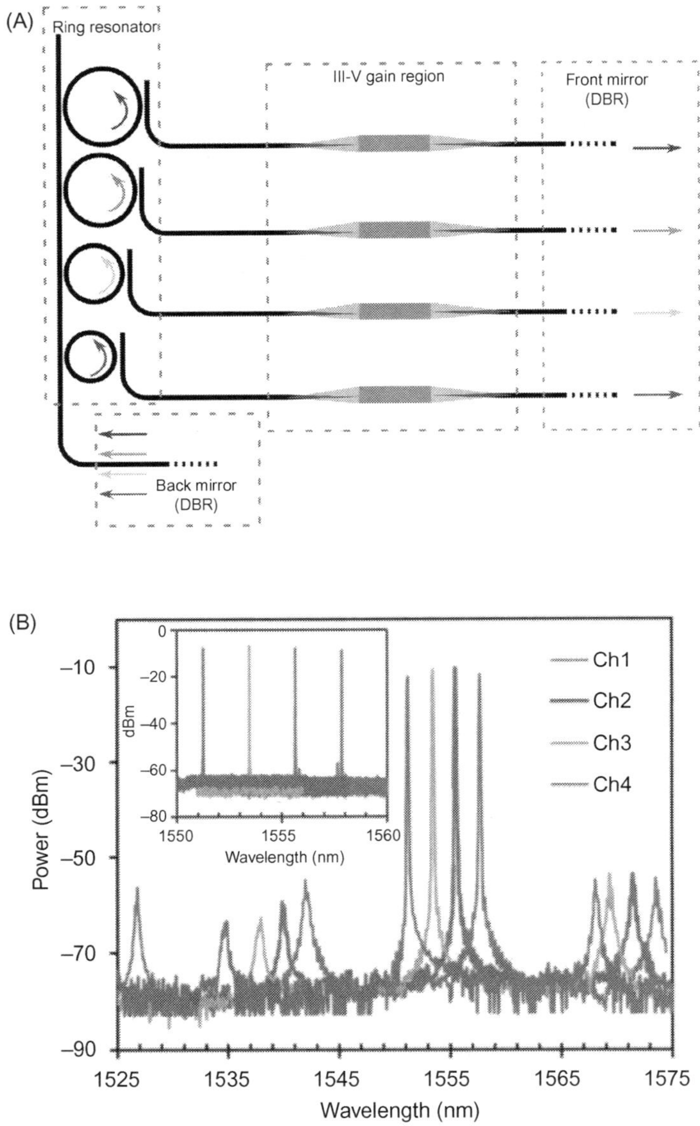

Figure 7.39

(A) Schematic four-channel laser array using ring resonators. (B) Spectral response for the four-channel laser array, measured with a resolution bandwidth of 0.2 pm. *From [52] S. Keyvaninia, S. Verstuyft, S. Pathak, F. Lelarge, G.-H. Duan, D. Bordel, et al., Iii-v-on-silicon multi-frequency lasers, Opt. Express 21 (11) (2013) 13675–13683.*

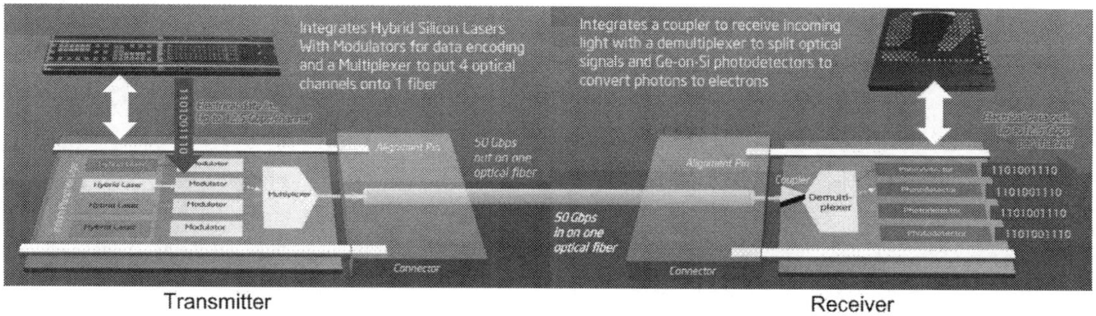

Figure 7.40
Schematic of 50 Gb/s data link. *From [54] https://www.intel.com.*

All these modulated lights are combined to a silicon waveguide by a wavelength multiplexer and finally coupled to an optical fiber.

The receiver chip consists of a four-channel Si demultiplexer and four Ge on Si photodetectors. In the receiver chip light couple to a silicon waveguide from the optical fiber, the Si demultiplexer separate the light into four wavelength channels. Finally, these wavelength channels connected to four Ge on Si photodetectors returns the data from optical domain back to electrical domain.

References

[1] V.R. Almeida, R.R. Panepucci, M. Lipson, Nanotaper for compact mode conversion, Opt. Lett. 28 (15) (2003) 1302–1304.

[2] T. Tsuchizawa, K. Yamada, H. Fukuda, T. Watanabe, J. Ichi Takahashi, M. Takahashi, et al., Microphotonics devices based on silicon microfabrication technology, IEEE J. Sel. Top. Quantum Electron. 11 (1) (2005) 232–240.

[3] G. Roelkens, P. Dumon, W. Bogaerts, D.V. Thourhout, R. Baets, Efficient silicon-on-insulator fiber coupler fabricated using 248-nm-deep uv lithography, IEEE Photonics Technol. Lett. 17 (12) (2005) 2613–2615.

[4] Q. Fang, J. Song, X. Luo, M. Yu, G. Lo, Y. Liu, Mode-size converter with high coupling efficiency and broad bandwidth, Opt. Express 19 (22) (2011) 21588–21594.

[5] N. Hatori, T. Shimizu, M. Okano, M. Ishizaka, T. Yamamoto, Y. Urino, et al., A hybrid integrated light source on a silicon platform using a trident spot-size converter, J. Lightwave Technol. 32 (7) (2014) 1329–1336.

[6] D. Taillaert, W. Bogaerts, P. Bienstman, T.F. Krauss, P.V. Daele, I. Moerman, et al., An out-of-plane grating coupler for efficient butt-coupling between compact planar waveguides and singlemode fibers, IEEE J. Quantum Electron. 38 (7) (2002) 949–955.

[7] D. Vermeulen, S. Selvaraja, P. Verheyen, G. Lepage, W. Bogaerts, P. Absil, et al., High-efficiency fiber-to-chip grating couplers realized using an advanced cmos-compatible silicon-on-insulator platform, Opt. Express 18 (17) (2010) 18278–18283.

[8] F.V. Laere, G. Roelkens, M. Ayre, J. Schrauwen, D. Taillaert, D.V. Thourhout, et al., Compact and highly efficient grating couplers between optical fiber and nanophotonic waveguides, J. Lightwave Technol. 25 (1) (2007) 151–156.

[9] X. Chen, C. Li, C.K.Y. Fung, S.M.G. Lo, H.K. Tsang, Apodized waveguide grating couplers for efficient coupling to optical fibers, IEEE Photonics Technol. Lett. 22 (15) (2010) 1156–1158.

[10] S. Pathak, M. Vanslembrouck, P. Dumon, D.V. Thourhout, W. Bogaerts, Compact soi-based polarization diversity wavelength de-multiplexer circuit using two symmetric awgs, Opt. Express 20 (26) (2012) B493–B500.

[11] K. Okamoto, Progress and technical challenge for planar waveguide devices: silica and silicon waveguides, Laser Photon. Rev. 6 (1) (2012) 14–23.

[12] Y. Hida, Y. Hibino, T. Kitoh, Y. Inoue, M. Itoh, T. Shibata, A. Sugita, and A. Himeno, "400-channel 25-GHz spacing arrayed-waveguide grating covering a full range of C- and L-bands," in Optical Fiber Communication Conference and International Conference on Quantum Information, 2001 OSA Technical Digest Series, paper WB2. https://www.osapublishing.org/abstract.cfm?uri = OFC-2001-WB2.

[13] F. M. Soares, W. Jiang, S.W. Seo, J.H. Baek, R.G. Broeke, J. Cao, et al., 20GHz Channel Spacing InP-based Arrayed Waveguide Grating, 33rd European Conference and Exhibition of Optical Communication, Berlin, Germany, 2007, pp. 1–2. http://ieeexplore.ieee.org/stamp/stamp.jsp?tp = &arnumber = 5758194 &isnumber = 5757122.

[14] M. Kohtoku, H. Sanjoh, S. Oku, Y. Kadota, Y. Yoshikuni, Y. Shibata, InPbased 64-channel arrayed waveguide grating with 50 ghz channel spacing and up to -20 db crosstalk, Electron. Lett. 33 (21) (1997) 1786–1787.

[15] W. Bogaerts, S. Selvaraja, P. Dumon, J. Brouckaert, K. De Vos, D. Van Thourhout, et al., Silicon-on-insulator spectral filters fabricated with CMOS technology, IEEE J. Sel. Top. Quantum Electron. 16 (1) (2010) 33–44.

[16] S. Pathak, K. Shang S. J. B. Yoo, Experimental demonstration of compact 16 channels-50 GHz Si3N4 arrayed waveguide grating, 2015 Optical Fiber Communications Conference and Exhibition (OFC), Los Angeles, CA, 2015, pp. 1–3. http://ieeexplore.ieee.org/stamp/stamp.jsp?tp = &arnumber = 7121837 &isnumber = 7121453.

[17] A. Malik, M. Muneeb, S. Pathak, Y. Shimura, J.V. Campenhout, R. Loo, et al., Germanium-on-silicon mid-infrared arrayed waveguide grating multiplexers, IEEE Photon. Technol. Lett. 25 (18) (2013) 1805–1808.

[18] S. Pathak, D.V. Thourhout, W. Bogaerts, Design trade-offs for silicon-oninsulator-based awgs for (de) multiplexer applications, Opt. Lett. 38 (16).

[19] S. Pathak, M. Vanslembrouck, P. Dumon, D. Van Thourhout, P. Verheyen, G. Lepage, et al., Effect of mask discretization on performance of silicon arrayed waveguide gratings, IEEE Photon. Technol. Lett. 26 (7) (2014) 718–721.

[20] S. Pathak, M. Vanslembrouck, P. Dumon, D.V. Thourhout, W. Bogaerts, Optimized silicon awg with flattened spectral response using an mmi aperture, J. Lightwave Technol. 31 (1) (2013) 87–93.

[21] C. Doerr, L. Stulz, R. Pafchek, S. Shunk, Compact and low-loss manner of waveguide grating router passband flattening and demonstration in a 64-channel blocker/multiplexer, IEEE Photon. Technol. Lett. 14 (1) (2002) 56–58.

[22] F. Horst, W.M. Green, S. Assefa, S.M. Shank, Y.A. Vlasov, B.J. Offrein, Cascaded Mach-Zehnder wavelength filters in silicon photonics for low loss and flat pass-band wdm (de-)multiplexing, Opt. Express 21 (10) (2013) 11652–11658.

[23] S. H. Jeong, Y. Tanaka, K. Morito, Silicon photonics optical demultiplexer technology for WDM optical interconnects, 2015 Opto-Electronics and Communications Conference (OECC), Shanghai, 2015, pp. 1–3. doi: 10.1109/OECC.2015.7340286, http://ieeexplore.ieee.org/stamp/stamp.jsp?tp = &arnumber = 7340286&isnumber = 7340059.

[24] W. Bogaerts, P. De Heyn, T. Van Vaerenbergh, K. De Vos, S. Kumar Selvaraja, T. Claes, et al., Silicon microring resonators, Laser Photon. Rev. 6 (1) (2012) 47–73.

[25] P.D. Heyn, J.D. Coster, P. Verheyen, G. Lepage, M. Pantouvaki, P. Absil, et al., Fabrication-tolerant four-channel wavelength-division-multiplexing filter based on collectively tuned si microrings, J. Lightwave Technol. 31 (16) (2013) 2785–2792.

[26] S. Park, K.-J. Kim, I.-G. Kim, G. Kim, Si micro-ring mux/demux wdm filters, Opt. Express 19 (14) (2011) 13531–13539.

[27] K.D. Vos, J. Girones, T. Claes, Y.D. Koninck, S. Popelka, E. Schacht, et al., Multiplexed antibody detection with an array of silicon-on-insulator microring resonators, IEEE Photon. J. 1 (4) (2009) 225–235.

[28] N.A. Yebo, D. Taillaert, J. Roels, D. Lahem, M. Debliquy, D.V. Thourhout, et al., Silicon-on-insulator (soi) ring resonator-based integrated optical hydrogen sensor, IEEE Photon. Technol. Lett. 21 (14) (2009) 960–962.

[29] A.W.-L. Fang, Silicon evanescent lasers, Ph.D. thesis, University of California, Santa Barbara, 2008.

[30] S. Stankovic, Hybrid iii-v/si dfb lasers based on polymer bonding technology, Ph.D. thesis, Ghent University, 2013.

[31] S. Stankovic, R. Jones, M.N. Sysak, J.M. Heck, G. Roelkens, D.V. Thourhout, 1310-nm hybrid iii-v/si fabry-perot laser based on adhesive bonding, IEEE Photon. Technol. Lett. 23 (23) (2011) 1781–1783.

[32] S. Stankovic, R. Jones, M.N. Sysak, J.M. Heck, G. Roelkens, D.V. Thourhout, Hybrid iii-v/si distributed-feedback laser based on adhesive bonding, IEEE Photon. Technol. Lett. 24 (23) (2012) 2155–2158.

[33] A.W. Fang, M.N. Sysak, B.R. Koch, R. Jones, E. Lively, Y.H. Kuo, et al., Single-wavelength silicon evanescent lasers, IEEE J. Sel. Top. Quantum Electron. 15 (3) (2009) 535–544.

[34] A.W. Fang, B.R. Koch, R. Jones, E. Lively, D. Liang, Y.H. Kuo, et al., A distributed bragg reflector silicon evanescent laser, IEEE Photon. Technol. Lett. 20 (20) (2008) 1667–1669.

[35] A. Masood, M. Pantouvaki, D. Goossens, G. Lepage, P. Verheyen, J.V. Campenhout, et al., Fabrication and characterization of cmos-compatible integrated tungsten heaters for thermo-optic tuning in silicon photonics devices, Opt. Mater. Express 4 (7) (2014) 1383–1388.

[36] W.M.J. Green, M.J. Rooks, L. Sekaric, Y.A. Vlasov, Ultra-compact, low rf power, 10 gb/s silicon mach-zehnder modulator, Opt. Express 15 (25) (2007) 17106–17113.

[37] L. Liao, D. Samara-Rubio, M. Morse, A. Liu, D. Hodge, D. Rubin, et al., High speed silicon mach-zehnder modulator, Opt. Express 13 (8) (2005) 3129–3135.

[38] H. Xu, X. Xiao, X. Li, Y. Hu, Z. Li, T. Chu, et al., High speed silicon mach-zehnder modulator based on interleaved pn junctions, Opt. Express 20 (14) (2012) 15093–15099.

[39] Z.-Y. Li, D.-X. Xu, W.R. McKinnon, S. Janz, J.H. Schmid, P. Cheben, et al., Silicon waveguide modulator based on carrier depletion in periodically interleaved pn junctions, Opt. Express 17 (18) (2009) 15947–15958.

[40] H. Yu, W. Bogaerts, A. De Keersgieter, Optimization of ion implantation condition for depletion-type silicon optical modulators, IEEE J. Quantum Electron. 46 (12) (2010) 1763–1768.

[41] Q. Xu, S. Manipatruni, B. Schmidt, J. Shakya, M. Lipson, 12.5 gbit/s carrier-injection-based silicon micro-ring silicon modulators, Opt. Express 15 (2) (2007) 430–436.

[42] A. Liu, L. Liao, D. Rubin, H. Nguyen, B. Ciftcioglu, Y. Chetrit, et al., High-speed optical modulation based on carrier depletion in a silicon waveguide, Opt. Express 15 (2) (2007) 660–668.

[43] J.W. Park, J.-B. You, I.G. Kim, G. Kim, High-modulation efficiency silicon mach-zehnder optical modulator based on carrier depletion in a pn diode, Opt. Express 17 (18) (2009) 15520–15524.

[44] H. Yu, M. Pantouvaki, J.V. Campenhout, D. Korn, K. Komorowska, P. Dumon, et al., Performance tradeoff between lateral and interdigitated doping patterns for high speed carrier-depletion based silicon modulators, Opt. Express 20 (12) (2012) 12926–12938.

[45] K. Kato, Ultrawide-band/high-frequency photodetectors, IEEE Trans. Microw. Theory Tech. 47 (7) (1999) 1265–1281.

[46] Z. Sheng, L. Liu, J. Brouckaert, S. He, D.V. Thourhout, Ingaas pin photodetectors integrated on silicon-on-insulator waveguides, Opt. Express 18 (2) (2010) 1756–1761.

[47] J. Brouckaert, G. Roelkens, D.V. Thourhout, R. Baets, Thin-film iii-v photodetectors integrated on silicon-on-insulator photonic ics, J. Lightwave Technol. 25 (4) (2007) 1053–1060.

[48] P.R.A. Binetti, X.J.M. Leijtens, T. de Vries, Y.S. Oei, L.D. Cioccio, J.M. Fedeli, et al., Inp/ingaas photodetector on soi photonic circuitry, IEEE Photon. J. 2 (3) (2010) 299–305.

[49] T.H. Loh, H.S. Nguyen, C.H. Tung, A.D. Trigg, G.Q. Lo, N. Balasubramanian, et al., Ultrathin low temperature sige buffer for the growth of high quality ge epilayer on si(100) by ultrahigh vacuum chemical vapor deposition, Appl. Phys. Lett. 90 (9) (2007) 092108.

[50] T. Yin, R. Cohen, M.M. Morse, G. Sarid, Y. Chetrit, D. Rubin, et al., 31ghz ge n-i-p waveguide photodetectors on silicon-on-insulator substrate, Opt. Express 15 (21) (2007) 13965–13971.

[51] H. Chen, P. Verheyen, P.D. Heyn, G. Lepage, J.D. Coster, S. Balakrishnan, et al., -1 v bias 67 ghz bandwidth si-contacted germanium waveguide p-i-n photodetector for optical links at 56 gbps and beyond, Opt. Express 24 (5) (2016) 4622–4631.

[52] S. Keyvaninia, S. Verstuyft, S. Pathak, F. Lelarge, G.-H. Duan, D. Bordel, et al., Iii-v-on-silicon multi-frequency lasers, Opt. Express 21 (11) (2013) 13675–13683.

[53] S. Keyvaninia, G. Roelkens, D.V. Thourhout, C. Jany, M. Lamponi, A.L. Liepvre, et al., Demonstration of a heterogeneously integrated iii-v/soi single wavelength tunable laser, Opt. Express 21 (3) (2013) 3784–3792.

[54] http://download.intel.com/pressroom/pdf/photonics/50G_Silicon_Photonics_Link.pdf?iid = pr_smrelease_vPro_materials1.

CHAPTER 8

Graphene Based Optical Interconnects

Xinbo Wang and Berardi Sensale-Rodriguez
The University of Utah, Salt Lake City, UT, United States

8.1 Introduction

The complexity and the bandwidth needs of network interconnect dramatically scale year after year. Data centers need more interconnection between machines. Access speeds to memory are a limiting factor in von Neumann computing [1]. To this end, a variety of options including 3D integration [2,3] and 2.5D system-in-package integration [4] are proposed to bring the memory access much closer to the computation hardware. In this context, optical interconnects have emerged as a solution for both conventional data networks and intra/interchip data links [5,6]. For optical communications, light needs to be actively manipulated, ideally at intrachip scales to bring memory closer to its target. Due to its compatibility with ubiquitous CMOS electronics, silicon photonics has turned into one of the leading optical technologies [7–11]. In addition, silicon photonics benefits from low manufacturing cost and the performance enhancement resulting from electronic–photonic integration. Today silicon photonics is a multimillion dollar industry, which is expected to exponentially grow up to the billion dollar level in the next decade [12]. However, the optical response of silicon, particularly in the near-infrared (IR) 1550-nm telecommunications range is relatively weak. Because of this, current optoelectronic technologies rely on the integration of other complex materials (e.g., III–V materials [13–15]) to fulfill the functions that silicon fails to deliver. Graphene, an intrinsic 2D material, can be an excellent platform for realizing silicon-compatible optoelectronic devices [16–19]. Although many proposed applications of graphene rely on its remarkable electronic properties [20–23], including large electron mobility [24], its optical properties are equally noteworthy [25]. For instance, at visible and near-IR wavelengths, graphene exhibits a nearly wavelength-independent optical absorption of about 2.3%, which is related to the fine structure constant [26]. When considering that graphene is an atom-thick material, such absorption is quite remarkable. Additionally, as a result of its high carrier mobility and broadband optical absorption, graphene has promised to realize fast, broadband silicon-compatible optoelectronic devices.

Nanoelectronics. DOI: https://doi.org/10.1016/B978-0-12-813353-8.00007-5

Recently, several exciting demonstrations have been made, including photodetectors [27–34], electro-optic modulators [35–39], saturable absorbers [40–42], and phase shifters [43]. During the last few years, dramatic progress has been made on improving the performance of graphene electro-optic modulators. Theoretical limits in these devices on waveguide configurations predict bandwidths ~30–40 GHz and modulation of ~0.2 dB/μm [44]. Recent experimental reports have already approached these performance limits (e.g., [38,45]). Thus, new paradigms in graphene modulator design are required to continue improving performance. In this chapter, we will discuss key structural and optical properties of graphene and how these can enable the realization of waveguide-integrated optical devices, furthermore we will discuss the operation principle behind these devices as well as potential future approaches that are predicted to deliver superior performance.

8.2 Graphene: Structure and Electrical Properties

As a result of its ease of integration, low cost, and extraordinary electrical/optical properties, graphene has attracted attention for many mechanical, electrical, and optical applications. Graphene is an intrinsically 2D semiconductor comprising sp^2-bonded carbon atoms that are arranged in a hexagonal lattice. A sketch of graphene's lattice in real space is depicted in Fig. 8.1A, where a unit cell, primitive lattice vectors, and vectors between nearest neighbor atoms are highlighted. In the reciprocal space, graphene is also described by a hexagonal lattice, depicted in Fig. 8.1B as its band structure. Near the

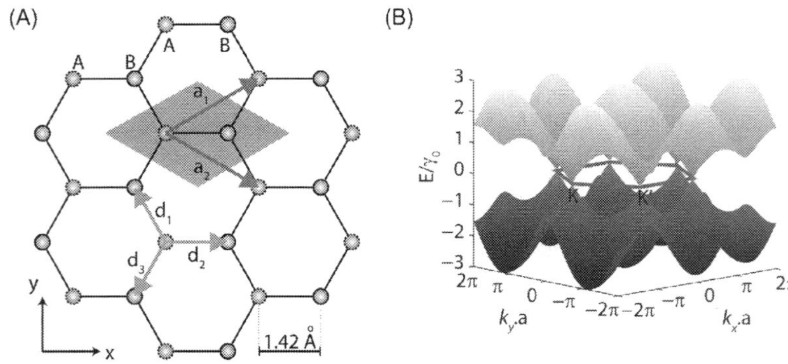

Figure 8.1

(A) Lattice structure of graphene in real space. A unit cell comprises two carbon atoms (A–B) is colored in gray. Primitive lattice vectors and vectors between near atoms are depicted in red and green, respectively. (B) Graphene band structure. Near the edges of the first Brillouin zone $K(K')$ point the E–k dispersion is linear. γ_0 is the hopping parameter between nearest neighbors, and its value is ~2.8 eV.

edges of the first Brillouin zone—$K(K')$ points—its tight binding Hamiltonian can be linearized as

$$H = \hbar v_F \begin{pmatrix} 0 & k_x + ik_y \\ k_x - ik_y & 0 \end{pmatrix}, \tag{8.1}$$

where $v_F \sim 10^6$ m/s is the Fermi velocity and $\hbar$ is the reduced Planck constant. The details of this derivation can be easily found in the literature e.g., Ref. [46]. From the linearized Hamiltonian, the following energy dispersion is obtained for charge carriers in graphene (i.e., electrons and holes):

$$E(k) = \hbar v_F |k|. \tag{8.2}$$

Because of the linear $E-k$ dispersion resulting in Eq. (8.2), which resembles that of massless particles such as photons, and since its Hamiltonian has the form of a "Dirac Hamiltonian" for "massless Dirac Fermions," charge carriers in graphene are usually called *massless Dirac Fermions*.

In terms of fabrication, the first production of a single-layer graphene was reported by mechanical exfoliation [21]. Lately, several chemical vapor deposition (CVD) methods to produce high-quality large-area graphene were proposed and demonstrated, including epitaxial growth on metallic substrates [47] and SiC [48]. However the material quality of CVD graphene, e.g., achievable mobility, is still much inferior to that of exfoliated graphene. Typical quality CVD graphene usually exhibits room temperature mobilities between 500 and 2000 cm^2/Vs over large areas ($\sim mm^2$), while the highest reported room temperature mobility in graphene exfoliated onto boron nitride substrates can approach 50,000–100,000 cm^2/Vs ($\sim \mu m^2$) [49].

8.3 Graphene: Optical Properties

Although so far, many applications of graphene, such as field effect transistors and transparent electrodes, relied on its electronic properties [50], its optical properties, which are inherited also from its singular conical band structure, are equally noteworthy [16,19]. A thorough description of its optical properties and their relation to its electrical properties is provided in this section.

Optical transitions in graphene include both intra and interband transitions, as depicted in Fig. 8.2A. Intraband transitions require change in momentum (k) thus phonon mediation. Based on the linearization of graphene's tight binding Hamiltonian near the edges of the Brillouin zone (8.1), its optical conductivity can be expressed as [25,51,52]

$$\sigma(\omega) = \sigma_{intra}(\omega) + \sigma_{inter}(\omega), \tag{8.3}$$

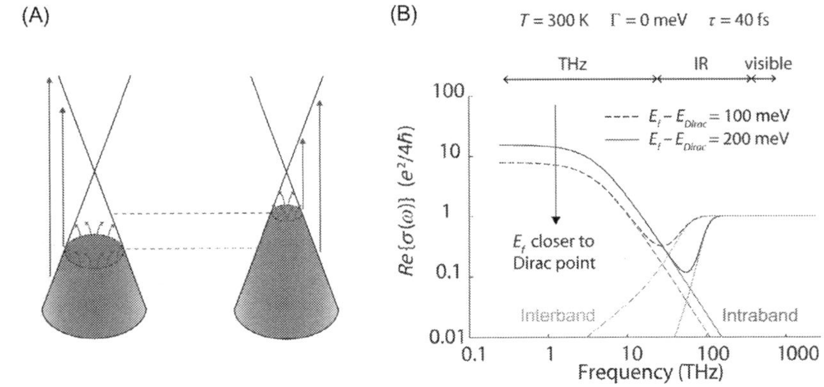

Figure 8.2
(A) Sketch of the possible optical transitions in graphene: interband (green -bright-) and intraband (red -dark-). (B) Real part of graphene's optical conductivity versus frequency. The contributions from intraband/interband transitions are plotted in red/green respectively.

where

$$\sigma_{intra}(\omega) = \frac{ie^2 E_f}{\pi h^2(\omega + 1/\tau)};\sigma_{inter}(\omega) = \frac{ie^2\omega}{\pi}\int_0^{\infty}\frac{f(\varepsilon - E_f) - f(-\varepsilon - E_f)}{(2\varepsilon)^2 - (h\omega + i\Gamma)^2}d\varepsilon. \tag{8.4}$$

Here, e is the electron charge, ω is the angular frequency, E_f is the Fermi level. f (.) is the Fermi distribution function, τ is the momentum relaxation time, and Γ is a parameter describing the broadening of interband transitions. In deriving Eq. (8.4), it has been assumed that E_f—$E_{Dirac} >> k_B T$, where k_B is the Boltzmann's constant and T represents temperature.

At low frequencies thus low photon energies (including terahertz wavelengths [53,54]), the optical conductivity of graphene is mainly determined by intraband transitions while at high frequencies the contribution of interband transitions becomes dominant as seen in Fig. 8.2B. By taking the limit when $\omega \rightarrow \infty$ in Eqs. (8.3) and (8.4), it follows that at very high frequencies, including the visible range, its optical conductivity approaches a constant universal value of $e^2/4\hbar$ thus $\sim$2.3% light absorption per graphene layer [26]. Based on this property, graphene has been proposed as a platform for transparent-flexible electrodes [55] and as UV-transparent electrode for internal photoemission measurements [56]. However, at higher wavelengths in the UV range, the band structure of graphene stops being linear thus its optical response becomes more complicated. Pi-plasmon resonances at wavelengths on the order of 4.7 eV can take place, which are evidenced by an increase in its UV optical conductivity [57,58]. Interaction between these pi-plasmons and surface plasmons in noble metals can provide for wavelength tunability in graphene-based UV plasmonic structures [59].

The dependence of the optical conductivity of graphene with its Fermi level in Eqs. (8.3) and (8.4) indicates that in the IR range, its optical conductivity can be modified by controlling the Fermi level but only weakly [60,61]. Nevertheless, this weak dependence is sufficient (in fact superior to that what is typically available in conventional semiconductors [62]) for realizing tunable devices such as ultrafast electro-absorption IR modulators [35,36]. These properties will be discussed in later sections in the context of waveguide-integrated graphene-based devices for optical interconnects. On the other hand, at long wavelengths, such as in the terahertz range [53,54,63], the contribution from interband transitions becomes negligible, and therefore the real part of its optical conductivity (8.3,8.4) can be rewritten employing a Drude-like dispersion:

$$Re\{\sigma(\omega)\} = \frac{\sigma_{DC}}{\left(1 + (\omega\tau)^2\right)}, \tag{8.5}$$

where, unlike in classical semiconductors, σ_{DC} the DC electrical conductivity of graphene (i.e., when $\omega \rightarrow 0$) is a nonlinear function of the charge density (n) since $E_f \propto \sqrt{n}$ [64]. The optical conductivity of graphene in the terahertz range, at frequencies such that $f < 1/2\pi\tau$ (where τ depends on the graphene sample structural quality), closely follows its DC electrical conductivity, and thus can be effectively electrostatically tuned [65–68] as depicted in Fig. 8.2B. As the Fermi level in graphene becomes closer to the Dirac point, the density of states available for intraband transitions decreases. As a result, its optical conductivity also decreases. As the frequency is increased, although the optical conductivity decreases, the terahertz absorption still can be effectively manipulated by employing structures allowing plasmonic resonances or by means of field enhancement [68–78].

8.4 Waveguide-Integrated Graphene Devices: Fundamental Operation Principles

Optical modulation in near-IR telecommunications band has relied on the integration of other materials into silicon to achieve better performance than silicon-based modulators. The primary electric field effects that are traditionally useful in semiconductor materials for causing either electro-absorption or electro-refraction are the Pockels effect, the Kerr effect, and the Franz–Keldysh effect. However, it has been shown that these are weak in pure silicon at the telecommunications wavelength of 1.55 μm [79,80]. Alternative methods are required to achieve effective optical modulation. One option is thermal modulation in silicon, owing to the large thermo-optic coefficient of silicon, but this is too slow for the high speed required by modern telecommunications systems [81]. The most common method of achieving modulation in silicon devices has been to exploit the plasma dispersion effect, in which the concentration of free charge in silicon changes the real and imaginary parts of the refractive index [80]. However, modulation performance is still poor in relation to what is possible in other materials (e.g., InP) [13–15]. Since alternate materials cannot be easily integrated using CMOS compatible fabrication processes, complex integration steps are required. In this context, resulting from its ease of integration with silicon in CMOS foundry compatible processes and its strong optical response, graphene has emerged as a promising material for modulation applications.

Resulting from the Fermi level dependence of its optical absorption at these IR frequencies, as discussed in Section 8.3, broadband optical modulators have been proposed in graphene. One of the earlier breakthroughs in the field was the work by Liu et al. [35], who reported the first experimental demonstration of graphene-based waveguide-integrated optical modulators. These devices consist of graphene on top of a Si/SiO_2 ridge waveguide, as depicted in Fig. 8.3A; here a single-layer graphene sheet, which is gated by the Si-substrate itself in a field-effect-transistor-like configuration, is the active element of the device. The waveguide was designed for operation at 1.53 μm. At the operation wavelength, a substantial part of the optical mode intensity lies in the region where the graphene layer is located as illustrated in Fig. 8.3B. The operation principle of this device is shown in Fig. 8.3C. In general, for a photon (energy $h\nu$) to be absorbed, momentum needs to be conserved during the optical transition. In the case of graphene, as a result of its linear dispersion, only electrons occupying the energy level

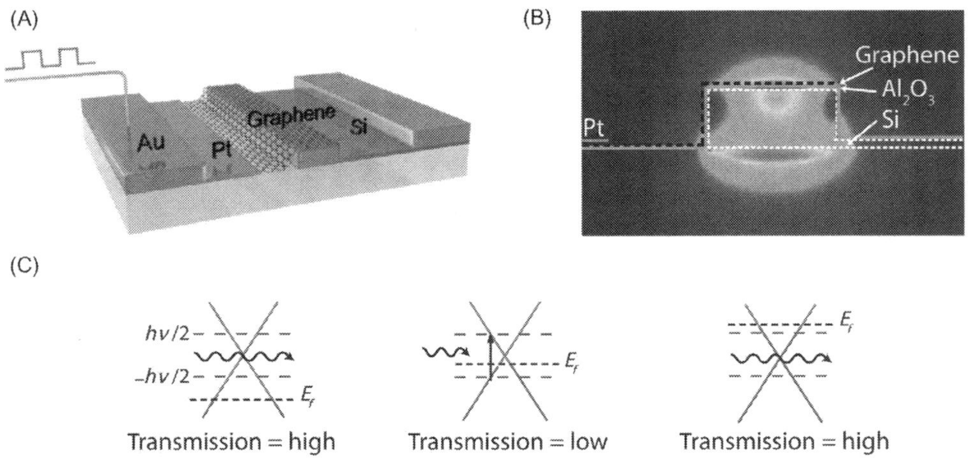

Figure 8.3

Early works on graphene-based electro-optic modulators reflecting the operation of this technology. (A) Sketch of a graphene-based waveguide-integrated broadband optical modulator. (B) Cross section of the device with an optical mode overlay. (C) Mechanism explaining the operation of the device. *Source: (A) and (B) extracted from M. Liu, et al., A graphene-based broadband optical modulator, Nature 474 (7349), 2011, 64–67.*

corresponding to $E - E_{DIRAC} = -h\nu/2$ can suffer such an optical transition guaranteeing momentum conservation. As a result, since optical transitions require: (1) the lower energy level to be filled with electrons, and (2) the higher energy level, which in this case corresponds to $E - E_{DIRAC} = +h\nu/2$, to be empty, the Fermi level in graphene has to lie in-between these two energy levels so that light can be absorbed (i.e., the following relation needs to be satisfied so that optical absorption is possible: $-h\nu/2 < E_F - E_{DIRAC} < +h\nu/2$). If the Fermi level lies below the $E - E_{DIRAC} = -h\nu/2$ level, then there are no electrons available to absorb the photon thus suffer an optical transition. Alternatively, if the Fermi level is above the $E - E_{DIRAC} = +h\nu/2$ level, then optical transitions cannot take place since the destination energy level is already completely filled by electrons. In these modulators, the switching speed is limited by an *RC* time constant. Experimentally, in the work by Liu et al. this was determined to be in the order of ~1 GHz. The modulation depth was found to be ~0.09 dB/μm. Furthermore, the intrinsic insertion loss in these modulators is negligible since the intraband absorption in graphene is extremely low at near-IR wavelengths. Loss in these devices is attributed to coupling loss of light into the waveguide.

Several other waveguide-integrated structures, including devices where the reconfigurable region is graphene–insulator–graphene, have been later proposed for optical modulation (see e.g., Ref. [36,37]). In these devices, the modulation depth can be increased due to a larger number of active graphene layers. Experimentally, the modulation reported in double-layer graphene-based waveguide optical modulators is ~0.16 dB/μm [36] with a reported 3-dB limited bandwidth of ~1 GHz at 1.53 μm. Later work by Youngblod et al. also in terms of double-layer structures showed modulation depth ~0.05 dB/μm with a larger bandwidth of ~2.5 GHz [82]. However, these devices exhibited a large insertion loss, in the order of 0.15 dB/μm, which might be owed to a less favorable Fermi level alignment in the stacked graphene layers (i.e., if the graphene layers are initially heavily doped with charges of same polarity, then there will be always a finite absorption by one of the layers). However, theoretical limits in graphene-based modulators predict bandwidths >30 GHz and modulations exceeding 0.2 dB/μm [44]. From this perspective, it is observed that although the modulation depth in these early experiments has already approached these limits, parasitics, doping control issues, etc., have heavily limited their experimental operational. This results from a large *RC* time constant in these devices. Two approaches need to be taken to improve this speed: (1) reduce the access and contact resistances in these devices, by a better control over the access region doping as well as better contacts to graphene, and (2) reduce the capacitance in these devices, by the use of low-*k* dielectrics, such as boron nitride, rather than Al_2O_3 and use of thicker gate oxide layers.

8.5 Waveguide-Integrated Graphene Devices: Recent Experimental Developments

Following some of the approaches discussed in Section 8.4, recent work by Phare et al. [38] reported graphene electro-optic modulator based on resonator loss modulation at critical coupling, showing a drastically increased speed with respect to the prior art (see Fig. 8.4). These devices operate with 30-GHz bandwidth and can attain ~20-dB modulation with a 50-V voltage swing. The active region of the device (a self-gated graphene pair on top of a buried SiN waveguide) has 30-μm length, thus one could define a remarkable ratio between modulation and active region length of 20 dB/30 μm = 0.67 dB/μm. However, the total footprint of the device is relatively large due to the large diameter (80 μm) of the ring resonator, which leads to a footprint of ~5000 μm^2 (excluding contacts). For reference, considering a ~0.09-dB/μm modulation depth (as reported in Ref. [35]), a length of 220 μm will be required so to obtain 20-dB modulation; however, the total footprint of such a device will be on the order of <5 μm × 220 μm = 1100 μm^2 (excluding contacts). From a footprint perspective, the advantages of ring-resonator-based modulators with respect to waveguide-only integrated modulators are not so obvious. Furthermore, although remarkable, the 30-GHz speed observed in these ring resonator structures results mainly from the decreased capacitance in the device due to a thick gate oxide (65 nm) in comparison to what has been employed in previous works (<10 nm) [35,36]. By applying similar oxide scaling approaches, as discussed in Section 8.4, it is fundamentally possible to attain similar speeds without a need of resorting to topologies relying on ring resonators, which compromises optical bandwidth. Furthermore, as discussed by Kim et al. [83], the optical absorption by graphene, thus modulation depth, could be further improved by means

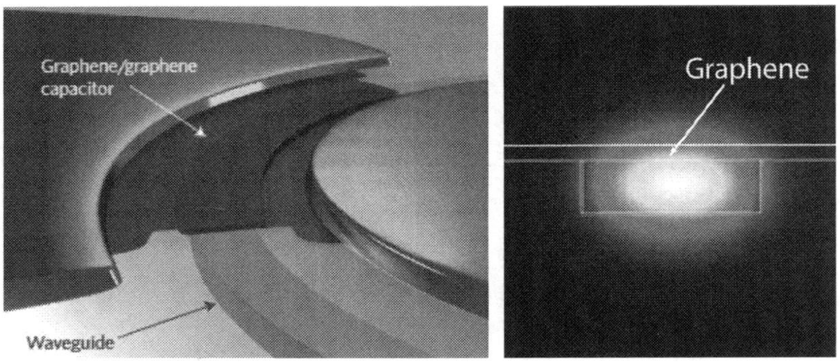

Figure 8.4
Sketch of a ring-resonator-based modulator. A SiN buried waveguide was employed and a graphene/graphene capacitor transferred on top. Source: *Extracted from C.T. Phare, Y.-H. Daniel Lee, J. Cardenas, and M. Lipson, Graphene electro-optic modulator with 30 GHz bandwidth, Nat. Photonics, 9 (8) (2015) 511–514.*

of placing graphene inside the buried waveguide, i.e., placing graphene in the region where the intensity of the optical mode is maximum (as in [35]). It is worth mentioning that all the device approaches described in Section 8.4 and this section could also be applied to other devices, such as graphene-based photodetectors. However, new paradigms leading to dramatically smaller footprints are required so to truly advance the art in terms of graphene optoelectronics. In the next section, we will describe one of such design approaches.

8.6 Emerging Research Trends in Graphene-Based Optical Devices

Free-form metamaterials offer a new degree of design freedom that can enable unique and efficient integrated photonic applications in very compact areas [84]. The premise of this approach is that via nanofabrication, one can control the local refractive index of the device. By spatial engineering of the refractive index, it is possible to design devices with much higher performance than what is otherwise achievable. Based on this design approach, several photonic structures such as beam splitters, waveguide bends, optical diodes, and polarizers exhibiting high performance (e.g., high efficiency) and the smallest reported footprints have been reported [85,86]. By employing this recently proposed free-form metamaterials concept, it is possible to overcome the fundamental limits of present, graphene-based, waveguide electro-optic modulators, and to develop highly efficient more complex devices such as phase shifters, active couplers, polarizers, etc. In this section, we discuss on the design principle of free-form metamaterial, modulators with graphene as reconfigurable element. These devices are demonstrated as being capable of achieving a superior modulation depth than what is otherwise possible with waveguide-integrated approaches. Overall, this idea extends further and aims to explore the feasibility of employing the recently proposed "designer metamaterial" concept as a design principle so to reduce the footprint of reconfigurable devices. In order to show the applicability of this design approach, we applied nonlinear optimization in order to design an optical modulator with very small footprint. A silicon ridge waveguide couples light in and out of the device, which consists of a silicon nanopattern with graphene on top. The nanopattern contains multiple "pixels," i.e., regions which may be comprised entirely of silicon or air. For a given nanopattern dimension, "seed" geometries were selected from a pool of randomly generated patterns. After that, a search algorithm was executed. Because of the enhanced local fields in the optimized nanopattern, the light–matter interaction in graphene is enhanced thus leading to a superior modulation performance and a reduced footprint. Simulation results show that modulation depth in excess of 0.3 dB/μm, that is $3\times$ larger than what is possible otherwise in ridge waveguide alone approaches. Furthermore, this modulation occurs in a broad wavelength range, i.e., from 1.55 to 1.6 μm, in contrast to what is possible in narrowband approaches such as those employing ring resonators.

The structure of the modulator is illustrated in Fig. 8.5A. A nanopattern with multiple pixels in a 3 × 3 μm^2 area connects two ridge silicon waveguides on the left and right sides. For the Si waveguide, the height is 250 nm and the width is 600 nm. The middle section of the device is composed of a nanopattern region consisting of 30 × 30 pixels, which is in the shape of a square. Every pixel has a side of 100 nm, generating a total footprint of 3 × 3 μm^2. A transverse magnetic (TM) excitation is set at the (left) input waveguide; the transmitted electric field is then found at the (right) output waveguide. For illustrative purposes, the whole nanopattern region is covered by a graphene layer, whose optical conductivity is properly varied. The optical conductivity of graphene can be effectively modified by electro-static control of its Fermi level. For this purpose, the Fermi level energy was varied from 0 to 1 eV, with 0 eV corresponding to the maximum absorption state (MAX state) and 1 eV corresponding to the minimum absorption state (MIN state). Furthermore, the surface-normal component of permittivity for graphene was set to be a constant of value equal to 2.5.

According to the change in Fermi level of graphene, the graphene-based optical modulator can attain a modulation depth given by the following expression:

$$MD = \frac{T(MAXstate) - T(MINstate)}{T(MIN\ state)}\%, \tag{8.6}$$

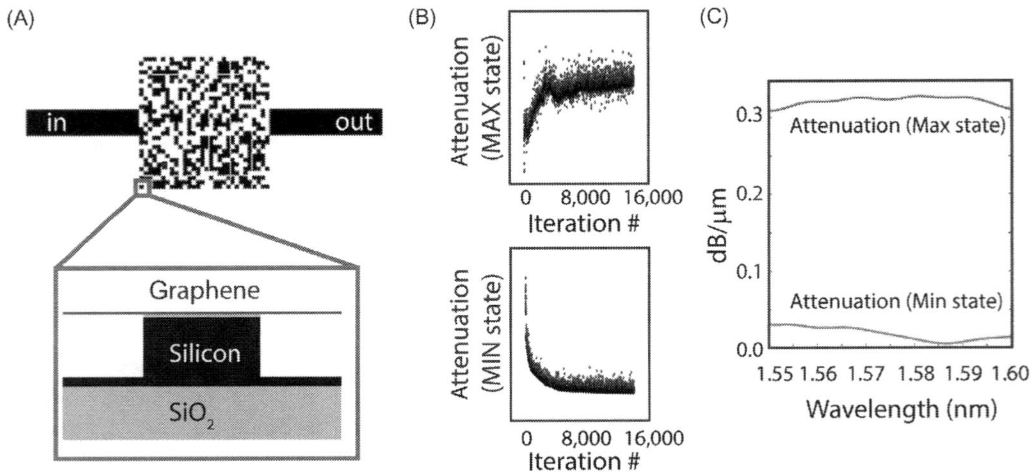

Figure 8.5

(A) Geometry of the analyzed modulator structure; graphene is placed on top of the nanopattern region, which is to be optimized. The nanopattern is composed of a 30 × 30 pixels, where each pixel can be either Si or etched-Si (air). (B) Evolution of loss and modulation depth with number of iterations. (C) Lumerical simulation results (attenuation dB/μm) for the optimized 30 × 30 pattern.

where T represents optical transmission. By utilizing Eq. (8.6), one can compare the performance of different designs. From this point of view, Eq. (8.6) will be employed as a target function to be maximized during the search algorithm. The nanopattern geometries were simulated using MEEP, which is an open-source software based on the finite-difference time-domain (FDTD) method. In order to speed the search algorithm to perform simulations and evaluate the modulation depth over thousands of different designs during the optimization loop, for the sake of computational cost, the graphene layer was defined as 20-nm thick during the optimization loop simulations. Moreover, 20 nm was also employed as unit cell in the mesh. After finishing the optimization loop, the designs providing best performances were simulated using Lumerical FDTD solutions; the thickness of graphene was assumed to be 1 nm in these simulations so to guarantee the accuracy of the simulations.

It is worth mentioning that the proposed device structure is a simplified geometry utilized just for illustrative purposes in order to show the feasibility of the free-form metamaterial design approach. In this regard, the following two assumptions were made: (1) graphene being suspended on the "pixels" where silicon is etched away, and (2) the Fermi level in graphene being uniform across the whole nanopattern. In order to verify the accuracy of our graphene model, first, the optical transmission for a ridge-waveguide structure with graphene on top was simulated with varying Fermi level. Such structure directly corresponds to the structure as analyzed in [35]; simulation results show transmission (per unit length) of 0.09 dB/μm for $E_F = 0$ eV and 0.01 dB/μm for $E_F = 1.0$ eV, which agrees well with the experimental results reported in [35] (~0.09-dB/μm modulation depth). For the 30×30 pixel nanopattern, a "seed" geometry was selected from a pool of randomly generated patterns. Since each pixel may be comprised entirely of silicon or of air (when the silicon is completely etched away), random patterns were chosen while varying the fraction of air-filled to silicon-filled pixels (f). For each value of f, with f ranging between 10% and 90%, 90 random patterns were generated, which were simulated in both graphene states, i.e., $E_F = 0$ eV and $E_F = 1$ eV, from here the pattern with the best figure of merit was selected as "seed" pattern. Then, the search algorithm (in this case simulated annealing) was executed so to maximize the figure of merit starting from the chosen "seed" pattern. Fig. 8.5B shows the evolution of modulation depth and insertion loss with number of iterations showing convergence into an optimal design. Finally, upon reaching convergence, in this case after ~14,000 iterations, the algorithm is stopped and the final pattern is chosen.

Upon convergence, simulations (Lumerical, modeling the graphene layer with appropriate thickness) show transmission (per unit length) of 0.32 dB/μm for $E_F = 0$ eV and 0.02 dB/μm for $E_F = 1.0$ eV. Such modulation depth is substantially larger than what was obtained for the ridge-waveguide structure with graphene on top, and is well above the theoretical limits for modulation discussed in Ref. [44]. Furthermore, as depicted in Fig. 8.5C, the optimal design provides for such a large modulation over a broad optical bandwidth (1.55–1.6 μm). Finally, in order to show that the design principle is scalable and it is possible to obtain high performance on free-form metamaterials regardless of the a priori chosen nanopattern extension, the algorithm was executed for a 60 × 30 pixel nanopattern structure. A pattern comprises two consecutive times the 30 × 30 pixel pattern obtained before as seed pattern; after running the search algorithm, the optimal design exhibited transmission (per unit length) of 0.36 dB/μm for $E_F = 0$ eV and 0.04 dB/μm for $E_F = 1.0$ eV. From this perspective, both, the modulation depth and the insertion loss levels seem to scale properly with respect to those obtained in the previous design. Moreover, broad bandwidth operation was also observed. Results show that the modulation depth of the designed metamaterial optical modulator is substantially improved with respect to that in graphene-on-waveguide structures. Overall, simulation results indicate the potential of the free-form metamaterial design approach to construct high-performance optoelectronic devices based on graphene with reduced footprint. However, it is worth noticing that employing the free-form design approach is expected to greatly reduce the footprint of graphene-based optoelectronic devices without having a significant impact on their speed. This is a result of the device length not really affecting its RC time constant since C scales proportional to L whereas R scales inversely proportional to L.

References

[1] M.D. Godfrey, D.F. Hendry, The computer as von Neumann planned it, IEEE Ann. Hist. Comput. 15 (1) (1993) 11–21.
[2] G.H. Loh, 3D-stacked memory architectures for multi-core processors, in Proceedings of the 35th Annual International Symposium on Computer Architecture, Washington, DC, USA, 2008, pp. 453–464.
[3] L. Li et al., Addressing bandwidth challenges in next generation high performance network systems with 3D IC integration, in 2012 IEEE 62nd Electronic Components and Technology Conference, 2012, pp. 1040–1046.
[4] R.V. Shenoy, K.-Y. Lai, E. Gusev, (Invited) 2.5D advanced system-in-package: processes, materials & integration aspects, ECS Trans. 61 (3) (2014) 183–190.
[5] D.A.B. Miller, Optical interconnects to silicon, IEEE J. Sel. Top. Quantum Electron. 6 (6) (2000) 1312–1317.
[6] N. Savage, Linking with light [high-speed optical interconnects], IEEE Spectr. 39 (8) (2002) 32–36.
[7] B. Jalali, S. Fathpour, Silicon photonics, J. Light. Technol. 24 (12) (2006) 4600–4615.
[8] M. Lipson, Guiding, modulating, and emitting light on silicon-challenges and opportunities, J. Light. Technol. 23 (12) (2005) 4222–4238.
[9] C. Gunn, CMOS photonics for high-speed interconnects, IEEE Micro 26 (2) (2006) 58–66.
[10] R. Soref, The past, present, and future of silicon photonics, IEEE J. Sel. Top. Quantum Electron 12 (6) (2006) 1678–1687.
[11] R. Soref, Mid-infrared photonics in silicon and germanium, Nat. Photonics 4 (8) (2010) 495–497.
[12] E. Mounier and C. Troadec, A view on the silicon photonics market. Yole Developpement Market Report, 2014.
[13] D. Liang, G. Roelkens, R. Baets, J.E. Bowers, Hybrid integrated platforms for silicon photonics, Materials 3 (3) (2010) 1782–1802.

[14] G. Roelkens, et al., III–V/silicon photonics for on-chip and intra-chip optical interconnects, Laser Photonics Rev 4 (6) (2010) 751–779.

[15] M.J.R. Heck, et al., Hybrid silicon photonics for optical interconnects, IEEE J. Sel. Top. Quantum Electron 17 (2) (2011) 333–346.

[16] F. Bonaccorso, Z. Sun, T. Hasan, A.C. Ferrari, Graphene photonics and optoelectronics, Nat. Photon. 4 (9) (2010) 611–622.

[17] P. Avouris, Graphene: electronic and photonic properties and devices, Nano Lett. 10 (11) (2010) 4285–4294.

[18] Q. Bao, K.P. Loh, Graphene photonics, plasmonics, and broadband optoelectronic devices, ACS Nano 6 (5) (2012) 3677–3694.

[19] B. Sensale-Rodriguez, Graphene-based optoelectronics, J. Light. Technol. 33 (5) (2015) 1100–1108.

[20] K.S. Novoselov, et al., Two-dimensional gas of massless Dirac fermions in graphene, Nature 438 (7065) (2005) 197–200.

[21] K.S. Novoselov, et al., Electric field effect in atomically thin carbon films, Science 306 (5696) (2004) 666–669.

[22] A.H. Castro Neto, F. Guinea, N.M.R. Peres, K.S. Novoselov, A.K. Geim, The electronic properties of graphene, Rev. Mod. Phys. 81 (1) (2009) 109–162.

[23] Y.-M. Lin, et al., 100-GHz transistors from wafer-scale epitaxial graphene, Science 327 (5966) (2010) 662. 662.

[24] K.I. Bolotin, et al., Ultrahigh electron mobility in suspended graphene, Solid State Commun. 146 (9–10) (2008) 351–355.

[25] L.A. Falkovsky, Optical properties of graphene, J. Phys. Conf. Ser. 129 (1) (2008) 12004.

[26] R.R. Nair, et al., Fine structure constant defines visual transparency of graphene, Science 320 (5881) (2008) 1308. 1308.

[27] F. Xia, T. Mueller, Y. Lin, A. Valdes-Garcia, P. Avouris, Ultrafast graphene photodetector, Nat. Nanotechnol. 4 (12) (2009) 839–843.

[28] T. Mueller, F. Xia, P. Avouris, Graphene photodetectors for high-speed optical communications, Nat. Photonics 4 (5) (2010) 297–301.

[29] B.Y. Zhang, et al., Broadband high photoresponse from pure monolayer graphene photodetector, Nat. Commun. 4 (2013) 1811.

[30] X. Wang, Z. Cheng, K. Xu, H.K. Tsang, J.-B. Xu, High-responsivity graphene/silicon-heterostructure waveguide photodetectors, Nat. Photonics 7 (11) (2013) 888–891.

[31] X. Gan, et al., Chip-integrated ultrafast graphene photodetector with high responsivity, Nat. Photonics 7 (11) (2013) 883–887.

[32] A. Pospischil, et al., CMOS-compatible graphene photodetector covering all optical communication bands, Nat. Photonics 7 (11) (2013) 892–896.

[33] C.-H. Liu, Y.-C. Chang, T.B. Norris, Z. Zhong, Graphene photodetectors with ultra-broadband and high responsivity at room temperature, Nat. Nanotechnol. 9 (4) (2014) 273–278.

[34] D. Schall, et al., 50 GBit/s photodetectors based on wafer-scale graphene for integrated silicon photonic communication systems, ACS Photonics 1 (9) (2014) 781–784.

[35] M. Liu, et al., A graphene-based broadband optical modulator, Nature 474 (7349) (2011) 64–67.

[36] M. Liu, X. Yin, X. Zhang, Double-layer graphene optical modulator, Nano Lett. 12 (3) (2012) 1482–1485.

[37] S.J. Koester, M. Li, High-speed waveguide-coupled graphene-on-graphene optical modulators, Appl. Phys. Lett. 100 (17) (2012) 171107.

[38] C.T. Phare, Y.-H. Daniel Lee, J. Cardenas, M. Lipson, Graphene electro-optic modulator with 30 GHz bandwidth, Nat. Photonics 9 (8) (2015) 511–514.

[39] Z. Sun, A. Martinez, F. Wang, Optical modulators with 2D layered materials, Nat. Photonics 10 (4) (2016) 227–238.

[40] Q. Bao, et al., Atomic-layer graphene as a saturable absorber for ultrafast pulsed lasers, Adv. Funct. Mater. 19 (19) (2009) 3077–3083.

[41] W.D. Tan, C.Y. Su, R.J. Knize, G.Q. Xie, L.J. Li, D.Y. Tang, Mode locking of ceramic Nd:yttrium aluminum garnet with graphene as a saturable absorber, Appl. Phys. Lett. 96 (3) (2010) 31106.
[42] I.H. Baek, et al., Efficient mode-locking of sub-70-fs Ti:Sapphire laser by graphene saturable absorber, Appl. Phys. Express 5 (3) (2012) 32701.
[43] J. Capmany, D. Domenech, P. Muñoz, Silicon graphene waveguide tunable broadband microwave photonics phase shifter, Opt. Express 22 (7) (2014) 8094.
[44] S.J. Koester, M. Li, Waveguide-coupled graphene optoelectronics, IEEE J. Sel. Top. Quantum Electron. 20 (1) (2014) 84–94.
[45] W. Li, et al., Ultrafast all-optical graphene modulator, Nano Lett. 14 (2) (2014) 955–959.
[46] R. Saito, G. Dresselhaus, M.S. Dresselhaus, Physical Properties of London, Imperial College, UK, 1999.
[47] X. Li, et al., Large-area synthesis of high-quality and uniform graphene films on copper foils, Science 324 (2009) 1312–1314.
[48] Luxmi, et al., Temperature dependence of epitaxial graphene formation on SiC(0001), Journal of Electronic Materials 38 (6) (2008) 718–724.
[49] C.R. Dean, et al., Boron nitride substrates for high-quality graphene electronics, Nature Nanotechnology 5 (2010) 722–726.
[50] Y.M. Lin, et al., Wafer-scale graphene integrated circuit, Science 332 (2011) 1294–1297.
[51] J. Dawlaty, et al., Measurement of the optical absorption spectra of epitaxial graphene from terahertz to visible, App. Phys. Lett. 93 (2008) 131905.
[52] T. Ando, Y. Zheng, H. Suzuura, Dynamical conductivity and zero-mode anomaly in honeycomb lattices, J. Phys. Soc. Jpn. 71 (5) (2002) 1318–1324.
[53] B. Sensale-Rodriguez, et al., Graphene for reconfigurable terahertz optoelectronics, Proceedings of the IEEE 101 (7) (2013) 1705–1716.
[54] K. Yang, et al., The linear and nonlinear THz properties of graphene. Terahertz, Sci. Technol. 6 (4) (2013) 223–233.
[55] K.S. Kim, et al., Large-scale pattern growth of graphene films for stretchable transparent electrodes, Nature 457 (2009) 706–710.
[56] R. Yan, et al., Determination of graphene work function and graphene-insulator-semiconductor band alignment by internal photoemission spectroscopy, Appl. Phys. Lett. 101 (2) (2012) 022105.
[57] K.F. Mak, H. Felipe, K. He, J. Deslippe, N. Petrone, J. Hone, et al., Tuning many-body interactions in graphene: the effects of doping on excitons and carrier lifetimes, PRL 112 (20) (2014) 207401.
[58] V.G. Kravets, A.N. Grigorenko, R.R. Nair, P. Blake, S. Anissimova, K.S. Novoselov, et al., Spectroscopic ellipsometry of graphene and an exciton-shifted van Hove peak in absorption, Physical Review B 81 (15) (2010) 155413.
[59] Y. Wang, et al., Modification of UV surface plasmon resonances in aluminum hole-arrays with graphene, In CLEO: Applications and Technology (2017). JTh2A-49.
[60] Z.Q. Li, et al., Dirac charge dynamics in graphene by infrared spectroscopy, Nat. Phys. 4 (2008) 532–535.
[61] F. Wang, et al., Gate-variable optical transitions in graphene, Science 320 (2008) 5876.
[62] G.T. Reed, G. Mashanovich, F.Y. Gardes, D.J. Thomson, Silicon optical modulators, Nature Photonics 4 (2010) 518–526.
[63] M. Hasan, et al., Graphene terahertz devices for communications applications, Nano Communication Networks 10 (2016) 68–78.
[64] L.A. Falkovsky, S.S. Pershoguba, Optical far-infrared properties of a graphene monolayer and multilayer, Phys. Rev. B 76 (2007) 153410.
[65] B. Sensale-Rodriguez, et al., Unique prospects for graphene-based terahertz Modulators, Appl. Phys. Lett. 99 (2011) 113104.
[66] B. Sensale-Rodriguez, et al., Broadband graphene terahertz modulators enabled by intraband transitions, Nat. Commun. 3 (2012) 780.
[67] J. Horng, et al., Drude conductivity of Dirac fermions in graphene, Phys. Rev. B 83 (2011) 165113.

[68] I. Maeng, et al., Gate-controlled nonlinear conductivity of Dirac fermion in graphene field-effect transistors measured by terahertz time-domain spectroscopy, Nano Lett. 12 (2012) 551–555.
[69] L. Ju, et al., Graphene plasmonics for tunable terahertz metamaterials, Nature Nanotech. 6 (2011) 630–634.
[70] R. Yan, et al., A new class of electrically tunable metamaterial terahertz modulators, Optics express 20 (27) (2012) 28664–28671.
[71] B. Sensale-Rodriguez, et al., Efficient terahertz electro-absorption modulation employing graphene plasmonic structures, Applied Physics Letters 101 (26) (2012) 261115.
[72] H. Yan, et al., Tunable infrared plasmonic devices using graphene/insulator stacks, Nature Nanotech. 7 (2012) 330–334.
[73] B. Sensale-Rodriguez, et al., Extraordinary control of terahertz beam reflectance in graphene electro-absorption modulators, Nano letters 12 (9) (2012) 4518–4522.
[74] K. Yang, et al., Graphene-based tunable metamaterial terahertz filters, Applied Physics Letters 105 (9) (2014) 093105.
[75] S. Arezoomandan, et al., Graphene-based electrically reconfigurable deep-subwavelength metamaterials for active control of THz light propagation, Applied Physics A 117 (2) (2014) 423–426.
[76] S. Arezoomandan, Berardi Sensale-Rodriguez, Geometrical tradeoffs in graphene-based deeply-scaled electrically reconfigurable metasurfaces, Scientific reports 5 (2015) 8834.
[77] R. Yan, et al., Exceptional terahertz wave modulation in graphene enhanced by frequency selective surfaces, ACS Photonics 3 (3) (2016) 315–323.
[78] S. Arezoomandan, et al., Graphene-based reconfigurable terahertz plasmonics and metamaterials, Carbon 112 (2017) 177–184.
[79] R. Soref, B. Bennett, Electrooptical effects in silicon, IEEE J. Quantum Electron. 23 (1) (1987) 123–129.
[80] G.T. Reed, A.P. Knights, Silicon Photonics: An Introduction, John Wiley & Sons, Ltd, Chichester, UK, 2004.
[81] G. Cocorullo, I. Rendina, Thermo-optical modulation at 1.5 mu m in silicon etalon, Electron. Lett 28 (1) (1992) 83–85.
[82] N. Youngblood, Y. Anugrah, R. Ma, S.J. Koester, M. Li, Multifunctional graphene optical modulator and photodetector integrated on silicon waveguides, Nano Lett. 14 (5) (2014) 2741–2746.
[83] K. Kim, et al., A role for graphene in silicon-based semiconductor devices, Nature, vo. 479 (2011) 338.
[84] J. Lu, J. Vučković, Nanophotonic computational design, Opt. Express 21 (11) (2013) 13351.
[85] B. Shen, P. Wang, R. Polson, R. Menon, Integrated metamaterials for efficient and compact free-space-to-waveguide coupling, Opt. Express 22 (22) (2014) 27175.
[86] B. Shen, P. Wang, R. Polson, R. Menon, Ultra-high-efficiency metamaterial polarizer, Optica 1 (5) (2014) 356.

PART IV

Plasmonics

CHAPTER 9

Hot Carrier Generation in Plasmonic Nanostructures: Physics and Device Applications

Ravi S. Hegde[1] and Saumyakanti Khatua[2]

[1]*Department of Electrical Engineering, Indian Institute of Technology, Gandhinagar, Gujarat, India*
[2]*Department of Chemistry, Indian Institute of Technology, Gandhinagar, Gujarat, India*

9.1 Introduction

Surface plasmon resonances [1] are collective oscillations of free electrons that are coupled to surface electromagnetic modes (modes confined to the conductor-dielectric interfaces). Due to the excitation of these resonances, the nanoparticles scatter and absorb light intensely (near the resonance wavelengths, the scattering cross sections in many cases exceeds the physical cross sections [2,3]). In addition to the efficient coupling of free space light waves to nanostructures, the excitation of localized surface plasmon resonances (LSPR) also leads to substantially enhanced and highly localized electromagnetic fields [4]. Plasmonic nanostructures are interesting in a variety of applications like sensing, imaging, and energy harvesting. Nanophotonic structures employing plasmonic resonators are particularly important for chemical and biological sensing [3] and various forms of surface-enhanced spectroscopies such as surface-enhanced Raman spectroscopy (SERS) [5,6], surface-enhanced infrared absorption spectroscopy (SEIRA) [7], and surface-enhanced nonlinear harmonic generation [8]. They have also been investigated in connection with ultra-high resolution color image sensors [9–11], multispectral sensors [12–14], low-power, low-cost, wide area, and high-performance displays [15–17], colorimetric chemical and biochemical sensing [18,19], building integrated photovoltaics [16,20], optical information security [21,22] devices, and high-density data storage [22,23].

Surface plasmon resonances in plasmonic metals are not long lived. Once excited, they are subject to decay and ultimate conversion to heat. Since the modes are spatially confined, the heating effect is also local in nature. Although this localized heating can be utilized for some applications [24,25], it is mostly a nuisance for most applications [26]. The decay of surface plasmons proceeds via the so-called Landau damping [27], whereby the energy of the

Nanoelectronics. DOI: https://doi.org/10.1016/B978-0-12-813353-8.00003-8

collective oscillation is transferred to individual charge carriers. These individual carriers then have higher energies in comparison with their peers, earning themselves the moniker of "hot" carriers ("hot" electrons or "hot" holes). These "hot" carriers are also themselves prone to give back their excess energy to the surrounding; but, if properly utilized in a timely fashion, it can enable processes not possible with thermalized carriers [24,25,28].

Plasmon-induced hot carrier effect thus provides a way to manage the loss-related problems in devices and that are currently of interest [29]. Utilizing plasmon-derived hot carriers enables us to combine efficient light capture and confinement with energy conversion and catalysis at the nanoscale [28,30–32]. The hot carriers have also been demonstrated to directly drive chemical reactions at the metal surface [33]. They could be transferred to a semiconductor for use in photovoltaics and photoelectrochemical systems [34]. Plasmon-enhanced energy conversion devices such as for water splitting have been demonstrated using composite metal-semiconductor (Ms) photocatalysts as well [35,36].

This chapter provides a tutorial review of the physics of hot carrier generation, transport, and extraction processes in plasmonic nanostructures. It then reviews the progress in two specific applications: hot carrier-driven surface chemistry and internal photoemission (IPE) photodetectors. The objective is to enable the reader to design applications suitably by considering the physical limitations.

9.2 The Physics of Hot Carrier Generation, Scattering, and Transport Processes

Fig. 9.1 provides a broad overview of plasmon decay-related hot carrier generation, scattering, transport, and collection. Four distinct steps can be considered: (i) the excitation of localized plasmon resonance in the nanoparticle; (ii) damping of the plasmon resonance and birth of hot carriers; (iii) various elastic and nonelastic scattering processes of the generated hot carriers; and finally (iv) the extraction of hot carriers through the surface of the nanoparticle. Step (i) typically can be worked out using classical electrodynamics with knowledge of the experimental dielectric permittivity tables. For qualitative and quantitative analysis of step (ii) and step (iii), a detailed understanding of the allowed electronic and phononic states, electron-electron and electron-phonon scattering, and various multiparticle states is needed. Step (iv) may involve either Ms, metal-molecule, or a metal-electrolyte interface. These four steps are described in detail in this section.

9.2.1 The Optical Properties of Plasmonic Nanoresonators

The first step involves light shining on a plasmonic nanoparticle and excitation of a localized plasmon resonance. This process is quite well understood; classical electrodynamics can explain the optical properties of nanoparticles even down to the size

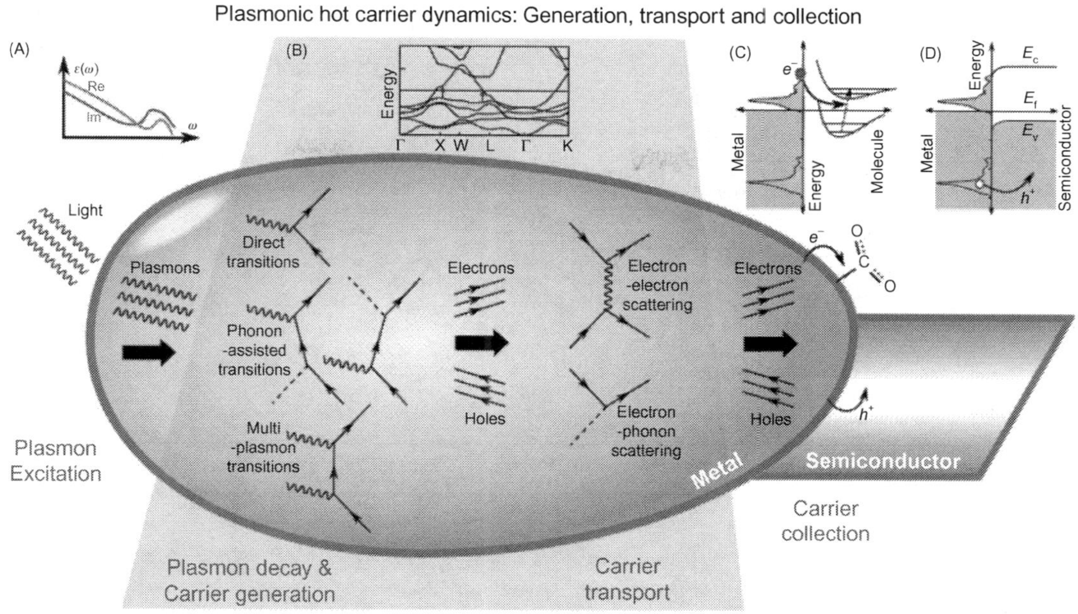

Figure 9.1

Overview of the light-driven plasmon decay-based hot carrier generation and extraction process. (A) For the light-driven plasmon excitation process, classical electrodynamics simulations utilizing the frequency dependent complex dielectric permittivity and the geometry is sufficient. (B) For the plasmon dephasing and carrier excitation as well as carrier scattering processes, details of the electronic and phononic states in the bulk and modification of these states due to surfaces, defects, and due to deviations from the bulk in very small nanoparticles are needed. (C) and (D) show the transport of excited carriers across metal-semiconductor and metal-molecule interfaces. To quantitatively understand the transport, detailed calculations concerning band-/energy-level alignment and consideration of traps and other nonidealities need to be considered. *Reprinted under the Creative Commons International License from Reference P. Narang, R. Sundararaman, H.A. Atwater, Plasmonic hot carrier dynamics in solid-state and chemical systems for energy conversion, Nanophotonics 5 (1) (2016) 96–111 [28], © 2016 DeGruyter.*

of a few nanometers [37]. Gustav Mie provided the first mathematical description of the spectral dependence of the scattering by a spherical nanoparticle [38] in 1908. With the advent of colloidal chemistry and advanced nanofabrication techniques, it is now possible to create a variety of nanostructures. However, only the simplest shapes remain analytically tractable; numerical tools [39] are essential to analyze and design compound and complex nanostructures. In designing a suitable plasmonic nanostructure for a particular application, the geometry and the dielectric environment are chosen to obtain the desired optical response in the form of precisely tailored line shapes [40–43] and controlled confinement of electromagnetic energy ("hot spots") by using numerical tools.

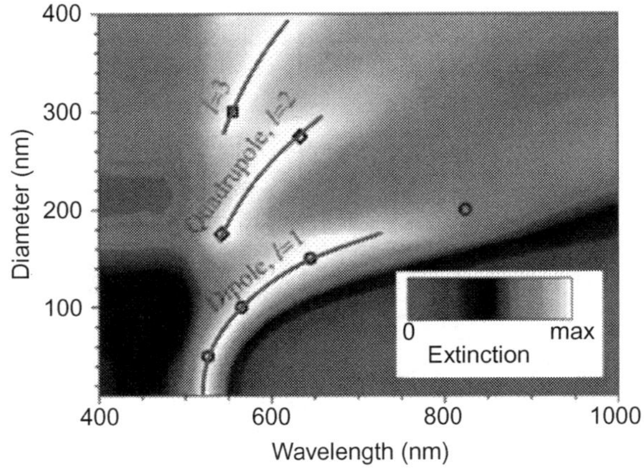

Figure 9.2

The extinction spectra of gold nanospheres in water for a wide range of particle sizes. The extinction values have been normalized so as to track the peak locations. *Reprinted with permission from Reference V. Myroshnychenko, J. Rodríguez-Fernández, I. Pastoriza-Santos, A.M. Funston, C. Novo, P. Mulvaney, et al., Modelling the optical response of gold nanoparticles., Chem. Soc. Rev. 37 (9) (2008) 1792–1805 [44], © 2008 Royal Society of Chemistry.*

First, let us consider the simplest case of the spectral response of an isolated spherical nanoparticle fully surrounded by a homogeneous medium. As seen in the nanosphere extinction spectra (Fig. 9.2), the scattering is a resonant phenomenon and shows a strong spectral variation. The size of the nanoparticle in relation to the excitation wavelength is seen to be the primary determinant of the resonant wavelength. A primary effect is the so-called retardation (the finite time it takes for light to traverse the full extent of a nanoparticle) which gives the observed strong size dependence. Retardation-related effects become noticeable when the largest length scale $L_{\max}$ associated with the nanoparticle begins to approach the wavelength of the light in the medium surrounding the nanoparticle (i.e., for $L_{\max} > 0.1\lambda$). For larger spheres, it is possible to observe the excitation of modes other than the fundamental mode (the dipolar mode). The effects associated with retardation manifest most strongly for dipolar scattering modes and successively less strongly for quadrupole and higher order multipolar modes. A phase delay of $4\pi R/\lambda$ is associated with the local fields at one end of the nanoparticle to changes in the charge distribution occurring at the opposite end in the dipolar mode of a spherical nanoparticle of radius R; the oscillation period will thus increase in proportion to this delay. The effective interaction distance is reduced due to the multiple nodes occurring in higher order scattering modes that reduce the influence of the retardation effect on the oscillation period.

For aspherical nanoparticles, the resonant wavelengths are influenced by both the retardation effects arising from the increased length scale and by the depolarization effects associated with the aspect ratio. Fig. 9.3 shows the influence of both these effects for ellipsoidal

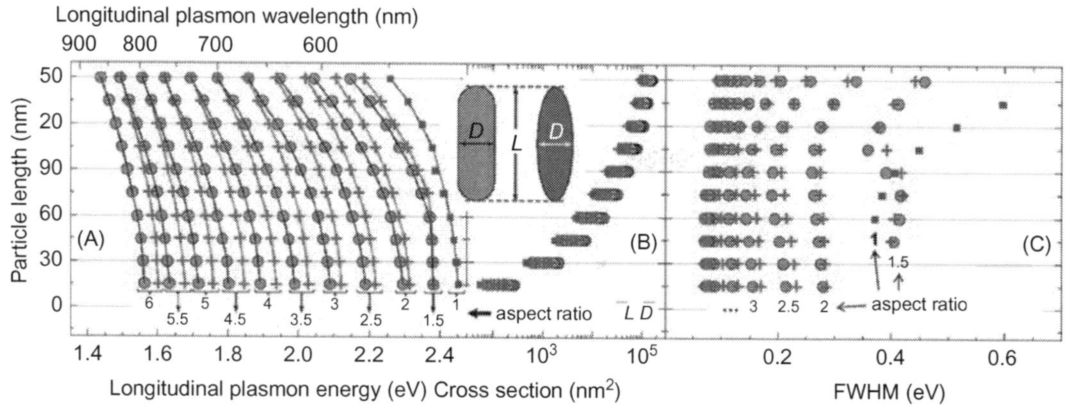

Figure 9.3

Influence of gold nanorod shape (ellipsoidal or hemispherical capped cylinders) and aspect ratio on the resonance wavelengths using numerical calculations. (A) Longitudinal resonance (with electric field polarized along the long edge of the ellipsoid) wavelength in gold ellipsoids (plus symbols) and rods with hemispherical caps (circles) as a function of particle length (vertical axis) and aspect ratio (labels at the bottom). Dipolar polaritons in spheres (unit aspect ratio) are shown as reference (squares). (B) Extinction cross-section at the resonant wavelength for illumination with incident electric field along the particle's axis of symmetry. (C) Full width at half maximum of the longitudinal eigenmode. *Reprinted with permission from Reference V. Myroshnychenko, J. Rodríguez-Fernández, I. Pastoriza-Santos, A.M. Funston, C. Novo, P. Mulvaney, et al., Modelling the optical response of gold nanoparticles., Chem. Soc. Rev. 37 (9) (2008) 1792–1805 [44], © 2008 Royal Society of Chemistry.*

nanoparticles. The size and aspect ratio of the ellipsoids are varied; the resonant wavelength of the dipolar mode is plotted for the incident wave polarized such that it excites the dipolar mode along the longest major axis of the ellipsoid (the other two semiaxes are equal in length). Keeping a fixed aspect ratio, an increase in the length L results in redshifting of the resonance wavelengths as a result of retardation effects. For a given length L, increasing the aspect ratio also leads to redshifted resonance as a result of the depolarization effects.

The presence of the substrate breaks the spatial symmetry and also alters the coupling strength between the nanoparticle eigenmodes and the free space propagating electromagnetic waves [45]. The most noticeable effect of an adjacent substrate is to redshift the resonance wavelength of the nanoparticle in close proximity to the interface in comparison to the resonance of the same particle far away from it. The influence of the mirror dipole is expected to be more pronounced for larger particles. Retardation effects become important in the substrate-nanoparticle interaction-induced changes in the optical spectra of nanoparticles. The breaking of spatial symmetry results in eigenmode hybridization [46] that may appear as an anomalous linewidth broadening [45] under unpolarized illumination conditions. In general, the presence of the substrate can induce interactions between the various multipoles of the nanoparticles and the corresponding mirror multipoles causing alterations in the resulting spectral response. To summarize, accurate numerical simulations must be performed to understand and exploit the spectral modifications of a metallic nanoparticle in close proximity to a supporting substrate.

The geometry and material permittivity dispersions are thus seen to have a strong influence on the spectral dependence of the optical properties of the nanostructure. At the resonance wavelength, we are also interested in the exact spatial distribution of electromagnetic energy. The so-called "hot spots" are created which confine electromagnetic energy. These hot spots can be probed experimentally; electron energy loss spectroscopy (EELS) is a well-established tool that allows us to probe the hot spots. In Fig. 9.4, EELS-based experimental study of the hot spots is shown and they compare well with the predictions of classical electrodynamics.

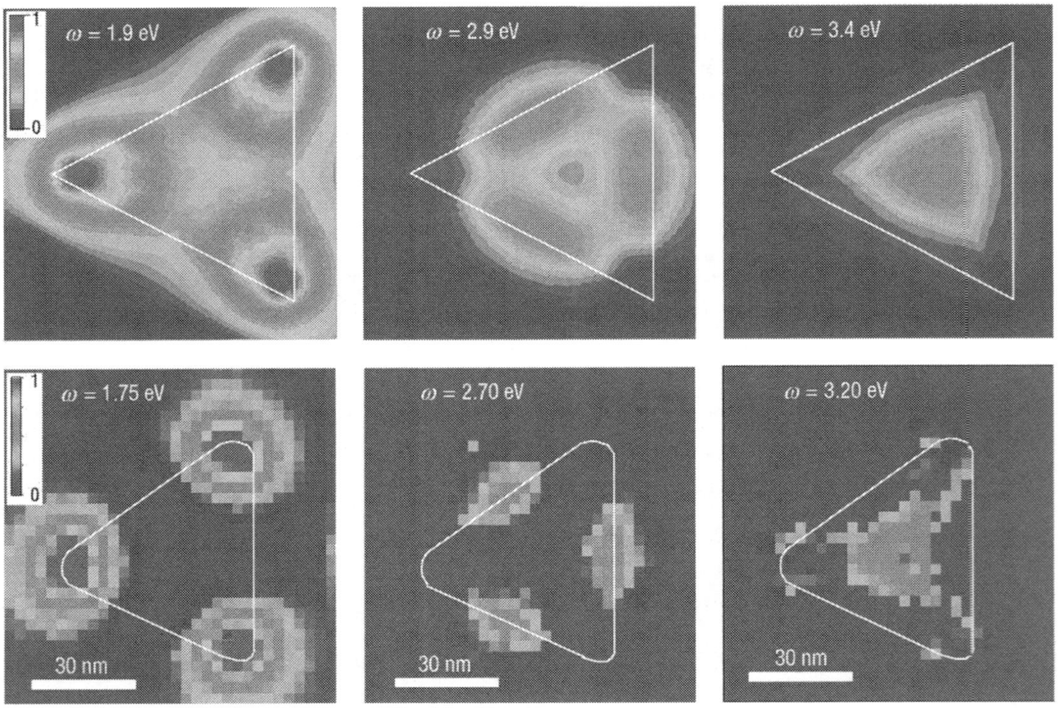

Figure 9.4

Simulated (a) and experimental (b) electron energy loss spectroscopy (EELS) amplitude maps demonstrate the strong localization of electromagnetic energy due to the excitation of surface plasmon resonances in a plasmonic nanoparticle. Distribution of the modes centered at 1.75, 2.70, and 3.20 eV, respectively, in the spectra of a triangular nanoparticle with 78nm-long sides on a mica substrate is shown (the outer contour of the particle is shown as a white line and the color scale is linear and in arbitrary units). *Reprinted with permission from Nelayah and coworkers F. Javier Garcia de Abajo, J. Nelayah, M. Kociak, O. Stephan, I. Pastoriza-Santos, M. Tence, et al., Mapping surface plasmons on a single metallic nanoparticle, Nat. Phys. 3 (2007) 348–353. [47]. Copyright © Nature Publishing Group.*

The rate of plasmon-derived hot carrier generation useful for transmission are located in a volume V adjoining the surface such that the path lengths from points in this volume to the interface are lesser than the mean free path of the carrier. For nanoparticles that are small enough that retardation effects become negligible, the rate of plasmon decay-related hot carrier generation will be approximately proportional to:

$$\frac{1}{2}\int_V d\mathbf{r}\omega|E(\mathbf{r})|^2\Im(\mathrm{e}), \tag{9.1}$$

where ω is the excitation frequency in radians, E is the spatially varying local electric field, and ϵ is the dielectric permittivity of the metal. The integral is carried out over the volume V. Certain plasmonic nanoparticle geometries are known to amplify the local electric field by orders of magnitude, and from Eq. (9.1) it is easy to see that this local field enhancement will improve the rate of production of hot carriers.

To summarize this subsection, knowing the frequency-dependent permittivity functions and by tuning the various geometrical parameters enables us to use well-established computational electromagnetics tools to optimize the optical absorption at desired excitation wavelengths that can, in turn, optimize the hot carrier generation rates.

9.2.2 The Generation of Hot Carriers and Their Energy Distribution

The coherent oscillation associated with the LSPR lasts for a very short time of the order of a few femtoseconds in most plasmonic metals. It then dephases through two channels: a radiative decay pathway involving the emission of photons that travel away and a nonradiative decay pathway that eventually releases heat. The so-called Landau damping is the mechanism behind the nonradiative process which mediates the conversion of coherently oscillating localized surface plasmons to hot charge carriers [27]. The plasmon dephasing in Landau damping primarily produces single electron–hole pairs in the metal and requires conservation of crystal momentum and total energy. Note that the radiative and nonradiative dephasing routes are competing mechanisms; it is not possible to entirely turn off the radiative decay [29] because that would prevent the efficient light-based excitation of the plasmon in the first place! The Landau damping processes occur at different rates depending on the material properties and the frequency of the plasmon resonance. In spectral regions, where the imaginary part of a metal's permittivity is higher, nonradiative damping mechanisms become important.

Landau damping shares similarity with the so-called internal photoemission (PE), whereby the direct absorption of photons at and near the surface of metals creates hot carriers. There has been significant work done in this connection that allows us to derive insights about the plasmon decay-derived hot carriers. The number density and energy distributions of photogenerated carriers can be obtained experimentally from PE and two-photon photoemission studies (2PPE) studies. Such experiments can also provide both time and spatially resolved data, and are thus invaluable in understanding the generation of photocarriers. The temporally resolved data can establish the rates of decay of photoexcited carriers and the dependence of lifetime on the initial energy. The spatially resolved data can help assess whether hot carrier generation is intimately related to the plasmonic hot spots.

Note that at thermal equilibrium, the electron energy distribution follows the well-known Fermi-Dirac statistics where the chemical potential is nearly independent of temperature and is well approximated by the Fermi energy. Following the plasmon decay, the equilibrium is disturbed and it is this nonequilibrium distribution that we wish to obtain. We will restrict attention to the four commonly used plasmonic metals: the three coinage metals Gold (Au), Silver (Ag), Copper (Cu), and Aluminum (Al). In order to understand the permitted electron transitions we need to examine the electronic band structures of these elements. Fig. 9.5 shows the band structure for four commonly studied plasmonic materials (the Fermi energy is marked as 0). In coinage metals plasmonic materials Au, Ag, and Cu, the valence d-band is located well below E_F. From below E_F to well above it, there exist hybridized s and p states giving the so-called sp-band. The sp-band electrons behave nearly like free electrons with parabolic dispersion. The valence d-band is characterized by a high density of states (DOS) and occurs for a narrow range of energies, while the sp-band gives nearly constant DOS over a wide range of energies surrounding the Fermi energy. The sp-band DOS is less compared to that of the d-band. The difference in energy between the Fermi energy and the d-band is the interband threshold. Note that Al has a markedly different band structure in comparison to the other three elements.

The possible electron transitions can be either interband (i.e., d to sp) or intraband (within sp). Since surface plasmons have negligible momenta in comparison with the crystal momenta of electrons, the direct decay of plasmons into an electron–hole pair in the bulk

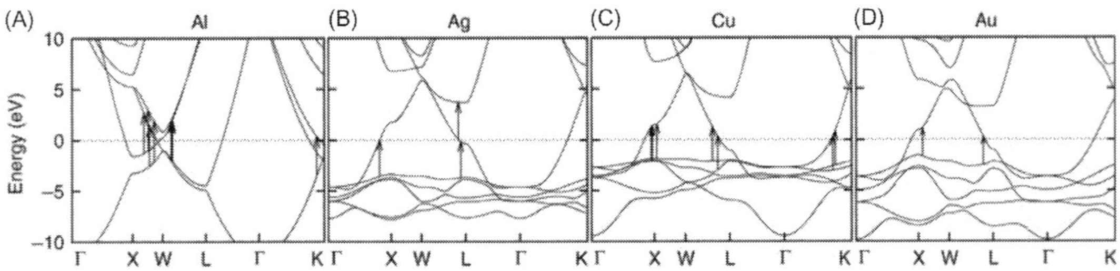

Figure 9.5

Theoretical electronic band structure for (A) aluminum, (B) silver, (C) copper, and (D) gold. (Fermi level at 0). The top panels show the band structure and blue arrows mark the allowed interband electron transitions. *Reprinted under the Creative Commons Attribution 4.0 International License from Reference M. Bernardi, J. Mustafa, J.B. Neaton, S.G. Louie, Theory and computation of hot carriers generated by surface plasmon polaritons in noble metals, Nat. Commun. 6 (2015) 1–9 [48],* crystals requires interband transitions. The blue arrows show the possible interband transitions in Fig. 9.5 for all the four metals. Since interband transitions in Ag, Au, and Cu occur from the *d* band located deep below the Fermi energy, the hot electrons have less energy in comparison to the hot holes. The prediction that hot holes have higher energy in comparison to hot electrons for plasmons with energy above the interband threshold in Au, Ag, and Cu is borne out by experiments [29].

Indirect decay pathways that involve other quasiparticles like phonons can create electron–hole pairs using intraband transitions. The crystal momentum conservation requirement becomes relaxed in the limit of very small particles as quantum confinement effects create localized states without well-defined crystal momentum. These so-called geometry-assisted transitions create an additional pathway that facilitates intraband transitions [49,50].

In Fig. 9.6, the normalized energy distributions were reported for Au, Ag, Cu, and Al planar interfaces (ignoring the geometry-assisted intraband transitions). For Au, Ag, and Cu, two distinct operational zones can be observed depending on whether the plasmon energy is above or below the interband threshold. The interband transitions create hot electrons, mostly located near the Fermi energy with much more energetic holes. Above the interband threshold, sharply peaked distributions are seen. As the DOS function is nearly constant for the sp-band and the probability of transitions is also nearly constant, it can be observed that the distribution of hot electron and hot hole energies is nearly uniform for intraband transitions. Below the interband threshold, uniform energy distributions are observed. The almost uniform energy distribution of electron and hole energies in aluminum and the lack of a transition threshold strongly contrasts with the hole-dominant energy distribution in copper and gold. A bimodal hot hole and hot electron distributions are observed in silver due to the position of the d-bands. Geometry-assisted transitions can be an additional knob to obtain favorable electron distributions for particular applications [28,48–50]. Geometry changes the relative contributions of interband and intraband contributions, but does not affect the carrier energy distributions within each contribution. Depending on the specific hot carrier-related application, traditionally used plasmonic metals may not always be the best material of choice. Munday and coworkers have recently surveyed a broader selection of materials that may be more suitable [51].

9.2.3 Scattering and Lifetimes of Hot Carriers

The hot carrier energy distributions upon birth have been discussed in the previous section. The hot carrier momentum distributions at birth are often anisotropic [28]. As time progresses, these hot carriers experience various scattering events. Accumulation of these events randomizes their momenta and redistributes their excess energy. These processes essentially seek to bring the hot carriers back into thermal equilibrium. Of primary interest are the electron-electron and hole-hole interactions, which occur on the timescale of a few

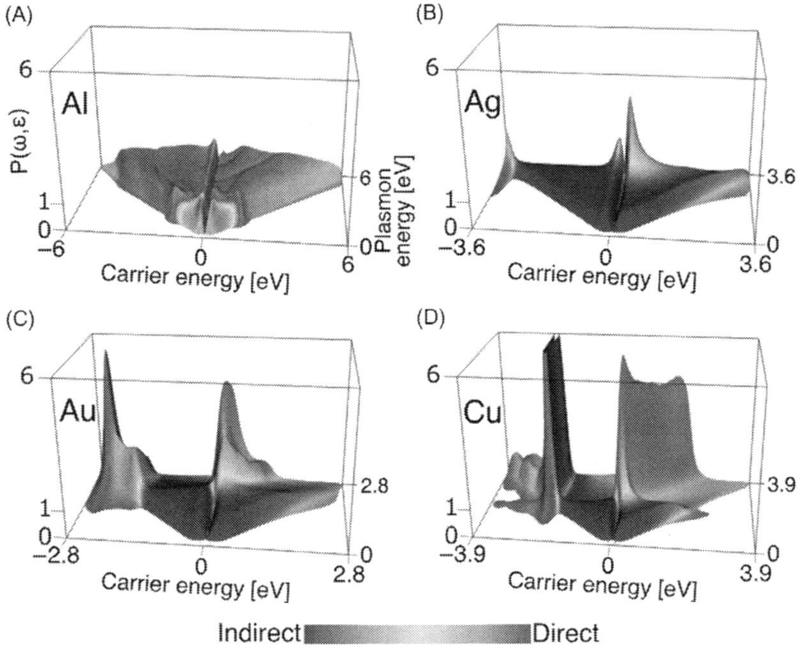

Figure 9.6

Theoretically calculated energy distributions of hot carriers generated at a planar interface by the decay of surface plasmons due to phonon-assisted and direct band transitions as a function of plasmon frequency and carrier energy. Four common plasmonic metals (A) Al, (B) Ag, (C) Au, and (D) Cu are shown. The color scale indicates the relative contributions of phonon-assisted (red) and direct (blue) transitions. *Reprinted with permission from A.M. Brown, R. Sundararaman, P. Narang, W.A. Goddard, H.A. Atwater, Nonradiative plasmon decay and hot carrier dynamics: effects of phonons, surfaces, and geometry, ACS Nano 10 (2016) 957–966 [49]. © 2016 American Chemical Society.*

femtoseconds, and electron-phonon and hole-phonon interactions, which occur on a much slower timescale of few picoseconds. Hot carriers with energy in excess of 1 eV in relation to the Fermi energy do not typically interact with phonons due to large energy mismatch [29]. Due to these reasons, electron-electron and hole-hole interactions are more important as it is usually the high energy charge carriers which we are primarily concerned with.

The investigation of hot carrier dynamics in metals has a long history predating the current interest in plasmon-induced hot carrier science and technology. Theoretical methods and first principle calculations are indispensable tools in this connection and excellent agreements have been found between experimentally determined lifetimes and first principle calculations [52]. This approach predicts [29,52] that the hot electron lifetime τ due to electron-electron scattering is proportional to the inverse square of the initial electron energy E,

$$\tau_{e-e}(E) \approx \left(\frac{E_f}{E-E_f}\right)^2 \tag{9.2}$$

where E_f is the Fermi energy. Unfortunately, this implies that the hottest charge carriers (i.e., those that are most likely to be collected) lose their energy most rapidly. Although this equation is derived for a free electron gas (the so-called Fermi liquid), the behavior of electron lifetimes in real metals show a similar energy dependence. Fig. 9.7 shows the first

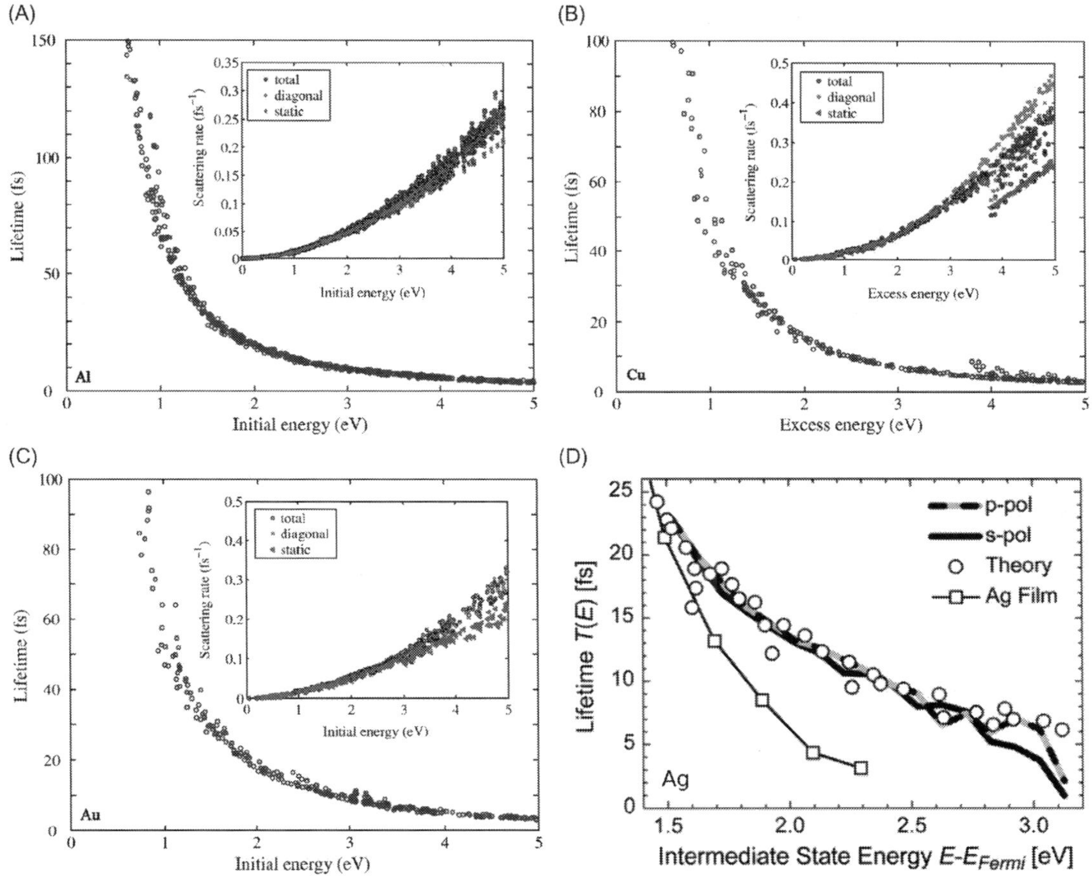

Figure 9.7

Ab-initio calculations of hot electron lifetimes as a function of initial electron energy in metal slabs of (A) Al (B) Cu (C) Au. In (D), the theoretical lifetimes are compared against experimentally observed ones for an Ag film. *Panels (a), (b), and (c) reprinted with permissions from Reference F. Ladstädter, U. Hohenester, P. Puschnig, C. Ambrosch-draxl, Firstprinciples calculation of hot-electron scattering in metals, Phys. Rev. B 70 (2004) 235125-10 [52]. © 2004 American Physical Society. Panel (d) reprinted with permissions from Reference M. Merschdorf, C. Kennerknecht, W. Pfeiffer, Collective and single-particle dynamics in time-resolved two-photon photoemission, Phys. Rev. B 70 (2004) 193401–193404 [53]. © 2004 American Physical Society.*

principle calculations of the lifetimes in the four elements: Al, Au, Ag, and Cu. Fig. 9.7(D) also shows the comparison of theoretical and experimentally determined (through 2PPE studies) lifetimes for Ag. The first observation is that the lifetime energy dependence shows good similarity with a free electron gas lifetimes. Especially for Al, the lifetimes are almost equal. For Ag, Au, and Cu, although the functional dependence is obeyed, the absolute lifetimes are enhanced in comparison to those in the free electron gas [52]. This discrepancy is attributed to d-band screening, where the high d-band DOS below the Fermi energy results in efficient screening contributions [52]. Additionally, we notice in Fig. 9.7(D) that the theory overpredicts the lifetimes, especially at high initial energies. The carrier dynamics are thus expected to deviate strongly from the Fermi liquid theory for plasmon energies exceeding the interband threshold where a large activity is expected from the d-band electrons [29].

The mean free path due to electron-electron scattering events, λ_{e-e}, can be determined analytically through the formula (attributed to Quinn [32]):

$$\lambda_{e-e}(E) = \frac{24a_0\sqrt{\frac{\alpha_e r_s}{\pi}}(3E_F/E^2 + 2/E)}{\arctan\sqrt{\frac{\pi}{\alpha_e r_s}} + \frac{\sqrt{\frac{\alpha_e r_s}{\pi}}}{1+\frac{\alpha_e r_s}{\pi}}} \tag{9.3}$$

where a_0 is the Bohr radius (0.0529 nm), $\alpha_e = \sqrt{4/(9\pi)}$, and r_s is the radius of a sphere equal to the volume of one conduction electron in units of the Bohr radius. The mean free path also follows an E^{-1} dependence. For gold with r_S of 3, the mean free paths are larger than 10 nm even for energies of 3.5 eV. Note that the mean free paths are often comparable to the dimensions of nanostructures.

9.2.4 Hot Carrier Injection Into Semiconductors

Once hot carriers are generated, they need to be extracted out of the metallic nanoparticles as soon as possible to prevent their decay. In semiconductors, the majority carriers typically are long lived; Ms Schottky photodiodes have been extensively studied and form the basis of many hot carrier collection setups. Similar to a solar cell, the Ms diode can be operated in the fourth quadrant to derive electrical power or can be operated in the second quadrant (Ms is reverse biased) as a photodetector. The seminal paper by Catchpole and coworkers [54] analyzed MS photodiodes and derived efficiency estimates. Several other reports have followed this work and have further refined the theoretical estimates of the device efficiencies [32,55–60]

Fig. 9.8(A) to (C) shows the schematic of the ideal MS contact analyzed by Catchpole and coworkers. Fig. 9.8(A) shows the basic geometry of a nanoparticle array resting on a semiconductor substrate and the resulting MS junction operating in the fourth quadrant as a power source. The hot carriers move to the semiconductor from the metal nanoparticles, whereby they drift to the contact and then on to drive a load. This circuit keeps replenishing the carriers in the metal. An Ms interface can form either a Schottky junction or an ohmic junction. Let us consider the case of a Schottky contact between the plasmonic metal and an n-doped semiconductor. The energy band diagram of this contact is shown in Fig. 9.8(C) with the barrier height Φ_B. The barrier height, in theory, depends only on the work functions of the metal and the semiconducting material [61]. In practice, the details of the interface layers may strongly influence the barrier height.

Catchpole and coworkers idealized the metal free electrons by considering them to be almost like the free electron gas. This assumption is somewhat true for the coinage metals for hot carrier generation below the interband threshold (the sp-band edge occurs about 5.5 eV below the Fermi energy for Gold, for instance). The assumption leads to the well-known $\sqrt{E}$ dependence for the electron and hole DOS. The energy distributions of the hot electrons is determined by the electron distribution joint density of states (EDJDOS) and that for the holes is determined by the corresponding hole distribution joint density of states (HDJDOS). The EDJDOS can be rigorously determined through first principles calculations. For the simple parabolic DOS function, the EDJDOS assumes a nearly flat dependence, as seen in Fig. 9.8(B).

At a given bias voltage V, the current density in the cell j_c can be written as $jsc - j_{thermionic}$, where j_{sc} is the short-circuit photocurrent and $j_{thermionic}$ is the thermionic emission of high energy majority carriers from the SC to the metal. In the case of forward bias, the thermionic component is large due to the reduction in the barrier for the flow from the SC to the metal. The hot carrier component of the current density is given by:

$$j_{sc} = e\int d\lambda \phi(\lambda)\eta_{abs}(\lambda)\eta_i(\lambda), \tag{9.4}$$

where e is the charge of the electron, ϕ denotes the photon flux, η_{abs} is the efficiency that captures the optical absorption process, and η_i is the efficiency that captures the gamut of processes beginning with the plasmon decay, the birth of the hot carrier, the transport of the hot carrier without scattering to the interface, and finally surmounting the barrier to reach the semiconductor. In order for a carrier to surmount the Schottky barrier, it needs to have energy in excess of the Schottky barrier height and, additionally, its momentum in the direction perpendicular to the interface must be sufficient [32]. Fig. 9.8(D) shows the acceptance cone of allowed momenta as reported by Leenheer and coworkers. In small nanoparticles and interfaces with roughness, the emission cone requirement may be relaxed [28].

Even with optimal conditions, the efficiencies of plasmonic hot carrier cells are only in the range of $\leq$ 10%, and there is a lot of room for further improvement [58]. In plasmonic hot carrier photodetectors, the observed responsivity values is quite low in comparison to the theoretical upper limits [56] (see panel (e) in Fig. 9.8). The main effect that reduces the

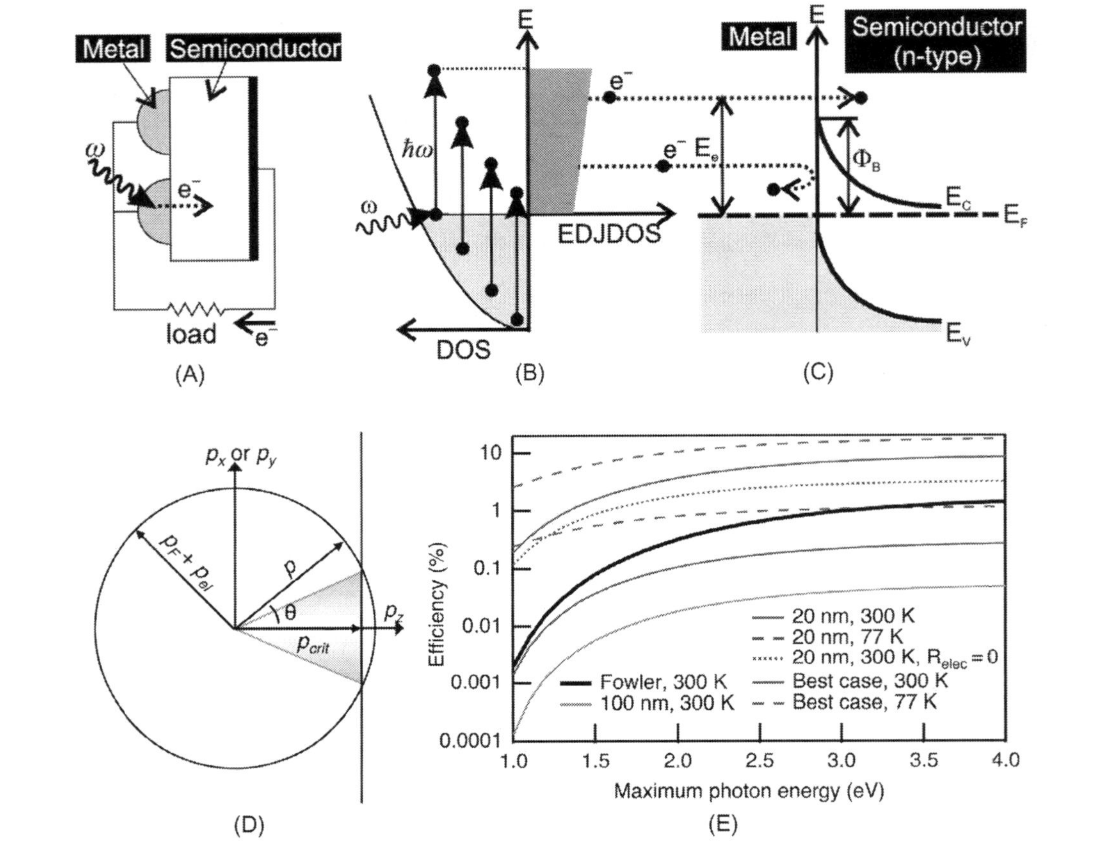

Figure 9.8

(A) Geometry of a plasmon-enhanced internal photoemission solar cell. (B) Excitation of electrons in the metal from occupied energy levels in the conduction band (shaded gray) to unoccupied levels above the Fermi energy. (C) Energy diagram of the Schottky junction at the metal (n-type) semiconductor interface. Only those hot electrons with energy $> \Phi_B$ can cross the barrier into the semiconductor. (D) Schematic of isotropic distribution of hot electron momentum on a sphere in momentum space with a limited escape cone. (E) Solar conversion efficiency for IPE sourced by a 5800 K blackbody spectrum (up to a given maximum photon energy) for different film thicknesses, operating temperatures, and escape cone limitations. *(C) Reprinted with permission from Reference T.P. White, K.R. Catchpole, Plasmon-enhanced internal photoemission for photovoltaics: theoretical efficiency limits, Appl. Phys. Lett. 101 (7) (2012) 073905 [54], © 2012 American Institute of Physics.(E) Reprinted with permission from Reference E. Cortes, W. Xie, J. Cambiasso, A.S. Jermyn, R. Sundararaman, P. Narang, et al., Plasmonic hot electron transport drives nano-localized chemistry, Nat. Commun. 8 (14880) (2017) 1–10 [32], © 2012 American Institute of Physics.*

short-circuit photocurrent turns out to be the flat EDJDOS functions as a significant amount of the hot carriers will simply never cross the barrier. Considerations of the exact EDJDOS distributions, including band-structure effects, Chang and coworkers have revised the efficiency estimates [59] further. Using accurate EDJDOS functions for gold, Boriskina and coworkers [56] predicted upper limits of 4% for hot electron-based cells and up to 10% for hot hole-based cells. Note that these estimates are absolute upper limits as they assume absolute optical absorption and ballistic transport of carriers to the interface. Govorov and coworkers have reported a detailed theoretical analysis and suggest that nanostructures for efficient carrier injection should be in the shape of slabs, platelets, cubes, and nanowires, with relatively small widths in the direction perpendicular to the interface [60]. EDJDOS engineering in nanostructures may be another pathway for improving the efficiencies of hot carrier cells [57].

9.3 Applications of Hot Carrier Generation

9.3.1 Photodetectors

Photodetectors are an integral part of optical systems and are continually seeing improvements in performance, cost, and speed over the years. Nanostructured materials, in particular, are very promising for performance improvements in photodetector technology [62]. Silicon photodetectors have an important limitation, as they are not suitable for operation beyond the visible region. In the short- and long-wave IR, silicon becomes nearly transparent, while at ultraviolet frequencies [63] they have large dark currents. Compound semiconductors dominate the photodetector landscape beyond the reach of silicon. Nanostructured materials can have a huge impact if they can widen the reach of silicon while maintaining its long-standing advantages [62,64].

Long before the current interest in plasmonic hot carriers, the IPE effect occurring in the metal side of a Ms Schottky diode has been explored for sub-bandgap photodetectors in silicon both in the free space and integrated configurations [65]. Plasmonics has breathed a new life into these devices [61] as it offers these key advantages: (1) plasmonic nanostructures can improve optical absorption cross-sections significantly [66], and this can be a solution to the low external quantum efficiencies (EQE) that have been plaguing IPE devices [65]; (2) the possibility of using very small nanoparticles opens up the possibility to manipulate the photogenerated hot carrier energy and momentum distributions [60], and thus allows a way to improve the internal quantum efficiencies (IQE); (3) tuning the geometrical parameters allow us to tailor the spectral [67,68] and polarization [69] response with high precision; and (4) using plasmonic nanostructures in one of the layers [70] of a metal-insulator-metal (MIM) triode configuration (this configuration has much higher speed [operational bandwidth] compared to the Schottky configuration) gives a high degree of asymmetry and improves the performance.

The interest in plasmon-enhanced Schottky photodetectors for photodetection of free space waves was triggered by the seminal work of Knight and coworkers [68]. In that contribution, Knight *et al.* demonstrated a free space hot electron photodetector based on a plasmonic nanoantenna that could detect near-infrared light below the silicon bandgap energy. The function of the gold nanoantenna is twofold: to perform optical absorption (silicon is transparent for this wavelength range) and to be the emitter of hot electrons (an ITO layer on the top provides the electrical contract that completes the circuit. The polarization and spectral response of the reported photocurrent is consistent with the excitation of the dipolar mode along the long axis of the nanostrips.

The linewidth of isolated nanostructures is usually quite large. Sobhani and coworkers [71] have reported narrowband hot electron photodetectors using plasmonic gratings in place of isolated nanoantenna (see panel B in Fig. 9.9). The extraordinary optical transmission phenomenon is responsible for narrowing the linewidth (see (b) in panel B of Fig. 9.9) as the free space light is coupled into long range propagating SPPs. In addition, the electrically connected metallic grating structures also facilitate electron transport without the need for the ITO layer. They achieved nearly 20 times improvement in the responsivity (0.6 mA W^{-1}) compared to Knight and coworkers.

Valentine and coworkers [72] further improved upon the responsivity by further optimizing both the optical absorption efficiency and the internal charge collection efficiencies (eta_{abs} and eta_i in Eq. (9.4)). The optical absorption improvements occur as a result of using the perfect absorber (see panel C in Fig. 9.9) geometry. This geometry not only significantly improves absorption even with ultrathin metal layers (only about 15 nm), but also confers omni directionality of optical absorption (see (c) and (d) in panel C of Fig. 9.9). Having an ultrathin (thin in comparison to the mean free paths of the carriers) metal layer improves the hot electron extraction across the barrier. In this optimal configuration, responsivity values reaches 3 mA W^{-1} at the wavelength of 1300 nm. As discussed in Section 2.4, the acceptance cone limits the injection efficiency. By using a conical shape, Desiatov and coworkers simultaneously reduce the dark current and improve responsivity [73]. The conical tip serves to orient the plasmons perpendicular to the interface and is believed to be the cause for the observed increase in responsivity of about 12 mA W^{-1} at the wavelength of 1300 nm. Nazirzadeh and coworkers [74] have reported that these devices can be fabricated through simple and scalable techniques, and thus this is an active area for low-cost near IR photodetector applications.

Melosh and coworkers [75] reported an alternate architecture based on SPP excitation within a MIM device that produces a photocurrent by asymmetrically exciting plasmons only on the top electrode. Plasmons excited primarily in the upper metal were absorbed and created a high concentration of hot electrons, which could tunnel through the thin insulating barrier to produce current. Chalabi and coworkers [70] replaced the planar top electrode with a plasmonic grating which enables simpler coupling to free space light. Using the perfect absorber geometry, the responsivity of MIM structures has been shown to scale up by a factor of 3 [76] in comparison to the simple grating geometry. Using ITO as one of the electrodes [77] is another improvement on the MIM geometry.

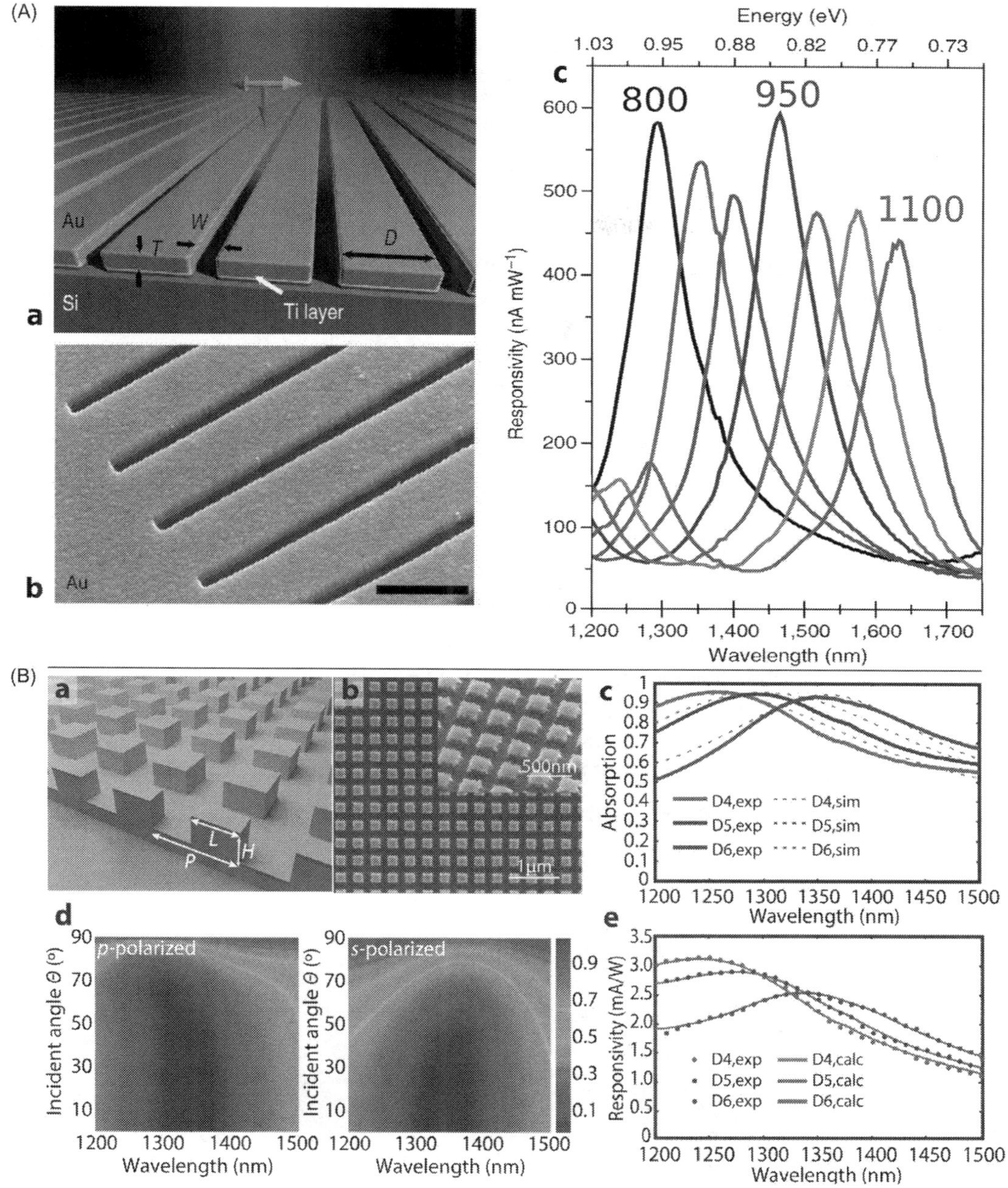

Figure 9.9

Silicon-gold Schottky interface-based free space hot electron photodetectors. A schematic (a) of a gold grating on silicon configuration and (b) the narrowband spectral response of the photocurrent for this device. (b) shows the spectra for different values of the strip widths *D*; *T* and *W* are kept constant at 200 nm and 250 nm, respectively. B schematic (a) and SEM (b) of a plasmonic perfect absorber-based geometry. Omnidirectional optical absorption is demonstrated for *s* and *p* polarizations in (d). Experimentally measured optical absorption in (c) is compared with simulations and (e) shows the photoresponsivity spectrum. *Figures reproduced with permission from: Panel B Reference A. Sobhani, M.W. Knight, Y. Wang, B. Zheng, N.S. King, L.V. Brown, et al., Narrowband photodetection in the near-infrared with a plasmon-induced hot electron device, Nat Commun. 4 (2013) 1643 [71], Qc 2013 Nature Publishing Group; Panel C Reference W. Li, J. Valentine, Metamaterial perfect absorber based hot electron photodetection, Nano. Lett. 14 (6) (2014) 3510–3514 [72], © 2014 American Chemical Society.*

An altogether different class of hot carrier-based photodetector is based on this effect in integrated waveguides [78]. The waveguide geometry is, in fact, slightly more conducive for plasmon-enhanced IPE photodetectors [65] owing to: (1) optical power is confined close to the Ms interface; (2) light can be absorbed over a longer length; and (3) the local electric fields at the Ms interface are perpendicular to the interface leading to a higher probability for the hot carrier momentum to fall in the acceptance cone. Both the Schottky and MIM configurations have been explored in this connection (see [64,65,78–83]).

9.3.2 Chemical Reactions Through Transfer of Charge Carriers

9.3.2.1 Mechanism of charge carrier transfer to adsorbed molecules

The hot carriers resulting from plasmon decay can transiently occupy available orbitals of molecules adsorbed onto the nanoparticle before their energy get dissipated via carrier-phonon interactions. This transient transfer of charge carriers to an accessible adsorbent orbital creates transient ions or excited nanoparticle-adsorbate system, and the excess energy is converted to nuclear motions, which may result in activation/dissociation of a chemical bond. In other words, the energy of charge carriers is converted to vibrational motion of the atoms in adsorbed molecules along the reaction coordinate. There are two proposed mechanisms for the charge transfer to take place between the plasmonic nanostructure and the adsorbed molecules: indirect charge transfer and direct charge transfer [84,85]. Schematic descriptions for these two processes are shown in Fig. 9.10. In the indirect charge transfer mechanism, the plasmon relaxes to produce hot electrons. Many of such electrons have

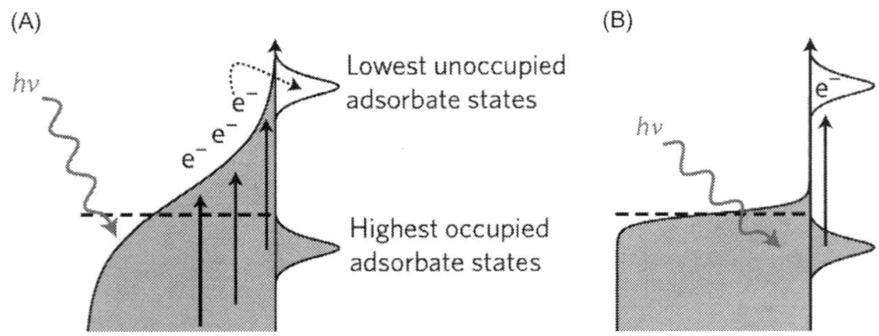

Figure 9.10

Charge transfer mechanisms between plasmonic nanostructures and adsorbed molecules. (A) Indirect charge transfer mechanism. In this scenario, plasmon relaxation creates an electron distribution and the electrons which have equal or higher energy than the lowest unoccupied adsorbate orbital can be transferred to those orbitals. (B) Direct charge transfer mechanism. In this scenario, plasmon decays can cause direct excitation of electron to an unoccupied adsorbate state. *Figures reproduced with permission from Reference S. Linic, U. Aslam, C. Boerigter, M. Morabito, Photochemical transformations on plasmonic metal nanoparticles, Nat. Mater. 14 (2015) 567–576 [84], © 2015 Nature publishing group.*

energies equal or higher than the energy of lowest unoccupied state of adsorbate molecules, and these electrons can be transferred to these orbitals. In direct transfer mechanism, on the other hand, plasmon decay causes direct excitation of an electron to an unoccupied adsorbent orbital without the formation of hot electrons [86,87]. The later mechanism is commonly known as chemical interface damping, as it provides additional nonradiative decay path for plasmons making the resonance weaker and broader [86,88].

9.3.2.2 Examples of charge carrier-driven chemical reaction

Charge carrier-induced chemistry has been studied on single crystal metal surfaces [89] as well as on metal nanoparticles. However, nanoparticles have better suitability compared to bulk surfaces owing to their physical characteristics such as: they support strong surface plasmon resonances, which enhances the optical excitation and concentrates the optical field within a tiny volume near the surface; the amount of light absorption can be optimized through geometry tuning; and they have higher concentration of surface electronic states compared with bulk materials, and therefore charge carriers will have higher probability of interacting with the adsorbed molecules [89]. Here, we focus our discussions on plasmonic nanoparticles.

Halas and coworkers [33] have demonstrated room temperature dissociation of H_2 molecules on small gold nanospheres (diameter of 2–10 nm) under excitation light intensity of $\approx$ 2.5 W cm^{-2}. When excited with the resonant light, the rate of reaction increased sixfold (see Fig. 9.11(E)) accompanied by a small increase of temperature (6 K). Increase of temperature under dark condition only had a minor effect on the reaction rate (see panel (f) of Fig. 9.11), confirming that the photocatalysis is not due to heating. The authors proposed that the reaction happened through formation of hot electrons, which were transferred to the vacant $1\sigma u*$ orbital of adsorbed H_2 molecule (see panels (a) to (d) of Fig. 9.11). The same group recently reported that hot electrons from Al nanoparticles under broadband visible excitation can also dissociate hydrogen molecules [90]. Linic et al. [91] have demonstrated charge carrier-induced oxidation reactions on silver nanocubes (50 nm length) under a visible broadband excitation with maximum power of 250 mW cm^{-2}. It was found that the rate of reaction was enhanced by energetic electrons adsorbed to oxygen molecules producing transient O_2^-, which dissociated to produce atomic oxygen. Maier and coworkers have demonstrated hot electron-induced reduction of nitro compounds using Ag nanoantenna under resonant excitation [31]. The authors have also shown that the reduction is more specific to the nanogaps of the bowtie antenna, indicating abundance of hot electrons at a region where field intensity is maximum.

Another interesting application of hot electron-induced chemistry is in conversion of solar energy to fuel such as hydrogen. There are two main advantages of plasmonic nanoparticles for this purpose: they can harvest the solar light more efficiently than organic dyes owing to their size and shape dependent optical properties (which can be tuned over the entire visible

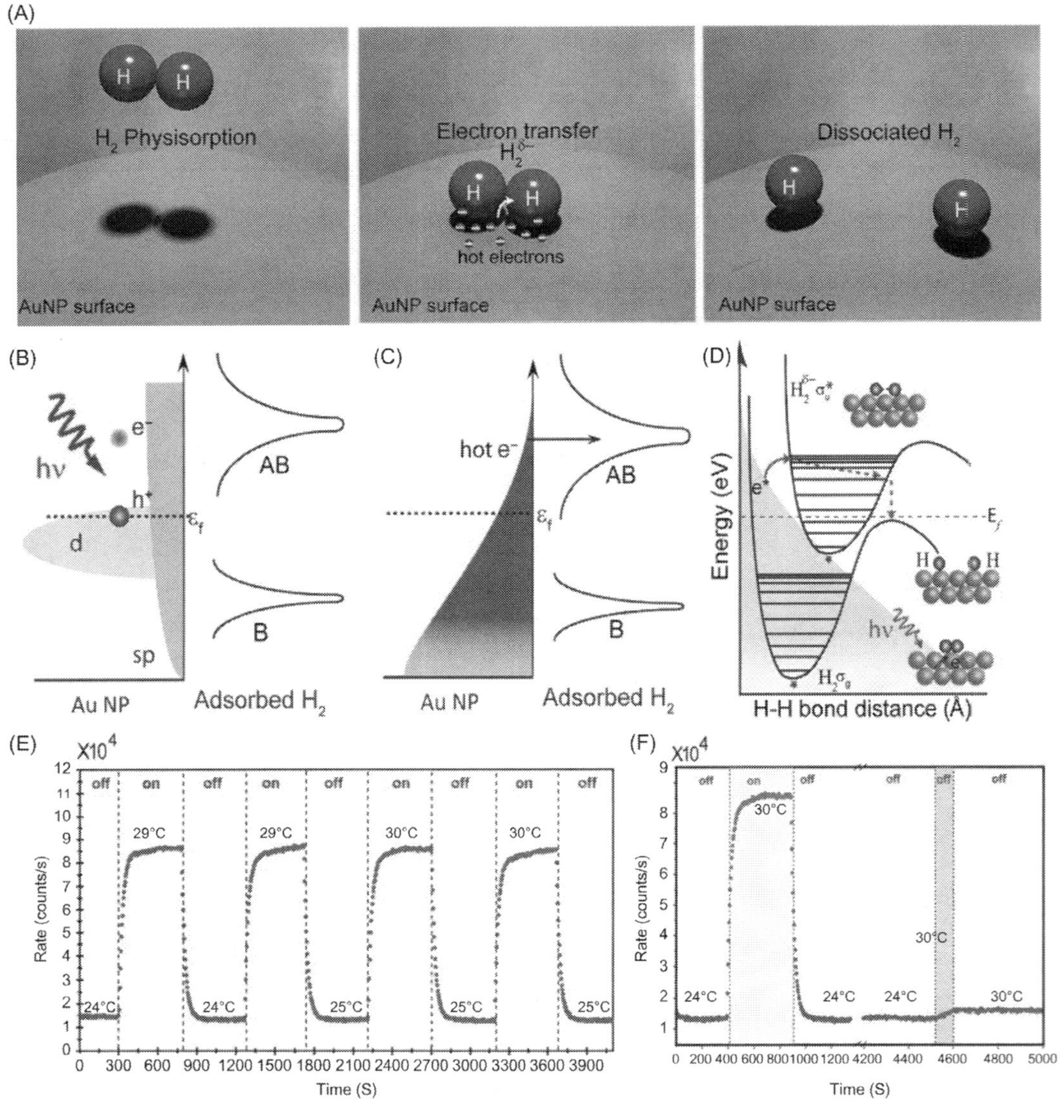

Figure 9.11

Hot electron induced dissociation of hydrogen on gold nanoparticles. (A) Schematic description of various steps involved in hydrogen dissociation process. (B) Schematics of hot electron excitation in Au nanoparticles after plasmon decay. Bonding and antibonding orbitals of adsorbed hydrogen are shown as B and AB. (C) Schematics of Fermi-Dirac distribution of hot electrons and hot electron transfer to antibonding orbital of hydrogen. (D) Proposed mechanism of hot electron induced dissociation of hydrogen. (E) Experimental results on hydrogen dissociation on Au nanoparticles under illumination and in dark conditions. Fast and reversible increase of rate is observed under laser excitation. A small increase of temperature from 24°C to 30°C was also observed. (f) Temperature increase from 24°C to 30°C in dark conditions shows much less enhancement of rate compared to (e). *Figures reproduced with permission from Reference S. Mukherjee, F. Libisch, N. Large, O. Neumann, L.V. Brown, J. Cheng, et al., Hot electrons do the impossible: plasmon-induced dissociation of H2 on Au, Nano Lett. 13 (1) (2013) 240–247 [33], © 2013 American Chemical Society.*

to NIR spectrum); hot carriers generated by the light absorption can catalyze splitting of water into hydrogen and oxygen. There are many reports of hot electron-induced water splitting [35,92,93]. In a particularly intriguing study, Moskovits and coworkers [93] used gold nanorods antenna covered with a thin layer of TiO_2 and demonstrated hydrogen production rates up to 2.8 mmol h^{-1} g^{-1} under 1 Sun illumination. In dark condition, no measurable amount of hydrogen was produced. This increase in the rate of hydrogen production under illumination was attributed to the hot electrons, which were generated through plasmon decay and were energetic enough to travel to the reaction site (Pt catalyst was used) through the TiO_2 layer.

Recent reports also demonstrated energetic charge carrier-induced chemistry with bimetallic nanoparticles consisting of a plasmonic and a more reactive (such as Pd, Pt, and Cu) nanoparticle in close proximity [90,94–96]. For example, Majima and coworkers have demonstrated hydrogen production reaction with gold nanorod-Pt nanoparticle bimetallic nanostructures (see Fig. 9.12) at room temperature under visible and near-IR excitations [94]. Using single particle spectroscopy, the authors show efficient transfer of electrons from gold nanorods to Pt nanoparticles, where the catalytic conversion takes place. Some other examples of such bimetallic nanoparticles-induced chemistry are Suzuki coupling with Au nanorods-Pd nanostructures [96]; CO_2 conversion with Al-Cu2O nanostructures [97]; and hydrogen dissociation on Al-Pd nanostructures [95].

9.4 Conclusion

This chapter covered the basics of hot carrier generation and scattering effects in plasmonic nanostructures and their collection across an interface. Two key applications of this include photodetectors and selectively powering chemical reactions with spatial localization. Although the field has witnessed several breakthroughs in the last few years, obtained device efficiencies fall short for practical applications. Current research is being directed at several aspects simultaneously: (1) exploration of various materials and interfaces to improve efficiencies [51,98]; (2) geometrical optimization of nanostructures to improve both the optical absorption and for optimizing the carrier energy distributions [99]; (3) quantum transport studies to improve the hot electron extraction efficiencies; and (4) alternate damping mechanisms [85,86,88] for the plasmon decay. Plasmon-induced hot carrier effects are also of interest in the context of novel two-dimensional materials [62,100–103].

Detailed understanding of various charge transfer mechanism and their relative contributions toward a chemical reaction still remains a challenge. One may assume that in a plasmon-assisted chemical reaction, charge transfer through indirect and direct pathways (as depicted in Fig. 9.10) as well as plasmon-induced heating will act together. Experimentally distinguishing these mechanisms and developing theoretical models to find their relative contribution still remains a challenge and future works need to address them. Recent results tend to suggest [87] that efficient charge transfer is possible through the

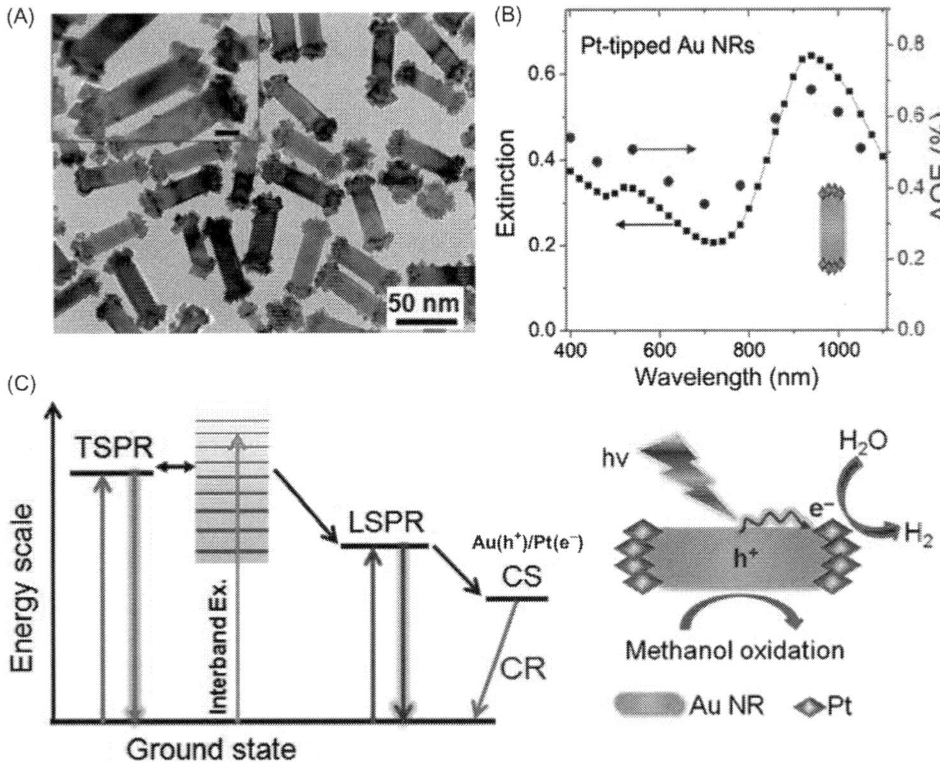

Figure 9.12

Catalytic reactions with bimetallic nanoparticles. (A) TEM image of Pt-attached gold nanorods. Note that Pt nanoparticles were selectively attached to the tips. (B) Apparent quantum efficiencies (ACQ) of hydrogen production match well with the extinction spectrum. (C) Proposed mechanism of hydrogen production. TSPR: Transverse surface plasmon resonance; LSPR: Longitudinal surface plasmon resonance; CS: Charge separation; CR: Charge recombination. *Figures reproduced with permission from Reference Z. Zheng, T. Tachikawa, T. Majima, Single-particle study of Pt-modified Au nanorods for plasmon-enhanced hydrogen generation in visible to near-infrared region, J. Am. Chem. Soc. 136 (2014) 6870–6873 [94].* *© 2014 American Chemical Society.*

direct transfer pathway (chemical interface damping); however, quantitative estimation of its contribution relative to other damping mechanisms and how the chemical nature of the adsorbate affects its efficiency is not well understood yet.

Acknowledgment

The authors wish to acknowledge financial support from the Department of Science and Technology, Government of India, under the DST Nanomission Project Proposal #SR/NS/NM-65/2016(G) and from startup funding from Indian Institute of Technology, Gandhinagar. RSH acknowledges support from the Department of Science and Technology, Government of India, Extramural Grant # SB/S3/EECE/0200/2015. SK wishes to acknowledge support from the Department of Science and Technology, Government of India, Proposal #EMR/2015/001013.

References

[1] S.A. Maier, Plasmonics: Fundamentals and Applications, Springer, New York, 2007.

[2] X. Fan, W. Zheng, D.J. Singh, Light scattering and surface plasmons on small spherical particles, Light Sci Appl 3 (e179) (2014) 1–14.

[3] J.N. Anker, W.P. Hall, O. Lyandres, N.C. Shah, J. Zhao, R.P. Van Duyne, Biosensing with plasmonic nanosensors, Nat Mater 7 (6) (2008) 442–453.

[4] C. Ciraci, R.T. Hill, J.J. Mock, Y. Urzhumov, A.I. Fernandez-Dominguez, S.A. Maier, et al., Probing the ultimate limits of plasmonic enhancement, Science 337 (6098) (2012) 1072–1074.

[5] M.I. Stockman, Electromagnetic Theory of SERS, Vol. 103, Springer-Verlag, Berlin Heidelberg, 2006.

[6] J.B. Jackson, N.J. Halas, Surface-enhanced Raman scattering on tunable plasmonic nanoparticle substrates, Proc Natl Acad Sci U S A 101 (52) (2004) 17930–17935.

[7] J. Kundu, F. Le, P. Nordlander, N.J. Halas, Surface enhanced infrared absorption (SEIRA) spectroscopy on nanoshell aggregate substrates, Chem Phys Lett 452 (1–3) (2008) 115–119.

[8] M. Scalora, M.A. Vincenti, D. De Ceglia, V. Roppo, M. Centini, N. Akozbek, et al., Secondand third-harmonic generation in metal-based structures, Phys Rev AMO Phys 82 (4) (2010) 33.

[9] Q. Chen, D. Chitnis, K. Walls, T.D. Drysdale, S. Collins, D.R.S. Cumming, CMOS photodetectors integrated with plasmonic color filters, IEEE Photon Technol Lett 24 (3) (2012) 197–199.

[10] S. Yokogawa, S.P. Burgos, H.A. Atwater, Plasmonic color filters for CMOS image sensor applications, Nano Lett 12 (8) (2012) 4349–4354.

[11] S.P. Burgos, S. Yokogawa, H.A. Atwater, Color imaging via nearest neighbor hole coupling in plasmonic color filters integrated onto a complementary metal-oxide semiconductor image sensor, ACS Nano 7 (11) (2013) 10038–10047.

[12] H. Park, K.B. Crozier, Multispectral imaging with vertical silicon nanowires, Sci Rep 3 (2013) 2460.

[13] M. Najiminaini, F. Vasefi, B. Kaminska, J.J.L. Carson, Nanohole-array based device for 2D snapshot multispectral imaging, Sci Rep 3 (2013) 2589.

[14] H. Aouani, M. Rahmani, H. Šípová, V. Torres, K. Hegnerová, M. Beruete, et al., Plasmonic nanoantennas for multispectral surface-enhanced spectroscopies, J Phys Chem C 117 (36) (2013) 18620–18626.

[15] J. Olson, A. Manjavacas, T. Basu, D. Huang, A.E. Schlather, B. Zheng, et al., High chromaticity aluminum plasmonic pixels for active liquid crystal displays, ACS Nano (2015). acsnano.5b06415.

[16] L. Wen, Q. Chen, S. Song, Y. Yu, L. Jin, X. Hu, Photon harvesting, coloring and polarizing in photovoltaic cell integrated color filters: efficient energy routing strategies for power-saving displays, Nanotechnology 26 (26) (2015) 265203.

[17] J.A. Gordon, R.W. Ziolkowski, Colors generated by tunable plasmon resonances and their potential application to ambiently illuminated color displays, Solid State Commun 146 (5–6) (2008) 228–238.

[18] M. Khorasaninejad, S. Mohsen Raeis-Zadeh, H. Amarloo, N. Abedzadeh, S. Safavi-Naeini, S.S. Saini, Colorimetric sensors using nano-patch surface plasmon resonators, Nanotechnology 24 (35) (2013) 355501.

[19] N.S. King, L. Liu, X. Yang, B. Cerjan, H.O. Everitt, P. Nordlander, et al., Fano resonant aluminum nanoclusters for plasmonic colorimetric sensing, ACS Nano 9 (11) (2015) 10628–10636.

[20] K.-T. Lee, J.Y. Lee, S. Seo, L.J. Guo, Colored ultrathin hybrid photovoltaics with high quantum efficiency, Light Sci Appl 3 (10) (2014) e215.

[21] Y. Cui, R.S. Hegde, I.Y. Phang, H.K. Lee, X.Y. Ling, Encoding molecular information in plasmonic nanostructures for anti-counterfeiting applications, Nanoscale 6 (1) (2014) 282–288.

[22] K. Kumar, H. Duan, R.S. Hegde, S.C.W. Koh, J.N. Wei, J.K.W. Yang, Printing colour at the optical diffraction limit, Nat Nanotechnol 7 (9) (2012) 557–561.

[23] Y. Cui, I.Y. Phang, R.S. Hegde, Y.H. Lee, X.Y. Ling, Plasmonic silver nanowire structures for two-dimensional multiple-digit molecular data storage application, ACS Photonics 1 (7) (2014) 631–637.

[24] M.L. Brongersma, N.J. Halas, P. Nordlander, Plasmon-induced hot carrier science and technology, Nat Nanotechnol 10 (1) (2015) 25–34.

[25] B.J.C. Ndukaife, M. Vladimir, Plasmonics—turning loss into gain, Science 351 (6271) (2016) 334–335.

[26] J.B. Khurgin, How to deal with the loss in plasmonics and metamaterials, Nat Nanotechnol 10 (1) (2015) 2–6.
[27] T.V. Shahbazyan, Landau damping of surface plasmons in metal nanostructures, Phys Rev B 94 (2015) 235431.
[28] P. Narang, R. Sundararaman, H.A. Atwater, Plasmonic hot carrier dynamics in solid-state and chemical systems for energy conversion, Nanophotonics 5 (1) (2016) 96–111.
[29] P. Christopher, M. Moskovits, Hot charge carrier transmission from plasmonic nanostructures, Annu Rev Phys Chem 68 (2017) 379–398.
[30] R. St-gelais, G.R. Bhatt, L. Zhu, S. Fan, M. Lipson, Hot carrier-based near-field thermophotovoltaic energy conversion, ACS Nano 11 (3) (2017) 3001–3009.
[31] E. Cortes, W. Xie, J. Cambiasso, A.S. Jermyn, R. Sundararaman, P. Narang, et al., Plasmonic hot electron transport drives nano-localized chemistry, Nat Commun 8 (14880) (2017) 1–10.
[32] A.J. Leenheer, P. Narang, N.S. Lewis, H.A. Atwater, Solar energy conversion via hot electron internal photoemission in metallic nanostructures : efficiency estimates, J Appl Phys 115 (2017) (2014) 134301.
[33] S. Mukherjee, F. Libisch, N. Large, O. Neumann, L.V. Brown, J. Cheng, et al., Hot electrons do the impossible: plasmon-induced dissociation of H_2 on Au, Nano Lett 13 (1) (2013) 240–247.
[34] C. Clavero, Plasmon-induced hot-electron generation at nanoparticle/metal-oxide interfaces for photovoltaic and photocatalytic devices, Nat Photonics 8 (2) (2014) 95–103.
[35] B. Hou, L. Shen, H. Shi, R. Kapadia, S.B. Cronin, Hot electron-driven photocatalytic water splitting, Phys Chem Chem Phys 19 (2017) 2877–2881.
[36] J.Y. Park, S.M. Kim, H. Lee, I.I. Nedrygailov, Hot-electron-mediated surface chemistry: toward electronic control of catalytic activity, Acc Chem Res 48 (8) (2015) 2475–2483.
[37] H. Duan, A.I. Fernandez-Dominguez, M. Bosman, S.A. Maier, J.K.W. Yang, Nanoplasmonics: classical down to the nanometer scale, Nano Lett 12 (3) (2012) 1683–1689.
[38] G. Mie, Beiträge zur Optik trüber Medien, speziell kolloidaler Metallösungen, Ann Phys 330 (3) (1908) 377–445.
[39] B. Gallinet, J. Butet, O.J.F. Martin, Numerical methods for nanophotonics: standard problems and future challenges, Laser Photonics Rev 9 (6) (2015) 577–603.
[40] H. Fischer, O.J.F. Martin, Engineering the optical response of plasmonic nanoantennas, Opt Express 16 (12) (2008) 9144–9154.
[41] B. Gallinet, O.J.F. Martin, Influence of electromagnetic interactions on the line shape of plasmonic fano resonances, ACS Nano 5 (11) (2011) 8999–9008.
[42] J. Becker, A. Trugler, A. Jakab, U. Hohenester, C. Sonnichsen, The optimal aspect ratio of gold nanorods for plasmonic bio-sensing, Plasmonics 5 (2) (2010) 161–167.
[43] R.S. Hegde, M. Mesch, H. Giessen, Quasistatic perturbation theory for the optimal design of plasmon induced transparency based sensors, in: META'14 Fifth International Conference on Metamaterials, Photonic Crystals and Plasmonics, Singapore, 2014, pp. 1–2.
[44] V. Myroshnychenko, J. Rodríguez-Fernández, I. Pastoriza-Santos, A.M. Funston, C. Novo, P. Mulvaney, et al., Modelling the optical response of gold nanoparticles, Chem Soc Rev 37 (9) (2008) 1792–1805.
[45] M.W. Knight, Y. Wu, J.B. Lassiter, P. Nordlander, N.J. Halas, Substrates matter: influence of an adjacent dielectric on an individual plasmonic nanoparticle, Nano Lett 9 (5) (2009) 2188–2192.
[46] H. Chen, T. Ming, S. Zhang, Z. Jin, B. Yang, J. Wang, Effect of the dielectric properties of substrates on the scattering patterns of gold nanorods, ACS Nano 5 (6) (2011) 4865–4877.
[47] F. Javier Garcia de Abajo, J. Nelayah, M. Kociak, O. Stephan, I. Pastoriza-Santos, M. Tence, et al., Mapping surface plasmons on a single metallic nanoparticle, Nat. Phys. 3 (2007) 348–353.
[48] M. Bernardi, J. Mustafa, J.B. Neaton, S.G. Louie, Theory and computation of hot carriers generated by surface plasmon polaritons in noble metals, Nat Commun 6 (2015) 1–9.
[49] A.M. Brown, R. Sundararaman, P. Narang, W.A. Goddard, H.A. Atwater, Nonradiative plasmon decay and hot carrier dynamics: effects of phonons, surfaces, and geometry, ACS Nano 10 (2016) 957–966.
[50] J.R.M. Saavedra, A. Asenjo-garcia, F.J. García de Abajo, Hot-electron dynamics and thermalization in small metallic nanoparticles, ACS Photonics 3 (2016) 1637–1646.

[51] T. Gong, J.N. Munday, Materials for hot carrier plasmonics [invited], Opt Mater Express 5 (11) (2015) 56–62.
[52] F. Ladstädter, U. Hohenester, P. Puschnig, C. Ambrosch-draxl, Firstprinciples calculation of hot-electron scattering in metals, Phys Rev B 70 (2004). 235125-10.
[53] M. Merschdorf, C. Kennerknecht, W. Pfeiffer, Collective and single-particle dynamics in time-resolved two-photon photoemission, Phys Rev B 70 (2004) 193401–193404. August.
[54] T.P. White, K.R. Catchpole, Plasmon-enhanced internal photoemission for photovoltaics: theoretical efficiency limits, Appl Phys Lett 101 (7) (2012) 073905.
[55] S.K. Cushing, A.D. Bristow, N. Wu, Theoretical maximum efficiency of solar energy conversion in plasmonic metal semiconductor, Phys Chem Chem Phys 17 (2015) 30013–30022.
[56] S.V. Boriskina, J. Zhou, W.-C. Hsu, B. Liao, G. Chen, Limiting efficiencies of solar energy conversion and photo-detection via internal emission of hot electrons and hot holes in gold, Arxiv (2015) 1–4 (arXiv:1509.02423).
[57] P. Reineck, D. Brick, P. Mulvaney, U. Bach, Plasmonic hot electron solar cells: the effect of nanoparticle size on quantum efficiency, J Phys Chem Lett 7 (2016) 4137–4141.
[58] Y. Zhang, C. Yam, G.C. Schatz, Fundamental limitations to plasmonic hot-carrier solar cells, J Phys Chem Lett 7 (2016) 1852–1858.
[59] Y.-J. Chang, K.-H. Shih, Y.-J. Chang, K.-H. Shih, Solar energy conversion via internal photoemission in aluminum, copper, and silver: band structure effects and theoretical efficiency estimates, J Appl Phys 119 (2016) 183101.
[60] A.O. Govorov, H. Zhang, Y.K. Gunko, Theory of photoinjection of hot plasmonic carriers from metal nanostructures into semiconductors and surface molecules, J Phys Chem C 117 (2013) 16616–16631.
[61] W. Li, J.G. Valentine, Harvesting the loss: surface plasmon-based hot electron photodetection, Nanophotonics 6 (1) (2017) 177–191.
[62] H. Chen, H. Liu, Z. Zhang, K. Hu, X. Fang, Nanostructured photodetectors: from ultraviolet to terahertz, Adv Mater 28 (2015) 403–433.
[63] A. Soni, S. Purohit, R.S. Hegde, Multilayered aluminum plasmonic metasurfaces for ultraviolet bandpass filtering, IEEE Photon Technol Lett 29 (1) (2017) 110–113.
[64] P.C. Eng, S. Song, B. Ping, State-of-the-art photodetectors for optoelectronic integration at telecommunication wavelength, Nanophotonics 4 (2015) 277–302.
[65] M. Casalino, Recent Advances in Silicon Photodetectors Based on the Internal Photoemission Effect, in: V. I. Talanin (Ed.), New Research on Silicon Structure, Propertie, Technology, InTech, 2017. Ch. 11.
[66] Z. Jakšić, M. Obradov, S. Vukovi, M. Beli, Plasmonic enhancement of light trapping, facta universitatis, Ser Electron Energ 27 (2) (2014) 183–203.
[67] T. Xu, H. Shi, Y.K. Wu, A.F. Kaplan, J.G. Ok, L.J. Guo, Structural colors: from plasmonic to carbon nanostructures, Small 7 (22) (2011) 3128–3136.
[68] M.W. Knight, H. Sobhani, P. Nordlander, N.J. Halas, Photodetection with active optical antennas, Science 332 (2011) 702–704.
[69] E. Panchenko, J.J. Cadusch, T.D. James, A. Roberts, Plasmonic metasurface-enabled differential photodetectors for broadband optical polarization characterization, ACS Photonics 2 (2016) 1833–1839.
[70] H. Chalabi, D. Schoen, M.L. Brongersma, Hot-electron photodetection with a plasmonic nanostripe antenna, Nano Lett 14 (2014) 1374–1380.
[71] A. Sobhani, M.W. Knight, Y. Wang, B. Zheng, N.S. King, L.V. Brown, et al., Narrowband photodetection in the near-infrared with a plasmon-induced hot electron device, Nat Commun 4 (2013) 1643.
[72] W. Li, J. Valentine, Metamaterial perfect absorber based hot electron photodetection, Nano Lett 14 (6) (2014) 3510–3514.
[73] B. Desiatov, I. Goykhman, N. Mazurski, J. Shappir, U. Levy, Plasmonic enhanced schottky detectors based on internal photoemission in nano pyramids for near ir regime, in: 2015 International Conference on Optical MEMS and Nanophotonics (OMN), Jerusalem, Israel, 2015.
[74] M.A. Nazirzadeh, F.B. Atar, B.B. Turgut, A.K. Okyay, Random sized plasmonic nanoantennas on silicon for low-cost broad-band near-infrared photodetection, Sci Rep 4 (7103) (2014) 1–5.

[75] F. Wang, N.A. Melosh, Plasmonic energy collection through hot carrier extraction, Nano Lett 11 (2011) 5426–5430.

[76] K. Wu, Y. Zhan, C. Zhang, S. Wu, X. Li, Strong and highly asymmetrical optical absorption in conformal grating system for plasmonic hot-electron photodetection application, Sci Rep 5 (2015) 14304.

[77] Y. Zhan, K. Wu, C. Zhang, S. Wu, Xiaofeng Li, Infrared hot-carrier photodetection based on planar perfect absorber, Opt Lett 40 (18) (2015) 4261–4264.

[78] R. Li, C. Zhang, X. Li, R. Li, C. Zhang, X. Li, Schottky hot-electron photodetector by cavity-enhanced optical Tamm resonance, Appl Phys Lett 110 (013902) (2017) 1–6.

[79] S. Muehlbrandt, A. Melikyan, T. Harter, K. Kohnle, A. Muslija, P. Vincze, et al., Silicon-plasmonic internal-photoemission detector for 40 Gbit/s data reception, Optica 3 (7) (2016) 741–747.

[80] M. Alavirad, A. Olivieri, L. Roy, P. Berini, High-responsivity sub-bandgap hot-hole plasmonic Schottky detectors, Opt Express 24 (20) (2016) 22544–22554.

[81] Z. Han, I.P. Radko, N. Mazurski, B. Desiatov, J. Beermann, O. Albrektsen, et al., On-chip detection of radiation guided by dielectric-loaded plasmonic waveguides, Nano Lett 15 (2014) 476–480.

[82] S. Ishii, S.-I. Inoue, R. Ueda, A. Otomo, Optical detection in a waveguide geometry with a single metallic contact, ACS Photonics 1 (2014) 1089–1092.

[83] L. Yang, P. Kou, J. Shen, E.H. Lee, S. He, Proposal of a broadband, polarization-insensitive and high-efficiency hot-carrier Schottky photodetector integrated with a plasmonic silicon ridge waveguide, J Opt 17 (12) (2015) 125010.

[84] S. Linic, U. Aslam, C. Boerigter, M. Morabito, Photochemical transformations on plasmonic metal nanoparticles, Nat Mater 14 (2015) 567–576.

[85] C. Boerigter, U. Aslam, S. Linic, Mechanism of charge transfer from plasmonic nanostructures to chemically attached, ACS Nano 10 (2016) 6108–6115.

[86] B. Foerster, A. Joplin, K. Kaefer, S. Celiksoy, S. Link, C. Sonnichsen, Chemical interface damping depends on electrons reaching the surface, ACS Nano 11 (2017) 2886–2893.

[87] K. Wu, J. Chen, J.R. Mcbride, T. Lian, Efficient hot-electron transfer by a plasmon-induced interfacial charge-transfer transition, Science 349 (6248) (2015) 632–635.

[88] P. Zijlstra, P.M.R. Paulo, K. Yu, Q.-H. Xu, M. Orrit, Chemical interface damping in single gold nanorods and its near elimination by tip-specific functionalization, Angew Commun 51 (2012) 8352–8355.

[89] D.N. Denzler, C. Frischkorn, C. Hess, M. Wolf, G. Ertl, Electronic excitation and dynamic promotion of a surface reaction, Phys Rev lett 91 (22) (2003) 226102.

[90] C. Zhang, H. Zhao, L. Zhou, A.E. Schlather, L. Dong, M.J. Mcclain, et al., Al Pd nanodisk heterodimers as antenna reactor photocatalysts, Nano Lett 16 (2016) 6677–6682.

[91] P. Christopher, H. Xin, S. Linic, Visible-light-enhanced catalytic oxidation reactions on plasmonic silver nanostructures, Nat Chem 3 (6) (2011) 467–472.

[92] D.B. Ingram, S. Linic, Photoelectrodes: evidence for selective plasmon induced formation, J Am Chem Soc 133 (2011) 5202–5205.

[93] S. Mubeen, J. Lee, N. Singh, S. Krämer, G.D. Stucky, M. Moskovits, An autonomous photosynthetic device in which all charge carriers derive from surface plasmons, Nat. Nanotechnol. 8 (4) (2013) 247–251.

[94] Z. Zheng, T. Tachikawa, T. Majima, Single-particle study of Pt-modified Au nanorods for plasmon-enhanced hydrogen generation in visible to near-infrared region, J. Am. Chem. Soc. 136 (2014) 6870–6873.

[95] S. Tian, F. Berg, H.O. Everitt, E.A. Carter, P. Nordlander, Aluminum nanocrystals as a plasmonic photocatalyst for hydrogen dissociation, Nano Lett. 16 (2016) 1478–1484.

[96] F. Wang, C. Li, H. Chen, R. Jiang, L.-D. Sun, Q. Li, et al., Plasmonic harvesting of light energy for Suzuki coupling reactions, J. Am. Chem. Soc. 135 (2013) 5588–5601.

[97] H. Robatjazi, H. Zhao, D.F. Swearer, N.J. Hogan, L. Zhou, A. Alabastri, et al., Plasmon-induced selective carbon dioxide conversion on earth-abundant aluminum-cuprous oxide antenna-reactor nanoparticles, Nat. Commun. 8 (27) (2017) 1–9.

[98] L. Krayer, J.N. Munday, Design concepts for hot carrier-based detectors and energy converters in the near ultraviolet and infrared, SPIE J Photonics Energy 6 (4) (2016) 042510.

[99] A.O. Govorov, H. Zhang, H.V. Demir, Y.K. Gun, Photogeneration of hot plasmonic electrons with metal nanocrystals: quantum description and potential applications, Nano Today 9 (1) (2014) 85–101.

[100] W. Wang, A. Klots, D. Prasai, Y. Yang, K.I. Bolotin, J. Valentine, Hot electron-based near-infrared photodetection using bilayer MoS2, Nano Lett. 15 (11) (2015) 7440–7444.

[101] U. Levy, M. Grajower, P.A.D. Gonçalves, N.A. Mortensen, J.B. Khurgin, U. Levy, et al., Plasmonic silicon Schottky photodetectors: the physics behind graphene enhanced internal photoemission, APL Photonics 2 (2017) 026103.

[102] Z. Fang, Z. Liu, Y. Wang, P.M. Ajayan, P. Nordlander, Graphene-antenna sandwich photodetector, Nano Lett. 12 (2012) 3808–3813.

[103] Z. Chen, X. Li, J. Wang, L. Tao, M. Long, S.-J. Liang, et al., Synergistic effects of plasmonics and electron trapping in graphene short-wave infrared photodetectors with ultrahigh responsivity, ACS Nano 11 (2017) 430–437.

CHAPTER 10

Plasmonic Metamaterial-Based RF-THz Integrated Circuits: Design and Analysis

Rahul Kumar Jaiswal, Nidhi Pandit and Nagendra Prasad Pathak

RF to THz Integrated Circuits, System & Sensors Laboratory, Department of Electronics and Communication Engineering, Indian Institute of Technology, Roorkee, Uttarakhand, India

10.1 Introduction

Semiconductor-based microelectronics and integrated circuits have been well developed and explored. This builds a strong building block for various applications, mainly in case of the information data processing and transmission of data. Since demand and development in the field of information transmission is increasing day by day, performance of electronic components is always limited because of less data supported by electrons and due to quantum limit provided by the Moore's law that predicts the doubling of transistor density in a particular area every 18 months. Henceforth bounding the miniaturization of the electronic circuitry. Additionally, other challenges like thermal dissipation, thermal noise, and leakage current necessitate the replacement of electronic circuitry by alternative sources [1–3]. A solution for these problems is provided by the use of photonics circuit, which use photons as an information carrier. For science and engineering background people, this came in reality when Charles K. Kao made successful use of optical fiber with large amount of information in the form of light, photons used for communication due to larger information capacity [4]. Unfortunately, as compared to the electronic component, photonic components like optical fiber are larger in size (due to the diffraction limit of light that actually requires that the optical waveguide wider dimension should be equal to at least half of the wavelength of the traveling light in the waveguide).

So there is a clear-cut tradeoff between circuit performance and miniaturization. Recently, solution for this tradeoff limitation has been provided that can be overcome by the surface plasmon-based photonics (SPPs) technique or plasmonics technique [5].

In the process of communication, signals may take several forms. For an instance, first, base information signals that is generated in the electrical form in an electronic circuit is converted to guided transmission waves before transmitting it as free electromagnetic (EM)

Nanoelectronics. DOI: https://doi.org/10.1016/B978-0-12-813353-8.00009-9

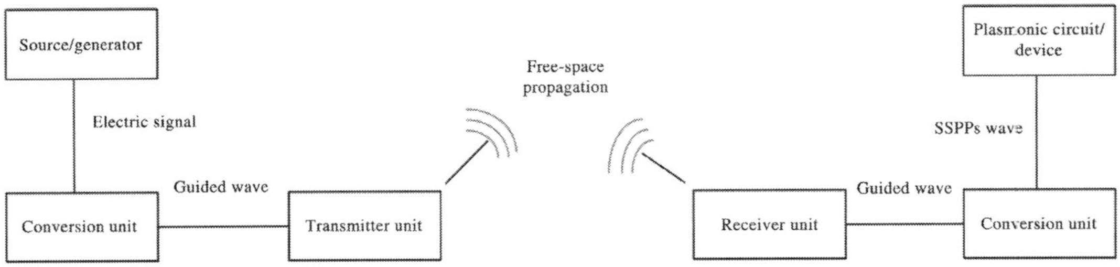

Figure 10.1
Conversion of signal during transmission.

propagating waves by a transmitting antenna, then receiver captures the signals as guided EM waves, and then conversion units may transform it into spoof surface plasmon polaritons (SSPPs) if necessary as shown in Fig. 10.1. The conversion is important because of instrument used in the microwave frequency regime that normally distribute the signal energies as guided waves along transmission lines. Hence, the conversion from guided microwave frequency waves to SPPs and vice versa would allow better integration of SSPPs in the conventional microwave circuitry [6–17].

10.1.1 Surface Plasmon Polaritons

The study of optical phenomena related to the EM response of metal is called as plasmonics, and it may involve the control of optical radiation on subwavelength scale that is manipulation of surface plasmon polaritons (SPPs) wave, which is a collective oscillations of surface conduction electrons of a metal which interact with the incident EM fields. Surface plasmon polaritons are unique EM waves found at the optical and infrared frequency regime that bound on the metal–dielectric interface and propagate parallel to the interface and the amplitude of this wave decays exponentially in the direction perpendicular to the interface because of negative permittivity behavior of metals. Since the confinement of SPPs occurs at the metal–dielectric interface (Fig. 10.2), therefore subwavelength design of circuitry could be realized [1].

In the metal, availability of large number of electrons gives rise to special dielectric response which is described by Drude model as given in Eq. (10.1)

$$\varepsilon_m = 1 - \left(\frac{\omega_p^2}{\omega^2 + j\gamma\omega}\right) \tag{10.1}$$

where ω_p is the plasma frequency and γ is the scattering rate of electron.

It is interesting that whenever dielectric response of metal is negative, light is allowed to transmit in the form of surface wave, which is called as SPP as discussed above. The EM

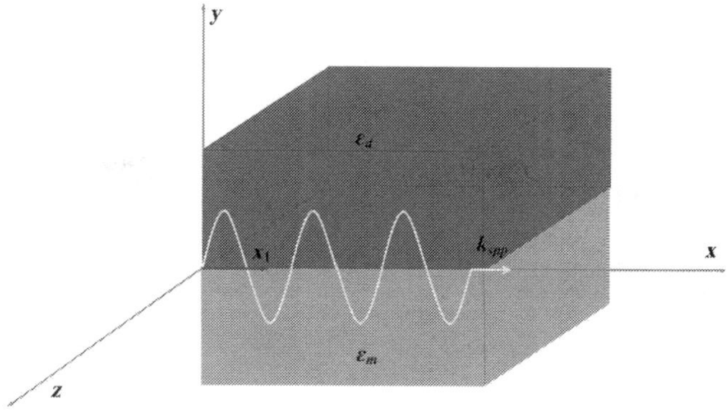

Figure 10.2
SPP wave at the interface of two materials (dielectric–metal).

field of an SPP, which interacts with the surface conduction electron of the metal at a metal–dielectric interface, is achieved from the solution of Maxwell's equations and associated boundary conditions in each medium. To derive the characteristics of the SPP mode, a metal–dielectric interface has been configured as shown in Fig. 10.2.

The solutions of Maxwell's equations in medium that have wave-like nature in the x_1-direction and whose amplitudes decay exponentially with the increasing distance into mediums from the interface $y = 0$ can be written as follows: [1]

Region $y > 0$,

$$H(x,t) = (0, C, 0)\ e^{jkx_1 - k_3(1)y - j\omega t} \tag{10.2}$$

$$E(x,t) = -C\frac{c}{j\omega\varepsilon_d}(k_3^{(1)}, 0, jk)\ e^{jkx_1 - k_3(1)y - j\omega t}. \tag{10.3}$$

Region $y < 0$

$$H(x,t) = (0, D, 0)\ e^{jkx_1 - k_3(m)y - j\omega t} \tag{10.4}$$

$$E(x,t) = -D\frac{c}{j\omega\varepsilon_m}(k_3^{(m)}, 0, jk)\ e^{jkx_1 - k_3(m)y - j\omega t} \tag{10.5}$$

${K_3}^{(1,m)}$ determine the decay of the EM field with increasing distance from the surface.

$$k_3^{(1)} = \left(k^2 - \varepsilon_d(\omega/c)^2\right)^{1/2} \tag{10.6}$$

$$k_3^{(m)} = \left(k^2 - \varepsilon_m(\omega/c)^2\right)^{1/2}. \tag{10.7}$$

Real parts of the ${k_3}^{(1,m)}$ must be positive so that Eqs. 10.2–10.5 describe an EM wave localized to the dielectric–metal interface at $y = 0$.

The boundary condition at the plane $y = 0$ yields a pair of equations

$$C = D, \; C\frac{k_3^{(1)}}{\varepsilon_d} = -D\frac{k_3^{(m)}}{\varepsilon_m}. \tag{10.8}$$

Condition that this system of equations has nontrivial solutions provide the dispersion relation connecting the frequency ω of the TM polarized wave and its wave number is given as

$$\frac{k_3^{(m)}}{k_3^{(1)}} = -\frac{\varepsilon_m}{\varepsilon_d}. \tag{10.9}$$

If, dielectric function response of metal ε_m is assumed to be real then in order that Eqs. 10.2–10.5 describe the surface EM wave, the ${k_3}^{(1,m)}$ must be positive and real. Hence to fulfill this purpose ε_m of metal must be negative for the existence of surface wave. The surface EM wave whose field vectors are given by Eqs. 10.1–10.4 with $C = D$ and whose frequency is calculated from the equation of dispersion relation (10.9) is known as SPPs, i.e., photons which are coupled to collective excitation of surface conduction electrons near a metal surface yield the following dispersion relation [1]:

$$k_{spp} = \frac{\omega}{c}\sqrt{\frac{\varepsilon_m \varepsilon_d}{\varepsilon_m + \varepsilon_d}}. \tag{10.10}$$

Here, ε_d is the permittivity of dielectric. To make k_{spp} real, $\varepsilon_m + \varepsilon_d < 0$, and thus $\omega < \omega_{sp}$ surface plasmon frequency and equals to$\omega_p/\sqrt{1 + \varepsilon_d}$. This is the frequency range for the existence of this surface mode.

But this exotic feature of SPP is found only at optical frequencies i.e., in the visible and near-IR field, and it is not supported at terahertz (THz) and microwave frequencies regime. Hence, to find the similar characteristics at lower frequency regimes, scientific community has find the solution in terms of plasmonic metamaterial that is spoof or designer SPP.

10.1.2 Spoof Surface Plasmon Polaritons

Spoof surface plasmons are bound EM waves at frequencies outside the plasmonic range mimicking ("spoofing") SPPs, which propagate on the periodically corrugated metal surfaces. The first experimental demonstration of the spoof plasmons was done using periodically arranged hollow brass tubes. As we know that metal surfaces support EM surface wave in a wide range from the microwave to visible frequencies. At visible regimes, the surface wave, which is referred as SPPs, can be tightly confined within distances of the order of wavelength in the dielectric and this is much less than the wavelength in metal. But at microwave and THz frequencies, SPPs surface wave cannot be supported by simple metal surface structure, which means SPP mode cannot be excited and supported in the microwave or THz wavelength at the metal–dielectric interface since metal behaves like perfect electric conductor (PEC) at these frequencies. Recently, researches have shown that metal surfaces drilled with subwavelength holes and grooves produce field confinement in the microwave and THz frequency. Further, to engineer surface plasmon at lower frequencies, Pendry et al. presented a corrugated plasmonic metamaterial that supports SPP-like bound surface mode called as spoof or designer SPPs. They analyze two cases: one-dimensional arrays of grooves and two-dimensional arrays of holes. In both cases, they find that these structures support surface bound states and that the dispersions of these modes have strong similarities with the dispersion of the SPP modes of real metals. The main advantage of this metamaterial is that the dispersion can be tuned at will by decorating the metal surfaces with subwavelength holes or grooves which provide an effective method to guide and manipulate EM waves in the form of SSPPs. Pendry and co-workers have shown that bound EM surface waves mimicking SPPs can be sustained even by a perfect conductor provided that its surface is periodically corrugated. If the size and spacing of the corrugations are much smaller than the wavelength λ_0, the photonic response of the surface can be described by an effective medium dielectric function $\varepsilon(\omega)$ of the plasma form, with ω_p determined by the geometry. Thus, the dispersion relation of the surface mode can be engineered via the geometry of the surface, allowing tailoring to particular frequencies. In the effective medium model, the establishment of surface waves can be physically understood by realizing that the surface modulations allow for an average finite field penetration into the effective surface layer, akin to the field penetration into real metals at visible frequencies leading to the formation of confined SPPs. A material with subwavelength structure exhibiting such an effective photonic response is also known as a metamaterial. The size and spacing of holes can be controlled, which allows the creation of designer surface plasmons with almost arbitrary dispersion in frequency and space [18–27].

Under the PEC condition, the dispersion equation of corrugated metal structure can be described as follows [23]:

$$k_x^2 - k_0^2 = \left(\frac{d^2}{p^2}\right) k_0^2 \tan^2(k_0 h). \tag{10.11}$$

Eq. (10.11) represents the relationship between propagation constant with the geometrical parameter of plasmonic metamaterial which is spoof SPP. It is observed that as the groove height increases, propagation constant of SSPP becomes larger than freely propagation constant k_0 which means SSPP wave travel slowly than light but confined more tightly along the corrugated metal surface.

10.2 Unit Cell Design and Dispersion Analysis

To study and demonstrate the behavior of plasmonic metamaterial, unit cell is designed and analysis of its dispersion curve is done. The designed spoof SPP strip is much like

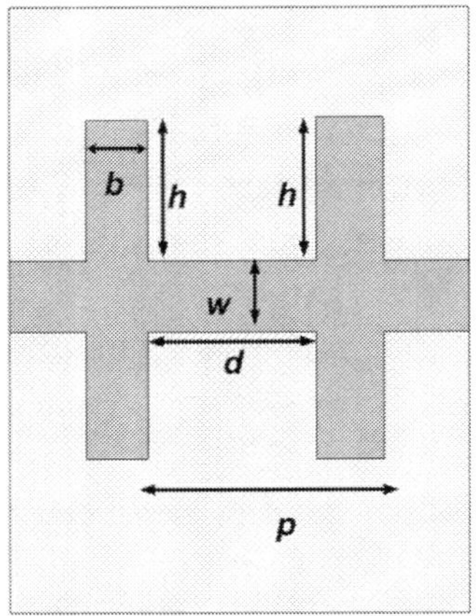

Figure 10.3
Schematic of unit cell (SSPP strip).

microstrip transmission line except that the signal line is changed to periodical grooves strip as shown in Fig. 10.3. The corrugation in metal surface, which is deposited on dielectric substrate, can be one sided and double sided. For symmetrical transmission, the SSPP strip is designed with double-sided groove. Since plasmonic metamaterials are geometry dependent hence guided-mode characteristics can be controlled by geometry parameter like period p, width d, depth h, and thickness t of metal. The simple corrugated metal rectangular groove unit cell is shown in Fig. 10.3 with their physical parameter.
The numerical simulation of dispersion curve has been performed using commercially available CST microwave studio. The design of unit cell and its dispersion is analyzed from RF to THz range of frequency.

10.2.1 Design and Analysis at Terahertz Frequency Regime

Figs. 10.4 and 10.5 depict the dispersion relation for the designed unit cell with varying the different parameters of unit cell at THz frequency. Fig. 10.4 illustrates the dispersion relation for fix depth h and varying gap d. To evaluate the impact of d on dispersion characteristics and performance device, the dispersion relation for structure with different d has been calculated. As gap d increases from 10 to 40 μm with fixed depth h equals to 40 μm, the dispersion curve deviate from the freely propagating wave vector line i.e., light line and thus cutoff frequency decreases significantly.

To study the effect of depth h on the performance of unit cell with different d, dispersion is calculated with varying h from 10 to 40 μm and d from 10 to 30 μm and shown in Fig. 10.5. The green color line is for freely propagating wave vector i.e., light line that is a line without any grooves.

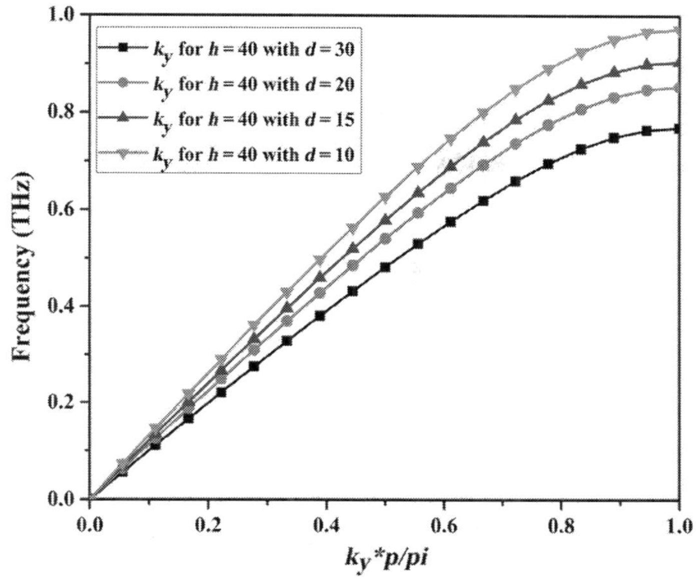

Figure 10.4
Dispersion diagram relating frequency and propagating wave vector for unit cell as shown in Fig. 10.3 with fix height $h = 40$ μm and varying d with 30, 20, 15, and 10 μm.

It is observed that when depth h is varied from 10 to 40 μm, the dispersion curve deviates from light line and cutoff frequency becomes lower. When the wave frequency is close to the cutoff frequency, the EM waves can be coupled into the grooves and confined at the surface, owing to the large mismatch between the wave vector of SPPs and that of the free-space waves. On the other hand, as frequency decreases, the dispersion curves are shifted closer to the light line, indicating weaker surface confinement of the propagating waves.

The cutoff frequency of designed unit cell, as shown in Fig. 10.3, can be brought to lower frequency i.e., at mm-wave from THz wave by just scaling up the dimensions of unit cell. It can be observed from Fig. 10.6 that the cutoff frequency of unit cell is nearly 65 GHz which is mm-wave frequency by scaling dimension up by a factor of 10 in THz frequency.

10.2.2 Design and Analysis at Microwave Frequency

Since in the lower frequency regimes of EM spectra i.e., at THz and microwave frequency bands, metals behave like perfect conductor (PEC). The scale invariance of classical electromagnetism indicates that the planar plasmonic metamaterial can be scaled down to the microwave frequency. By scaling the dimensions up 100 times at THz frequency, excellent propagation characteristics of SPP at microwave frequency are observed. Here, various designs of unit cell have been explored and analyzed by adding stubs in the

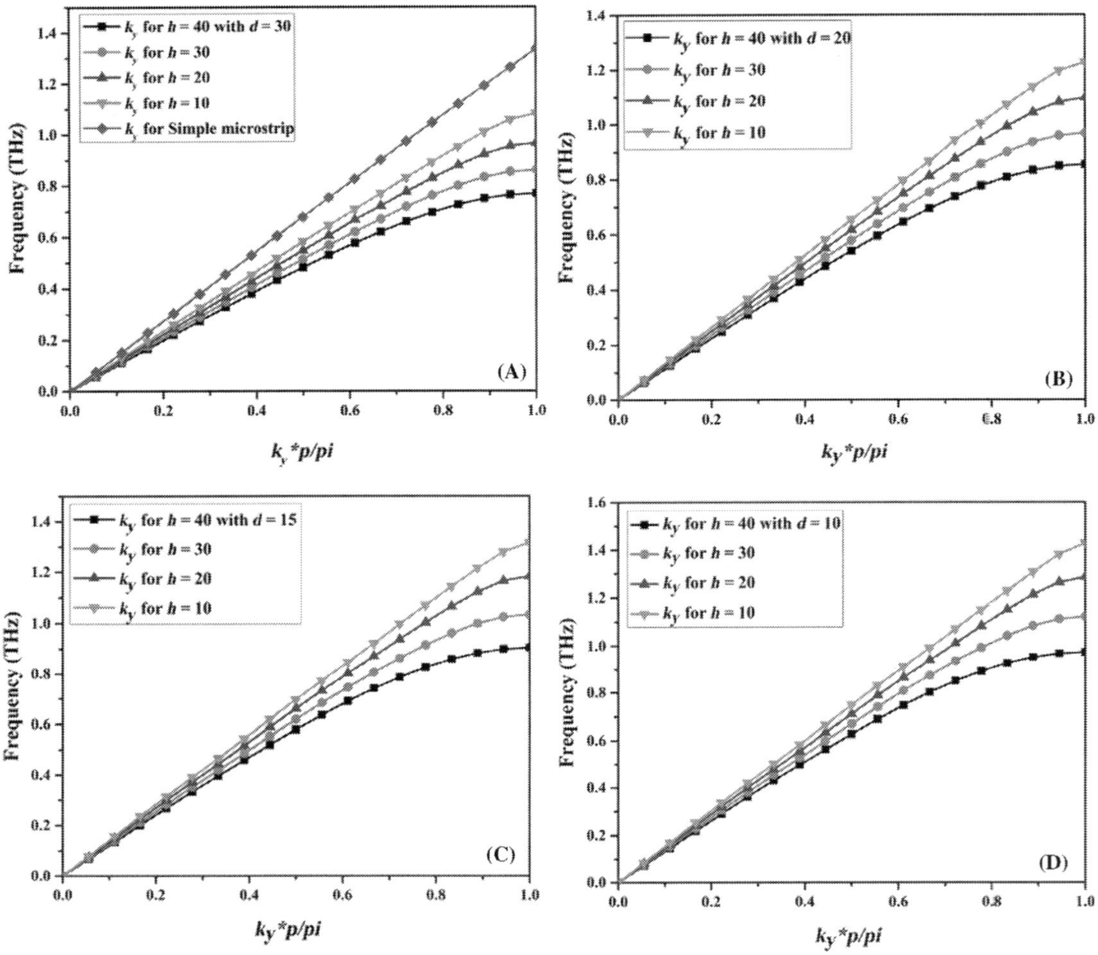

Figure 10.5

Dispersion diagram relating frequency and propagating wave vector for unit cell as shown in Fig. 10.3 with fix period d and varying height h with 30, 20, 15, and 10 μm. (A) $d = 30$ μm, (B) $d = 20$ μm, (C) $d = 15$ μm, and (D) $d = 10$ μm.

conventional unit cell. Fig. 10.7 shows the different types of unit cell design and its dispersion behavior analysis has been shown in Fig. 10.8. By analyzing the dispersion curve, it has been seen that the dispersion characteristics and cutoff frequencies are geometry dependent and can be tuned according to our requirements. The dispersion relation for stub-loaded geometry is lower than the conventional geometry. Also the propagation constant for these geometry is higher and hence results in higher confinement of EM field (Fig. 10.8).

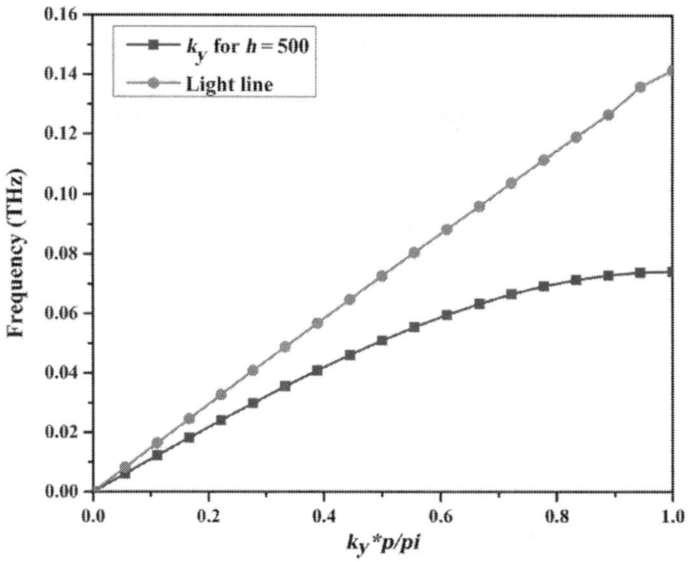

Figure 10.6
Dispersion diagram relating frequency and propagating wave vector for unit cell as shown in Fig. 10.3 at mm-wave frequency.

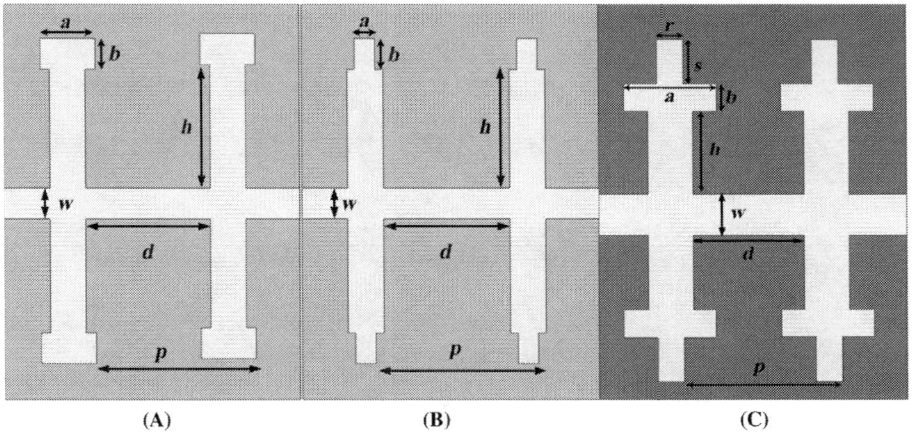

Figure 10.7
Schematic of unit cell (SSPP strip) with different shapes added to conventional unit cell. (A) stub type 1, (B) stub type 2, and (C) stub type 3.

Fig. 10.9 shows the geometry which is of different shapes than conventional rectangular one. Design of unit cell with trapezoidal shape has been explored and analyzed. Fig. 10.9A is conventional trapezoidal shape unit cell and Fig. 10.9B shows trapezoidal shape groove loaded stepped impedance unit cell. Fig. 10.10 shows that dispersion relation comparing the

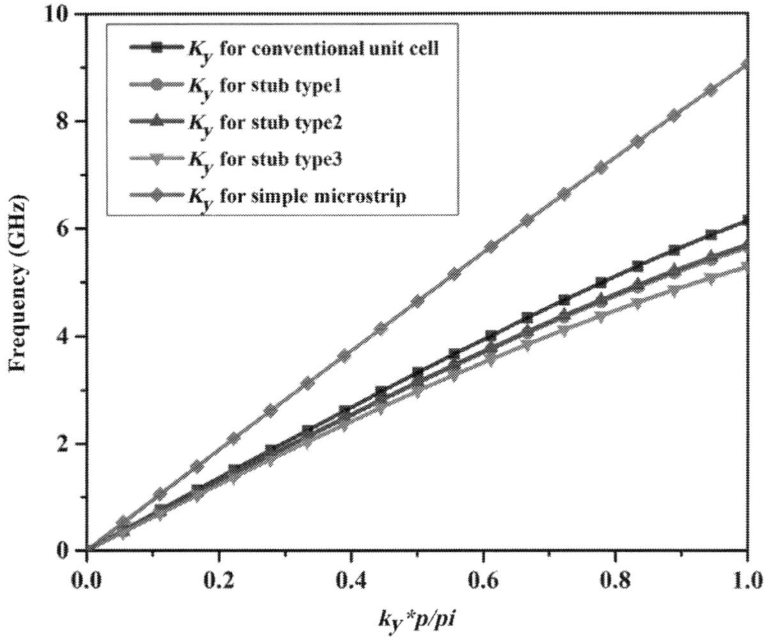

Figure 10.8
Dispersion diagram for unit cell given in Fig. 10.3 and 10.7 at microwave frequency.

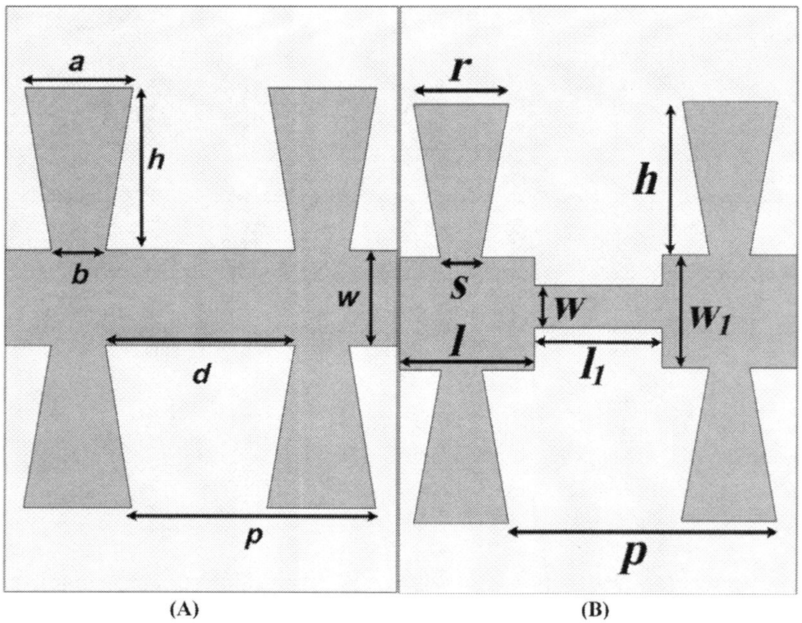

Figure 10.9
Schematic of unit cell (SSPP strip) with (A) conventional trapezoidal shape and (B) stepped stub trapezoidal shape.

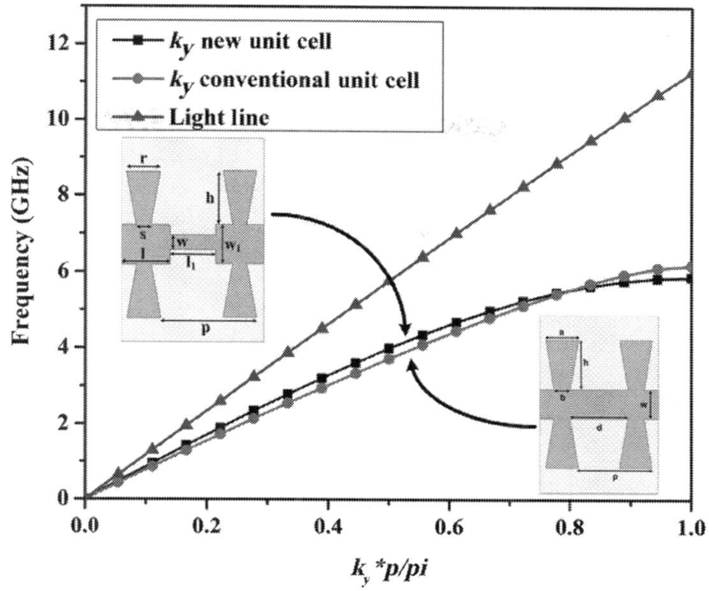

Figure 10.10
Dispersion diagram for unit cell given in Fig. 10.9.

both conventional and stepped type unit cell. The dispersion for unit cell (Fig. 10.9B) deviate more than the conventional (Fig. 10.9A) thus propagation constant is higher and cutoff frequency is lower than the conventional, hence EM field is more confined in case of stepped type unit cell. At lower frequency, the propagation constant is similar but as the frequency increases the propagation constant for SSPP strip becomes higher and deviate from the light line. Dimensions used at microwave frequency are as follows: $w = 2$ mm, $p = 6$ mm, $d = 4$ mm and $h = 4$ mm, $b = 1$ mm and $a = 0.5$ mm.

To design this type of unit cell at microwave frequency, dielectric substrates with dielectric constant of 3.38, loss tangent 0.0016, and height 1.524 mm with both sides copper layer of thickness 0.035 mm are used

10.2.3 Conversion and Momentum Matching (at Microwave, mm Wave and THz Frequencies)

From the dispersion curve discussed above at THz, mm wave, and microwave frequencies, it can be observed that propagation wave vectors of SSPP strip are mismatched from the wave vector of freely propagating wave i.e., light line. This mismatching causes very low transmission efficiency. Here, to match this propagation wave vector, a conversion is needed so that a gradual conversion of wave vectors can be achieved between conventional microstrip line and corrugated metal strip (SSPP strip). Figs. 10.11 and 10.12 give schematic of one unit section of this conversion part for rectangular as well as trapezoidal shape

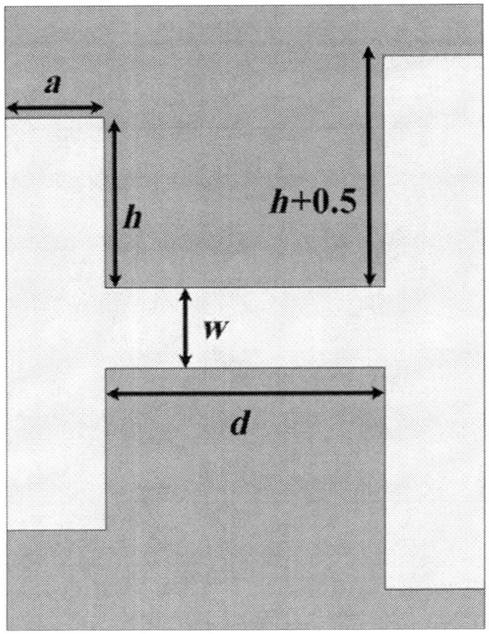

Figure 10.11
One unit cell with gradient groove of conversion and matching section for conventional rectangular shape unit cells.

respectively. Fig. 10.12 demonstrates the dispersion curve for rectangular shape unit cell at THz, millimetre wave and microwave frequency. Similarly Fig. 10.14 shows dispersion curve for trapezoidal unit cell at microwave frequency. In the conversion sections as shown in Fig. 10.15, height of grooves are increased with equal steps, labeled as $h_1 - h_7$. Also, the same layout and concept as shown in Fig. 10.15 is used for complete conversion of wave vectors at THz, mm wave, and microwave frequencies, respectively. At microwave frequency, heights vary from 0.5 to 3.5 mm with step equal to 0.5 mm. As height increases from h_1 to h_7, the wave vectors gradually deviate from the microstrip line and cutoff frequency becomes lower and thus SSPP wave is more confined at grooves. Thus, this gradient corrugated grooves provide very efficient conversion from traditional microstrip line to spoof SPP.

10.3 Plasmonic Metamaterial-Based Transitions and RF-Microwave Components

10.3.1 Transitions

It is impractical to operate designed SSPP strcutures independently. They need to be connected to the conventional existing microwave circuits which consist of two conductor transmission lines. Hence signal feeding into and signal extraction from such transmission

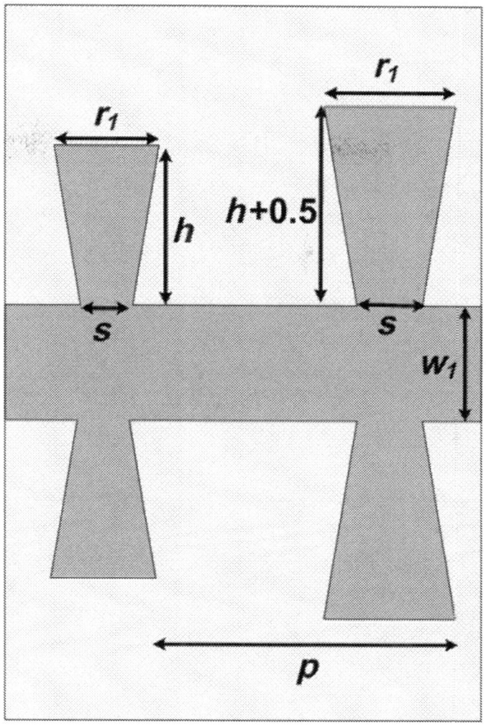

Figure 10.12
Schematic of one unit cell with gradient groove of conversion and matching section for trapezoidal shape unit cells.

lines are very difficult. Therefore, an efficient conversion of guided waves into spoof SPP wave is necessary. Hence, back-to-back transition from quasi-tranverse electro magnetic (QTEM) mode supported by microstrip-to-spoof SPP (TM) mode has been designed. Fig. 10.16 shows the schematic of back-to-back transition structure which consists of three parts: (1) a conversion section that converts the direction of the electric field from perpendicular to parallel to the strip, (2) a matching section with gradient corrugated grooves to match momentum and impedance, and (3) an SSPP waveguide to support SPP waves.

Fig. 10.17 shows the simulated S-parameter characteristics of broadband back-to-back transition at THz frequency. S_{11} parameters are below 10 dB and S_{21} charactristics are less than 5 dB in the frequency range from 0.1 to 0.9 THz. It shows that efficient conversion has been achieved from guided wave to SSPP wave.

Since nonavailabilty of fabrication and characterization facility at such high frequency, the design of such back-to-back transition has been done at microwave frequency. Fig. 10.18 shows the fabricated prototype for back-to-back transition strcutures. Fig. 10.18A–E shows

Figure 10.13
Schematic of complete section of conversion and matching with gradient grooves height *h* vary from 0.5 to 4 mm for rectangular and trapezoidal shape unit cells.

the fabricated prototype of back-to-back transition employing the unit cell described in Figs. 10.7 and 10.9.

The simulated and measured reflection and transmission characteristics of designed transition as shown in Fig. 10.18A–E, at microwave frequency, have been shown in Fig. 10.19 respectively. As it is observed that both simulated and measured S_{11} and S_{21} are in good agreement and less than 10 and 5 dB, respectively, in frequency range around from 1 to 7 GHz thus showing broadband and efficient conversion of waves.

10.3.2 Filters

Filters is one of the main building block for RF and microwave communication system applications. Nowadays, filters with more stringent requirements such as higher performance, smaller physical size, low profile, and low cost are required to satisfy modern applications. Researchers have reported various advancement in filters in terms of materials,

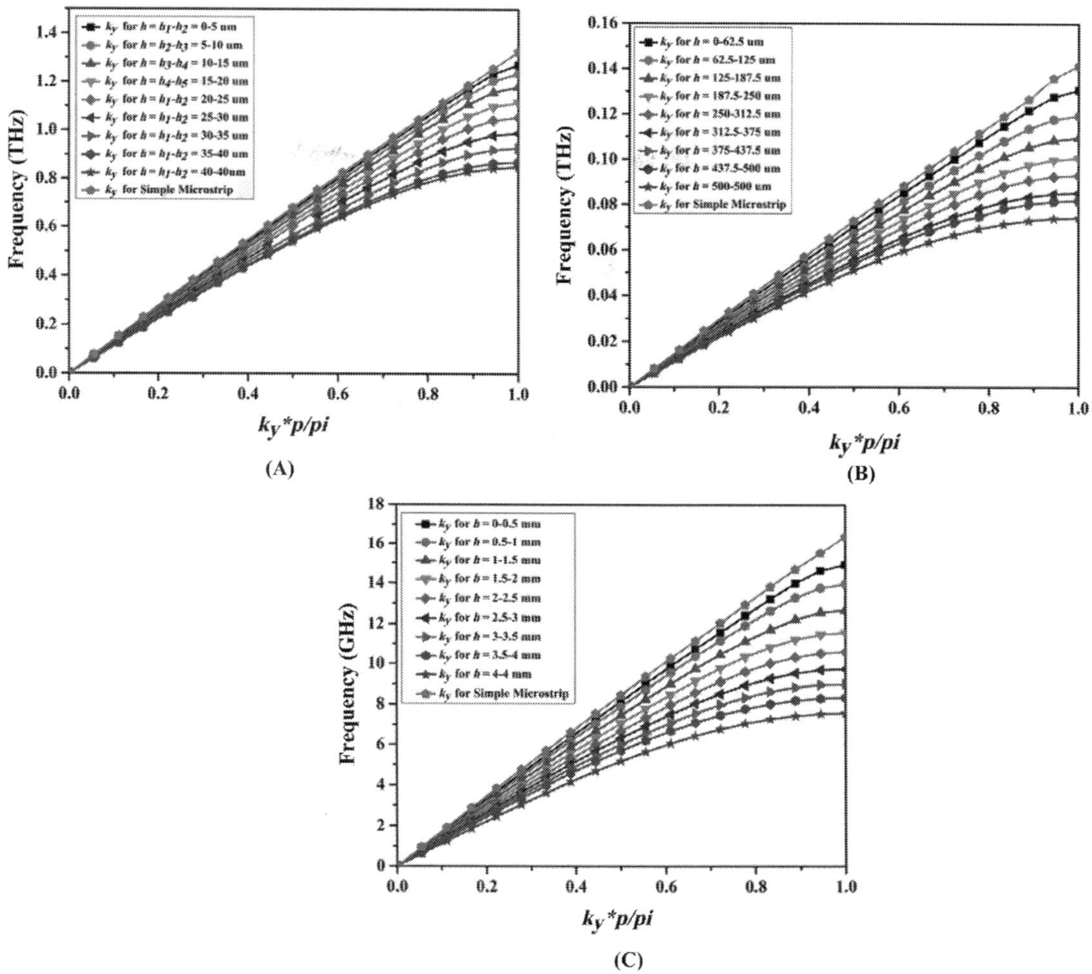

Figure 10.14

Dispersion relation showing the conversion from QTEM mode to SSPP (TM) mode, at (A) THz frequency, (B) mm-wave frequency, and (C) microwave frequency, respectively.

fabrication technologies, and in terms of miniaturization, like high-temperature superconductors (HTS) filters, low-temperature cofired ceramics (LTCC) filters, monolithic microwave-integrated circuits (MMIC) filters, microelectromechanic system (MEMS) filters, filters based on multimode structure, and metamaterials. Due to all this advancement in material and fabrication technologies, a rapid development of new microstrip and other filters for RF/microwave applications have been seen over the years.

The design of filter using plasmonic metamaterial has been done by employing complementory split-ring resonator (CSRR) at the bottom of the substrate. Split-ring

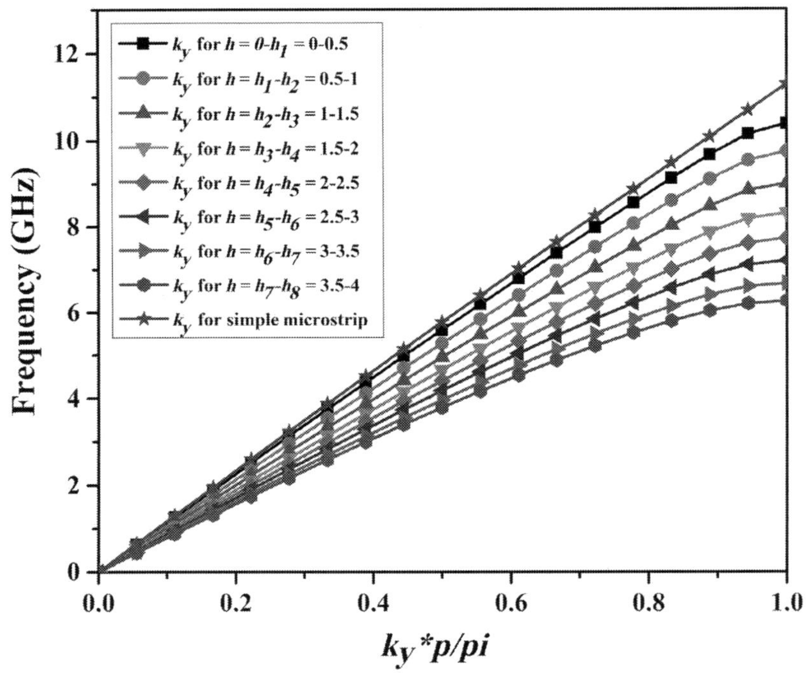

Figure 10.15
Dispersion relation showing the conversion from QTEM mode to SSPP (TM) mode at microwave frequency.

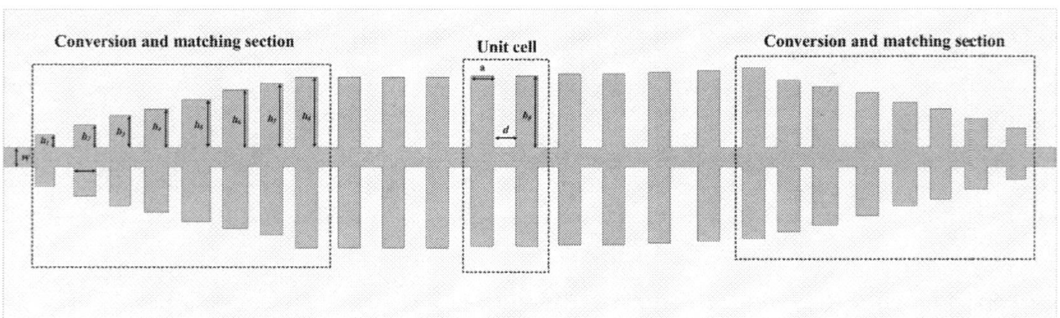

Figure 10.16
Schematic of back-to-back transition from QTEM mode of microstrip to SSPP mode.

resonators are normally used in periodic configurations to design metamaterial structures, slow-wave transmission line, and different types of microstrip filters. Circuit model of SRR consists of parallel L and C components that are placed in series form in a transmission line, therefore used in the design of band-stop filters. On the opposite side, according to

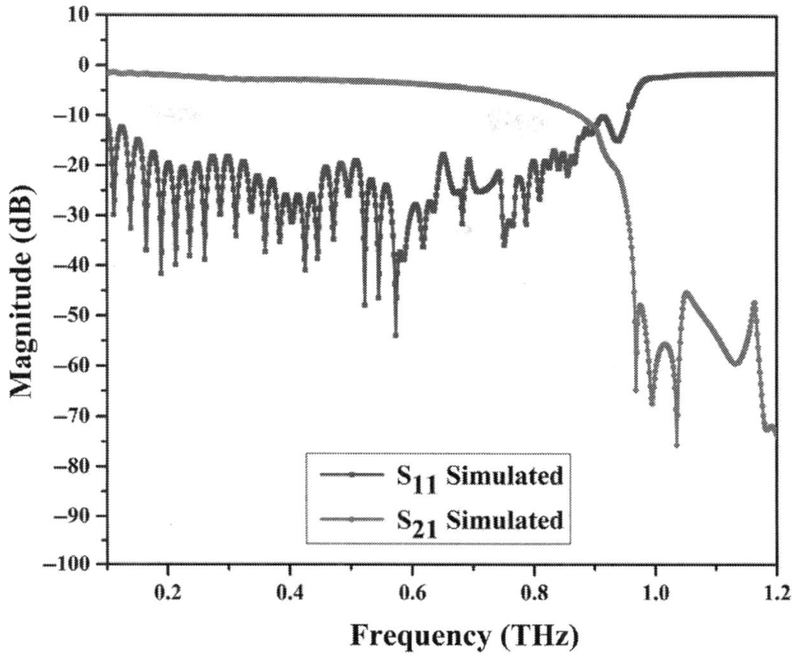

Figure 10.17
Simulated S-parameter characteristics of back-to-back transition at THz frequency.

duality theorem and Babinet's principles, complementary split-ring resonators (CSRRs) can be interpreted as bandpass filters.

Fig. 10.20 shows the schematic of a U-shaped CSRR ring with its equivalent lumped circuit model which consists of a parallel L and C components and situated in shunt form in a transmission line sections.

In order to obtain the bandpass filter (BPF) characteristics, two CSRR rings, symmetric and asymmetric shown in Fig. 10.21 are inserted in the ground plane of the back-to-back transition SSPP strip as shown in Fig. 10.18A–C. Two U-shaped CSRRs have been etched from the ground plane to achieve bandpass characteristics. The resonant frequencies of the CSRR have been calculated using the method as suggested in [28–31].

Fig. 10.22 shows the fabricated hardware to evaluate and validate the design concept where each of the transitions shown in Fig. 10.18A–C is etched with asymmetric and symmetric U-shaped CSRR shown in Fig. 10.21 in the ground plane. In continution to this, Fig. 10.23 depicts the simulated and measured reflection and transmission coefficient for designed SSPP-based filter for each of the case mentioned above. From the measured and simulated

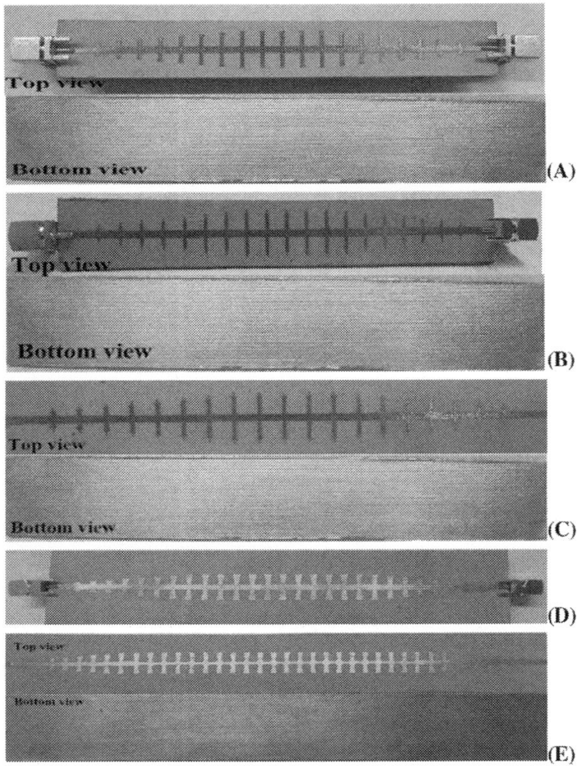

Figure 10.18
(A)–(E) Fabricated prototype of back-to-back transition for unit cell shown in Figs. 10.7 and 10.9.

response, it has been observed that S_{21} response in case of asymmetric CSRR is far apart from each other, while response in the case of symmetric CSRR is closed to each other.

10.3.3 Planar Ring Resonators

Ring resonators are simply a transmission line formed in a closed loop. The basic circuit of ring resonators consists of a resonator, coupling gap, and feed lines. Power is coupled into and out of resonators through feed lines and coupling gaps. The resonant frequencies are mainly dependent on the length of the ring resonator. Whereas if coupling gap is large then it does not affect the resonant frequencies of ring because of very less capacitance of gap usually called as loose coupling. However, if distance between feed lines and ring are small then it properly excite the resonator due to tight coupling. Thus, we need to consider the capacitance of coupling gap to accurately model the ring resonators. Various applications like antenna and frequency selective surfaces use the ring circuits and also microwave

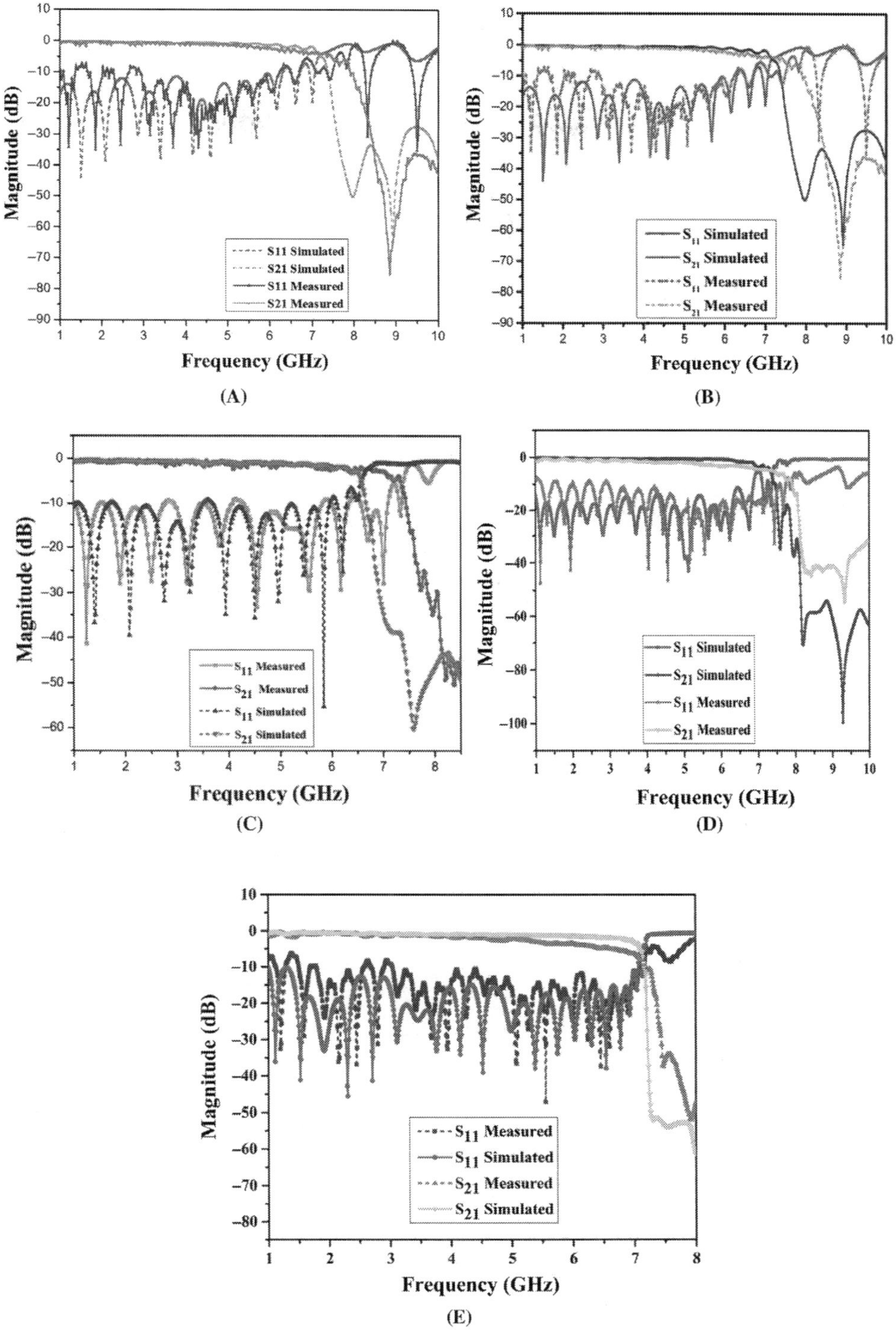

Figure 10.19

Simulated and measured S-parameter characteristics of back-to-back transition at microwave frequency for Fig. 10.18A–E.

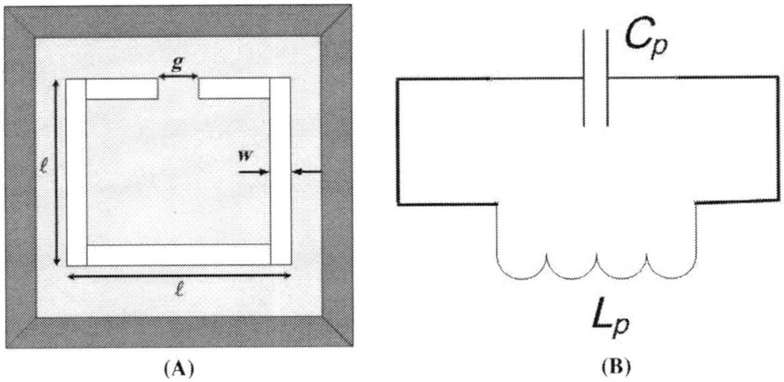

Figure 10.20
Schematic of (A) CSRR of U shape and (B) equivalent lumped circuit.

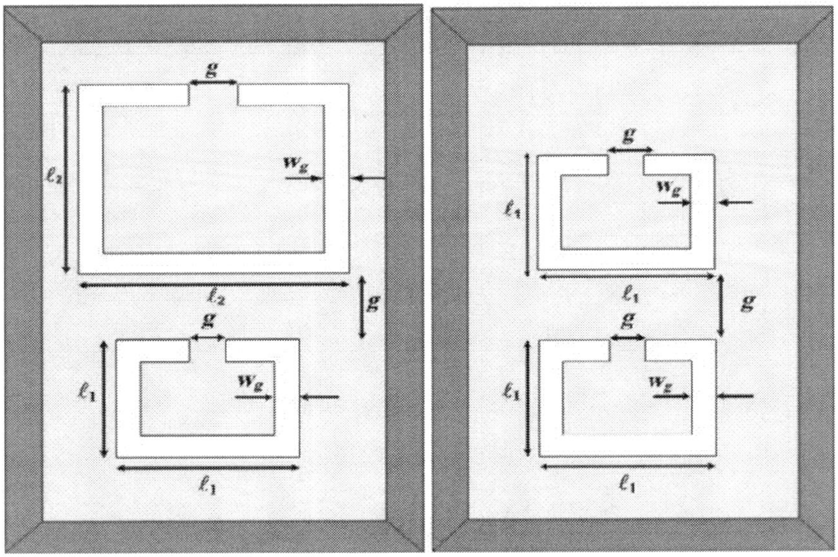

Figure 10.21
Schematic of asymmetric and symmetric U-shaped CSRR inserted in ground plane of the designed transition.

circuits like, filters, oscillators, mixers, baluns, and couplers are made using ring circuits. The integration of solid-state devices with ring was also realized to perform tuning, amplification, switching, and other optoelectronic functions [32].

The ring resonator kind of structure would only resonate if mean circumference of resonator will equivalent to an integral multiple of the guided wavelength [32].

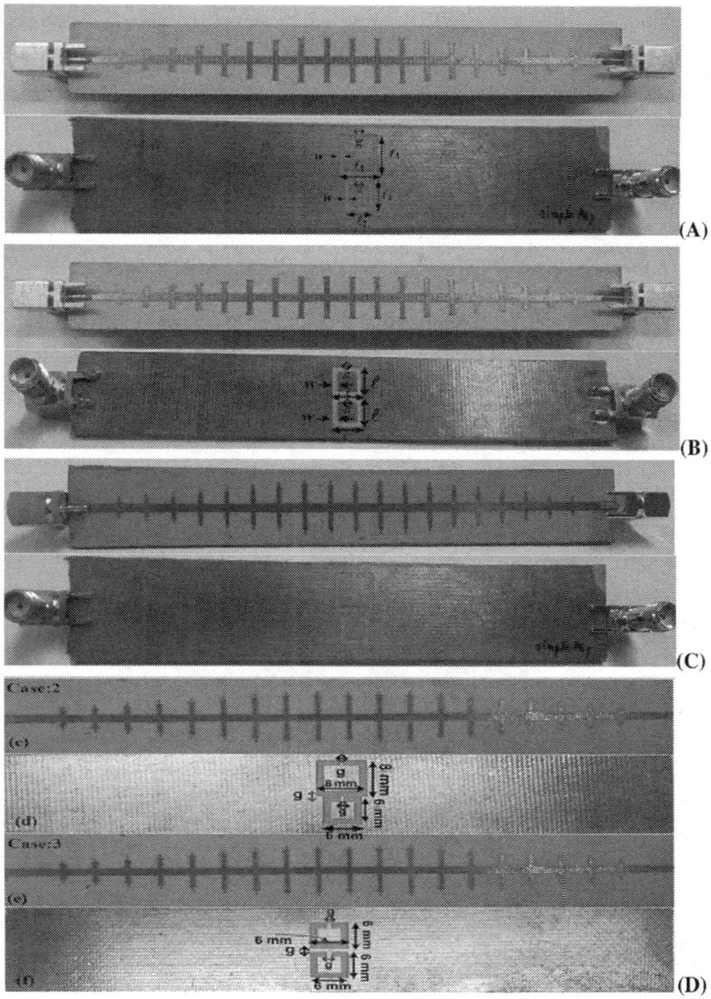

Figure 10.22
(A)–(D) Fabricated prototype of multiband filters using two U-shaped CSRRs in the ground plane in transition shown Fig. 10.18A–C.

$$2\pi r = n\lambda_g \quad \text{for } n = 1, 2, 3, \ldots.$$

Here, r belongs to mean radius of ring (average of outer and inner radius of ring), λ_g stands for the guided wavelength, and n is the mode number.

Planar plasmonic metamaterials have the property to support wideband THz waves in two orthogonal direction that decays exponentially at subwavelength scale (around 1/15 wavelength) and also provide low-loss propagation of waves [33,34]. Thus, EM wave can

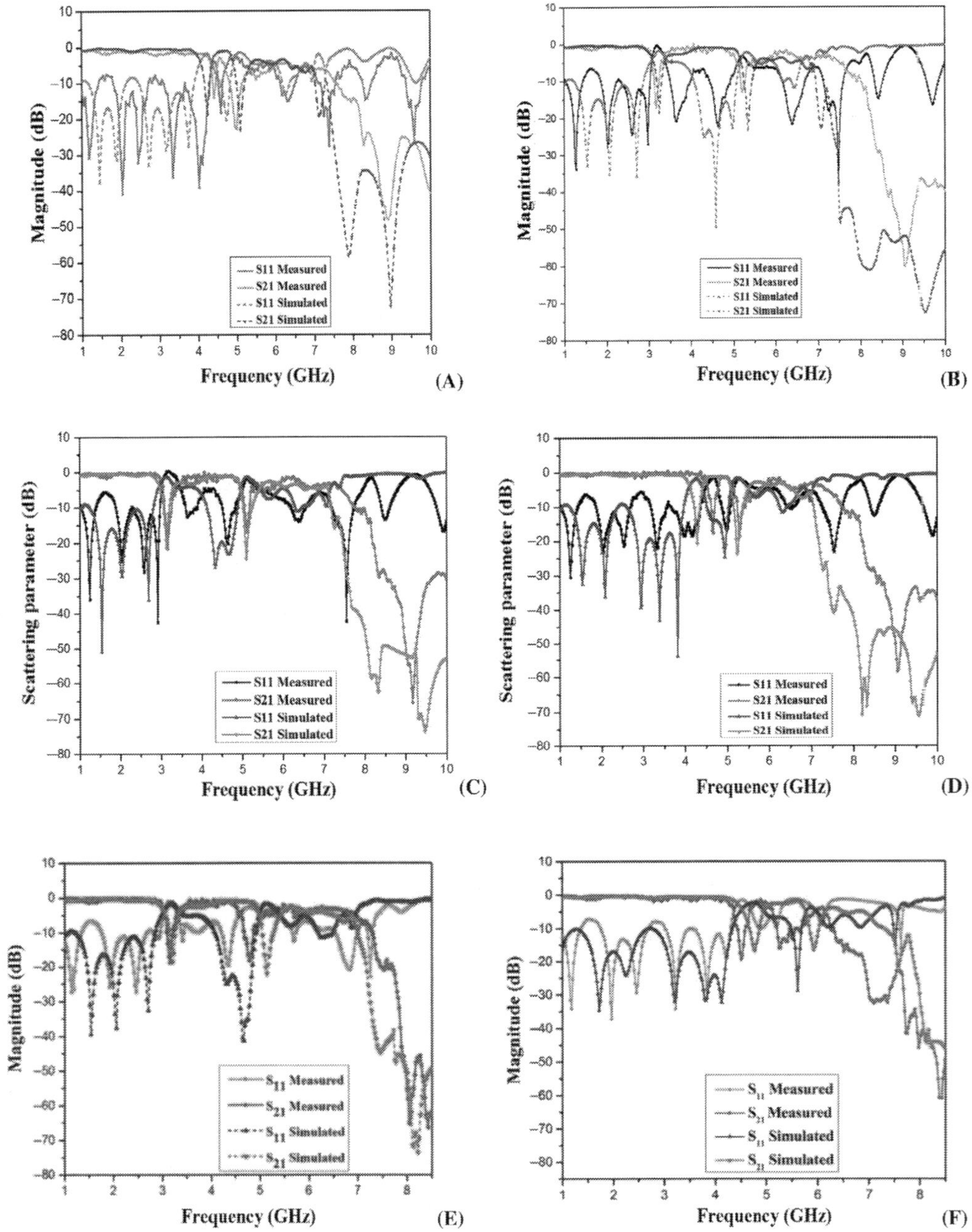

Figure 10.23
Simulated and measured S-parameter characteristics of multiband filter at microwave frequency for structures shown in Fig. 10.22A–D.

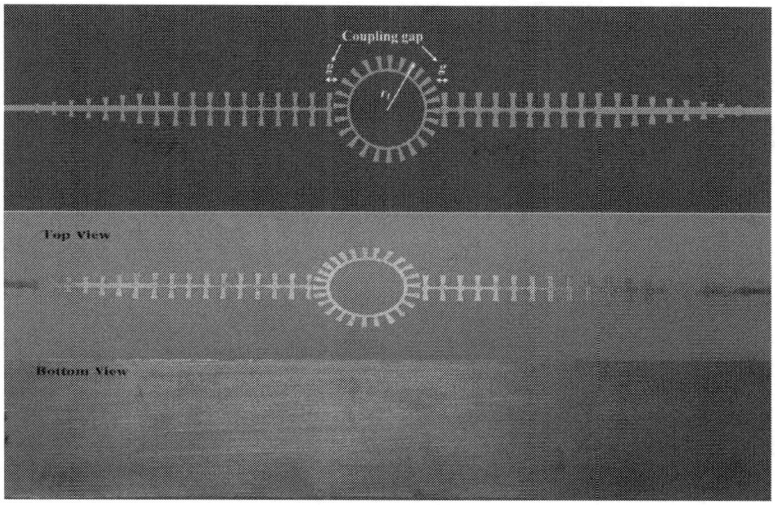

Figure 10.24
Schematic and fabricated prototype of corrugated planar ring resonator fed by spoof SPP (in-line coupling).

be bend efficiently in plasmonic metamaterial with this high orthogonal confinement of SPP and hence propagation of EM waves is possible to control in a planar surface. Figs. 10.24 and 10.27 are the designed planar ring resonators based on corrugated metal surface. Design (Fig. 10.24) consists a two-port corrugated ring fed by spoof SPP transitions at its both of the two port with gap g and while design (Fig. 10.27) consists of a straight plasmonic waveguide and corrugated ring above it. The output signal from such type of structure can be modulated by efficiently choosing ring radius r and coupling gap g. Here, ring radius is to be chosen equal to half of operating wavelength.

Here, in Fig. 10.24, we have chosen symmetric gap i.e., gap between feed lines and ring with both sides equal. When the ring is excited symmetrically, the maximum field points in both the field lines and ring are collinear. The proposed structure consists of a two-port planar corrugated metallic ring fed by a simple straight spoof plasmonic waveguide at input and output ports. The ring has a outer radius of 15.92 mm and the number of grooves has been set to 25. The grooves are of trapezoidal shape and their parameters are same as the above mentioned in transition part (Fig. 10.18E). Dielectric substrate with thickness of 1.52 mm and dielectric constant (ε_r) of 3.38 has been used for hardware prototyping. Coupling gap g, which is a gap between straight waveguide and ring, and ring radius (r) are the important structural parameter which determine the resonance spectrum of ring resonator unit. Resonance condition is achieved at particular frequencies when circumference of ring becomes an integral multiple of guided wavelength. Fig. 10.26 depicts the simulated electric field at first three resonating modes i.e., for mode number $n = 1$, 2, and 3. Blue color marked

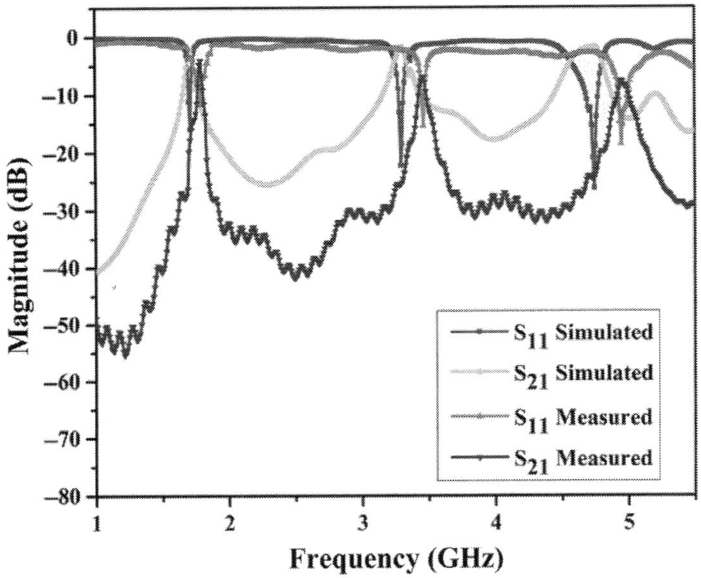

Figure 10.25
Simulated and measured S-parameter characteristics of bandpass filter using ring resonator shown in Fig. 10.24.

area illustrates the maximum field region for first three resonating modes in the proposed structure. We have fabricated the above-mentioned ring resonator-based bandpass filter with coupling gap $g = 0.15$ mm at both side of ring (Fig. 10.24).

The simulated and measured S-parameter results are shown in Fig. 10.25. A shift in measured result has been observed as compared to simulated S-parameter which is due to fabrication tolerance and use of two SubMiniature Version A (SMA) connectors at the input and output ports. From the simulated electric field response as shown in Fig. 10.26, it is observed that for $n = 1$, two field maxima are obtained at feeding position with 180° phase shift. Similarly for $n = 2$, four field maxima are obtained with 90° shift and for $n = 3$, six maxima of field are obtained with 45° phase difference. The characterization of aforementioned device has been performed using keysight field-fox microwave analyzer N9918A (Fig. 10.25).

In Fig. 10.27, a part of signal fed to input waveguide has been coupled to ring and then coupled to waveguide again after a traveling to ring normally termed as offline coupling. The ouput signal is modulated when these two waves interfere with each other at the right part of the ring on output port side. This depends on phase conditon around the ring, whenever the intereference is constructive at the right side of ring, the ring is at the resonance state and signal is transmitted to the output and when intereference is destructive, ring works at closed state and signal becomes cutoff. The simulated S-parameter for

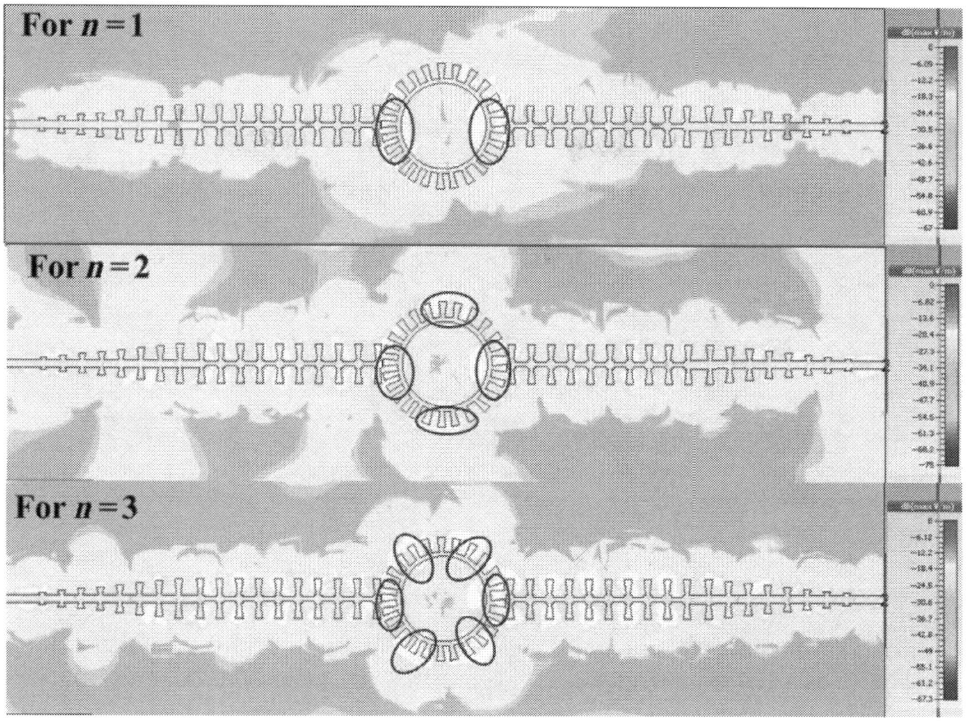

Figure 10.26
Simulated electric field distribution for designed band-pass filter (BPF) using planar ring resonator for mode $n = 1$, 2, and 3 for Fig. 10.24.

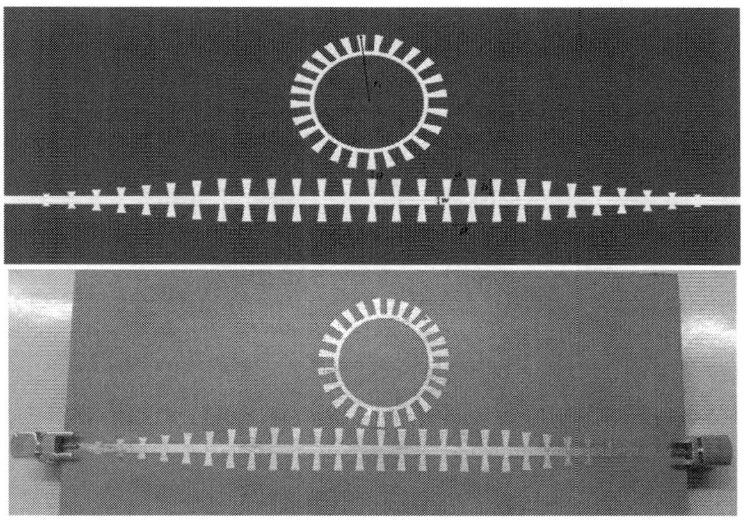

Figure 10.27
Schematic and fabricated prototype of corrugated planar ring resonator coupled with spoof SPP transition structure (offline coupling).

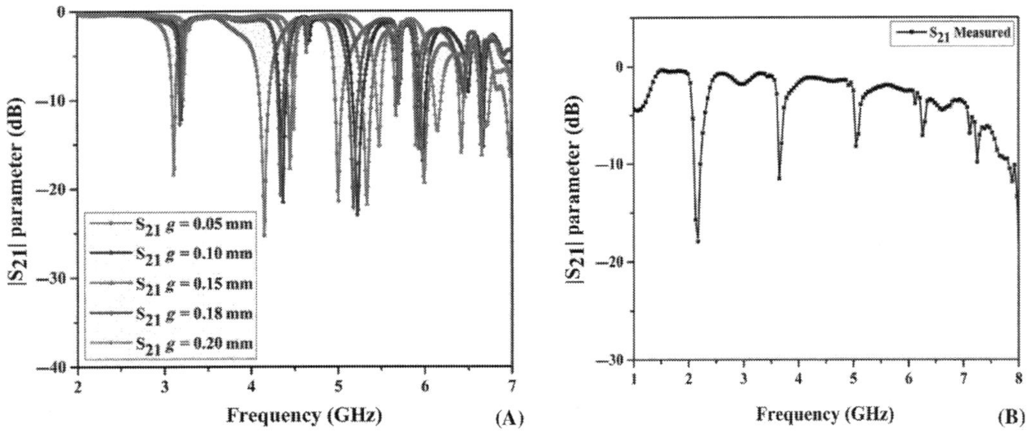

Figure 10.28

(A) Simulated S-parameter characteristics for the structure shown in Fig. 10.27 with different coupling gaps *g*. (B) Measured S-characteristics for fabricated prototype with $g = 0.18$ mm.

designed ring circuit (Fig. 10.27) has been shown in Fig. 10.28 with varying gap *g*. Measured results show the resemblance to simulated results.

10.3.4 Spoof SPP-Fed Antenna Design

As the demand for the high gain, high efficiency, and wide bandwidth antenna in wireless systems are increasing due to the extension of the frequency spectrum of wireless instrument. The wideband antenna has wide impedance bandwidth to reduce the number of antennas and the volume of communication instrument by using the same antenna among several communication systems. Current ultra-wide bandwidth (UWB) and high-gain antennas include mainly log-periodic antenna, traditional transverse electromagnetic (TEM) horn antenna, spiral antenna, bow-tie antenna, tapered slot antenna, and microstrip patch antenna. However, the traditional Yagi antenna and microstrip patch antenna have narrow bandwidth. TEM horn antennas are wideband antenna and have high gain over operating bandwidth and is used for UWB communication [35–39]. Planar horn antennas are highly desirable for integration with active and passive circuits [40]. It has wide application in radar systems, mainly in ground penetrating radar (GPR) and other such application [41,42]. The traditional TEM horn is three-dimensional and stereo structure with relative big volume and heavy weight usually requiring continuous or discrete resistive loading which results in low radiation efficiency. Horn maintains constant phase center thus they can transmit and receive impulse signals with minimal phase distortion. They are relatively compact and tend to have higher aperture efficiency than comparable reflector antenna and offer more directivity and gain for a given physical dimension.

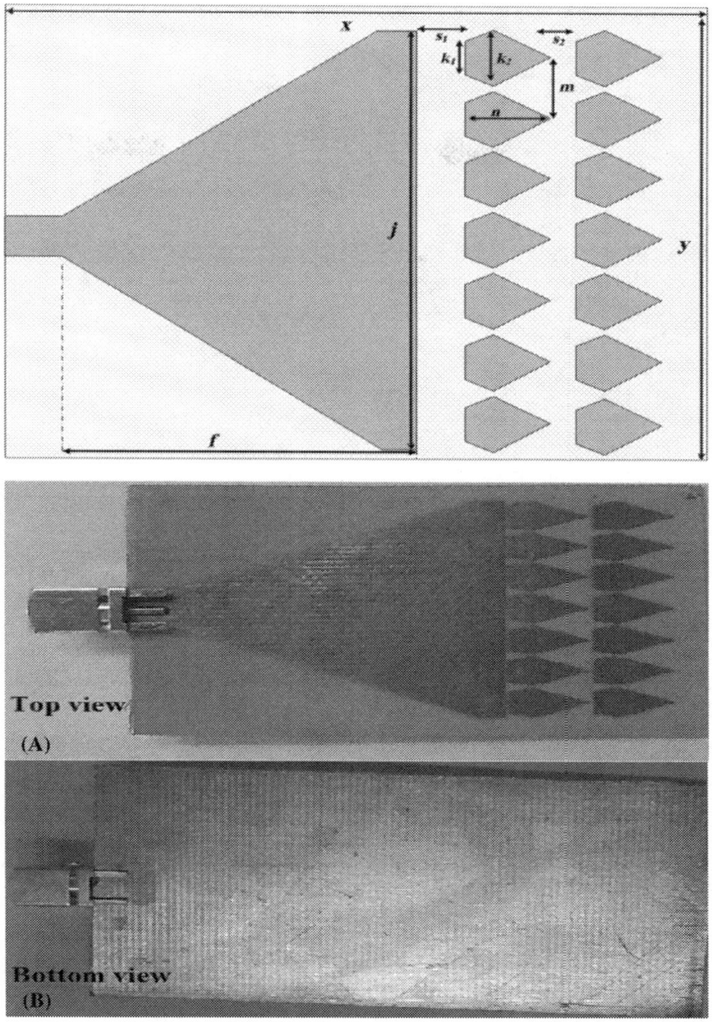

Figure 10.29
Schematic and fabricated prototype of the planar horn antenna fed by microstrip feed line.

There are different feeding methods that are used to feed the TEM horn antenna, i.e., Yagi antenna feeding, a long metallic horn and microstrip patch feeding and planar monopole-fed compact, and surface mounted horn antennas are reported [43–45]. Here, SPP-based feeding becomes a new approach to obtain the gain relatively high as compared to other feed. Thus, a novel method to feed the planar TEM horn antenna has been demonstrated. The developed QTEM mode to SSPP mode transition is used to feed the designed planar horn antenna.

Fig. 10.29 shows the schematic and fabricated design of planar horn antenna. It consists of a conventional microstrip patch of triangular shape and an array of patches with pentagons

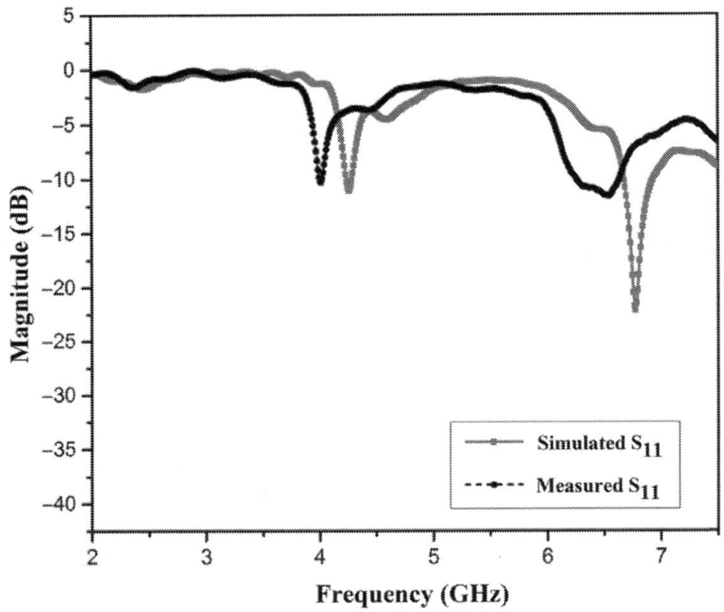

Figure 10.30
Simulated and measured reflection coefficient magnitude for planar horn antenna.

shapes that are arranged in front of the main patch. Advantage of adding small patches is better radiative properties such as high gain and radiation efficiency. Microstrip feeding technique is used to feed the designed planar horn. The dimensions used to design this antenna are described as follows: $x = 67$ mm, $y = 60$ mm, $f = 37$ mm, $j = 42.084$, $n = 10$ mm, $m = 6.1$ mm, $s_1 = 0.25$ mm, $s_2 = 0.35$ mm, $k_1 = 4$, and $k_2 = 6$ mm.

The simulated and measured S-parameter characteristics for this microstrip-fed planar horn antenna has been shown in Fig. 10.30. It is observed that the resonant frequencies of antenna are 4.267 and 6.778 GHz, and the S_{11} parameter at these frequencies are nearly -11 and -22 dB, respectively. Deviation in simulated and measured results is due to loss of SMA not included in simulation and fabrication tolerance. Far-field radiation patterns are shown in Figs. 10.31 and 10.32.

Realized gain of antenna is shown in Fig. 10.33, are 5.2 and 6.5 dB at 4.267 and 6.778 GHz of frequencies, respectively. Further, this designed antenna is fed by designed transition (Fig. 10.18C) and its simulation and measures results are recorded.

Schematic of spoof SPP transition-fed planar horn antenna has been shown in Fig. 10.34 with their dimensions. Fabricated protoype for this designed antenna is also developed

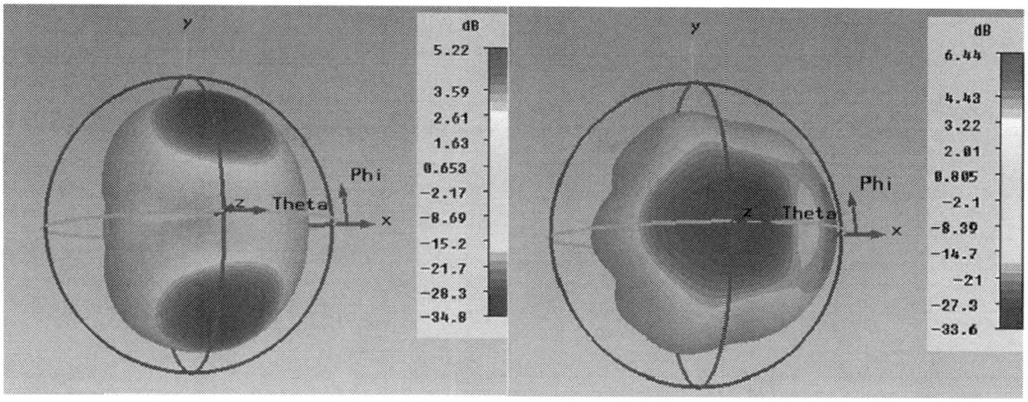

(A) Radiation pattern at 4.267 GHz **(B) Radiation pattern at 6.778 GHz**

Figure 10.31
Simulated far-field radiation plot for designed planar horn antenna as shown in Fig. 10.29.

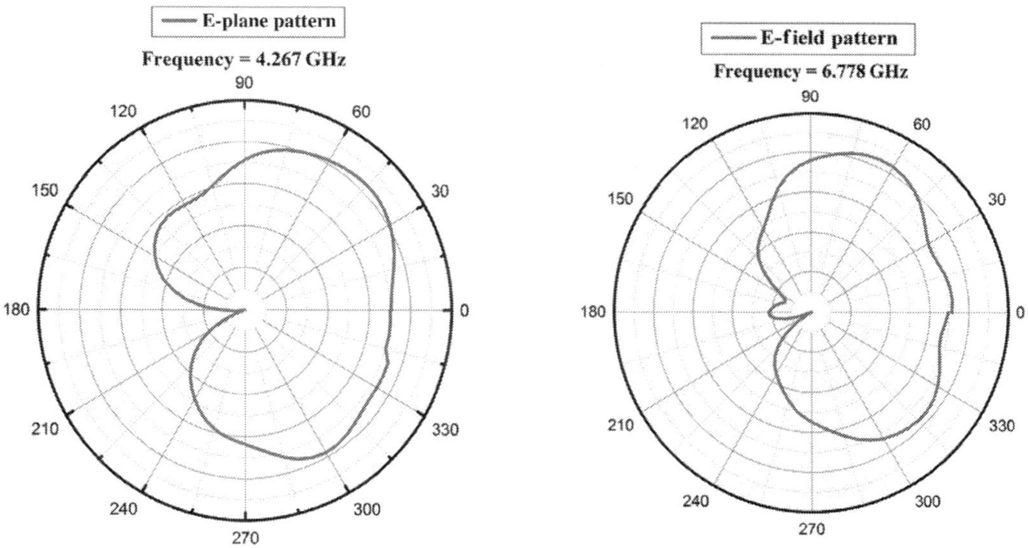

Figure 10.32
Measured E-plane pattern for designed planar horn antenna at 4.267 and 6.778 GHz frequency.

and shown in Fig. 10.34. The substrate with dielectric constant of 3.38, dissipation factor 0.0016, and height of 1.524 mm has been used to develop this structure.

The simulated and measured S-parameter characteristics for this spoof SPP-fed planar horn antenna has been shown in Fig. 10.35. It is observed that the resonant frequencies of

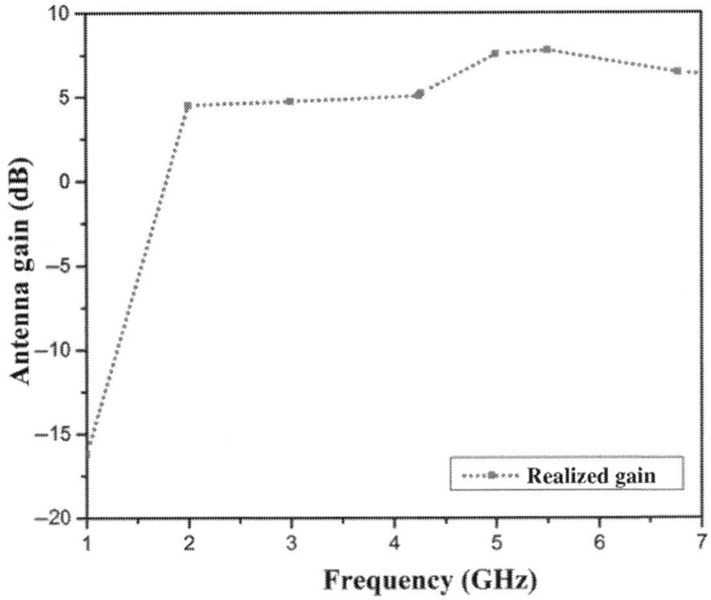

Figure 10.33
Realized gain for designed microstrip-fed planar horn antenna.

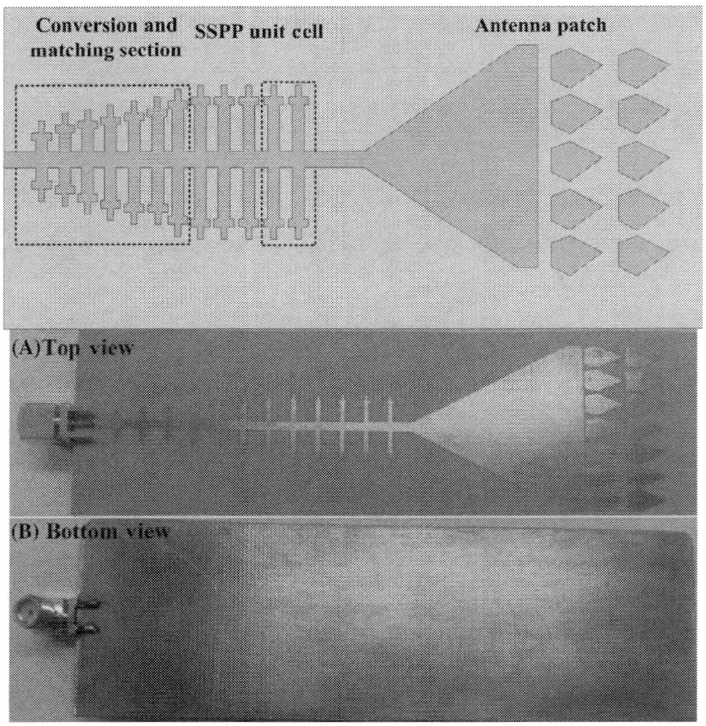

Figure 10.34
Schematic and fabricated prototype of designed spoof SPP-based planar horn antenna.

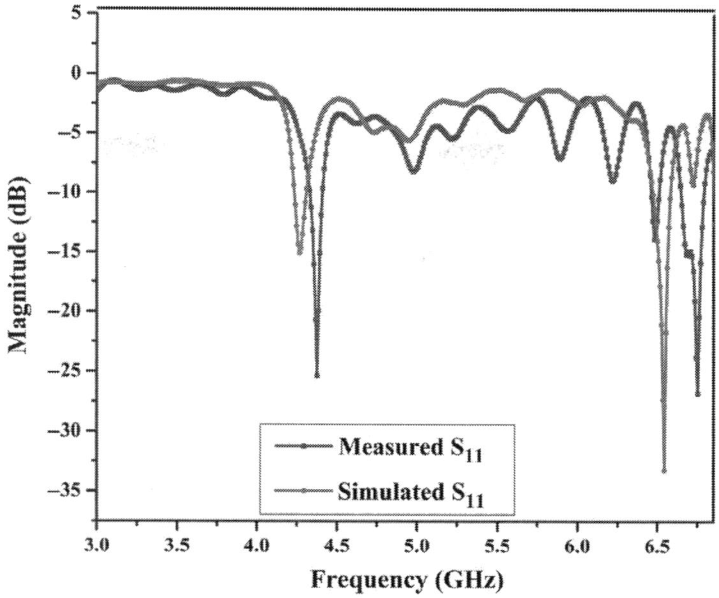

Figure 10.35
Simulated and measured reflection coefficient magnitude for spoof SPP-fed planar horn antenna.

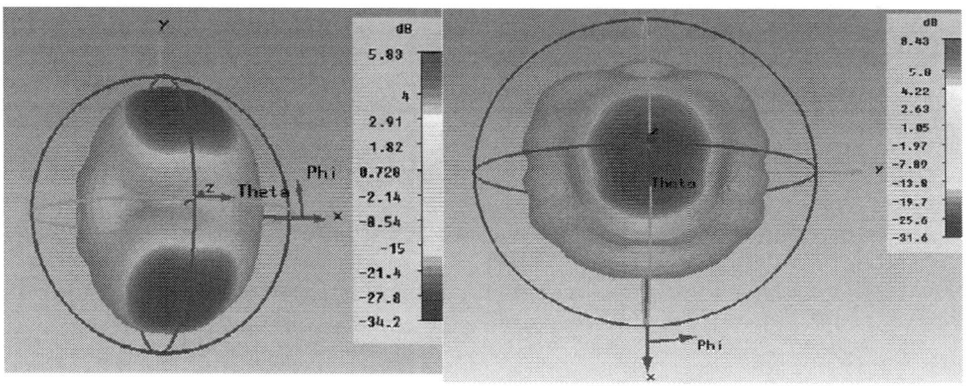

Figure 10.36
Simulated far-field radiation plot for designed spoof SPP-fed planar horn antenna.

this antenna are 4.267 and 6.544 GHz, and the S_{11} parameter at these frequencies are nearly —15 and −33 dB, respectively. Deviation in simulated and measured results is due to loss of SMA not included in simulation and fabrication tolerance. Far-field radiation patterns and surface charge current density and electric field distribution are shown in Figs. 10.36–10.39, respectively.

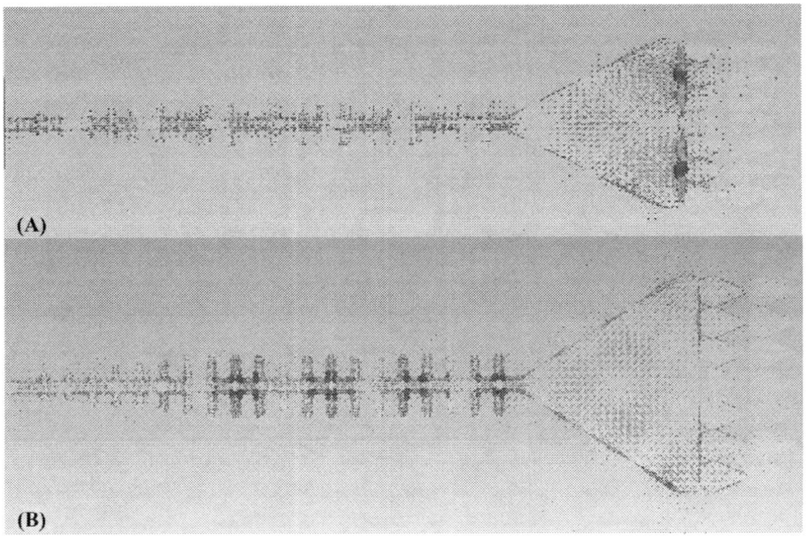

Figure 10.37
Surface current distribution of designed SSPP-based planar horn antenna (A) at 4.267 GHz and (B) at 6.544 GHz, respectively.

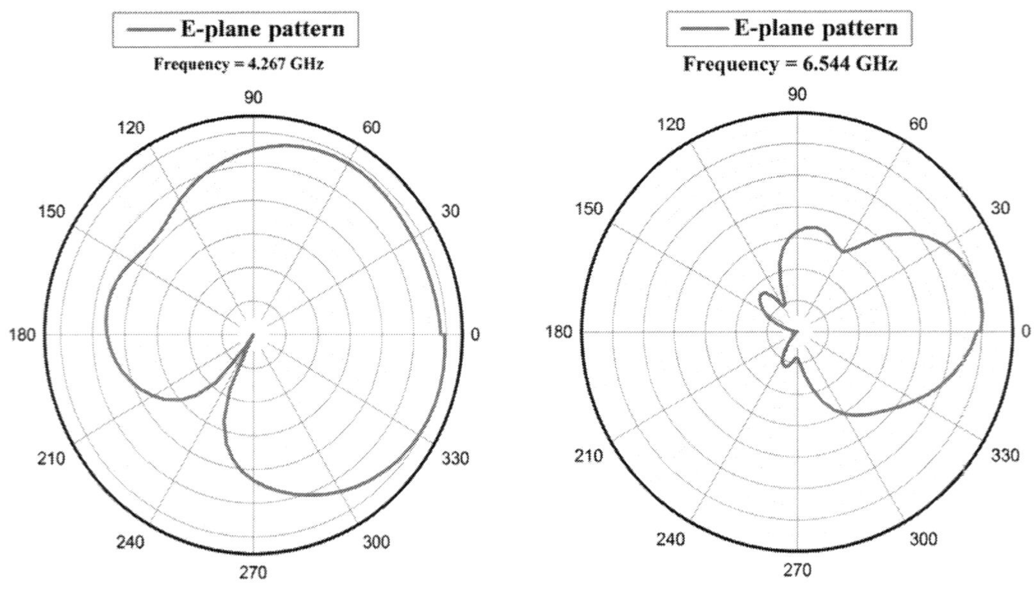

Figure 10.38
Measured E-plane pattern for designed planar horn antenna at 4.267 and 6.544 GHz frequency.

From Fig. 10.38, we observe that the designed spoof SPP-fed antenna has omnidirectional characteristic with one null at 4.267 GHz of frequency and directive characteristics at 6.544 GHz of frequency of designed antenna are shown.

Realized gain of SSPP-fed antenna is shown in Fig. 10.40, which are 5.82 and 8.43 dB at 4.267 and 6.544 GHz of frequencies, respectively. Thus, good radiation characteristics are obtained for an antenna when using spoof SPP feeding.

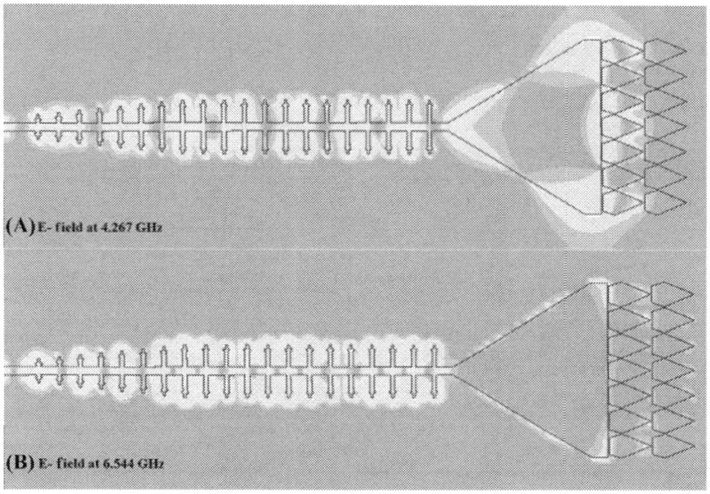

Figure 10.39
E-field distribution for designed spoof SPP-based planar horn antenna.

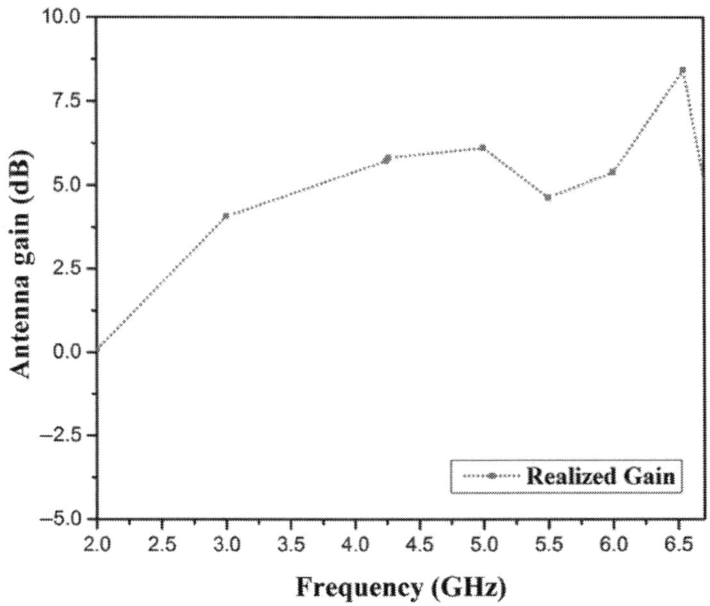

Figure 10.40
Realized gain for designed spoof SPP-based planar horn antenna.

10.4 Conclusion

Characteristics of plasmonic metamaterial have been demonstrated from RF-THz frequency regime. The dispersion relation and cutoff frequency of such geometry-dependent structure can be engineered at our will, hence named spoof or designer SPP. So, we can explore frequencies from microwave to THz in the EM spectrum. Also, transitions, antenna, and microwave circuits such as filter, resonator circuits using the concept of plasmonic metamaterial have been presented. Such a plasmonic technology plays a very vital role in designing miniscale structures.

References

[1] A.V. Zayats, I.I. Smolyaninov, A.A. Maradudin, Nano-optics of surface plasmon polaritons, Phys Report 408 (2005) 131–314.

[2] J.M. Pitarke, V.M. Silkin, E.V. Chulkov, P.M. Echenique, Theory of surface plasmons and surface-plasmon polaritons, Rep Prog Phys 1 (1) (2006) 54.

[3] J.R. Powell, The quantum limit to Moore's law, Proc. IEEE 96 (8) (2008) 1247–1248.

[4] Nobel Media AB, The 2009 nobel prize in physics-press release, 2009.

[5] V.J. Sorger, R.F. Oulton, R.-M. Ma, X. Zhang, Toward integrated plasmonic circuits, Materials Research Society 37 (2012) 728–738.

[6] H.C. Zhang, Y. Fan, J. Guo, X. Fu, T.J. Cui, Second-harmonic generation of spoof surface plasmon polaritons using nonlinear plasmonic metamaterials, ACS Photonics 3 (2015) 139–146.

[7] J.J. Xu, J.Y. Yin, H.C. Zhang, T.J. Cui, Compact feeding network for array radiations of spoof surface plasmon polaritons, Sci Rep 6 (22692) (2016). Available from: https://doi.org/10.1038/srep22692.

[8] H.A. Atwater, The promise of plasmonics, Scientific American 296 (4) (2007) 56–62.

[9] I.S.I. Web, S. This, H. Press, N. York, A. Nw, Plasmonics: merging photonics and electronics at nanoscale dimensions, Science Rev 311 (2006) 189–194.

[10] H. Liu, B. Wang, L. Ke, J. Deng, C.C. Chum, S.L. Teo, et al., High aspect subdiffraction-limit photolithography via a silver superlens, Nano Lett 12 (3) (2012) 1549–1554.

[11] B. Wang, X. Zhang, X. Yuan, J. Teng, Optical coupling of surface plasmons between graphene sheets, Appl Phys Lett 100 (131111) (2012). Available from: https://doi.org/10.1063/1.3698133.

[12] H. Atwater, A. Polman, E. Kosten, D. Callahan, P. Spinelli, C. Eisler, et al., Nanophotonic design principles for ultrahigh efficiency photovoltaics, AIP Conf Proc 1519 (2013) 17–21.

[13] M.A. Schmidt, D.Y. Lei, L. Wondraczek, V. Nazabal, S.A. Maier, Hybrid nanoparticle-microcavity-based plasmonic nanosensors with improved detection resolution and extended remote-sensing ability, Nat Commun 3 (2012). Available from: https://doi.org/10.1038/ncomms2109.

[14] S. Zhang, Y. Xiong, G. Bartal, X. Yin, X. Zhang, Magnetized plasma for reconfigurable subdiffraction imaging, Phys Rev Lett 106 (243901) (2011) 1–4.

[15] J.N. Anker, W.P. Hall, O. Lyandres, N.C. Shah, J. Zhao, R.P. Van Duyne, Biosensing with plasmonic nanosensors, Nat Mater 7 (6) (2008) 442–453.

[16] D.K. Gramotnev, S.I. Bozhevolnyi, Plasmonics beyond the diffraction limit, Nature Photonics (2010). Available from: https://doi.org/10.1038/nphoton.2009.282.

[17] N. Liu, F. Wen, Y. Zhao, Y. Wang, P. Nordlander, N.J. Halas, et al., Individual nanoantennas loaded with three-dimensional optical nanocircuits, Nano Lett 13 (1) (2013) 142–147.

[18] J.B. Pendry, L. Martín-Moreno, F.J. Garcia-Vidal, Mimicking surface plasmons with structured surfaces, Science 305 (5685) (2004) 847–848.

[19] A.P. Hibbins, B.R. Evans, J.R. Sambles, Experimental verification of designer surface plasmons, Science Report 308 (5722) (2005) 670–672.

[20] A. Rusina, M. Durach, M.I. Stockman, Theory of spoof plasmons in real metals, Appl Phys A Mater Sci Process 100 (2) (2010) 375–378.

[21] Gao, J. Hui Shi, X. Shen, H. Feng Ma, W. XiAng Jiang, L. Li, et al., Ultrathin dual-band surface plasmonic polariton waveguide and frequency splitter in microwave frequencies, Appl Phys Lett 102 (15) (2013) 9–13.

[22] X. Shen, T. Jun, D. Martin-cano, F.J. Garcia-vidal, Conformal surface plasmons propagating on ultrathin and flexible films, PNAS 110 (1) (2013) 40–45.

[23] H.F. Ma, X. Shen, Q. Cheng, W.X. Jiang, T.J. Cui, Broadband and high-efficiency conversion from guided waves to spoof surface plasmon polaritons, Laser Photonics Rev 8 (1) (2014) 146–151.

[24] B. Cao Pan, Z. Liao, J. Zhao, T. Jun Cui, M.E. Stewart, Controlling rejections of spoof surface plasmon polaritons using metamaterial particles, Chem. Rev 108 (2) (2014) 494–521.

[25] A. Kianinejad, Z.N. Chen, C.-W. Qiu, Design and modeling of spoof surface plasmon modes-based microwave slow-wave transmission line, IEEE Trans. On Microwave Theory and Tech 63 (6) (2015) 1817–1825.

[26] W. Zhang, G. Zhu, L. Sun, F. Lin, Trapping of surface plasmon wave through gradient corrugated strip with underlayer ground and manipulating its propagation, Jour App Phys 106 (021104) (2015). Available from: https://doi.org/10.1063/1.4905675.

[27] X. Liu, L. Zhu, Q. Wu, Y. Feng, Highly-confined and low-loss spoof surface plasmon polaritons structure with periodic loading of trapezoidal grooves, AIP Advances 5 (2015) 077123.

[28] A. Radkovskya, M. Shamonin, C.J. Stevens, G. Faulkner, D.J. Edwards, E. Shamonina, et al., Resonant frequencies of a combination of split rings: experimental, analytical and numerical study, Microwave and Optical Technology Letters 46 (5) (2005) 473–476.

[29] J.D. Baena, J. Bonache, F. Martin, R.M. Sillero, F. Falcone, T. Lopetegi, et al., Equivalent-circuit models for split-ring resonators and complementary split-ring resonators coupled to planar transmission lines, IEEE Trans. on Microw. Theory and Techn. 53 (4) (2005) 1451–1460.

[30] M.D. Sindreu, A. Velez, F. Aznar, G. Siso, J. Bonache, F. Martín, Applications of open split ring resonators and open complementary split ring resonators to the synthesis of artificial transmission lines and microwave passive components, IEEE Trans. on Microw. Theory and Techn. 57 (12) (2009) 3395–3403.

[31] J. Bonache, F. Martin, F. Falcone, J.D. Baena, T. Lopetegi, J. Garcia, et al., Application of complementary split-ring resonators to the design of compact narrow band-pass structures in microstrip technology, Microwave and Optical Technology Letters 46 (5) (2005) 508–512.

[32] K. Chang, L.-H. Hsieh, Microwave ring circuits and related structures, John Wiley & Sons, 2004.

[33] Y. Liu, J. Yan, Y. Shao, J. Pan, C. Zhang, Y. Hao, et al., Spoof surface plasmon polaritons based on ultrathin corrugated metallic grooves at terahertz frequency, Applied physics 55 (7) (2016).

[34] X. Shen, T.J. Cui, Planar plasmonic metamaterial on a thin film with nearly zero thickness, Applied physics letters 102 (2013) 211909.

[35] Y. Li, X. Yin, H. Zhao, L. Wang, M. Yang, Radiation enhanced broadband planar tem horn antenna., IEEE Asia Pacific Microwave Conference (2014) 720–722.

[36] M. Esquius-Morote, B. Fuchs, J.F. Zürcher, J.R. Mosig, Novel thin and compact H-Plane SIW horn antenna, IEEE Trans. on Ant. and Prop 61 (6) (2013) 2911–2920.

[37] H.G. Schantz, J. Jeon, Standard gain UWB planar horn antennas, IEEE International Conference on UltraWideband (2012). Available from: https://doi.org/10.1109/ICUWB.2012.6340489.

[38] L.-C.T. Chang, W.D. Burnside, An ultrawide-bandwidth tapered resistive TEM horn antenna, IEEE Trans. on Antennas and Prop 48 (12) (2000) 1848–1857.

[39] S. Esfandiarpour, A. Mallahzadeh, Wideband planar horn antenna using substrate integrated waveguide technique, IEEE APMC (2011) 1969–1972.

[40] L. Shu, Z. Xin-yue, Z. Xing-qi, Z. Xue-ying, C. Run-nan, H. Guan-long, et al., High-gain planar TEM horn antenna fed by balanced microstrip line, CHINACOM (2011). Available from: https://doi.org/10.1109/ChinaCom.2011.6158284.

[41] A.S. Turk, Ultra-wideband tem horn design for ground penetrating impulse radar systems, Microwave and optical technology letters 41 (5) (2004) 333–336.

[42] A.S. Turk, H. Nazli, Hyper-wide band tem horn array design for multi band ground penetrating impulse radar, Microwave and optical technology letters 50 (1) (2008) 76–81.

[43] M. Sironen, Y. Qian, T. Itoh, A 60 GHz conical horn antenna excited with quasi-Yagi antenna, IEEE MTT-S Int. Symp. Digest 1 (2001) 547–550.

[44] A.A. Rahman, A.K. Verma, A.S. Omar, High gain wideband compact microstrip antenna with quasi-planner surface mount horn, IEEE MTT-S Int. Microwave Symp. Digest 1 (2003) 571–574.

[45] Y. Ranga, A.K. Verma, K.P. Esselle, Planar monopole fed, surface mounted quasi-TEM horn antenna for UWB systems, IEEE Trans on Antenna and Prop 58 (7) (2010) 2436–2439.

PART V

Emerging Materials

CHAPTER 11

Advances in InSb and InAs Nanowire Based Nanoelectronic Field Effect Transistors

Suprem R. Das

Kansas State University, Manhattan, KS, United States

11.1 Introduction

11.1.1 Search for Better Materials and Devices

Since its invention, semiconductor logic and computing [complementary metal oxide semiconductor (CMOS)] are based on basic inverter circuit consisting of *p*-MOSFET (metal-oxide-semiconductor field-effect transistor) and *n*-MOSFET devices. High output gain, input output isolation, and fan out of the devices in the micro/nanoelectronic circuit are the key factors necessary for the development of microprocessor chips. Silicon has been unanimously accepted as the material of choice on which entire semiconductor foundry is developed. Continuous demand for faster and efficient computation typically involves an improved switching of an input signal between *two states* of an inverter (conventionally, 0 and V_{DD}, the supply voltage). As the physical device scales following *Moore's law* (i.e., the size of a basic modern-day transistor should scale down) [1] and power density scales following *Dennard scaling* (i.e., power density needed for such a transistor should remain constant) [2,3], the logic devices face an aggressive miniaturization. The uncertain future of electronics, on other hand, has provided tremendous new opportunities in micro/nanoelectronic in terms of architectures, materials, devices, and even system level [4]. In addition, the transport physics and reliability requirements in a present-day device technology and its integration are greatly threatened by the enormous heat dissipation in the chip. Since 2011, the technology is being aggressively scaled by leading industries by exploiting FINFET technology with unveiling of 7 and 5 nm technology nodes recently [5,6]. Yet it is reasonable to say that the "conventional scaling" will come to an end within next one decade. Number of new materials, device concepts, and integration techniques have been proposed. Some of them are still at the early stages of study (such as spin-based

Nanoelectronics. DOI: https://doi.org/10.1016/B978-0-12-813353-8.00005-1

electronic devices) but some are with great promises such as tunnel field-effect transistors (TFETs) [7,8]. The scope of the present chapter is outside of device scaling, but provide an alternative material approach in considering III–V nanowires (NWs).

Materials, in the reduced dimensionality, have become increasingly important to further explore the emerging technologies due to their unique quantum mechanical properties. In the context of CMOS devices, in particular, the ideal materials should be ones that could potentially provide platforms for the complex transport mechanism while mitigating chip's enormously large power dissipation problem. To give an example of the figures, each square millimeter in a 32 nm technology node Intel chip has ~382 million transistors in the chip [9]. Typically, average thermal dissipation power of an intel core i3-530 processor is 73 W although a basic *p*-MOS or *n*-MOS in the processor dissipates an extremely small energy of ~450 aJ. III–V materials quantum wells (QWs) (such as InGaAs and InSb) have shown great promises as alternate materials/devices of choice since last decade due to their high electron mobility and low power consumption [10]. Leading industries such as Intel, IBM/Global Foundries, Inter University Microelectronics Center (IMEC), Taiwan Semiconductor Manufacturing Company (TSMC), etc. are being successfully engaged in monolithic integration of III–V devices onto 300 mm wafer scale processes. Furthermore, more emerging technologies such as broken FINFET technologies, NW field-effect transistors (NWFETs) technologies (omega-type), and gate-all-around (GAA) type technologies are being researched and manufactured for future directions. Therefore, it is necessary to discuss various physical properties of III–V NWs and their transistor device characteristics. The present chapter brings a comprehensive review on the current and future materials and device explorations based on two types of NWs: InSb and InAs NWs. Based on their unique physical properties, some of the emerging device concepts are discussed in this chapter.

11.1.2 InSb and InAs materials and their nanowires

Low-dimensional III–V semiconductors are known for their superior transport properties (such as high speed and low power applications). InSb and InAs both share many common features such as (cubic) zinc blend crystal structure, with a direct bandgap of 0.17 eV and 0.35 eV, respectively, at room temperature. Due to low bandgap energies and low effective masses of the carriers (e.g., $m_e^*/m_0 = 0.0116$ at 300 K in case of InSb), InSb and InAs possess one of the highest known room temperature electron mobilities [~77,000 cm^2/(V*s) in case of InSb and ~40,000 cm^2/(V*s) in case of InAs]. Besides, both have unique optical and spin–orbit parameters (such as refractive index, optical phonon energy, Lande *g*-factor) that make the two III–V binary semiconductors technologically important materials. As a

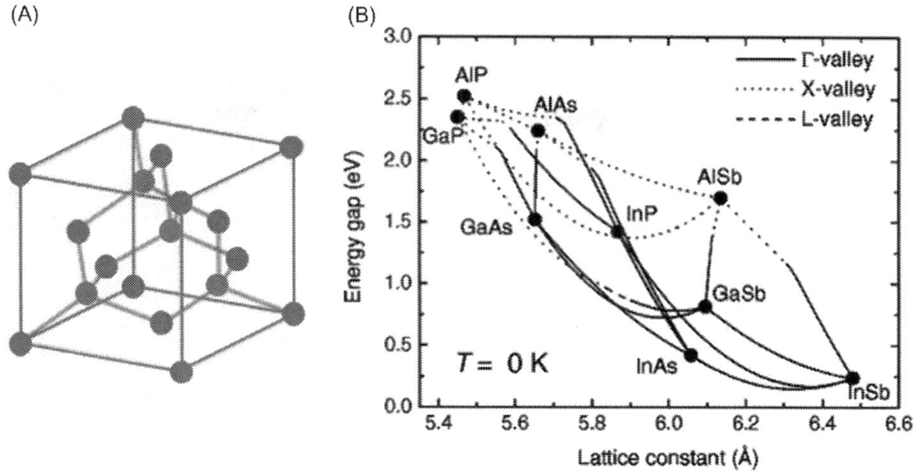

Figure 11.1

(A) Zinc blende crystal structure of InSb and InAs crystals, where the red and green circles denote the two different elements present in the semiconductor. (B) The energy gap (eV) vs lattice constant of bulk InSb and InAs with other well-known III–V semiconductors along various crystallographic directions. Source: *Figure (B)—Copyright American Institute of Physics. Reprinted with permission from I. Vurgaftman, J.R. Meyer, Band parameters for III–V compound semiconductors and their alloys, J. Appl. Phys. 89 (11) (2001) 5815.*

direct consequence of above parameters, they are potential candidates for high speed and low power transistors, infrared photodetectors, magnetic sensors, and spintronic devices applications. Fig. 11.1, showing bandgap energy vs lattice constant graph (along different crystallographic directions), provides the lowest bandgap values and their lattice mismatch/alignment with other III–V semiconducting materials [11]. Crystal designers often use these type of diagrams as guidelines to even tailor materials and form alloys of suitable bandgap (such as indium–antimony–arsenide ternary alloy) with tunable properties. Although InSb and InAs NWs grow in cubic zinc blende crystal structure, some hexagonal wurzite structures during the growth have been reported [12,13]. This change in the structure is associated with change in the energy bandgap and thus could be leveraged as a technique for bandgap and material engineering. Therefore, it is important to understand the growth mechanism to control the structure and property of the material to design desired properties in the device structure.

The various important materials properties of III–V semiconductors as well as recently discovered graphene are compared in the following table. Note that these are all bulk material properties and could substantially change with reduced dimensionality.

Bulk materials	Bulk mobility (cm^2/V s)	Bulk bandgap (eV)	Direct/indirect	Effective mass (bulk) (m_e^*/m_0)	Bulk crystal structure
Si	1400	1.12	Indirect	0.19	Diamond structure
Ge	3900	0.661	Direct	0.082	Diamond structure
GaN	2000	3.47		0.2	Wurtzite
GaAs	8500	1.424	Direct	0.067	Cubic zinc blende
InGaAs	12,000	0.74		0.041	Cubic zinc blende
InAs	40,000	0.354	Direct	0.023	Cubic zinc blende
InSb	80,000	0.17	Direct	0.014	Cubic zinc blende
Graphene (suspended)	200,000	0	Gapless	0	2D atomic thin layer

From Fig. 11.1B, it is apparent that an epi-layer of InSb on silicon, for example, would undergo significant lattice mismatch making it unsuitable for reliable device fabrication. In fact, the lattice mismatch issue constitutes one of the major challenges of InSb and III–V, in general, in the integration onto silicon electronics. In order to deal with the problem along with availing number of other advantages, the notion of thin film QWs type heterostructures and NWs have received widespread attention [10,14]. As shown in Fig. 11.2, leading chipmakers have realized the potential of III–V materials in reduced dimensional device structures (QWs, NWs GAA, TFET, etc.) for future nanoelectronic applications [15]. While the QW structures are beyond the scope of this chapter, the present chapter is limited to the recent development in InSb and InAs NW transistors. These NWs and their FETs could provide motivations to study more complex InSb and InAs NW devices.

11.2 InSb and InAs Nanowire Growth

Typically, NWs are wire-shape materials with circular cross-sections but sometimes they are reported to be triangular or square in shape. For simplicity, we will only consider wires with circular cross-section. When a whisker with a diameter 100 nm or less and a length much larger than its diameter, typically in micrometers (therefore the NWs must possess a high aspect ratio), is grown/fabricated, it constitutes a NW. However, more importantly, when the underlying physics in the materials' physical properties (such as electronic properties, photonic properties, thermal, or any such properties) behave differently from its bulk counterpart, the shape should define a NW structure. Due to the reduced structural

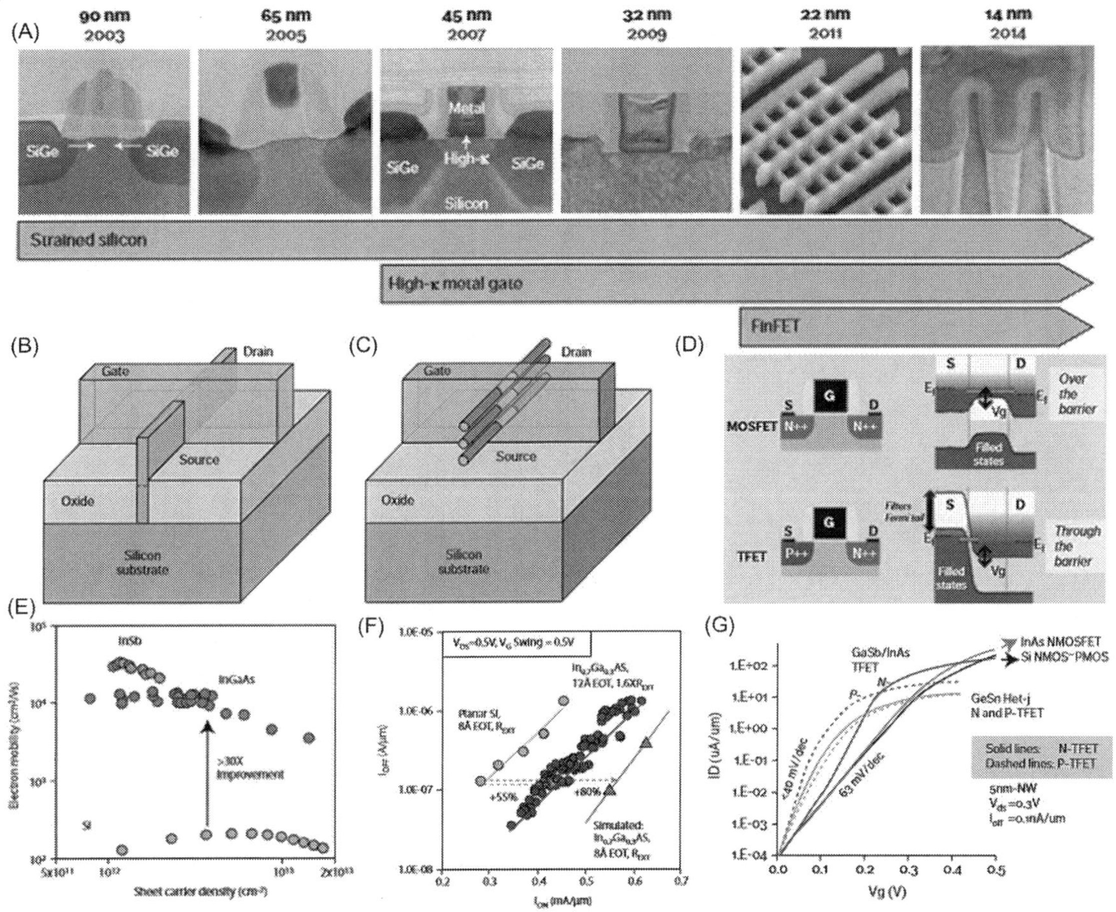

Figure 11.2

(A) Device scaling and Intel's six generations of transistors. (B) Schematic of a present day FinFET. (C) Schematic of a GAA structure. (D) MOSFET and TFET structures and energy diagram. (E) Electrical mobility vs carrier density in III−V and silicon. (F) OFF-current vs ON-current in III−V and silicon. (G) Drive current vs gate voltage for various transistor structures. Source: *Copyright: IEEE. Reprinted with permission from M.T. Bohr, I.A. Young, CMOS scaling trends and beyond, IEEE Comput. Soc. (2017) 20−29.*

geometry in two dimensions compared to the third one, carriers in these tailored structures undergo quantum confinement leading to the changes in the properties due to change in the electronic states. The change in the energy−momentum dispersion relation of charge carriers and the density of states, are best described mathematically and graphically by following (see Figure 11.3):

Dispersion relation

$$E_{3D} = h^2(k_x^2 + k_y^2 + k_z^2)/[8\pi^2 m^*] \tag{11.1}$$

$$E_{2D} = h^2(k_x^2 + k_y^2)/[8\pi^2 m^*] \tag{11.2}$$

$$E_{1D} = h^2 k_x^2/[8\pi^2 m^*] \tag{11.3}$$

$$E_{0D} = k^0. \tag{11.4}$$

Density of states

$$g_{DOS-3D} = \frac{1}{2\pi^2}\left(8\pi^2 m^*\right)^{\frac{3}{2}}(E-E_C)^{1/2} \tag{11.5}$$

$$g_{DOS-2D} = (\frac{4\pi m^*}{h^2})\sigma(E - E_C) \tag{11.6}$$

$$g_{DOS-1D} = \left(\frac{2m^*}{h}\right)[m^*/(2(E-E_C))]^{1/2} \tag{11.7}$$

$$g_{DOS-0D} = 2\delta(E - E_C). \tag{11.8}$$

In general, a top-down approach and a bottom-up approach are used to fabricate/grow low-dimensional materials, such as quantum wires or NWs. Each of these techniques comes with its unique advantages and disadvantages but the later one, where the materials could be grown with molecular and atomic control, provides numerous flexibilities in designing artificial materials. However, due to its much complex nature with underlying unknowns, the bottom-up growth of NWs possesses active research in every electronic material, including InSb and InAs NWs (Fig. 11.3).

Growth of InSb NWs is reported using two types of bottom-up techniques—an electrochemically deposited technique and vapor deposition technique [16−22]. InAs NWs were reported much earlier than InSb NWs using vapor-phase epitaxy method [23]. Over the last decades, high quality of these NWs and epitaxial NWs is mostly grown using chemical vapor deposition (CVD) techniques, such as a metal-organic vapor-phase epitaxy (MOVPE) and vapor-phase epitaxy methods [24−28]. Precursors such as a

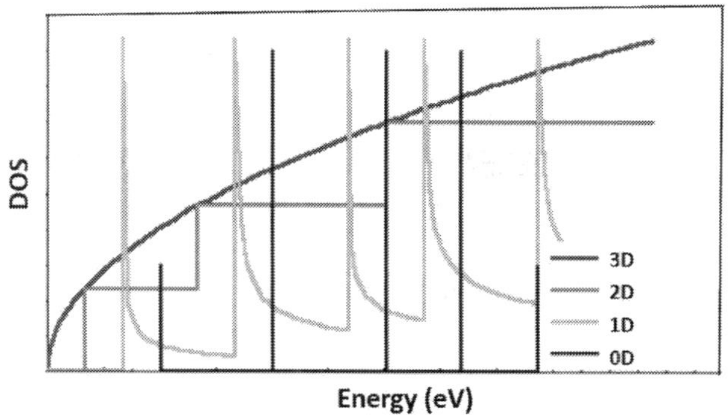

Figure 11.3
A qualitative diagram of the density of states vs the electronic energy showing the electronic properties in solids. The effect of dimensionality and quantum mechanical confinement is shown as the dimension reduces from 3D solids to lower dimensional materials. Source: *Figure modified originally from web.eecs.umich.edu/~fredty.*

metal-organic compound (e.g., trimethylindium, TMIn) and hydrides (such as AsH_3, arsine) are passed along with a carrier gas (nitrogen or hydrogen) and they undergo chemical reactions on a growth substrate in the reactor chamber. Typically, the reaction occurs at 600–700°C temperature and a vapor–liquid–solid (VLS) mechanism is often discussed to explain the growth process although the universal validity of such a mechanism is recently questioned [28].

11.3 InSb and InAs Materials and Their Nanowire Field-Effect Transistors

Physical properties of NWs due to quantum confinement and large surface to volume area make them interesting for various device applications. From a simple particle in a quantum well (QW) type of model, it could be realized that the energy spacing is inversely proportional to effective mass and diameter of the wire. As a result, a ~ 20 nm to 30 nm diameter InAs and InSb NW would demonstrate a similar sub-band spacing as required by a ~5 nm silicon NW. This, along with high mobility of carriers, provides a better opportunity to use InAs and InSb NWs for 1D electronic devices. FET, a voltage-controlled resistor, is the classic device for many applications such as logic devices to sensors, to name few. At low drain-to-source voltage (V_{ds}), the device is linear (in current–voltage relation) and could be well controlled with gate-to-source voltage (V_{gs}). At high field (V_{ds} high), the characteristics become nonlinear with the FET acting as a current source. The current (electron flow) is characterized by the formation of an inversion layer, leading to a condition that ($V_{gs}-V_{th}$) should exceed zero (where V_{th} is the threshold gate-source voltage). InSb and InAs, due to their high bulk electronic mobilities and low effective mass, are suitable candidates for high mobility/speed NWFETs. However, due to their low bandgap energies, a hole conduction is likely to appear simultaneously leading to lower ON/OFF ratio and high OFF-current in the device. However, their NW could positively influence the OFF current due to reduced dimensionality-induced quantum confinement and is suitable to engineer a tunable bandgap. Indeed, Chen et al. [29] have shown a tunable bandgap with diameter for InSb NWs, which significantly increases with wire diameter below 65 nm. This also leads to the absorption of electromagnetic radiation with longer wavelengths, providing enhanced optical properties [30]. An ultrathin diameter InSb and InAs NW could therefore support lower number of sub-bands leading to their potential application for linear devices such as radio-frequency communication [31]. A typical NWFET could be fabricated in many different configurations. Fig. 11.4 provides three commonly found device types studied in the literature. While type-I demonstrates a back-gate FET, type-II and type-III are top-gated FETs and could be operated in dual-gated configurations. A fourth type, a GAA, will be discussed below separately.

The NWFETs are fundamentally different from traditional silicon MOSFETs in the sense that they possess metal source and drain (S/D) contacts rather than doped silicon S/D region. A back-gated NWFET possesses a heavily doped silicon as gate electrode rather than a metal gate. Metal S/D contact electrodes form Shcottky barriers with channel NW material based on the Fermi level alignment. However, a NWFET could provide low parasitic S/D resistance, low temperature processing, and scaling considerations. Therefore, the NWFETs are otherwise known as Schottky barrier FETs (SB-FETs). There is no doping profile in these SB-FETs (except intrinsic doping of the wires obtained during the NW growth) and the transport is fully controlled by the two Schottky barriers controlled carrier injection (and collection) into the NW channel and the gate control of electrical potential in the wire. The following section discusses briefly transport model in these types of devices.

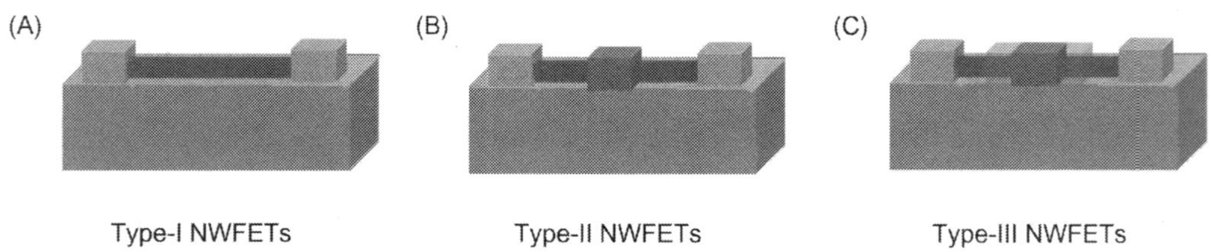

Figure 11.4

Schematic diagram of various generation of NWFETs studied in the literature. (A) Type-I NWFETs uses a back-gate configuration, (B) type-II NWFETs uses a top-gate but un-passivated device, and (C) type-III NWFETs uses a top-gate and passivated device.

11.4 Diffusive Transport Model Within the Channel

Assuming electrons to be majority carriers, a charge per unit length of NW channel (q/l) and drift velocity of each electrons (v_d), the drain−source current within the channel is simply

$$I_{ds} = \left(\frac{q}{L}\right)v_d. \tag{11.9}$$

Furthermore, the drift velocity in terms of mobility and field strength in the channel is given by

$$v_d = -\mu_e E = -\mu_e V_{ds}/L. \tag{11.10}$$

Combining Eqs. (11.9) and (11.10),

$$\frac{I_{ds}L}{q} = -\mu_e V_{ds}/L \tag{11.11}$$

$$I_{ds} = -\frac{\mu_e V_{ds} q}{L^2} = -\frac{\mu_e V_{ds}\left(\frac{q}{L}\right)}{L}. \tag{11.12}$$

Furthermore, the total charge per unit length is related to the gate capacitor and inversion layer as

$$q = C_{ox}\left(V_{gs} - V_{th}\right). \tag{11.13}$$

Combining Eqs. (11.12) and (11.13) and assuming I_{ds} to be electronic current,

$$I_{ds} = -\frac{\mu_e C_{ox}\left(V_{gs} - V_{th}\right)V_{ds}}{L^2} = \mu_e\left(\frac{C_{ox}}{L^2}\right)\left(V_{gs} - V_{th}\right)V_{ds}. \tag{11.14}$$

Note that, these relations hold for small V_{ds} (linear regime), as we discussed earlier.

11.4.1 Electrostatics and Channel Potential with Schottky Barrier at the Metal-nanowire Junctions

The three key physical phenomena that are responsible for transport at the two Schottky barriers are thermionic emission (TE), field emission (FE), and thermionic field emission (TFE). While the TE classically explains the thermally assisted cross over of the barriers by the charge carriers giving rise to a purely thermal current, the FE is responsible for quantum mechanical carrier tunneling through the barrier at the Fermi energy. At higher temperatures (including room temperature), the TFE, associated with thermally assisted tunneling at Fermi energy and FE of the barrier, plays the dominant role for the transport. Therefore, the transport in the device becomes strongly dependent on metal-semiconductor contacts (work function of the metal and nature of the semiconductor including its Fermi energy, etc.), and the height and thickness of the Schottky barrier. Solving for the channel-dielectric potential $\varphi_f(x)$ for the NWFETs, simply it could be approximated by using 1D Poisson's equation as follows [32]:

$$\frac{d^2\varphi_f(x)}{dx^2} - \frac{(\varphi_f(x) - \varphi_g - \varphi_{bi})}{\lambda^2} = -\frac{q(\rho \pm N)}{\epsilon_0 \epsilon_{nw}}, \tag{11.15}$$

where x, φ_g, φ_{bi}, λ, ρ, N, ϵ_0, ϵ_{nw}, and q denote the direction perpendicular to the NW-dielectric interface, gate-potential, built-in potential, a characteristic screening length, density of mobile charges in the NW, constant charge background due to NW doping, free space dielectric constant, dielectric constant of NW, and electronic charge, respectively. In the above equation, positive sign before N is if the charge carriers are donors and negative if they are acceptors. For a constant charge density, above equation could be solved analytically for a solution

$$\varphi_f(x) \propto \exp\left(-\frac{x}{\lambda}\right). \tag{11.16}$$

At the NW-metal electrode interface (i.e., at the Schottky barrier), the above solution (16) bends the conduction band down from the top of the Schottky barrier exponentially to the NW channel with the characteristic screening length λ. λ is a function of NW diameter (d_{nw}), oxide thickness (d_{ox}), NW dielectric constant (ε_{nw}), and oxide dielectric constant (ε_{ox}). For type-I (Fig. 11.4), type-II & III (Fig. 11.4), and GAA type NWFETs they are as follows, respectively:

$$\lambda = [\frac{\epsilon_{nw}}{\epsilon_{ox}} d_{nw} d_{ox}]^{1/2} \tag{11.17}$$

$$\lambda = [\epsilon_{nw} d_{nw}^2 \ln\left(1 + \frac{2d_{ox}}{d_{nw}}\right) / 8\epsilon_{ox}]^{1/2}. \tag{11.18}$$

Therefore, controlling the diameter and oxide thickness, the Schottky barrier could be controlled. The analytical solution for maximum channel potential [i.e., solution for Eq. (11.15), called the surface potential], at OFF state and when channel length is larger than screening length, is given by [32]

$$\varphi_f^{max} \approx \left[-\left(\varphi_{bi} + \varphi_g\right)\left(\varphi_d - \left(\varphi_{bi} + \varphi_g\right)\right)\right]^{\frac{1}{2}} \exp\left(-\frac{L/2}{\lambda}\right) + \varphi_g + \varphi_{bi}, \tag{11.19}$$

where $\varphi_d = -qV_{ds}$ is the drain potential.

For $L >> \lambda$, Eq. (11.19) gives

$$\varphi_f^{max} \approx \varphi_g + \varphi_{bi} \tag{11.20}$$

$$\frac{\partial \varphi_f^{max}}{\partial \varphi_g} \approx 1. \tag{11.21}$$

Under these ideal conditions, the surface potential maximum, that decides the carrier injection into the channel in the NW SB-FET, becomes one-to-one electrostatically controlled by the gate voltage.

11.4.2 1D Transport and Landauer Formalism

The conventional SB-FET subthreshold current, current due to the source−drain injected carrier flow into the channel, is given by

$$I_{ds} = \int_{E_C}^{\infty} g_{DOS-1D}(E) v(E) \{f_1 - f_2\} T(E) dE. \tag{11.22}$$

The above relation includes both electronic current and hole current and represents the total drain-to-source effective current. Therefore, irrespective of a n-type, p-type, or an ambipolar device, Eq. (11.22) should provide the overall device transfer characteristics. $g_{DOS-1\mathrm{D}}$ is the density of states in 1-D structure (11.7) and v represents the carrier velocity, f_1 and f_2 represent the fermi functions at source and drain, and T represents the transmission probability across the device channel. Depending on the nature of the device geometry and the metal-NW materials, contact transmission probability could be modeled and and its values could be used in Eq. (11.22) to calculate the device current.

11.5 InSb and InAs NW SB-FETs

Room temperature InSb NW SB-FETs with NWs of diameter ~100 nm and sub-100 have been recently experimentally fabricated and characterized. The wires are grown using two bottom-up approaches discussed earlier. For example, Das et al. [33] have grown InSb NWs using an electrodeposition technique with subsequent controlled environment thermal annealing producing phase pure and crystalline NWs. Furthermore, they have fabricated back-gated InSb NW SB-FETs with 500 nm channel length on 20 nm oxide and with Ni S/D electrodes. The devices showed high ON current (I_{ON} ~ 40 μA/wire) with ON/OFF ratio of ~20 and mobility of ~1200 cm^2/V s. While the average saturated electron velocity was shown to be comparable to bulk InSb saturation velocity, the reduced mobility is argued to be due to various scattering events (such as boundary scattering, electron−phonon, and electron−electron scattering). The device structure, crystallinity of the NW, and the results are summarized in Fig. 11.5 (upper row). Indeed, the NWs grown using electrodeposition method with a porous anodic alumina template by Das et al. produce NWs with rough interfaces and

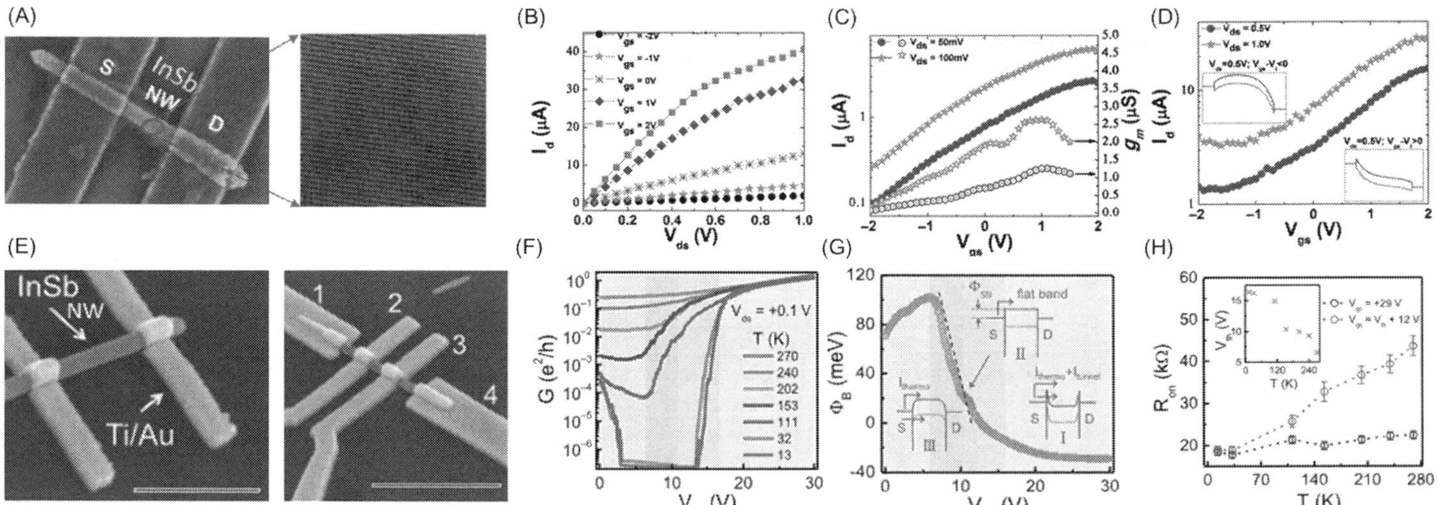

Figure 11.5

InSb NW SB-FETs studied using two growth techniques discussed (i.e., the electrodeposition and the vapor deposition techniques). All the devices shown have a diameter ~90–110 nm. Source: *Copyright: AIP and IoP. Reprinted with permission from S.R. Das, Room temperature device performance of electrodeposited InSb nanowire field effect transistors, Appl. Phys. Lett. 98 (2011) 243504; D. Fan, et al., Schottky barrier and contact resistance of InSb nanowire field-effect transistor. Nanotechnology 27 (2016) 275204.*

boundaries that could be significantly responsible for the carrier scattering in the transport. InSb NWs with very similar diameters but grown via vapor-phase method are recently used by Gul et al. [34] to fabricate bottom-gated SB-FETs on 285 nm gate oxide and with Ti/Au S/D contact metal of thickness of 5/145 nm. A low temperature, electronic mobility of $\sim 25{,}000 \text{ cm}^2/\text{V s}$ is reported at 4.2 K, showing the highest mobility achieved on NW FETs. However, the NWFET mobility is still three times smaller than bulk InSb mobility value and is largely attributed to the surface sensitivity of NWs toward the foreign species' adsorption as well as to the surface cleanliness of the substrate materials. Two-terminal and four-terminal devices were studied to probe the effect of contact resistance and the barrier heights (Fig. 11.5).

The choice of metals and the nature of contacts between the metal and the NW channel are directly related to the Schottky barrier heights and the contact resistance in the NWFETs. Fan et al. [35] have recently reported InSb NWFETs (both bottom-gated two-terminal and top-gated four-terminal devices) with Ti/Au contact electrodes and reported a gate tunable Schottky barrier height of ~ 20 meV. The temperature-dependent conductance, Schottky barrier height, and the ON-state contact resistance are all shown in Fig. 11.5F–H, respectively. The contact resistance severely affects the measured/extrinsic transconductance of the device and hence the field-effect mobility as follows [36]:

$$g_{int} = \frac{\mu C}{L^2} V_{ds}, \tag{11.23}$$

and the extrinsic transconductance is given by

$$g_{ext} = g_{int}/[1 + g_{int}R_s + (R_S + R_D)R_{int}, \tag{11.24}$$

where R_S, R_D, and R_{int} are source contact resistance, drain contact resistance, and the intrinsic (NW) channel resistance, respectively. Substituting Eq. (11.23) in (11.24), one could extract the extrinsic mobility of the device

$$\mu_{ext} = \left[\frac{\partial g_{ext}}{\partial V_{gs}}\right]_{V_{ds}}. \tag{11.25}$$

11.6 GAA NWFETs

The GAA NWFET architecture is being presently actively implemented by several leading chip industries for consideration of next generation CMOS beyond FinFET technology. While undoubtedly, the GAA structure has superior electrostatic gate control compared to three NWFET types mentioned earlier, issues such as dealing with parasitic capacitances and further fabrication–performance optimization remain to be addressed. III–V NWs, with additional advantages of high mobility and low power consumption, have great potential for the GAA NWFET technology. Fig. 11.6 shows the fabrication and transistor characterization of *n*-InAs NW GAA FETs reported by Egard et al. [37], where the wires were grown epitaxial on semi-insulating InP substrates and source, drain, and gate electrodes were fabricated with wrap-around design. NW arrays (70 NWs) with GAA structure with 100 nm gate length showed unity current gain cut-off frequency of 7.4 GHz and maximum frequency of oscillation of 20 GHz.

11.7 Transport in NW Tunnel FETs

Due to their inverse subthreshold slope below the thermionic limit of 60 mV/decade [ln(10) **kT/q*], TFETs are one of the most promising devices that have lower power consumption and thus they are potential candidates for future CMOS logic applications. Although the TFET signatures have been observed in few material systems, they are burdened by low ON-currents. III–V materials, with low effective tunneling mass, could be designed to overcome this constraint by primarily growing homogeneous III–V wires or hybrid systems (e.g., InAs/GaSb NW, InAs/InGaAsSb/GaSb NW, etc.) with narrow bandgap. Typically a type-III staggered bandgap or a type-III broken bandgap along the device channel is suitable [38–40]. Such NW materials could enable a GAA NWTFET which could not only increase the ON-current but enhance the electrostatic gate control. Three columns in Fig. 11.7 corresponding to three different configurations of III–V NW system, such as InAs/Si, InAs/GaSb, and InAs/InGaAsSb/GaSb NWs, respectively [38–40], show the TFETs with near 60 mV/decade or below 60 mV/decade subthreshold swing as well as a μA or above ON-current.

11.8 Emerging Non-CMOS Nanoelectronic Devices and Quantum Devices

In addition to the above-described NW devices, InSb and InAs NWs offer very rich platforms for studying quantum effects in lower dimensions. 1-D sub-band manipulation and ballistic transport achievements in these NW devices are some of the characteristics that are already observed at room temperature [31,41–43]. Both electron and hole quantum dots formation and their spin spectroscopy were studied recently in InSb NW quantum dot devices [44,45]. Although far from room temperature realization, recently there have been many exotic phenomena observed and reported in InSb and InAs NW devices with metals such as niobium and aluminum. Such systems support the existence of an elusive particle, called a Majorana fermion, a fermion that has been long proposed as its own antiparticle [46–48]. Such an entity and system are envisioned for constructing Majorana quantum bits for quantum computation [49].

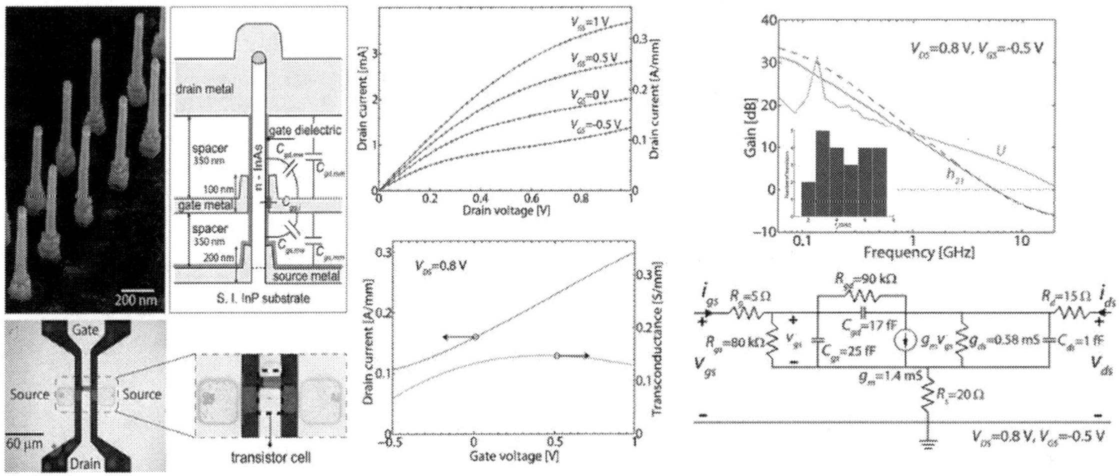

Figure 11.6

InAs NW-GAA device fabrication (left-most column): *n*-InAs were grown on semi-insulating InP substrates. Scanning electron micrograph (SEM) of grown InAs NWs on silicon and covered with 5-nm HfO_2 and a 20-nm SiO_2 spacer is shown along with the schematic of GAA structure. Lower images show top view optical images of the fabricated devives. Images in the middle column show the output characteristics, transfer characteristics, and the transconductance of the GAA NWFETs. The right column shows the current gain (h_{21}) and power gain (U) of ground-signal-ground coplanar device structure measured between 60 MHz and 20 GHz. The histogram (inset) shows the statistics of the cut-off frequency on 222 nominally identical devices on same chip. The circuit diagram shows the equivalent small signal model of the InAs GAA NWFET. Source: *Copyright: ACS (Nano Letters). Reprinted with permission from M. Egard, et al., Vertical InAs nanowire wrap gate transistors with ft > 7 GHz and fmax > 20 GHz, Nano Lett. 10 (2010) 809–812.*

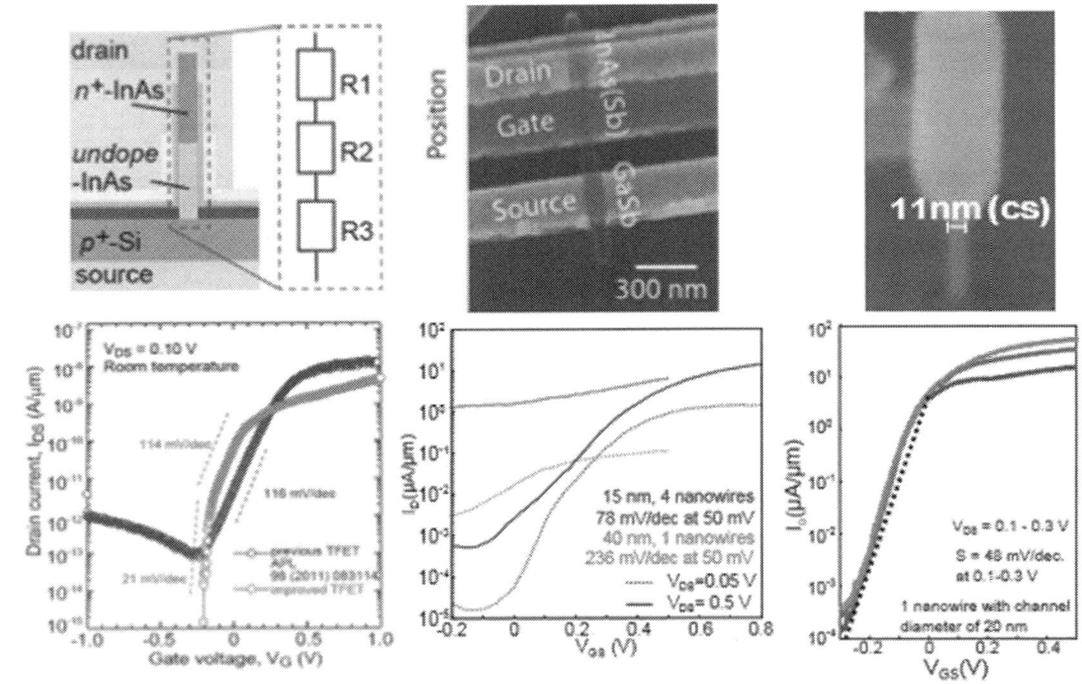

Figure 11.7

TFETs with III—V GAA NWFETs with three different configurations: left column represents the device schematic and the transfer characteristics of InAs NW/Si heterostructure at room temperature. Middle column corresponds to the TFET scanning electron microscope image and transfer characteristics of InAs/GaSb hybrid NW device. The right column shows the e ectron microscope image of a InAs/InGaAsSb/GaSb hybrid NW device and its transfer characteristics. Source: *Copyright: IEEE. Reprinted with permission from K. Tomioka, et al., Steep-slope tunnel field-effect transistors using III—V nanowire/Si heterojunction, 2012 Symposium on VLSI Technology Digest of Technical Papers, pp. 47—48; A.W. Dey, et al., High-current GaSb/InAs(Sb) nanowire tunnel field-effect transistors, IEEE EDL 34 (2013) 2; E. Memisevic et al., Individual defects in InAs/InGaAsSb/GaSb nanowire tunnel field-effect-transistors operating below 60 mV/decade, Nano Lett. 17 (2017) 4373—4380.*

11.9 Conclusion and Outlook

In conclusion, the present chapter discusses the promises of InSb and InAs NW transistors as future high speed and low power nanoelectronic devices. Summary of their growth techniques and underlying transport processes is made. Their assessment as potential future nanoelectronic devices and efforts to integrate with existing CMOS platforms is presented by summarizing numerous progress made in the recent years. Much of the electronic transport, size-dependent quantum confinement, and device physics-related work have provided further insight into their manipulation in the device structures. Furthermore, recent developments in the hybrid quantum devices (such as NW quantum dots as single electron and single hole devices) and quantum heterostructures (such as semiconductor wire with superconducting electrodes) bring numerous promises for future electronic devices for quantum computations and unconventional electronics. Overall, the III–V NWs and devices have great opportunities for exploring nanoscience and nanotechnology.

Acknowledgments

S.R.D. acknowledges Department of Industrial and Manufacturing Systems Engineering and Electrical and Computer Engineering at Kansas State University, Manhattan, Kansas.

References

[1] G.E. Moore, Cramming more components onto integrated circuits, Electronics 38 (8) (1965) 114–117.
[2] R. Dennard, et al., Design of ion-implanted MOSFETs with very small physical dimensions, IEEE Journal of Solid State Circuits, SC 9 (5) (1974) 256–268.
[3] M. Bohr, A 30 year retrospective on Dennard's MOSFET scaling paper, IEEE SSCS NEWSLETTER (2007) 11–13.
[4] K. Bernstein, et al., Device and architecture outlook for beyond CMOS switches, Proceedings of the IEEE 98 (12) (2010) 2169–2184.
[5] R. Xie, et al., A 7 nm FinFET technology featuring EUV patterning and dual strained high mobility channels, IEDM 978-1-5090-3902-9/16 (2016).
[6] Editorial Note: Moore's deviation, Nature Nanotechnology 12 (2017) 1105.
[7] S.A. Wolf, et al., Spintromics: a spin-based electronics vision for the future, Science 294 (2001) 1488–1495.
[8] A.C. Seabaugh, et al., Low-voltage tunnel transistors for beyond CMOS logic, Proceedings of the IEEE 98 (12) (2010) 2095–2110.
[9] https://ark.intel.com/products/46472/Intel-Core-i3-530-Processor-4M-Cache-2_93-GHz.
[10] M. Cooke, Digital III–Vs review, The Adv. Semicond. Magazine. 19 (3) (2006) 18–22.
[11] I. Vurgaftman, J.R. Meyer, Band parameters for III–V compound semiconductors and their alloys, J. Appl. Phys. 89 (11) (2001) 5815.
[12] S.Q. Wang, H.Q. Ye, A plane-wave pseudopotential study on III–V zinc-blende and wurtzite semiconductors under pressure, J. Phys. Condens. Matter 14 (2002) 957.
[13] K. Takahashi, T. Moriizumi, Growth of InAs whiskers in wurzite structure, Jpn. J. Appl. Phys. 5 (1966) 657.
[14] R. Chau, et al., Benchmarking nanotechnology for high-performance and low-power logic transistor applications, IEEE Trans. Nanotechnology 4 (2) (2005) 153–158.
[15] M.T. Bohr, I.A. Young, CMOS scaling trends and beyond, IEEE computer society (2017) 20–29.
[16] S.V. Zaitsev-Zotov, et al., Luttinger-liquid-like transport in long InSb nanowires, J. Phys.: Condens. Matter 12 (2000) L303.
[17] H.D. Park, et al., Growth of high quality, epitaxial InSb nanowires, J. Crystal Growth 304 (2007) 399–401.
[18] Y. Wang, et al., Field effect transistor based on single crystalline InSb nanowire, J. Mater. Chem. 21 (2011) 2459–2462.
[19] X. Zhang, et al., Fabrication of highly ordered InSb nanowire arrays by electrodeposition in porous anodic alumina membranes, J. Electrochem. Soc. 152 (2005) C664–C668.

[20] S.R. Das, et al., Electrodeposition of InSb branched nanowires: Controlled growth with structurally tailored properties, J. Appl. Phys. 116 (2014) 083506.
[21] M. Khan, et al., Electrochemical growth of InSb nanowires and report of a single nanowire field effect transistor, Journal of Nanoelectronics and Optoelectronics 3 (199-202) (2008) 4.
[22] X. Yang, et al., Ultrathin single crystal indium antimonide nanowires, Cryst. Growth Des. 10 (2010) 2479–2482.
[23] M. Yazawa, et al., Heteroepitaxial ultrafine wire-like growth of InAs on GaAs substrates, Appl. Phys. Lett. 58 (1991) 1080.
[24] M. Yazawa, et al., Effect of one monolayer of surface gold atoms on the epitaxial growth of InAs nanowhiskers, Appl. Phys. Lett. 61 (1992) 2051.
[25] K. Tomioka, et al., Control of InAs nanowire growth directions on Si, Nano Lett. 8 (2008) 3475–3480.
[26] B. Mandl, et al., Au-free epitaxial growth of InAs nanowires, Nano Lett. 6 (2006) 1817–1821.
[27] L.E. Jensen, et al., Role of surface diffusion in chemical beam epitaxy of InAs nanowires, Nano Lett. 4 (2004) 1961–1964.
[28] K.A. Dick, et al., Failure of the vapor – liquid – solid mechanism in Au-assisted MOVPE growth of InAs nanowires, Nano Lett. 5 (2005) 761–764.
[29] Chen, H., Sun, X., Lai, K.W.C., Meyyappan, M., Xi, N. Infrared detection using an InSb nanowire. Nanotechnology Materials and Devices Conference, 2009. NMDC '09. IEEE, 2-5 June 2009. DOI: 10.1109/NMDC.2009.5167558.
[30] A. Mohammad, et al., Wavelength-dependent absorption in structurally tailored randomly branched vertical arrays of InSb nanowires, Nano Lett. 12 (2012) 6112–6118.
[31] A. Razavieh, D.B. Janes, J. Appenzeller, Transconductance linearity analysis of 1-D, nanowire FETs in the quantum capacitance limit, IEEE Transactions on Electron Devices 60 (2013) 2071–2076.
[32] J. Appenzeller, et al., Toward nanowire electronics, IEEE Transaction on Electron Devices 55 (2008) 2827–2845.
[33] S.R. Das, et al., Room temperature device performance of electrodeposited InSb nanowire field effect transistors, Appl. Phys. Lett. 98 (2011) 243504.
[34] O. Gul, et al., Towards high mobility InSb nanowire device, Nanotechnology 26 (2015) 215202.
[35] D. Fan, et al., Schottky barrier and contact resistance of InSb nanowire field-effect transistor, Nanotechnology 27 (2016) 275204.
[36] W. Lu, P. Xie, C.M. Lieber, Nanowire transistor performance limits and application, IEEE Transactions on Electron Devices 55 (2008) 2859–2876.
[37] M. Egard, et al., Vertical InAs nanowire wrap gate transistors with $f_t > 7$ GHz and $f_{max} > 20$ GHz, Nano Lett. 10 (2010) 809–812.
[38] Tomioka, K. et al., Steep-slope tunnel field-effect transistors using III–V nanowire/Si heterojunction. 2012 Symposium on VLSI Technology Digest of Technical Papers, pp. 47-48.
[39] A.W. Dey, et al., High-current GaSb/InAs(Sb) nanowire tunnel field-effect transistors, IEEE EDL 34 (2) (2013).
[40] E. Memisevic, et al., Individual defects in InAs/InGaAsSb/GaSb nanowire tunnel field-effect-transistors operating below 60 mV/decade, Nano Lett. 17 (2017) 4373–4380.
[41] S. Chuang, et al., Ballistic InAs nanowire transistors, Nano Lett 13 (2013) 555–558.
[42] I. van Weperen, et al., Quantized conductance in an InSb nanowire, Nano Lett. 13 (2013) 387–391.
[43] E. Halpern, et al., Room temperature observation of quantum confinement in single InAs nanowires, Nano Lett. 15 (2015) 481–485.
[44] S. Nadj-Perge, et al., Spectroscopy of spin–orbit quantum bits in indium antimonide nanowires, Phys. Rev. Lett. 108 (2012) 166801.
[45] V.S. Probiag, et al., Electrical control of single hole spins in nanowire quantum dots, Nature Nanotechnology 8 (2013) 170–174.

[46] V. Mourik, et al., Signatures of Majorana Fermions in hybrid superconductor-semiconductor nanowire devices, Science 336 (2012) 1003–1007.
[47] M.T. Deng, et al., Anomalous zero-bias conductance peak in a Nb–InSb nanowire–Nb hybrid device, Nano Lett. 12 (2012) 6414–6419.
[48] L.P. Rokhinson, et al., The fractional a.c. Josephson effect in a semiconductor–superconductor nanowire as a signature of Majorana particles, Nature Physics 8 (2012) 795–799.
[49] A.Y. Kitaev, Unpaired Majorana fermions in quantum wires, Phys.-Usp. 44 (2001) 131.

CHAPTER 12

Carbon Nanotube and Nanowires for Future Semiconductor Devices Applications

Saurabh Chaudhury and Sanjeet Kumar Sinha

Department of Electrical Engineering, National Institute of Technology Silchar, Assam, India

12.1 Introduction

Metal oxide semiconductor field-effect transistor (MOSFET) has been the main building block in high-performance and low-power very large-scale integration (VLSI) chips for the last several decades. Device scaling is fundamental to technological advancements, which allows more devices to be integrated on a single die providing greater functionality per chip. Ultimately, the goal of scaling is to build an individual transistor that is smaller, faster, cheaper, and consumes less power. Scaling continued following Moore's law [1] initially and now we see an exponential growth in today's nanoscaled chip. However, device scaling to deep nanometer regime leads to exponential increase in leakage currents and excessive heat generation. Moreover, fabrication process variability causes a limitation to further scaling. The basic metal oxide semiconductor (MOS) transistor size has shrunk from a feature size of several microns to less than 32 nm during the last two decades. The short channel effects, threshold voltage roll-off, and drain-induced barrier lowering become increasingly significant as the channel length of semiconductor devices is reduced to deep nanometer range. Researchers believe that with a mix of chemistry, physics, and engineering, nano-electronics may provide a solution to increasing fabrication costs and may allow integrated circuits to be scaled beyond the limits of the modern transistor. As the oxide thickness scales down to 1.5 nm, leakage current increases sharply, leading to high power consumption and reduced device reliability. When the channel length of a MOSFET will go down below 10 nm, silicon-based technology will soon reach its limits [2] and Moore's law will no longer hold true. Researchers and engineers are continuously harnessing for new materials, devices, and fabrication technologies to mitigate the issue of scaling in the post-silicon era to meet increasing demand of the market. In fact semiconductor industry is looking for different materials and devices to integrate with the current silicon-based technology in a long-term future. Carbon nanotubes (CNTs) and nanowires (NWs) are found to be the most promising candidate in this context, which can fit into the traditional MOSFET structure as channel

Nanoelectronics. DOI: https://doi.org/10.1016/B978-0-12-813353-8.00014-2

material and facilitate further scaling. CNT- and NW-based FETs (CNTFET and NWFET, respectively) have been analyzed and characterized in laboratory and also been demonstrated as prototypes. The new devices, such as Silicon on insulator (SOI) MOSFET, FinFET, or Tunnel FET, have also attained additional performance improvements that can overcome the device scaling difficulty [3]. Much progress has been made in recent years showing that CNTFETs and NWFET can outperform the state-of-the-art silicon FETs in many ways [4,5].

This chapter at first presents various structures of CNTFET and NWFET devices and then an extensive simulation-based study and analysis is carried out for CNTFET and NWFET devices together with a comparison of the results with conventional MOSFET. Further, an exhaustive simulation study has been carried out to see the effect of variations of device parameters on the electrical characteristics of these new devices, especially the CNTFET. From this study, it can be concluded that these new devices have got some excellent properties and favorable characteristics which will make them the most suitable candidates for future semiconductor devices in post-silicon era.

12.2 Structure of CNTs

CNTs have some excellent properties which make them a promising candidate of future electronic systems. The performance of CNT transistors has been illustrated in Refs. [6–8]. Diamond and graphite are well-known allotropes of carbon. Allotropy or allotropism is the property of some elements so that they can exist in two or more different forms, known as allotropes of these elements. Allotropes are different structural modifications of an element, where the atoms of the element are bonded together in a different manner. In diamond, the carbon atoms are bonded together in a tetrahedral lattice arrangement, whereas in graphite, the carbon atoms are bonded together in a hexagonal lattice arrangement (Fig. 12.1).

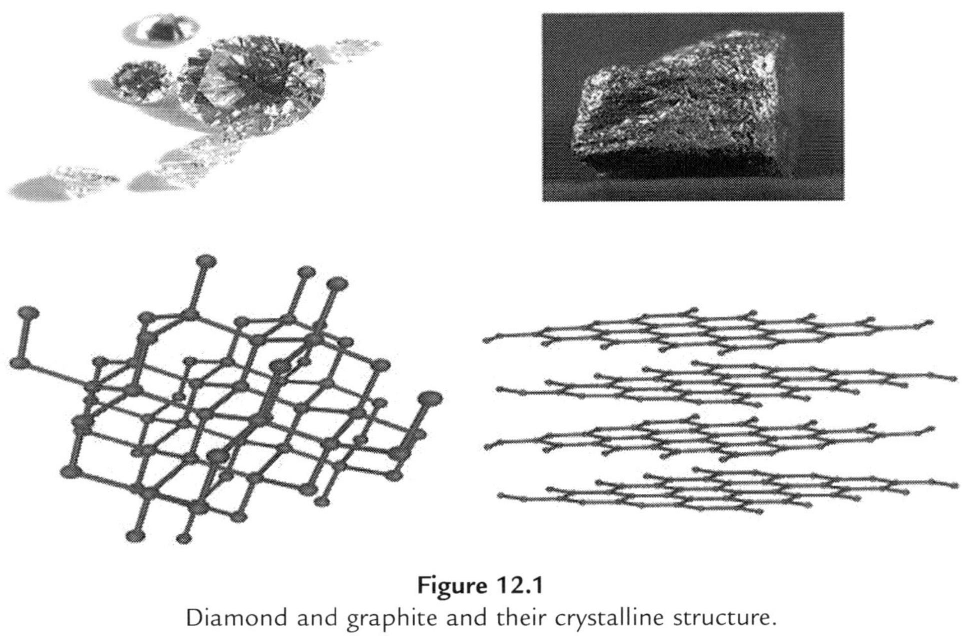

Figure 12.1
Diamond and graphite and their crystalline structure.

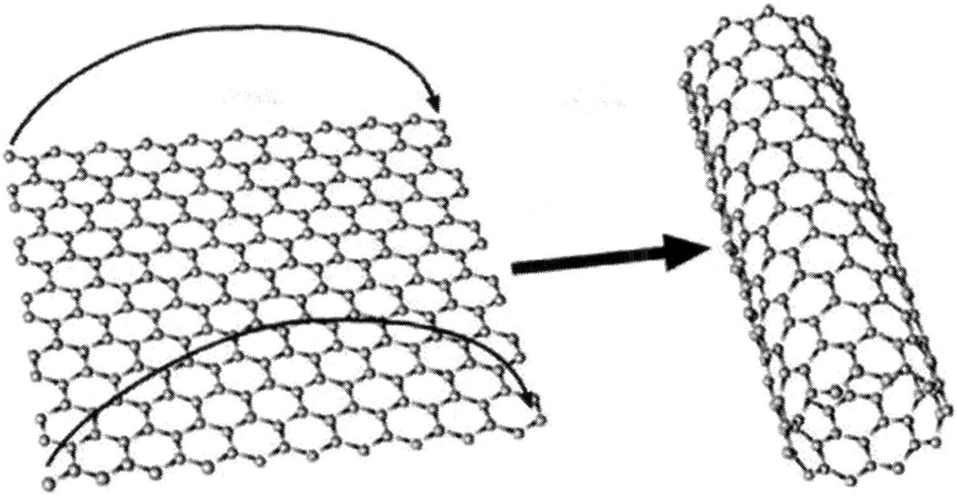

Figure 12.2
Graphene sheet rolled up to form CNT.

Graphite is formed by the carbon atoms when arranged in a honeycomb pattern. These honeycomb layers are stacked one above the other. A single sheet of graphite is very stable, strong, and flexible. CNTs were first observed by Japanese electron microscopist Sumio Iijima in 1991.

A CNT is one-atom-thick sheet of graphite (which is called graphene) rolled up into a seamless cylinder as shown in Fig. 12.2. In other words, CNTs are extended tubes of rolled graphene sheets. Even if every technology node is guided by Moore's law, from now onward, we can expect that by 2019 it will result in transistor which will be just a few atoms in width. We have already seen a progression from a micron, to submicron and then down to 45 nm scale. CNTs, whose walls are just one atom thick, with diameters of only 1–2 nm, seem to be one of the perfect candidates to take us right to the end of Moore's law curve. We possibly cannot go beyond that, so certainly CNTs have a promising future.

12.2.1 Symmetry Structure of Nanotubes

The structure of a CNT is formed by a layer of carbon atoms that are bonded together in a hexagonal (honeycomb) mesh. This one-atom-thick layer of carbon is called graphene, and it is wrapped in the shape of a cylinder and bonded together to form a CNT. Nanotubes can have a single outer wall of carbon, or they can be made of multiple walls (cylinders inside other cylinders of carbon). CNTs have a range of electric, thermal, and structural properties that can change based on the physical design of the nanotube. Single-walled CNTs can be formed in three different designs: armchair, chiral, and zigzag. The design depends on the way the graphene is wrapped into a cylinder. For example, imagine a rolling a sheet of

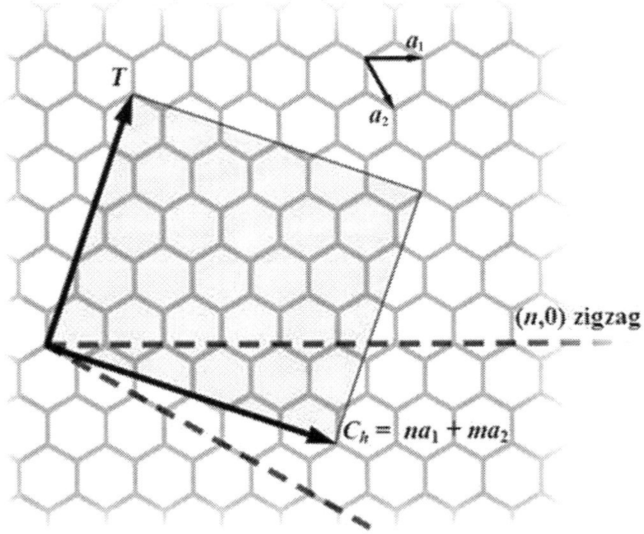

Figure 12.3
Chiral vectors.

paper from its corner, which can be considered one design and a different design can be formed by rolling the paper from its edge. A single-walled nanotube structure is represented by a pair of indices (m,n) called the chiral vector (Fig. 12.3).

Most CNTs do not have zigzag and armchair symmetric forms but have the chiral symmetry structure. Through this symmetry structure we can classify how the carbon sheet is wrapped into a tube. The way the graphene sheet is wrapped is represented by a pair of indices (m,n). The integers n and m denote the number of unit vectors along two directions in the honeycomb crystal lattice of graphene. If $m = 0$, the nanotubes are called zigzag nanotubes, and if $m = n$, the nanotubes are called armchair nanotubes. Otherwise, they are called chiral (Figs. 12.4–12.6).

The structural design has a direct effect on the electrical properties of nanotube. When $m - n$ is a multiple of 3, then the nanotube is described as "metallic" (highly conducting), otherwise the nanotube is a semiconductor. The armchair design is always metallic while other designs can make the nanotube a semiconductor.

CNTs have an important role especially when the devices are in nanometer regime, because of their novel chemical and physical properties. They are mechanically very strong and can conduct electricity extremely well. All of these remarkable properties give CNTs a potential application for nano-electronics. There are two structural models of multiwalled nanotubes. In the Russian Doll model, a CNT contains another nanotube inside it (the inner nanotube has a smaller diameter than the outer nanotube). In the Parchment model, a single graphene sheet is rolled around itself multiple times, resembling a rolled up scroll of paper.

Figure 12.4
Zigzag symmetry structure of CNT.

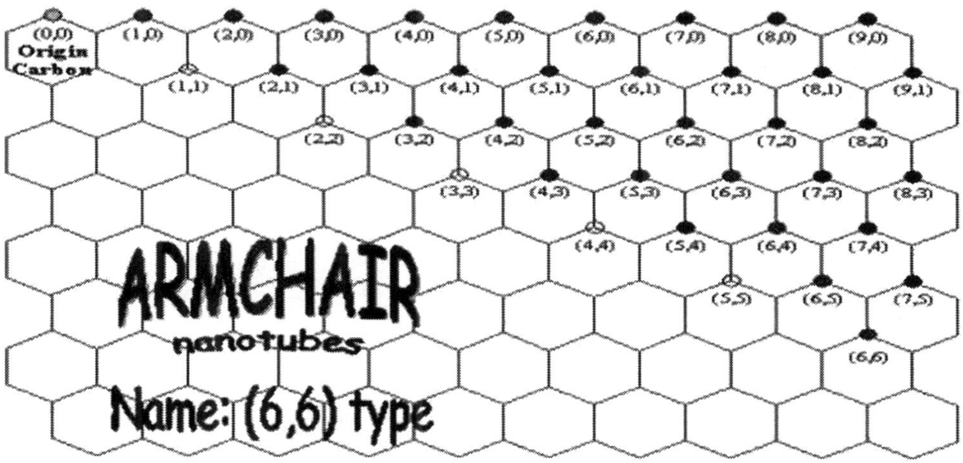

Figure 12.5
Armchair symmetry structure of CNT.

12.2.2 Electronic Characteristics

A free atom of carbon has the electronic structure of $1s^2 2s^2 2p^2$. In order to form covalent bonds, one of the 2s electrons is promoted to 2p and form sp^2 hybridization. Due to other single electron in s orbital it has semimetal characteristics. The overlap of orbitals on adjacent atoms in a given plane provides the electron bond network causing electrical conductivity of the graphite. The origin of the versatility of carbon is the possibility of hybridization of the orbitals of the valence electrons. Carbon has four valence electrons, two in s orbitals and two in p orbitals. Since it is energetically favorable to maximize the

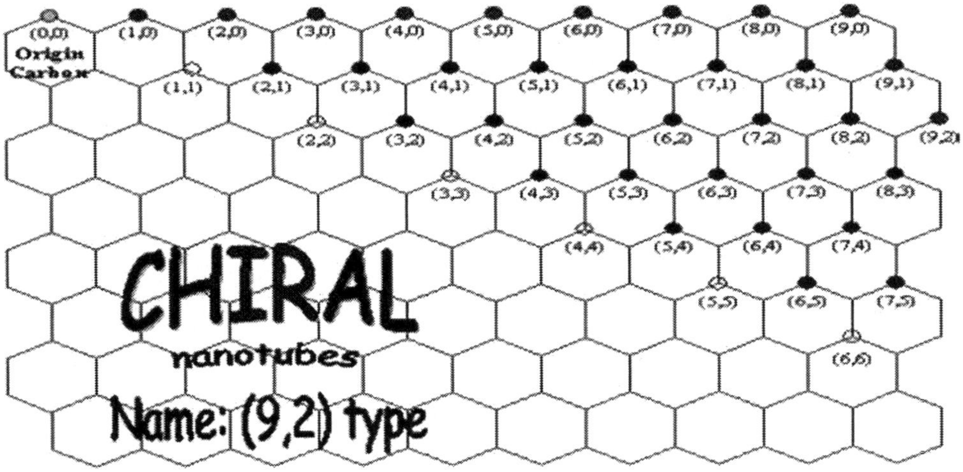

Figure 12.6
Chiral symmetry structure of CNT.

number of bonds, one s electron can be excited to the empty p orbital and one, two, or three of the p orbitals mix with the remaining electron in the s orbital. In diamond which has an sp^3 configuration one s and three p orbitals hybridize to give rise to four new orbitals. Graphite has an sp^2 configuration, only two of the p orbitals hybridize with an s orbital to give three orbitals that lie in a plane leaving the p_z orbital unaltered.

The electronic properties of CNTs make them not only suitable for transistors but also for applications such as sensors [9], high-frequency diodes [10], light-emitting diodes (LEDs) [11], and efficient photovoltaic devices [12]. Due to their high strength and low density, CNTs are attractive for nano-electromechanical systems [13]. Even though both metallic and semiconducting CNTs have excellent electrical properties, the coexistence of both types has been a serious obstacle for realization of many electronic applications that require use of only one type. There are techniques for separating CNT material [14] and for selective growth [15] of one type of CNTs, but none of these approaches have been successful enough to make CNTs competitive for large-scale integration in electronics.

12.2.3 The CNTFET Device Structures

Several CNTFET structures have been fabricated and demonstrated in literature. However, the following four structures have been widely illustrated. These are:

1. Top-gated CNTFET
2. Back-gated CNTFET
3. Gate all-around CNTFET (or wraparound gate CNTFETs)
4. Suspended CNTFET.

12.2.3.1 Top-gated CNTFET

It is one of the most recent and advanced fabrication processes [16]. This is the structure which we have taken into consideration for our all simulation and analysis purposes. Because of the presence of ultra-thin gate dielectric, a larger electric field can be generated with respect to the nanotube using a lower gate voltage. This is the reason why top-gated devices are generally preferred over back-gated CNTFETs.

The device structure is fabricated in the following way: Single-walled CNTs in the form of solution are deposited on the top of a silicon dioxide substrate. Individual nanotubes are then identified with the help of atomic force microscope or scanning electron microscope. Once an individual tube is isolated, source and drain contacts are defined and patterned using high-resolution electron beam lithography. A fine layer of top-gated dielectric is then deposited over the nanotube, via either via evaporation or atomic layer deposition. Finally, the gate contact is deposited on the gate dielectric, thus completing the fabrication process. In a similar manner, array of top-gated CNTFET can be fabricated.

12.2.3.2 Back-gated CNTFET

This is one of the oldest techniques for fabricating CNT field-effect transistors (FETs) involving pre-patterning parallel strips of metal across a silicon dioxide substrate, and then depositing the CNTs on top in a random pattern [17]. Benefit of the technique is that it is a simple process. However, it suffers from many drawbacks. First of all it gives a poor metal contact with nanotubes and because of the semiconducting nature of CNT it forms Schottky barrier at the metal–semiconductor interface. Secondly, because of back-gated geometry, the on–off becomes difficult at low gate voltage.

12.2.3.3 Gate all-around CNTFET

Gate all-around CNTFET, which is also known as wraparound gate CNTFET, is a further advancement over the top-gated device geometry. In this device, instead of gating just the part of the CNT which is closer to the metal gate contact, the entire circumference of the nanotube is gated [18]. This would ideally improve the electrical performance of the CNTFET, thus reducing the leakage current and improving the device on/off ratio.

12.2.3.4 Suspended CNTFET

Suspended CNTFET is yet another device geometry which involves suspending the nanotube over a trench to reduce contact with the substrate and gate oxide [19]. This technique has the advantage of reduced scattering at the CNT–substrate interface, thus improving the device performance.

12.3 Semiconductor NWs

Semiconductor exhibits an energy band gap, which represents an energy interval in which no electron states are allowed. In a semiconductor the states below the gap are fully occupied by electrons, and the states above the gap are empty. Since completely filled bands carry no net current, these materials are insulators at absolute zero. However, at room temperature these materials have a finite conductance that can be controlled by introducing impurity atoms and applying electrical fields. In a semiconductor transistor, doping is introduced during fabrication processes, and electrical fields are then applied during operation to control the conductance, so called complementary metal oxide semiconductor (CMOS) technology, and it is the core of today's semiconductor technology.

In Fig. 12.7, the elements of the most commonly used semiconductors are shown. Silicon is the most important semiconductor material and is found in all electronic devices today. The III–V semiconductors are compound materials composed of elements from groups III and V of the periodic table, e.g., GaAs and InAs. These are more expensive than Si, and the degree of perfection of the material is lower. However, these materials have other functionality that Si does not possess. Perhaps the most important difference is they exhibit a direct band gap material, meaning that electrons can relax to lower energy states with conserved momentum and emit light in the process. Thus, III–Vs are used in light-emitting applications such as LEDs and lasers.

Semiconductor NWs have a diameter less than 100 nm, and they have unique properties. In recent years substantial effort has been devoted to understanding their synthesis, and electrical and mechanical properties, and how these properties can be exploited in fields

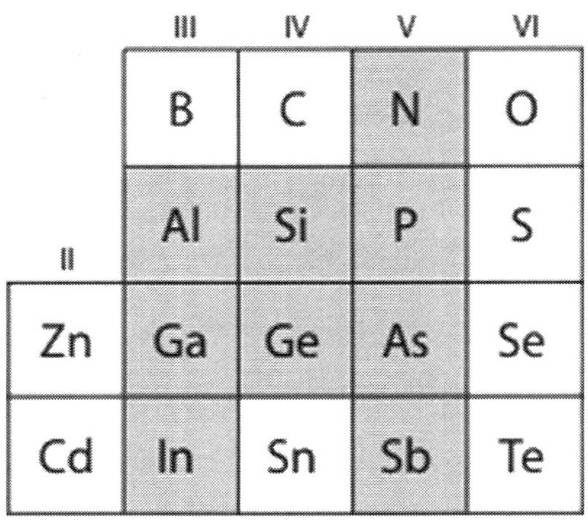

Figure 12.7
Periodic table showing the elements of group IV and III–V semiconductors.

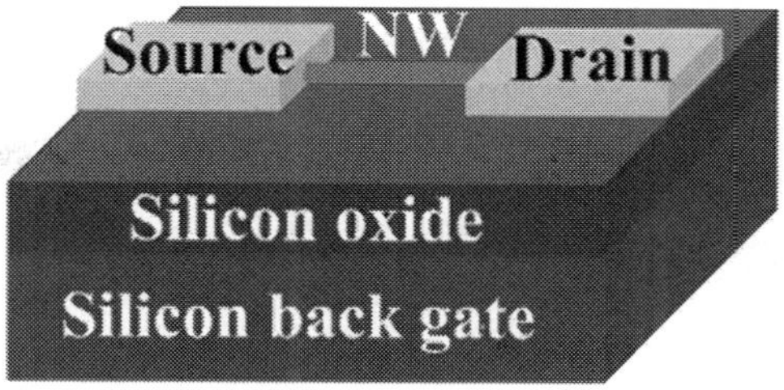

Figure 12.8
Schematic of NWFET.

such as electronics and photonics. NWs provide an ideal geometry for FETs with a wrap-gate around the NW active channel [20–22]. Although the III–V elements appear to be the preferred semiconductor materials for many applications, Si will probably remain the most preferred platform for many applications for several reasons. Si wafers are available with extremely high-purity and crystal-quality material which has good mechanical stability and has a stable SiO_2 oxide bond.

12.3.1 NWFET Structures

NWFETs have been proposed and studied by many research groups around the world, because they are promising candidate to sustain the relentless progress in scaling for MOS devices. Several key factors have contributed to the boom of NW research. First, semiconductor NWs can be prepared with high yield together with reproducible electronic properties as required for large-scale integrated systems. NWFETs, unlike planar MOSFETs, have metal source and drain contacts. That is, the source and drain contacts are made from metals instead of doped semiconductors. A schematic of NWFET is shown in Fig. 12.8.

NWFETs have several advantages over other devices. The ability of the suppression of the short channel effects and the suppression of the leakage current of NWFETs are very good, because of the gate surrounding configuration [23]. Semiconductor NWs are cylindrical single crystal structures with a diameter of a few nanometers that exhibit several interesting and novel properties.

12.4 Effect of Oxide Thickness on Gate Capacitance in Nanodevices

A simulation-based study on the effect of oxide thickness in gate capacitance of various nanoscale devices such as single gate and double gate MOSFET, CNTFET, and NWFET has been presented in this section using nanoHUB tools [24]. Table 12.1 provides the simulation results of gate capacitance in terms of quantum capacitance (QC) [25] in case of single gate MOSFET at different gate voltages, oxide thickness, and at a constant drain

Table 12.1: Effect of variation of gate voltage and oxide thickness on quantum capacitance in single gate MOSFET

Single Gate MOSFET				
At Drain Voltage = 1 (V)				
	Quantum Capacitance (F/cm^2)			
Gate Voltage (V)	T_{ox} = 0.7 nm	T_{ox} = 0.9 nm	T_{ox} = 1.2 nm	T_{ox} = 1.5 nm
0	2.10E − 10	2.10E − 10	2.10E − 10	2.10E − 10
0.083333	3.58E − 09	3.58E − 09	3.57E − 09	3.57E − 09
0.166667	6.00E − 08	5.98E − 08	5.95E − 08	5.93E − 08
0.25	8.31E − 07	8.03E − 07	7.65E − 07	7.32E − 07
0.333333	4.41E − 06	4.01E − 06	3.56E − 06	3.21E − 06
0.416667	8.47E − 06	7.74E − 06	6.86E − 06	6.17E − 06
0.5	1.08E − 05	1.02E − 05	9.25E − 06	8.47E − 06
0.583333	1.19E − 05	1.15E − 05	1.07E − 05	1.00E − 05
0.666667	1.24E − 05	1.21E − 05	1.16E − 05	1.11E − 05
0.75	1.26E − 05	1.25E − 05	1.21E − 05	1.17E − 05
0.833333	1.27E − 05	1.26E − 05	1.24E − 05	1.21E − 05
0.916667	1.28E − 05	1.27E − 05	1.26E − 05	1.24E − 05
1	1.28E − 05	1.27E − 05	1.27E − 05	1.25E − 05

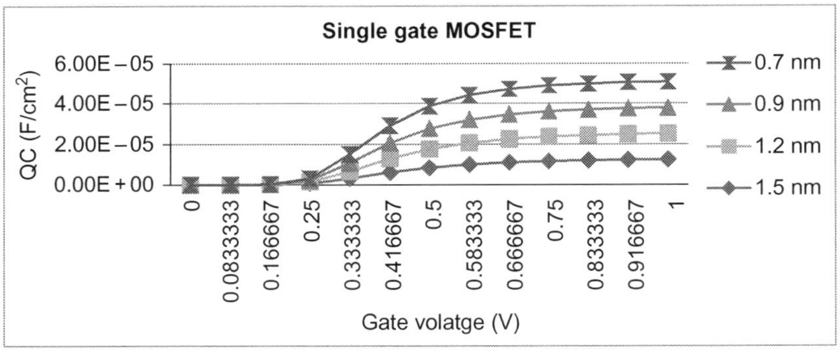

Figure 12.9
Plot showing QC versus gate voltage for single gate MOSFET.

voltage of 1 V. It can be observed that in single gate MOSFET as the oxide thickness goes down from 1.5 to 0.7 nm, the QC increases significantly as the gate voltage increases. This increment in QC is observed up to a gate voltage of 1 V. However, at a very low gate voltage such as at 0 and 0.083 V the value of QC is same for all oxide thicknesses considered here. This is also represented in the plot of Fig. 12.9, from where it can be concluded that with the decrease in gate oxide thickness the QC increases at different gate voltages. Table 12.2 provides the simulation results of double gate MOSFET under identical simulating conditions (such as MOSFET body thickness of 10 nm, insulator dielectric constant of 3.9, and a temperature of 300 K). A plot of gate voltage versus QC for double gate MOSFET is shown in Fig. 12.10, which shows that as the gate oxide thickness

Table 12.2: Effect of variation of gate voltage and oxide thickness on quantum capacitance in double gate MOSFET

Double Gate MOSFET				
At Drain Voltage = 1 (V)				
	Quantum Capacitance (F/cm^2)			
Gate Voltage (V)	T_{ox} = 0.7 nm	T_{ox} = 0.9 nm	T_{ox} = 1.2 nm	T_{ox} = 1.5 nm
0	2.10E − 10	2.10E − 10	2.10E − 10	2.10E − 10
0.083333	3.58E − 09	3.58E − 09	3.58E − 09	3.58E − 10
0.166667	6.03E − 08	6.02E − 08	6.01E − 08	5.99E − 08
0.25	8.88E − 07	8.71E − 07	8.46E − 07	8.24E − 07
0.333333	5.42E − 06	5.07E − 06	4.64E − 06	4.30E − 06
0.416667	1.02E − 05	9.61E − 06	8.89E − 06	8.27E − 06
0.5	1.20E − 05	1.17E − 05	1.12E − 05	1.07E − 05
0.583333	1.26E − 05	1.24E − 05	1.21E − 05	1.18E − 05
0.666667	1.27E − 05	1.27E − 05	1.25E − 05	1.24E − 05
0.75	1.28E − 05	1.27E − 05	1.27E − 05	1.26E − 05
0.833333	1.28E − 05	1.28E − 05	1.27E − 05	1.27E − 05
0.916667	1.28E − 05	1.28E − 05	1.28E − 05	1.27E − 05
1	1.28E − 05	1.28E − 05	1.28E − 05	1.28E − 05

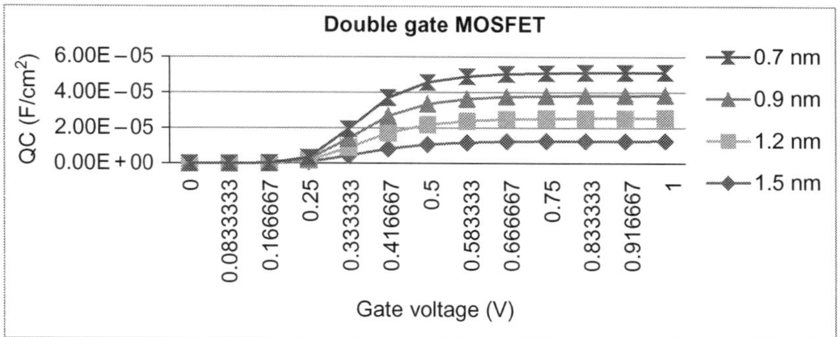

Figure 12.10
Plot showing QC versus gate voltage for double gate MOSFET.

decreases, the QC increases at different gate voltages. However it saturates beyond 0.75 V for very small oxide thickness such as 0.7 and 0.9 nm. Whereas in case of CNTFET, we observe that as the oxide thickness decreases from 1.5 to 0.7 nm, the QC also decreases (Table 12.3). This happens when we apply a gate voltage of 0.5 V and above, which is represented in the plot of Fig. 12.11. However, the change in QC is observed up to gate voltage of 1 V. Table 12.4 provides the simulation results in case of silicon NWFET. It also provides a characteristic similar to CNTFET. It can be observed that QC goes on increasing as we decrease the oxide thickness but up to a gate voltage of 0.5 V. However, it is clear from Fig. 12.12 that as the gate voltage is increased above 0.5 V it shows a drooping trend, i.e., QC decreases with the decrease in oxide thickness similar to CNTFET devices. Thus

Table 12.3: Effect of variation of gate voltage and oxide thickness on quantum capacitance in CNTFET

CNTFET				
At Drain Voltage = 1 (V)				
	Quantum Capacitance (F/cm)			
Gate Voltage (V)	T_{ox} = 0.7 nm	T_{ox} = 0.9 nm	T_{ox} = 1.2 nm	T_{ox} = 1.5 nm
0	1.37E − 16	1.37E − 16	1.37E − 16	1.37E − 16
0.083333	2.32E − 15	2.32E − 15	2.32E − 15	2.32E − 15
0.166667	3.88E − 14	3.87E − 14	3.86E − 14	3.85E − 14
0.25	5.09E − 13	4.96E − 13	4.82E − 13	4.70E − 13
0.333333	2.30E − 12	2.18E − 12	2.04E − 12	1.94E − 12
0.416667	3.67E − 12	3.57E − 12	3.42E − 12	3.20E − 12
0.5	3.37E − 12	3.78E − 12	3.82E − 12	3.81E − 12
0.583333	3.21E − 12	3.38E − 12	3.56E − 12	3.67E − 12
0.666667	2.74E − 12	2.91E − 12	3.12E − 12	3.28E − 12
0.75	2.43E − 12	2.56E − 12	2.74E − 12	2.89E − 12
0.833333	2.24E − 12	2.33E − 12	2.47E − 12	2.60E − 12
0.916667	2.11E − 12	2.18E − 12	2.29E − 12	2.39E − 12
1	2.02E − 12	2.08E − 12	2.16E − 12	2.24E − 12

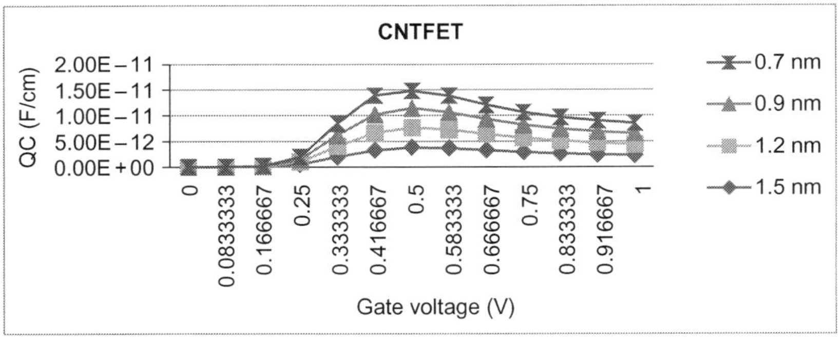

Figure 12.11
Plot showing QC versus gate voltage for CNTFET.

we can see that the effect of QC is quite prominent in case of silicon NWFET and CNTFET devices.

Such a decrease in QC can be best explained if we go back to the physics of semiconductor devices where we normally consider C_{ins} is much smaller than C_Q, so the energy associated with CQ can be considered negligible. However, as the device scaling approaches to a few nanometer, C_{ins} becomes very much comparable to or even larger than C_Q, so the effect of C_Q should be considered carefully while calculating the gate capacitance in scaled-down devices. Thus, we see that while gate capacitance in MOSFET increases with the decrease in oxide thickness in scaled-down devices, it reduces in

Table 12.4: Effect of variation of gate voltage and oxide thickness on quantum capacitance in silicon NWFET

Silicon NWFET				
At Drain Voltage = 1 (V)				
	Quantum Capacitance (F/cm)			
Gate Voltage (V)	T_{ox} = 0.7 nm	T_{ox} = 0.9 nm	T_{ox} = 1.2 nm	T_{ox} = 1.5 nm
0	2.07E − 16	2.07E − 16	2.07E − 16	2.07E − 16
0.083333	3.52E − 15	3.52E − 15	3.52E − 15	3.51E − 15
0.166667	5.85E − 14	5.83E − 14	5.81E − 14	5.78E − 14
0.25	7.12E − 13	6.89E − 13	6.62E − 13	6.42E − 13
0.333333	2.99E − 12	2.80E − 12	2.60E − 12	2.45E − 12
0.416667	5.04E − 12	4.78E − 12	4.48E − 12	4.25E − 12
0.5	5.60E − 12	5.62E − 12	5.52E − 12	5.37E − 12
0.583333	4.90E − 12	5.18E − 12	5.47E − 12	5.60E − 12
0.666667	4.64E − 12	4.70E − 12	4.87E − 12	5.10E − 12
0.75	4.07E − 12	4.50E − 12	4.67E − 12	4.72E − 12
0.833333	3.09E − 12	3.74E − 12	4.37E − 12	4.60E − 12
0.916667	2.55E − 12	2.90E − 12	3.61E − 12	4.14E − 12
1	2.37E − 12	2.50E − 12	2.87E − 12	3.39E − 12

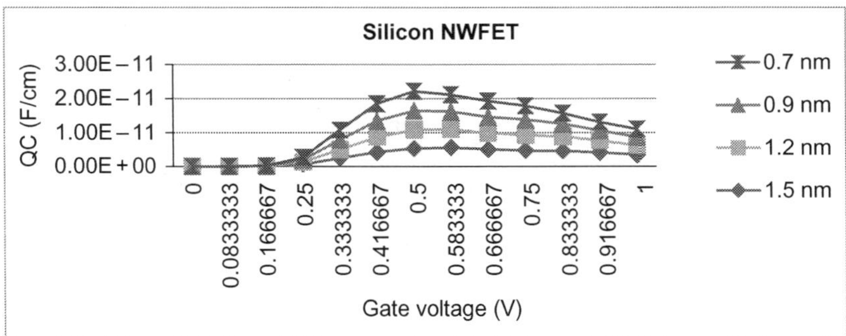

Figure 12.12
Plot showing QC versus gate voltage for silicon NWFET.

CNTFET devices. This makes NWFET and CNTFET excellent semiconductor devices in deep nanometer regime

12.5 Effect of Device Parameters on Threshold Voltage in CNTFET Devices

Here in this section, we will see the effect of device parameters such as chiral vector, temperature, metal work function, dielectric constant, and channel length on threshold voltage in CNTFET devices. Some insight into these facts will also be given.

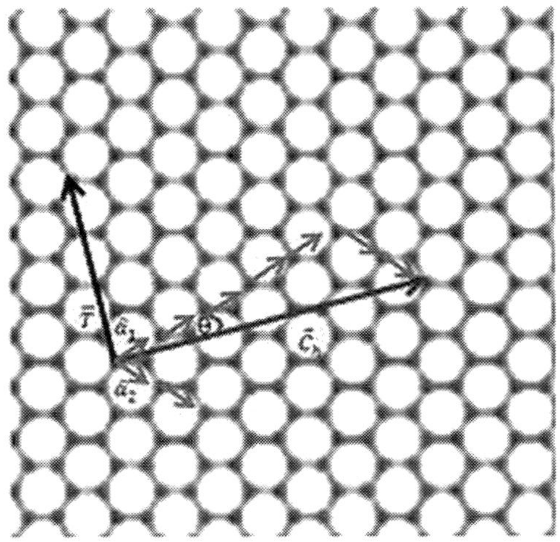

Figure 12.13
Graphene atomic structure with chiral vector.

12.5.1 Effect of Chiral Vector

A CNT can act as either a conductor or a semiconductor, depending on the angle of the atom arrangement along the tube. This is referred to as the chirality vector and is represented by the integer pair (*n*,*m*). The circumference of CNT can be expressed in terms of a chiral vector, $c = na_1 + ma_2$, which connects two equivalent sites of the two-dimensional graphene sheet, as shown in Fig. 12.13, where n and m are integers and a_1 and a_2 are the unit vectors of the hexagonal honeycomb lattice. A simple method to determine if a CNT is metallic or semiconducting is to consider its indexes (*n*,*m*). The nanotube is metallic if $n = m$ or $n - m = 3i$, where i is an integer. Otherwise, the tube is semiconducting [26].

The geometry of the graphene lattice and the chiral vector of the tube determine the structural parameters like diameter, unit cell, and its carbon atoms, as well as the size and shape of the brillouin zone. The diameter of the CNT can be calculated using the following equation:

$$D_{\mathrm{CNT}} = \frac{\sqrt{3}a_0}{\pi}\sqrt{n^2 + m^2 + mn} \tag{12.1}$$

where $a_0 = 0.142$ nm is the interatomic distance between each carbon atom and its neighbor. The direction of the chiral vector is measured by the chiral angle θ. The chiral angle θ can be calculated using the equation:

$$\cos\theta = \frac{(n+m)/2}{\sqrt{n^2 + m^2 + mn}} \tag{12.2}$$

Table 12.5: Threshold voltage w.r.t. different chiral vector

CNTFET Parameters		
Chiral Vector (*m*,*n*)	(*m* − *n*)	Threshold Voltage (V)
(4,0)	4	0.69
(6,2)	4	0.66
(16,12)	4	0.17
(8,0)	8	0.59
(9,1)	8	0.51
(17,9)	8	0.18
(11,0)	11	0.45
(13,2)	11	0.33
(24,13)	11	0.12
(14,0)	14	0.33
(17,3)	14	0.24
(26,12)	14	0.12

The differences in the chiral angle and the diameter cause the differences in the properties of the CNTs.

$$V_{\text{th}} = a(V_{II})/\sqrt{3}qD_{\text{CNT}} \tag{12.3}$$

where q = electronic charge, $a = 2.49$ Å is the lattice constant, and $V_{\text{II}} = 3.033$ eV is the carbon π to π bond energy.

CNTFET provides a unique opportunity to control threshold voltage by changing the chirality vector, or the diameter of the CNT as it can be understood from Eqs. (12.1) and (12.3). The gate-to-source voltage that generates the same reference current is taken as the threshold voltage for the transistor that has different chirality. Chiral vector has an important role in deciding the diameter of the CNT. Table 12.5 shows the result of simulation of varying chiral vectors and its effect on threshold voltage of CNTFET using HSPICE from Synopsys. From the simulation results it is very much clear that the threshold voltage of a CNTFET device can be controlled by varying the chiral vector. We have considered 12 such combinations of chiral vector pairs starting from (4,0) to (26,12) for verifying the fact.

Further, we find that when the difference of the chiral vector (m,n) is small, e.g., (m,n) = (4,0) the threshold voltage is larger (0.69 V), whereas when the difference is small and the chiral vector (m,n) is large, e.g., (m,n) = (16,12), then the threshold voltage is small.

Thus when the chiral vector (m,n) goes on increasing, the threshold voltage goes on reducing as shown in Fig. 12.14.

12.5.2 Effect of Temperature

The MOSFET device characteristics and circuit behavior change with the change (increase) in temperature. The maximum tolerable operating temperature for silicon devices is 150°C.

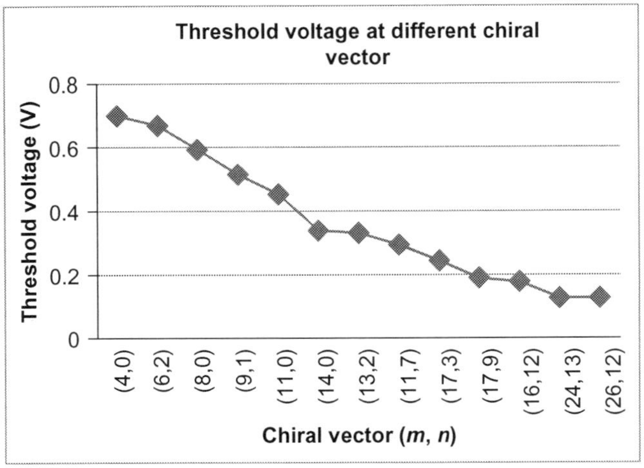

Figure 12.14
Threshold voltage decreases with increase in chiral vector (*m*,*n*).

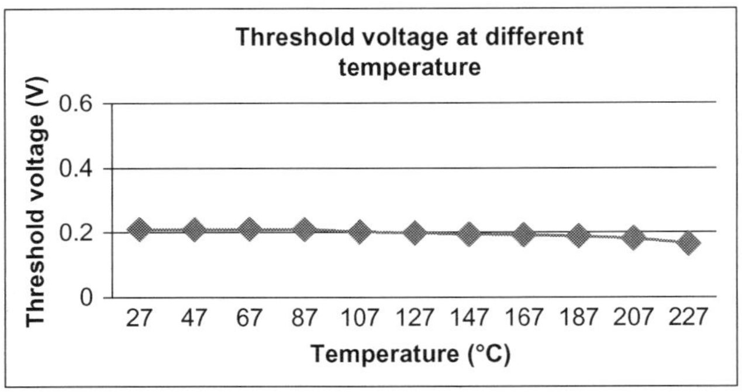

Figure 12.15
Effect of temperature (increase) on threshold voltage.

Because of the strong covalent carbon–carbon bonding in the sp^2 configuration, CNTs are chemically inert and are able to transport large amounts of electric current [27,28]. CNTs are able to conduct heat nearly as good as diamond. In CNTFET devices effect of temperature on threshold voltage is negligibly small. Plot of Fig. 12.15 shows the simulation results with a temperature variation of 27–227°C. From the simulation results we can observe that temperature hardly affects the threshold voltage as there is only a 4.6% variation in *V*th while raising the temperature from 27°C to 227°C. This is not possible in case of MOSFET as it can operates at maximum temprature of 150°C and the device characteristics change abruptly at higher temperature and especially leakage current goes very high.

The plot of Fig. 12.16 shows the simulation results when the operating temperature is varied between 0°C to −60°C. It is observed that there is a nominal increment of 3.5% in

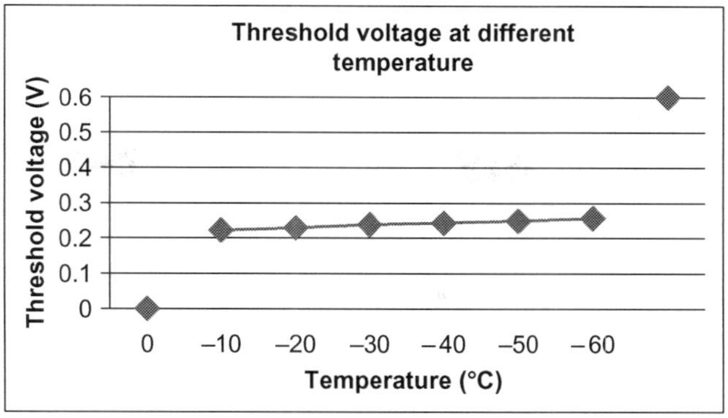

Figure 12.16
Effect of temperature (decrease) on threshold voltage.

Table 12.6: Threshold voltage w.r.t. work function

Effect of Work Function on Threshold Voltage		
Metal	**Work Function (eV)**	**Threshold Voltage (V)**
N + polysilicon	4.10	0.240
Aluminum (Al)	4.28	**0.244**
Copper (Cu)	4.65	0.231
Gold (Au)	5.10	0.210

Even a low cost, low work function metal such as aluminum can provide higher threshold (essential for low leakage).

threshold voltage which negligibly affects the characteristics of CNTFET device.
In both Figs. 12.8 and 12.9, we have represented each threshold voltages by observing the drain current against gate voltages characteristics of CNTFET at different drain voltages.

12.5.3 Effect of Metal Gate Work Function

Metal work function plays an important role in the design of solid-state electronics.
The work function of the metal gate should be given the most important consideration in selecting the gate material, because the work function determines the threshold voltage of the semiconductor devices [29]. The conventional polysilicon gate is limited by the well-known gate depletion effect, which reduces drive current. High resistivity and thermal instability with high-K dielectric materials also limit the scalability of polysilicon gate. In semiconductor devices the work function difference between the semiconductor and gate influences the threshold voltage. The work function is to be smaller for metals with an open lattice and larger for metals whose atoms are closely packed (Table 12.6).

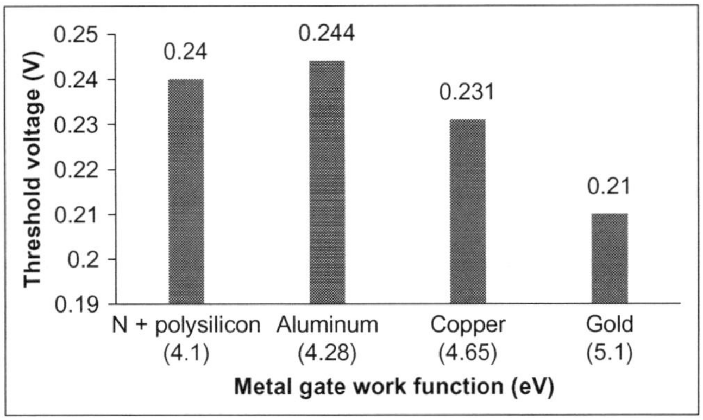

Figure 12.17
Threshold voltage versus work function.

Polysilicon has been utilized as gate material since the evolution of MOSFET device. However, due to polysilicon depletion, the equivalent oxide thickness increases [30], so the aggressive scaling of devices requires the implementation of a metal gate in place of conventional polysilicon.

As shown in Fig. 12.17, as work function of metal gate increases the threshold voltage of CNTFET device decreases and the maximum threshold voltage is for aluminum instead of other metals considered here. Moreover, it is even better than N + polysilicon with a work function of 4.1 eV.

*12.5.4 Effect of High-*K *Dielectric*

Silicon oxide (SiO_2) has been used as gate dielectric material since the inception of MOSFET in 1960, because of its electrical stability at the interface between Si and SiO_2 and thermal stability. High-speed and compact integrated circuits are achieved by scaling down of physical thickness of the SiO_2 gate dielectrics and gate length. But, the gate leakage current due to direct tunneling of electrons through the SiO_2 will be too high, and circuit power dissipation will increase to unacceptable values, when the gate dielectric goes ultra-thin and less than 1.5 nm [31]. While MOSFETs have been realized at 1.5 nm oxide thickness, a major obstacle to overcome is the high level of direct tunneling current through the gate. One possible solution is to incorporate the use of a high-K gate dielectric.

Table 12.7 provides the threshold voltage of CNTFET with different high-K materials.

Table 12.8 provides the required drain voltage for drain currents. It is clear from the table that the requirement of drain voltage for a particular fixed drain current is very low when a high-K dielectric material is used.

Table 12.7: Threshold voltage w.r.t. high-*K*

Effect of High-*K* Dielectric Oxide on Threshold Voltage		
Compound	K_{ox}	**Threshold Voltage (V)**
SiO_2	3.9	0.240
$Al2O_3$	9.0	0.231
HfO_2	15.6	0.224
ZrO_2	19	0.218

Table 12.8: Drain current and drain voltage w.r.t. high-*K*

	Drain Current (μA) at Different Drain Voltage (V)			
Kox	**20 (μA)**	**40 (μA)**	**60 (μA)**	**80 (μA)**
3.9	0.30	0.56	0.84	–
9.0	0.15	0.28	0.44	0.56
15.6	0.11	0.22	0.33	0.44
19.0	**0.12**	**0.18**	**0.30**	**0.40**

With high-*K* dielectric, the drain voltage required is the relatively low for a particular drain current as compared to low-*K* dielectric.

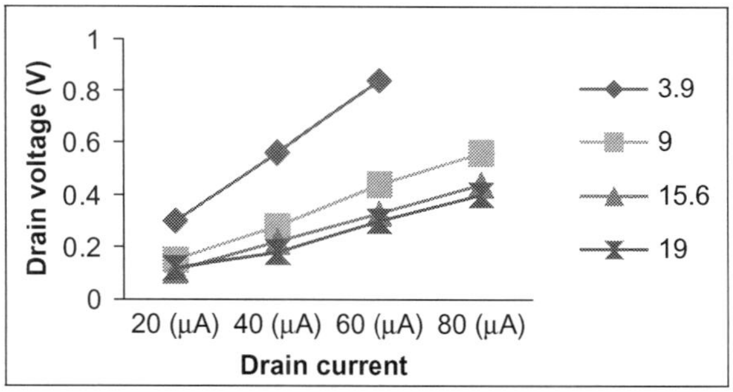

Figure 12.18
Drain current versus drain voltage of different high-*K* materials.

Fig. 12.18 shows the plot of the results as given in Table 12.8, of drain voltage at fixed drain current for different high-*K* materials. It is very much clear that in CNTFET device, at high-*K* materials the requirement of drain voltage is very less for a particular drain current.

12.5.5 Effect of Channel Length

Channel length is the key parameter in MOSFET devices. In 1960, when the MOSFET was first developed, the channel length was longer than 20 μm. Today channel lengths less than 0.1 μm have been fabricated in volume production, and channel lengths of 0.01 μm or so have been created in research laboratories. In integrated circuit MOSFET device has become smaller and smaller in order to achieve higher packing density. Minimization of leakage currents is enormously important in VLSI circuits these days. As the device is

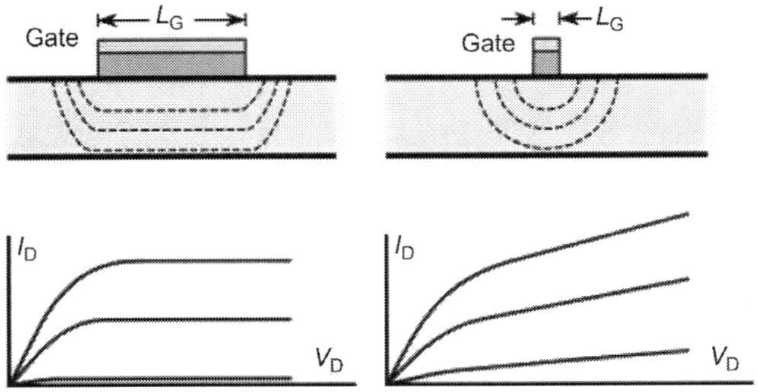

Figure 12.19
Characteristics of MOSFET at different channel length.

scaled down to nanometer level the threshold voltage is also scaled down to increase the performance, and consequently the leakage power increases [32–35]. Leakage current has become the limiting factor for oxide thickness thinner than 1.5 nm. Decreasing the channel length increases the current drive capability of the transistor. Much of the scaling is therefore driven by decrease in channel length. If only this parameter is scaled many problems are encountered, such as increased electric field. If the channel length becomes too short, the depletion region from the drain can reach the source side and reduces the barrier for electron injection. This is known as punch through, and because of this, devices characteristics degrade as shown in Fig. 12.19.

In long channel length devices, the gate is completely responsible for depleting the semiconductor. In short channel devices, part of the depletion is accomplished by the drain and source bias. Since less gate voltage is required for the transistor to deplete, the barrier for electron injection from source to drain decreases. This is known as drain-induced barrier lowering. For both the devices CNTFET and MOSFET, we have considered the channel length from 100 to 10 nm. An interesting observation can be obtained from Table 12.9 that, in case of CNTFET as the channel length goes down from 20 to 10 nm threshold voltage increases rapidly. This is due to the contribution of QC as described in Section 12.4. A characteristic plot of threshold voltage with respect to different channel length of CNTFET devices is shown in Fig. 12.14, from which also clarifies the fact that in case of MOSFET devices the threshold voltage rolls off sharply when channel length reduces from 20 to 10 nm, while in case of CNTFET it rises abruptly.

Comparison of threshold voltage with respect to channel length of CNTFET and MOSFET devices is shown in Fig. 12.20, where we can see that below 20 nm channel length, the threshold voltage increases rapidly in case of CNTFET device and the threshold voltage goes on decreasing beyond 20 nm channel length in case of MOSFET that causes increased leakage power dissipation and device degradation in MOSFET. The advantage of using CNTFET device in nanometer regime is that we have an increased threshold voltage at

Table 12.9: Threshold voltage w.r.t. different channel length

Channel Length Effect of Different Devices		
	Threshold Voltage (V)	
Channel Length (nm)	CNTFET	MOSFET
100	0.227	0.241
90	0.227	0.241
80	0.227	0.241
70	0.228	0.242
60	0.228	0.242
50	0.229	0.243
40	0.231	0.245
30	0.239	0.247
20	0.263	0.248
10	**0.605**	**0.168**

Threshold voltage (Vth) rolls-off as the channel length reduces, that is, in short channel MOSFET devices. On the contrary, Vth increases in CNTFET devices due to quantum capacitance effect as explained in Tables 12.1–12.4.

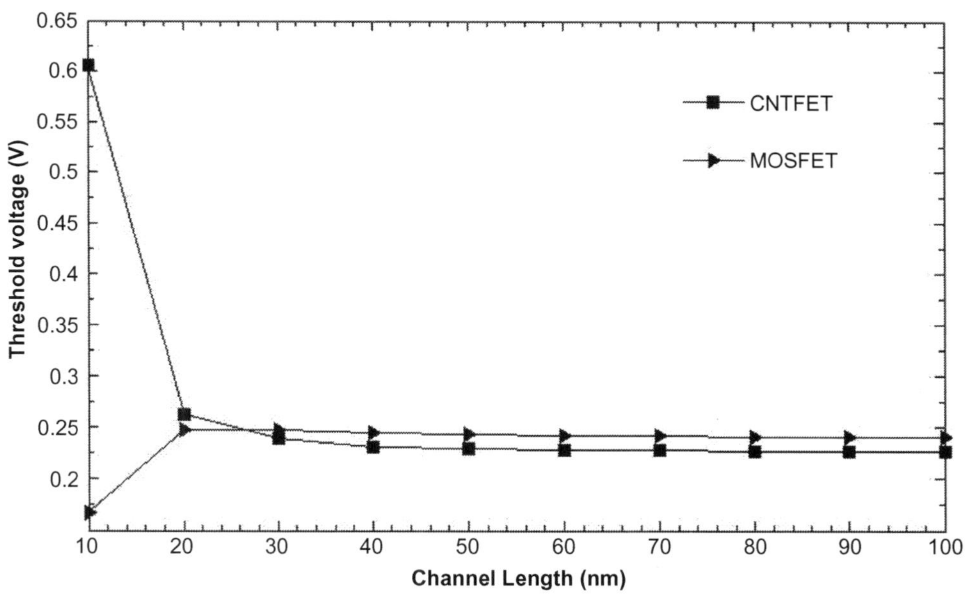

Figure 12.20
Threshold voltage versus channel length in CNTFET and MOSFET devices.

10 nm channel length or below. In case of MOSFET while reducing the channel length the threshold voltage is also reduced which lead to more leakage power.

12.6 Conclusion

Shrinking of the device dimensions with improved technologies is the driving force toward circuit miniaturization, portability, and low cost. International technology roadmap for semiconductors (ITRS) has predicted that in deep nanometer regime, the expected high-density integration will encounter substantial difficulties in terms of physical phenomena and technology limitations, possibly preventing the continued improvements in figure of merit, such as low power and high performance.

CNTFET and NWFET devices are getting more and more important today because of their high channel mobility and improved gate capacitance against gate voltage. It is seen that in nanometer regime QC is the deciding factor in calculating the gate capacitance of an FET device. CNTFET and silicon NWFET have a favorable characteristic of decreasing gate capacitance with the decrease in oxide thickness in deep nanometer regime, which is not possible to get in a single gate or a double gate MOSFET. CNTs have some unique features and special properties that offer a great potential for nano-electronic devices. In this work, we have analyzed first the effect of dielectric thickness on gate capacitance in nanodevices. Next we have seen the effect of chiral vector, metal work function, channel length, and high-K dielectric on threshold voltage of CNTFET devices. It is also observed that the temperature has a negligible effect on threshold voltage of CNTFET thus making it a more reliable device.

Fabrication of CNTFET and NWFETs is of course the real bottleneck for today's researchers and engineers. Finding out a suitable technology for the mass-scale production of it is still a challenge. Efforts are on, but still it is at the level of development stage. It has been reported to fabricate a CNTFET device experimentally [36].

References

[1] G. Moore, Progress in digital electronics, IEDM Tech. Digest (1975) 11–13.

[2] H.-S. P. Wong, D.J. Frank, P.M. Solomon, C.H.J. Wann, J.J. Welser, Nanoscale CMOS, Proc. IEEE 87 (4) (1999).

[3] ITRS. <http://www.itrs.net/>, 2011.

[4] Z. Zhang, X. Liang, S. Wang, K. Yao, Y. Hu, Y. Zhu, et al., Doping-free fabrication of carbon nanotube based ballistic CMOS devices and circuits, Nano Lett. 7 (12) (2007) 3603–3607.

[5] C. Thelander, P. Agarwal, S. Brongersma, J. Eymery, L.F. Feiner, A. Forchel, et al., Nanowire-based one-dimensional electronics, Mater Today 9 (10) (2006) 28–35.

[6] S.J. Wind, J. Appenzeller, R. Martel, V. Derycke, P. Avouris, Fabrication and electrical characterization of top gate single-wall carbon nanotube field-effect transistors, J. Vacuum Sci. Technol. B 20 (2002) 2798–2801.

[7] A. Javey, J. Guo, D.B. Farmer, Q. Wang, E. Yenilmez, R.G. Gordon, et al., Self-aligned ballistic molecular transistors and parallel nanotube arrays, Nano Lett. 4 (7) (2004) 1319–1322.

[8] A. Javey, J. Guo, Q. Wang, M. Lundstrom, H.J. Dai, Ballistic carbon nanotube field-effect transistors, Nature 424 (2003) 654–657.

[9] J. Suehiro, G. Zhou, M. Hara, Fabrication of a carbon nanotube-based gas sensor using dielectrophoresis and its application for ammonia detection by impedance spectroscopy, J. Phys. D Appl. Phys. 36 (2003) 109–114.

[10] H.M. Manohara, E.W. Wong, E. Schlecht, B.D. Hunt, H. Siegel Peter, Carbon nanotube Schottky diodes using Ti-Schottky and Pt-ohmic contacts for high frequency applications, Nano Lett. 5 (2005) 1469–1474.

[11] T. Mueller, M. Kinoshita, M. Steiner, V. Perebeinos, A.A. Bol, D.B. Farmer, et al., Efficient narrow-band light emission from a single carbon nanotube p–n diode, Nature Nanotechnol. 5 (2010) 2731.

[12] N.M. Gabor, Z. Zhong, K. Bosnick, J. Park, P.L. McEuen, Extremely efficient multiple electron–hole pair generation in carbon nanotube photodiodes, Science 325 (2009) 1367–1371.

[13] V. Sazonova, Y. Yaish, H.Ä. UstÄunel, D. Roundy, T.A. Arias, P.L. McEuen, A tunable carbon nanotube electromechanical oscillator, Nature 431 (2004) 284–287.

[14] M.S. Arnold, A.A. Green, J.F. Hulvat, S.I. Stupp, M.C. Hersam, Sorting carbon nanotubes by electronic structure using density differentiation, Nature Nanotechnol. 1 (2006) 60–65.

[15] A.R. Harutyunyan, G. Chen, T.M. Paronyan, E.M. Pigos, O.A. Kuznetsov, K. Hewaparakrama, et al., Preferential growth of single-walled carbon nanotubes with metallic conductivity, Science 326 (2009) 116–120.

[16] S.J. Wind, J. Appenzeller, R. Martel, V. Derycke, Ph Avouris, Vertical scaling of carbon nanotube field-effect transistors using top gate electrodes, Appl. Phys. Lett. 80 (20) (2002) 3817.

[17] C. Dekker, S.J. Tans, A.R.M. Verschueren, Room-temperature transistor based on a single carbon nanotube, Nature 393 (1998) 6680.

[18] Z. Chen, D. Farmer, S. Xu, R. Gordon, P. Avouris, J. Appenzeller, Externally assembled gate-all-around carbon nanotube field-effect transistor, IEEE Electron. Device Lett. 29 (2) (2008) 183.

[19] J. Cao, Q. Wang, H. Dai, Electron transport in very clean, as-grown suspended carbon nanotubes, Nature Mater. 4 (10) (2005) 745–749.

[20] T. Bryllert, L.E. Wernersson, L.E. Fröberg, L. Samuelson, Vertical high mobility wrap-gated InAs nanowire transistor, IEEE Electron Device Lett. 27 (2006) 323.

[21] D. Moon, S.-J. Choi, J.P. Duarte, Y.-K. Choi, Investigation of silicon nanowire gate—all-around junctionless transistors built on a bulk substrate, IEEE Transact. Electron Devices 60 (4) (2013) 1355–1360.

[22] S. Lee, J.-S. Noh, J. Kim, M.G. Kim, S.Y. Jang, J. Park, et al., The optoelectronic properties of PbS nanowire field-effect transistors, IEEE Transact. Nanotechnol. 12 (6) (2013) 1135–1138.

[23] S. Jin, T.W. Tang, M.V. Fischetti, Simulation of silicon NW transistors using Boltzmann transport equation under relaxation time approximation, IEEE Transact. Electron. Devices 55 (3) (2008) 727–736.

[24] nanoHUB Tools. <www.nanohub.org>.

[25] S.K. Sinha, S. Chaudhury, Impact of oxide thickness on gate capacitance—a comprehensive analysis on MOSFET, nanowire FET and CNTFET devices, IEEE Transact. Nanotechnol. 12 (6) (2013).

[26] M. Dresselhaus, G. Dresselhaus, R. Saito, Carbon fibers based on C60 and their symmetry, Phys. Rev. B 45 (11) (1992).

[27] B.Q. Wei, R. Vajtai, P.M. Ajayan, Reliability and current carrying capacity of carbon nanotubes, Appl. Phys. Lett. 79 (8) (2001) 1172–1174.

[28] J. Appenzeller, Carbon nanotubes for high-performance electronics: progress and prospect, Proc. IEEE 96 (12) (2008) 201–211.

[29] I. De, D. Johri, A. Srivastave, C.M. Osburn, Impact of gate work function on device performance at the 50 nm technology node, Solid-State Electron. 44 (2000) 1077–1080.

[30] C.J. Choi, M.Y. Jang, Y.Y. Kim, M.S. Jun, T.Y. Kim, B.C. Park, et al., Effective metal work function of high-pressure hydrogen postannealed Pt–Er alloy metal gate on HfO2 film, Jpn. J. Appl. Phys. 46 (1) (2007) 125–127.

[31] C.G. Parker, G. Lucovsky, J.R. Hauser, Ultrathin oxide–nitride gate dielectric MOSFET's, IEEE Electron Device Lett. 19 (4) (1998).

[32] K. Terada, H. Muta, A new method to determine effective MOSFET channel length, Jpn. J. Appl. Phys. 18 (1979) 953.

[33] J.G.J. Chern, P. Chang, R.F. Motta, N. Godinho, A new method to determine MOSFET channel length, IEEE Electron Device Lett. EDL-1 (1980) 170.

[34] Y. Taur, Y.J. Mii, R. Logan, H.S. Wong, On effective channel length in 0.1 m MOSFET's, IEEE Electron Device Lett. 16 (1995) 136.

[35] S. Jain, Measurement of threshold voltage and channel length of submicron MOSFET's, Proc. Inst. Elect. Eng. 135-1 (1988) 162.

[36] A. Javey, R. Tu, D.B. Farmer, J. Guo, R.G. Gordon, H. Dai, High performance carbon nanotube field-effect transistors with chemically doped contacts, Nano Lett. 5 (2) (2005) 345–348.

CHAPTER 13

Role of Nanocomposites in Future Nanoelectronic Information Storage Devices

Vaishali Shukla[1], Bhargav Raval[1], Sachin Mishra[2,3] and Man Singh[4]

[1]School of Nanoscience, Central University of Gujarat, Gandhinagar, India [2]Biosensors and Bioelectronics Centre, IFM, Linköping University, Linköping, Sweden [3]Department of Physics, University of the Free State, Bloemfontein, South Africa [4]School of Chemical Sciences, Central University of Gujarat, Gandhinagar, India

13.1 Introduction

Science requires measurement, and measurement is the language of science. Nanoscale implies a scale of measurement that exists at the level of the nanometer (1 nm = 10^{-9} m), and it can design the materials for its size-based properties. Nanotechnology originates from the Greek word "dwarf." Nanomaterials (NMs) get accumulated for maximum surface area with least potential energy and it becomes a natural barrier on the full surface of targeted material. Therefore, NMs incorporate the methodology, designing, strength to transfer the potential energy into the kinetic energy due to nonaccumulated particles of the surface which play a role in the relative properties; in general manner nanoparticles (NPs) do not have much energy so that it can easily enter into bulk materials which make it uses in the on-going process. Two principal factors cause the properties of NMs to differ significantly from other materials. The increase in surface-to-volume ratio in nanometer-sized systems can be efficiently explained considering that the greater of atoms lie at the surface of a nanostructure differently to macroscopic solid matter and the dominance of quantum effects creates fascinating new possibilities for tuning the physical and chemical properties of materials by controlling the size of constituent domains and the manner in which they are assembled. Hence, these factors can change or enhance properties such as reactivity, strength, electrical properties, and optical characteristics; make them more reactive and more sensitive for the innovative properties-based applications in the different technological fields [1].

Nanotechnology is a broad and interdisciplinary field dealing with structures and particles at the nanoscale. Nanotechnology can be defined as research and technology development

Nanoelectronics. DOI: https://doi.org/10.1016/B978-0-12-813353-8.00011-7

at nanometer scale of approximately 1–100 nanometers in any dimension [2]. Thus, nanotechnology involving in the application of different nanoscale materials in different potential areas provide novel technological advances mainly in the field of nanoelectronics, which involves utilization of nanotechnology for the benefit of upcoming more potential-based devices. These specified disciplines are overlapping in many ways, like use of nanotechnology in various applications of nanoelectronics. With the help of nanotechnology, it is now possible to provide fabrication at a molecular level. Recently various novel and advance methods of device fabrication based on NPs are being developed. These designed nanostructures are used as conducting or semiconducting materials, fluorescent materials, energy and memory storage materials for molecular research tools. Recent modifications of NMs systems such as paramagnetic NPs, quantum dots (QDs), core/shell NPs and nanocomposites are widely used for nanoelectronics.

Optimization of hybrid-nanocomposite-based nonvolatile memory devices offers great potential to implement future-integrated nanoelectronic systems for both on-board computing and information storage. Conventional data memory works on the basis of electrons that are moved around and stored. However, even by atomic standards, electrons are extremely small. It is very difficult to control them. This does not only limit storage density, it also costs a great deal of energy. For this reason, researchers are working feverishly all over the world on nanoelectronic components that make use of ions, i.e., charged atoms, for storing data. Ions are some thousands of times heavier than electrons and therefore much easier to "hold down." In this way, the individual storage elements can almost be reduced to atomic dimension, which enormously improve the storage density.

Among the various types of nonvolatile memory devices, organic bistable devices (OBDs) fabricated utilizing hybrid organic/inorganic nanocomposites have currently been receiving broad attention because of their excellent performance with high-mechanical flexibility, simple fabrication, and low cost. The prospect of potential applications of nonvolatile memory devices fabricated utilizing hybrid nanocomposites has led to substantial research and development efforts to form various kinds of nanocomposites by using various methods. Generally, hybrid nanocomposites are composed of organic layers containing metal NPs, semiconductor QDs, core-shell semiconductor QDs, fullerenes, carbon nanotubes (CNTs), graphene or graphene oxides (GOs).

This chapter summarizes the most recent development of this new area of research including the investigations of developments in nonvolatile memory devices based on hybrid nanocomposites. The device structure, fabrication, and electrical characteristics of nonvolatile memory devices are discussed, and the switching and carrier transport mechanisms in the hybrid nonvolatile memory devices are reviewed. Furthermore, various flexible memory devices fabricated utilizing hybrid nanocomposites are described and their future prospects are discussed.

13.2 Classification of Nanomaterials

There are two methods that are used to produce NMs. One is called top-down approach and another one is bottom-up approach. In the top-down approach very small components are produced using large-scale parts of the materials and in the bottom-up approach NMs are produced by processing at atomic or molecular level. NMs are classified based on their origin, dimensions, and structural configuration relevant [3].

13.2.1 Origin Relevant Nanomaterials

According to their origin NMs are classified as follows:

Natural NMs: NMs which are belonging to resource of nature are defined as natural NMs. As per examples biological antibody, amino acid molecules originated from nature are some natural nanostructured materials. In addition, mineralized materials such as clays, natural colloids, fog (aerosol type), gelatine (gel type), etc., are categorized under natural NMs.

Synthetic NM: Synthetic NMs are those which are prepared deliberately through a well-defined physical, chemical, or biological route. The examples of such materials are semiconducting NPs, QDs, nanotubes, graphene, etc.

13.2.2 Dimension Relevant Nanomaterials

Fig. 13.1 presents dimension relevant classification and provides examples of each form of nanostructure, including clusters/spheres, nanowires, thin films, and bulk specimens. Since the surface properties of NMs dominate bulk, by controlling the surface/volume aspect (particle radius, film thickness, or grain size), it is possible to develop materials whose physical properties will be related to surface [4].

Zero dimensional (0-D): These NMs have the entire three dimensions in the nanometer range. Metallic NPs including gold and silver NPs and semiconductor such as QDs are the perfect example of this kind of NPs. Most of these NPs are spherical in shape and the diameter of these particles will be in the range less than 100 nm. Cubes and polygons shapes are also found for this kind of NMs.

One dimensional (1-D): In these nanostructures, two dimensions of the nanostructure will be in the nanometer range. Nanowires, nanorods, and nanotubes are the best examples of 1-D nanostructure materials. These materials are long (several micrometer in length), but with diameter of only a few nanometers. Nanowire and nanotubes of metals, oxides, and other materials are few examples of this kind of materials.

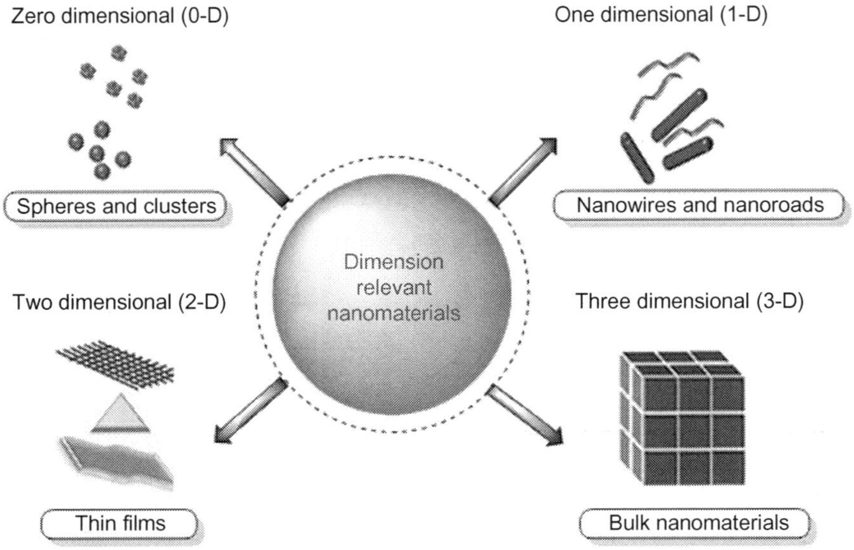

Figure 13.1
Dimension relevant classifications of NMs.

Two dimensional (2-D): In this type of NMs, only one dimension is in the nanometer range. These include different kinds of nanofilms such as coatings and thin-film multilayers, nanosheets, or nanowalls. The area of the nanofilms can be large (several square micrometers), but the thickness is always in nanoscale range.

Three dimensional (3-D): In these nanostructure materials, entire three dimensions are outside the nanometer range. These include bulk materials composed of the individual blocks which are in the nanometer scale (1–100 nm).

13.2.3 Structural Configuration Relevant Nanomaterials

Structural configuration relevant NMs can be categorized into four types:

Carbon-based NMs: This kind of NMs is hollow spheres, ellipsoids, sheet, or tubes in structures. Spherical and ellipsoidal configured carbon NMs are defined as fullerenes; sheet-like structure is defined as graphene, whereas cylindrical ones are described as nanotubes.

Metal-based materials: Metal is the key component of this type of NMs. These types of NMs include metal NPs, metal-oxide NPs, closely packed semiconductors such as QDs.

Dendrimers: Dendrimers are highly branched macromolecules with the nanometer scale dimensions. The surfaces of a dendrimer possess numerous chains which can be modified to perform specific chemical functions.

Nanocomposites: Nanocomposite can be described as a multiphase material where at least one of the phases has one, two, or three dimensions in nanoscale. The most common examples of these materials are colloids and copolymers.

13.3 Trends and Future Applications

NMs are extremely reactive because of their higher surface area and they have remarkable properties as compared to other bulk materials which makes them more suitable in the trend-based applications. In the nanolevel, materials utilized smartly to produce innovative products due to quantum effect and then electrical, optical, and magnetic properties can be effectively exploited to certain applications.

Energy: nanotechnologies provide essential improvement in the development of both conventional energy sources and renewable energy sources like geothermal energy, sun, wind, water (H_2O), tides, or biomass. The future generations will be striving to explore the alternative energy sources like hydrogen-based fuel to fulfil their requirements. Nanotechnologies will play a decisive role in the intensified use of solar energy through photovoltaic systems. Recently, graphene-based nanostructures could lead to emerge as more efficient solar cells. The nanostructures are fabricated on top of graphene samples to concentrate the electromagnetic field in the region of the material where light is converted to electrical current, so as to drastically increase the generated photovoltage. Compared to silicon which is a material currently being used in solar cells, graphene is more conductive and can absorb longer range spectrum of sunlight, and that may boost the efficiency of solar cells. There is a considerable research program is going on to develop alternative source of energy in the form of hydrogen fuel by splitting H_2O using sunlight in the presence of titanium dioxide (TiO_2) NMs. In case of conventional nanocrystalline material solar cells, for instance, increases in efficiency are achievable by antireflection layers for higher light yield. The evolving hydrogen generated by this process can itself become a very good source of fuel for automobiles and other purposes [5].

Electronics: Application of NMs in electronics is significantly increased in recent years. The most common NM which is enormously investigated for the application of electronic science is carbon-based NMs and semiconductor nanocrystal (QDs). Due to their superior electrical properties carbon NMs such as 1-D CNTs and 2-D graphene are evolving as significant tools which provide a faster and more power-efficient electronics. In addition, high surface-to-volume ratio, robust, and durable mechanical properties had established them a highly sensitive and lower energy consuming building block for nanosensors. All the CNTs have an ability to emit electrons in the presence of an electric field. This property has been utilized in the field emission devices like television, computer screens, and other instruments which are based on cathode-ray emission. The chemical and physical state of the environment may alter the electrical conductivity of CNTs and developed as chemical and physical sensors. The efficiency of CNTs to detect any changes in chemical composition of the environment is three times higher in magnitude than conventional solid-state sensors. The other advantage of these nanoelectronic device is smaller in size and can be used at room temperature [6].

Aerospace: Applications of nanotechnology in aerospace industry are mainly based on the concept of using high strength and low weight nanocomposites. Space vehicles need high-performance multifunctional materials which can withstand harsh and extreme environments. Materials should also sustain at high or low temperature and pressure. Polymer nanocomposites using silica fibers and NPs have high stiffness, low temperature coefficient of expansion, and high impact strength. In addition, the radiation protections of NP-based composites are better as compared to microparticles-based composites. This makes polymer nanocomposites as a good material for aerospace application [7].

Automobile: In automobile industry, polymers have been mixed with NMs (mainly CNTs and metal NPs) to develop robust and durable parts of vehicles. Cars are spray painted with tiny size NPs that provide smooth, fine, and thin attractive coating. In addition, nanocoating-based paints can also alter the heat-reflecting properties depending on the intensity of the incident sunlight, which helps to regulate the temperature of the vehicle, making the air conditioning system more efficient, easier, and less fuel consuming. Self-cleaning glass can be developed by dissolving small amount of TiO_2 NPs with glass manufacturing ingredients during the melting time. TiO_2 NP is able to dissociate organic dust in the presence of UV light which is found in normal sunlight. The dissociated organic compound may fall down or simply evaporate and gives a clear sight [8].

Medical: Nanotechnology can be effectively used for diagnosis and drug delivery. Special sensor of nanorobots can be inserted into the blood under the skin to check different biological and chemical contents present in blood and trace out any possible diseases. They can also be used to check the sugar level present in the blood. Advantage at using such nanorobots is that they are very cheap to produce and easily portable. Nanotechnology is already used for the treatment and diagnosis of cancer, autoimmune diseases, and parasitic infections. Some of NPs have antibacterial properties thus they are used in hospital equipment, and magnetic NPs are used to diagnose conditions to avoid the risks of surgery. Further research investigation may explore the role of nanotechnology in tissue engineering and can be used as replacement of body parts. Ligand-based target drug therapy for cancer patients also becomes effective after the invention of nanotechnology [9]. Some commercial products on the market today utilizing NMs include stain-resistant textiles and reinforced tennis rackets. Food will stay fresh longer if the packaging is less permeable to atmosphere [10].

13.4 Nanocomposite: A Brief Overview

The concept of enhancing mechanical, thermal, and physical properties of materials through addition of reinforcement material is being practiced for over long decades. The concentrated efforts in this direction resulted in the development of exotic materials both as reinforcements such as whiskers, organic, and inorganic-based high-performance fibers as well as matrix systems [11]. The current nanotechnology era has equally attracted the structural scientists, engineers, and industries with simple promise that using building blocks with dimensions in the nanosized range makes it possible to design and develop new multifunctional materials, the so-called nanocomposite. A nanocomposite is a multiphase solid material where one of the phases covered in less than 100 nm dimensions. Nanocomposite materials have emerged as suitable alternatives to overcome limitations of microcomposites and monolithic, while posing preparation challenges related to the control of elemental composition and stoichiometry in the nanocluster phase [12].

The methods used for preparing nanocomposites may vary from mechanical and chemical routes to vapor phase deposition, mainly depending on the type of matrix used. The key issues in the development of nanocomposites are production of nano-reinforcements, control over distribution in size and their dispersion, tailoring interfaces at nanoscale between the constituents and understanding physical phenomena, stress transfer and fracture in composites [13].

13.4.1 Ceramic–Matrix Nanocomposites

In this type of composites the main part of the volume is occupied by a ceramic, i.e., a chemical compound from the group of oxides, nitrides, borides, silicides, etc. In most of the cases, ceramic–matrix nanocomposites encompass a metal as the second component. Ideally both components, the metallic one and the ceramic, are finely dispersed into each other in order to elicit the particular nanoscopic properties. Nanocomposite from this combination was demonstrated in improving their optical, electrical, and magnetic properties [14] as well as corrosion-resistance and other protective properties [15]. The binary phase diagram of the mixture should be considered in designing ceramic–metal nanocomposites and measures have to be taken to avoid a chemical reaction between both components. At the last point, a metallic component may easily react with the ceramic and thereby lose its metallic character. This is not an easily obeyed constraint, because the preparation of the ceramic component generally requires high process temperatures. Thus for safety there are carefully chosen immiscible metal and ceramic phases. A good example for such a combination is represented by the ceramic–metal composite of TiO_2 and Cu, which were found immiscible over large areas in the Gibbs' triangle of Cu–O–Ti [16].

The concept of ceramic–matrix nanocomposites was also applied to thin films i.e., solid layers of a few nanometers (nm) to some tens of micrometer (μm) thickness deposited upon an underlying substrate and that play an important role in the functionalization of technical surfaces. Gas flow sputtering by the hollow cathode technique turned out as an effective technique for the preparation of nanocomposite layers. The process operates as a vacuum-based deposition technique and is associated with high deposition rates up to some μm/s and the growth of NPs in the gas phase. Nanocomposite layers in the ceramics range of composition were prepared from TiO_2 and Cu by the hollow cathode technique [17] that showed a high-mechanical hardness, small coefficients of friction, and a high-resistance of corrosion.

13.4.2 Metal–Matrix Nanocomposites

Metal–matrix nanocomposites can be classified as continuous and discontinuous reinforced materials. One of the important nanocomposites is CNT metal–matrix composites (MMCs) which is emerging new materials to take advantage of the high tensile strength and electrical conductivity of CNT materials. Critical to the realization of CNT–MMC possessing optimum properties in these areas is the development of synthetic techniques, like; economically producible, provide for a homogeneous dispersion of nanotubes in the metallic matrix, and lead to strong interfacial adhesion between the metallic matrix and the CNTs.

In addition to CNT MMCs, boron nitride reinforced MMC and carbon nitride MMCs are the focused research areas on metal matrix nanocomposites [18]. Another kind of nanocomposite is the energetic nanocomposite, generally as a hybrid sol–gel with a silica base, which combined with metal oxides and nanoscale aluminum powder, can form supethermite materials [19–21].

13.4.3 Polymer–Matrix Nanocomposites

In the simplest case, NPs add into a polymer matrix can enhance its performance, often in very dramatic degree, by simply capitalizing on the nature and properties at the nanoscale [22]. This strategy is particularly effective in yielding high-performance composites, when good dispersion of the filler is achieved and the properties of the nanoscale filler are substantially different or better than those of the matrix, for example, reinforcing a polymer matrix by much stiffer NPs [23] of ceramics, clays, or CNTs. It should be noted that the improvement in mechanical properties may not be limited to stiffness or strength. Time-dependent properties could be improved by addition of the nanofillers [24]. Alternatively, the enhanced properties of high-performance nanocomposites may be mainly due to the high aspect ratio and/or the high surface area of the fillers [25,26], since NPs have extremely high surface area to volume ratios when good dispersion is achieved. Nanoscale dispersion of filler or controlled nanostructures in the composite can introduce new physical properties and novel behaviors that are absent in the unfilled matrices, effectively changing the nature of the original matrix [22]. Some examples of such new properties are fire resistance or flame retardancy and accelerated biodegradability [27].

13.5 Nanoelectronic: Information Storage Devices

Nanoelectronics refer to the use of nanotechnology in electronic components. The term covers a diverse set of devices and materials, with the common characteristic like interatomic interactions and quantum mechanical properties need to be studied extensively. The aim of nanoelectronics is to process, transmit, and store information by taking advantage into properties of matter that are distinctly different from macroscopic properties. The relevant length scale depends on the phenomena investigated for materials which act like transistors or information storage devices [28].

Semiconductor electronics have seen a sustained exponential decrease in size and cost, on the other side increase in performance and level of integration (known as Moore's law). The silicon roadmap is laid out for the next 10 years. After that, either economical or physical barriers will pose a huge challenge. The former is related to the difficulty of making a profit in view of the exorbitant costs of building the necessary manufacturing capabilities if recent technologies are extrapolated. The latter is a direct consequence of the shrinking device size, leading to physical phenomena impeding the operation of current devices. Quantum and coherence effects, high electric fields creating dielectric breakdowns, heat dissipation problems in closely packed structures as well as the nonuniformity of dopant atoms, and the relevance of single atom defects are all roadblocks along the current road of miniaturization [29]. These phenomena is even clear that electrons will be the method of choice for signal processing or computation in the long-term quantum computing, spin electronics, optics, or even computing based on (nano-) mechanics are actively being discussed. Nanoelectronics thus needs to be understood as a general field of research aimed at developing an understanding of the phenomena characteristic of nanometer-sized objects with the aim of exploiting them for information processing purposes [30].

13.5.1 Genesis of the Concept

Although an information storage device based on nanoelectronics can be made fully functional, its workload is restricted to its size. The basic principle is that the power of a machine will increase according to the increase in volume, but the amount of friction, i.e., machine's bearings hold will depend on the surface area of the machine. Information storage is the recording (storing) of information (data). Storing is accomplished by virtual form of energy. Electronic information storage requires electrical power to store and retrieve data. Without a significant amount of memory, a nanoelectronic device would just be able to perform fixed operations and immediately output the result based on switching and carrier transport mechanism in the hybrid nonvolatile information storage device.

13.5.2 Approaches of Information Storage Devices

There are several areas of nanoscience that demand novel approaches to nanofabrication. For example, the aim in storage of energy and information devices is cost and convenience (particularly in academic research, or devices for resource-poor settings), rather than state-of-the-art performance. In addition, the ability to pattern materials requires the combination of "top-down" fabrication with "bottom-up" processes.

13.5.2.1 Evaporation method

A typical structure for a hybrid organic/inorganic nonvolatile information storage device consists of composite organic molecules: metal/semiconductor NP layer sandwiched between two metal electrodes. The device area is defined by the overlap between the top and the bottom electrodes; therefore, a very high memory density can be easily achieved by using crossbar arrays [31–35]. The organic material used can consist of small molecules or monomers. Small organic molecules have low molecular weight and can be deposited under high vacuum by using thermal evaporation method. On the other hand, polymer consists of much larger molecules with long chains of repeating monomer units and will decompose before evaporation begins. Multiple layers of organic thin films can be deposited uniformly over a large surface area and this is a remarkable advantage of thermal evaporation method.

As a result, device characteristics can be controlled by changing the materials and thickness. Ma et al. first demonstrated the resistive switching that could be achieved by depositing a layer of aluminum NPs between two organic layers [34]. The depositions of other NPs and the organic layer are controlled independently in the devices. Fig. 13.2 shows a typical structure for such devices. Usually the metal cluster layer is formed by depositing a discontinuous thin metal film at very slow evaporation rate ($<0.1\ \text{As}^{-1}$) [36–38], but for the aluminum oxide it formed spontaneously because of residual oxygen in the vacuum chamber and would be important to improve the charge retention properties of the NPs. Oxygen plasma has also been used to form an oxide around the metal NPs [39]. Semiconducting materials, for example, MoO_3 and so on [40], with suitable energy band levels can also be used in place of the metal NPs.

However as shown by the vast differences in the electrical characteristics of the fabricated devices in different laboratories [34,36–38], the reproducibility of the device depends on the vacuum chamber and deposition conditions. Therefore, a more repeatable fabrication process is required for the large-scale manufacture of such devices. Besides depositing a

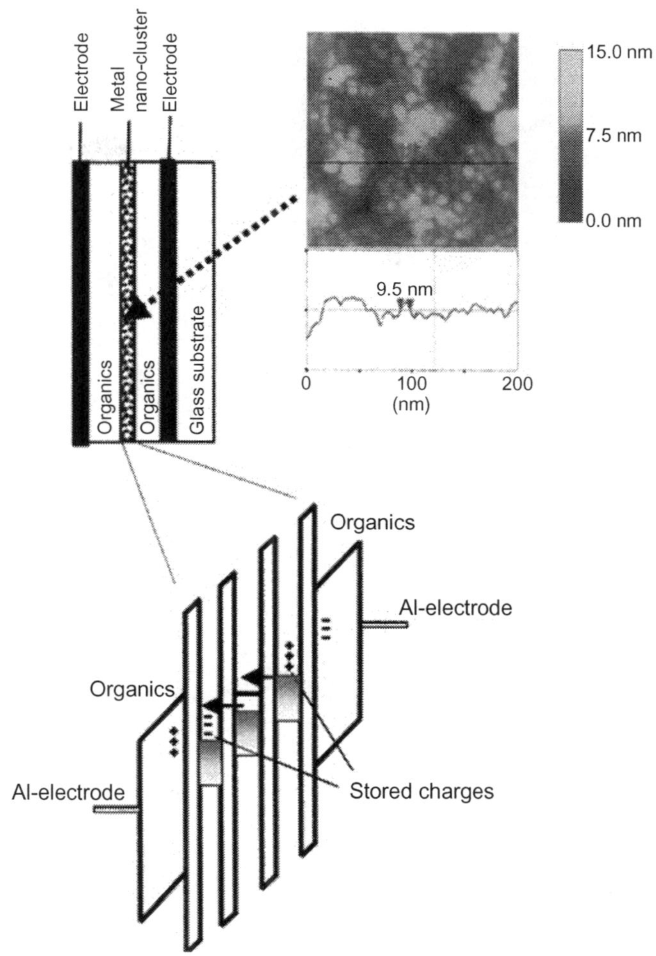

Figure 13.2

Typical device structure and atomic force microscope measurement of the metal cluster layer. *Source: L. Ma, et al. (2003), Nonvolatile electrical bistability of organic/metal-nanocluster/organic system, Appl. Phys. Lett., 82(9), pp. 1419–1421. doi: 10.1063/1.1556555. Copyright © 2003 American Institute of Physics.*

separate layer of metal/semiconductor clusters, electrons from metal electrodes can be driven into the organic layer by applying a large voltage across the device during the forming process [41]. Such processes are difficult to control and often result in nonuniform distributions of NPs due to local field enhancements at the rough edges of the electrodes (Fig. 13.3) [42]. Metals like gold can also be introduced into the organic layer during deposition. Gold is deposited at high temperature and does not react with the organic film. As a result, it readily diffuses into the organic layer, forming clusters within the organic layer [43]. Care has to be taken in the fabrication of such devices so that a short circuit does not form between the top and the bottom electrodes.

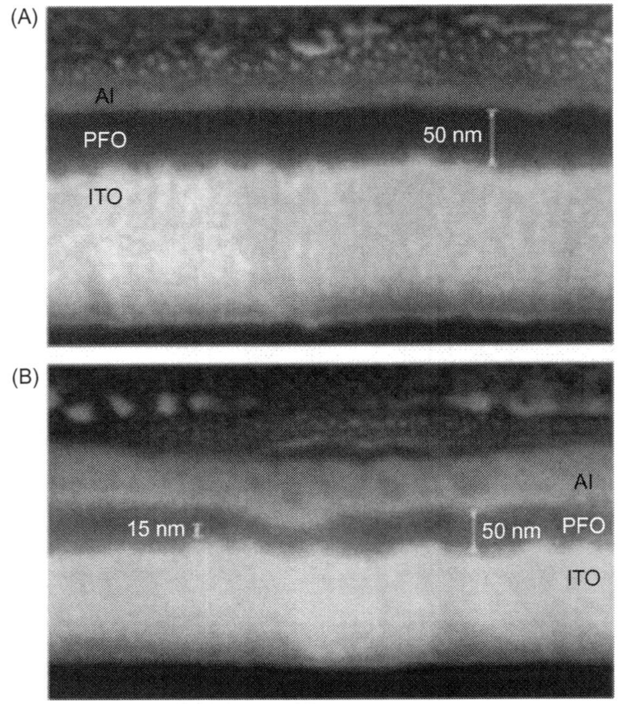

Figure 13.3

Cross-sectional scanning electron microscope images of the device at different spots: (A) uniform regions and (B) nonuniform regions. *Source: B. Lei, et al. (2009), Statistical characterization of the memory effect in polyfluorene based non-volatile resistive memory devices, Org. Electron. Phys. Mater. Appl., 10(6), pp. 1048–1053. doi: 10.1016/j.orgel.2009.05.015. Copyright © 2009 Elsevier.*

13.5.2.2 Solution method

In this method, one of the main advantages provided by organic materials and devices is low fabrication cost and simplicity. The fabrication of these devices through thermal evaporation requires high vacuum and stringent conditions to control the morphology of the organic active layer [44]. As alternatives solution processes including spin-coating, spray coating, dip-coating, roller-coating and ink-jet printing are well known and can be used to deposit polymer materials on a variety of substrates, such as glasses, plastics, metal foils, and wafers [45–51].

Single-layer structure: The simplest structure for a hybrid memory device fabricated utilizing the solution method is a single-polymer layer embedded with inorganic NMs and sandwiched between two metal electrodes, as shown in Fig. 13.4.

Generally, the hybrid nanocomposites are formed by dissolving inorganic NMs and a polymer matrix simultaneously in a certain organic solvent with a relatively high volatility [52]. To the date, fabrications of single-layer-structured nonvolatile information storage

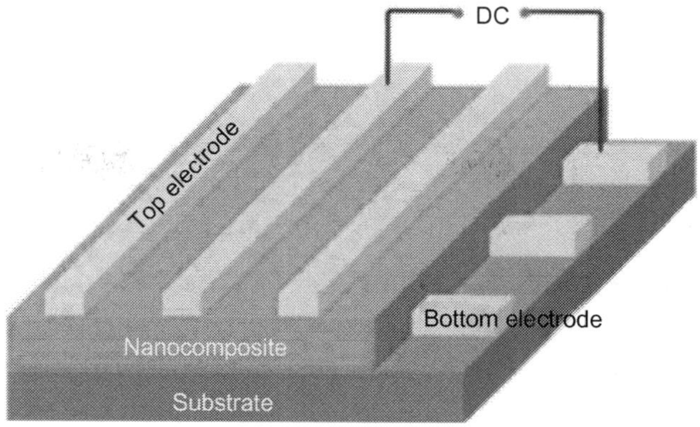

Figure 13.4
Schematic of a single-layer hybrid memory device.

based on various organic/inorganic hybrid nanocomposites have been reported. Inorganic species used in these devices include semiconductor NPs (ZnO, CdSe, Si, CuO, and so on) and metal NPs (Au, Ag, FeNi, and so on) [52–57]. Both insulating polymers, such as polyimide (PI), poly (methyl methacrylate), polystyrene, and conducting polymers, such as poly (*N*-vinylcarbazole) and poly [2-methoxy-5-(2-ethyhexoxy)-1, 4-phenylene vinylene], are used as a matrix for the inorganic NPs [58–62]. Most of these hybrid thin films are formed on a certain substrate by using a spin-coating technique, where the inorganic NPs are randomly dispersed in the polymer layer. The morphology and uniformity of the hybrid films can be controlled by varying the precursor concentration and spin conditions.

Multilayer structure: The single-layer structure fabricated via the solution process offers the simplest way to realize a nonvolatile information storage device. However, several key issues need to be addressed for this type of information storage device. One of the problems that single-layer memories face is the crosstalk of neighboring memory cells due to the presence of parasitic conduction paths, which causes misreading during the operating process [63]. A possible solution to this problem is to use one diode and one resistor or one transistor and one resistor structures to construct an individual memory cell with high reading accessibility [64,65]. Therefore, multilayer structures might be required to realize one diode and one resistor or one transistor and one resistor cells. On the other hand, device reproducibility may not be ensured because of the contact uncertainty at the interface between the hybrid layer and the metal electrode. Thus, the multilayer structure, where inorganic NPs are sandwiched between organic layers, has been employed in order to obtain memories with high stability and reproducibility [66]. The main challenge to obtain such a structure is the dissolving of the underlying polymer layer caused by the solvent used to form the upper polymer layer. Thus, the key point in realizing a multilayer structure is the availability of an appropriate polymer material. It has been shown that a polymer thin film made of polyimide (PI), after annealing at a certain temperature, can well withstand the organic solvent used to form the polymer layer. Some works concerning the formation of multilayer-structured nonvolatile information storage devices by using PI materials have been reported, and their memory effects have been demonstrated as well [67,68]. It was shown that the cell reproducibility and the cell-to-cell uniformity of the hybrid information storage devices could be significantly improved by employing a multilayer structure.

An ultra-high storage density, which can be achieved by using a vertically stacked 3-D structure, has been shown to be required for next-generation nonvolatile information storage devices [69–71]. Therefore, a hybrid resistive information storage device is developed with a 3-D stacked 8×8 crossbar array by using a composite of PI and 6-phenyl-C_{61} butyric acid methyl ester, which may offer a promising approach to future ultra-high-density memory [71]. The results indicated that the memory cells in each layer could be programmed or erased with a uniform parameter.

13.5.3 Fabrication of Information Storage Devices

13.5.3.1 Fabrication of nonvolatile information storage devices utilizing graphene materials embedded in a polymer matrix

Graphene, comprised of one or several monolayers of carbon atoms packed into a 2-D honeycomb lattice, is predicted to have remarkable properties, such as superior mobility, thermal conductivity, current-carrying capability, and room-temperature ballistic transport, which make it an attractive material for electric, optoelectronic, and photonic devices [72–79]. The two dimensionality of graphene allows scaling beyond the conventional semiconductor technology, which could be instrumental for high-density storage-memory applications. Furthermore, if graphene is integrated with polymers and flexible substrates, a spectrum of graphene-based memory applications, including transparent, flexible, and wearable electronics, is possible [80]. Some works concerning graphene-based memory devices have been reported recently [38,81–85]. For instance, Jeong et al. [84] reported a reliable and reproducible bipolar resistive switching effect in a GO film spin-casted on a flexible substrate, as shown in Fig. 13.5. The device exhibited a long retention time and a low switching voltage, but the ON/OFF current ratio was relatively modest, B_{100}, while 10^5-10^6 has been typically achieved in complementary metal-oxide semiconductor devices.

Graphene sheets, acting as charging and discharging media, have been particularly interesting because of their being promising candidates for hybrid-nanocomposite-based resistive-switching devices for potential applications in low-cost flexible memories. However, the pristine graphene materials are unsuitable for intercalation in large species, such as polymer chains, because graphene has a pronounced tendency to agglomerate in a polymer matrix, which leads to difficulty in material processing and to a high cost of device

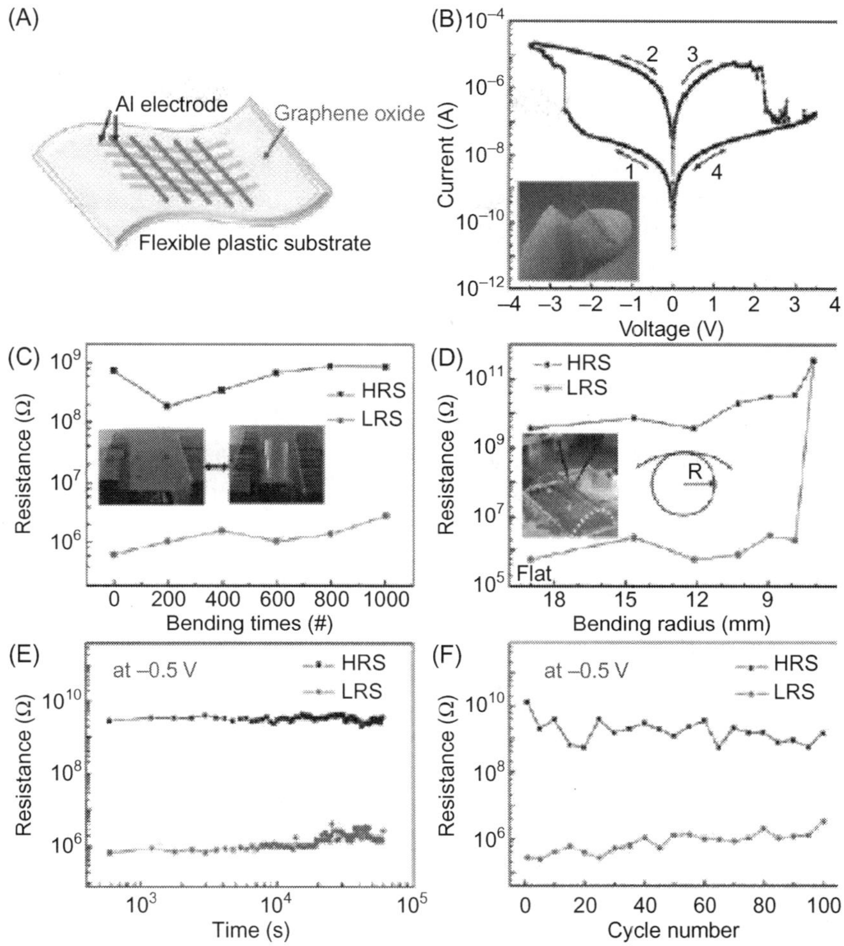

Figure 13.5

(A) Schematic of a GO-based flexible cross-bar memory device. (B) Typical I–V curve of an Al/GO/Al/PES device plotted on a semilogarithmic scale. The arrows indicate the voltage sweep direction. The left inset is a real photo image of a device. (C) Continuous bending effect of an Al/GO/Al/PES device. The insets show photographs of two repeated bending states. (D) The resistance ratio between the high-resistance state and the low-resistance state as a function of the bending radius (R). The inset is a photograph of an I–V measurement being performed under a flexed condition. (E) Retention test of an Al/GO/Al/PES device reads at −0.5 V. (f) Endurance performance of an Al/GO/Al/PES device measured during 100 sweep cycles. *Source: H.Y. Jeong, et al. (2010), Graphene oxide thin films for flexible nonvolatile memory applications, Nano Lett., 10(11), pp. 4381–4386. doi: 10.1021/nl101902k. Copyright © 2010 American Chemical Society.*

fabrication. One of the approaches for improving the solubility of the graphene material is to modify GO by using a chemical method [86–88].

For example, Zhuang et al. [89] demonstrated a single-layer hybrid bistable memory containing GO sheets covalently grafted with soluble conjugated polymer triphenylamine-based polyazomethine (TPAPAM). The device was fabricated by using a simple spin-coating technique, and the TPAPAM-GO-based information storage device exhibited typical bistable electrical switching and a nonvolatile rewritable memory effect, with a turn-on voltage of about −1.0 V and an ON/OFF current ratio of about 10^3. This is thought to be the first example of a polymer-modified graphene-based information storage device. However, multilayer-structured memory devices still need to be developed in order to improve device reproducibility. A nonvolatile hybrid bistable device fabricated with graphene sandwiched between two insulating poly(methyl methacrylate) polymer layers was reported by Son et al. [90] An ON/OFF ratio of 4.4×10^6 was maintained for retention times longer than 10^5 s, and the endurance number of the ON/OFF switching was 1.5×10^5 cycles. The bistable behavior for the poly (methyl methacrylate)-graphene-based device was attributed to conducting filaments having been formed in the hybrid active layer. Wu et al. [91] demonstrated a nonvolatile hybrid information storage device employing a tri-layer structure, PI/PI: GO/PI, in which insulating polymer PI was used as the matrix for the GO sheets. Apart from the high ON/OFF ratio ($\sim 10^5$) and the long retention time (> 1400 s), the device exhibited excellent cell-to-cell uniformity (70 out of 81 cells, 86%), as shown in Fig. 13.6.

Recent results clearly show that graphene/polymer hybrid nanocomposites are able to present a superior memory effect with a large ON/OFF ratio, good endurance, long retention time, and high cell-to-cell uniformity. These advantages may enable graphene/polymer nanocomposites to serve as storage media in low-cost flexible nonvolatile memories. However, for practical applications, some issues like in situ synthesis of graphene with polymer matrix and the crosstalk between the adjacent memory cells still need to be addressed. These problems may be overcome in the near future as more and more research is focused on graphene-based hybrid information storage devices.

13.5.3.2 Fabrication of nonvolatile information storage devices utilizing hybrid nanocomposites

One of the main advantages of a nonvolatile memory device based on hybrid nanocomposites is the possibility of realizing flexible memories at extremely low fabrication cost. These hybrid devices can be processed at low temperatures and therefore they are regarded as one of the most important types of information-storage components for future stretchable/portable electronic device applications [92−95].

For example, Kim and Lee [92] developed nonvolatile information storage devices on plastic substrates (PES) based on a pentacene layer embedded with self-assembled gold NPs, as shown in Fig. 13.7. The nanofloating gate memory exhibited a large memory window. The hybrid memory was reliable after several hundreds of repeated programming/erasing operation cycles. Additionally, the mechanical flexibility was confirmed using the bending-cycle test over 1000 times. Nonvolatile electrical bistability of a hybrid device fabricated utilizing CdSe/ZnS NPs embedded in a conducting poly(N-vinylcarbazole)

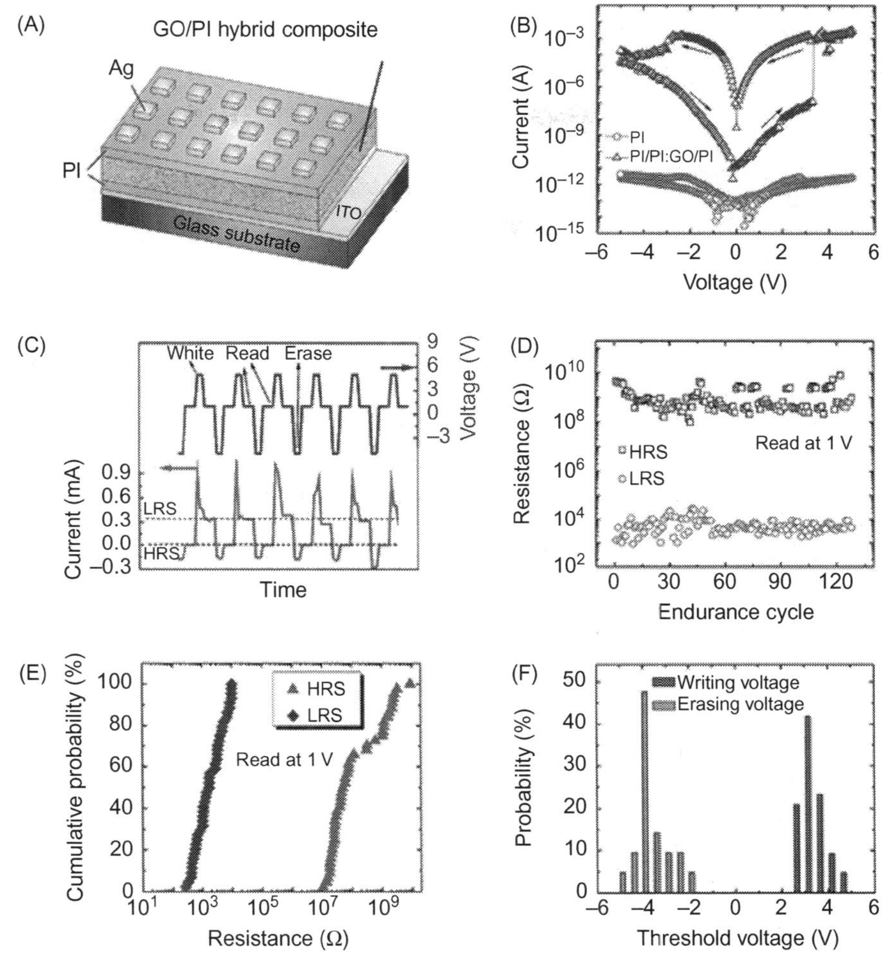

Figure 13.6

(A) Schematic of the hybrid memory device containing GO sheets embedded in a PI layer. (B) I–V characteristics for Ag/PI/GO: PI/PI/indium tin oxide (ITO) and Ag/PI/ITO devices. (C) Write–read-erase–read sequence test for an Ag/PI/GO: PI/PI/ITO memory device. (D) Resistances of the low-resistance state and high-resistance state at 1 V for 130 cycles. (E) Cumulative probability data for the resistances of the low-resistance state and high-resistance state. (F) Threshold voltage distributions for the "write" and the "erase" processes. *Source: C. Wu, F. Li, Y. Zhang, et al. (2011), Highly reproducible memory effect of organic multilevel resistive-switch device utilizing graphene oxide sheets/polyimide hybrid nanocomposite, Appl. Phys. Lett., 99(4), pp. 2011–2014. doi: 10.1063/1.3619815. Copyright © 2011 American Institute of Physics.*

polymer layer on flexible polyvinylidene fluoride and polyethylene terephthalate (PET) substrates has also been reported. The results indicate that the hybrid bistable memory devices are prospects for potential applications in flexible nonvolatile flash memory devices [95].

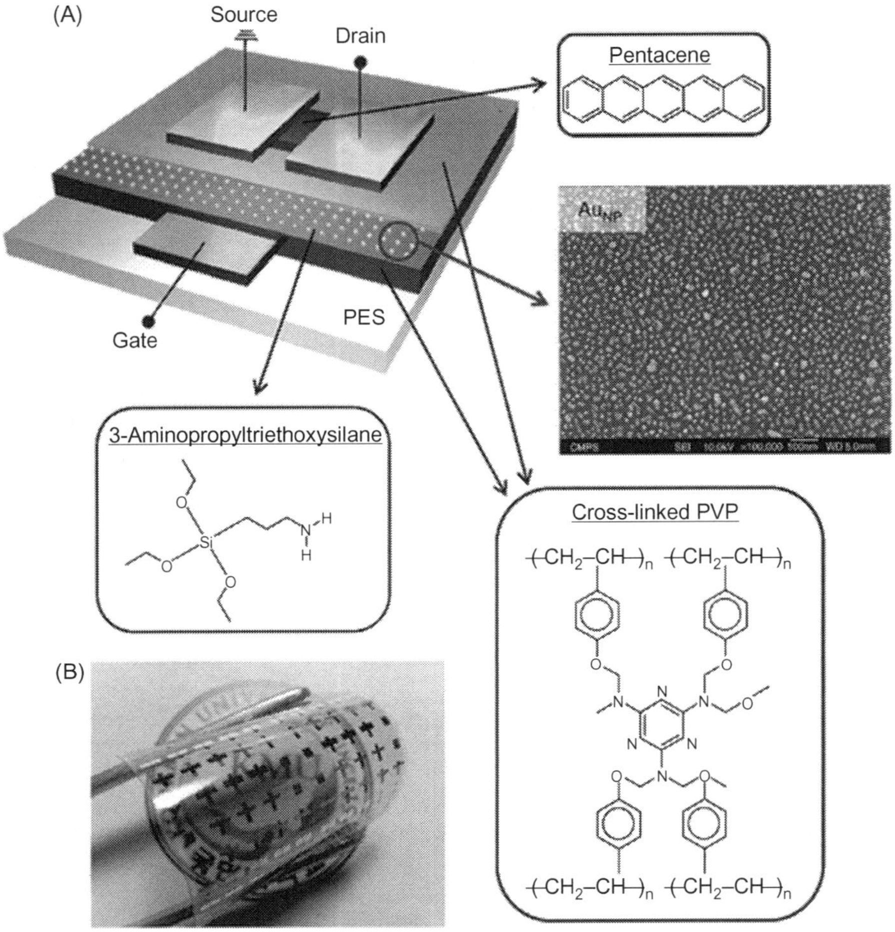

Figure 13.7
(A) 3-D schematic diagram of the flexible hybrid memory device architecture. Patterned Ti/Au on a PES plastic substrate was used as the gate electrode, and cross-linked Poly vinyl pyrrolidone (PVP) layers were used as the blocking and the tunnelling organic dielectric layers. The self-assembled gold NP charge storage layers were formed on the (3-Aminopropyl) triethoxysilane (APTES)-coated PV blocking dielectric layer. A pentacene active layer and gold source/drain contacts were formed to make the organic transistor-based memory devices. The chemical structures of the organic layers are shown in the figure. (B) Photograph of a fabricated flexible organic memory device. *Source: S.J. Kim, J.S. Lee (2010), Flexible organic transistor memory devices, Nano Lett., 10(8), pp. 2884–2890. doi: 10.1021/nl1009662. Copyright © 2010 American Chemical Society.*

Electrical stability of a hybrid memory during bending is important for practical applications in nonvolatile information storage. The memory should be reliable after thousands of bending-cycle tests. However, frequent folding of the information storage device could cause physical and electrical damage at the conducting substrate. Metals and indium tin oxide are widely used as electrode materials for hybrid information storage devices [96,97]. However, their brittle properties under bending may become obstacles for application in flexible devices. For this reason, Ji et al. [98] developed a flexible hybrid resistive information storage device with transparent multilayer graphene electrodes on a PET substrate (Fig. 13.8). The active layer of the memory is a hybrid nanocomposite of PI and 6-phenyl-C_{61} butyric acid methyl ester. The information storage devices show typical write-once read many-time characteristics with an ON/OFF ratio of over $\sim 10^6$. The current levels can be maintained upon bending for up to 10,000 times.

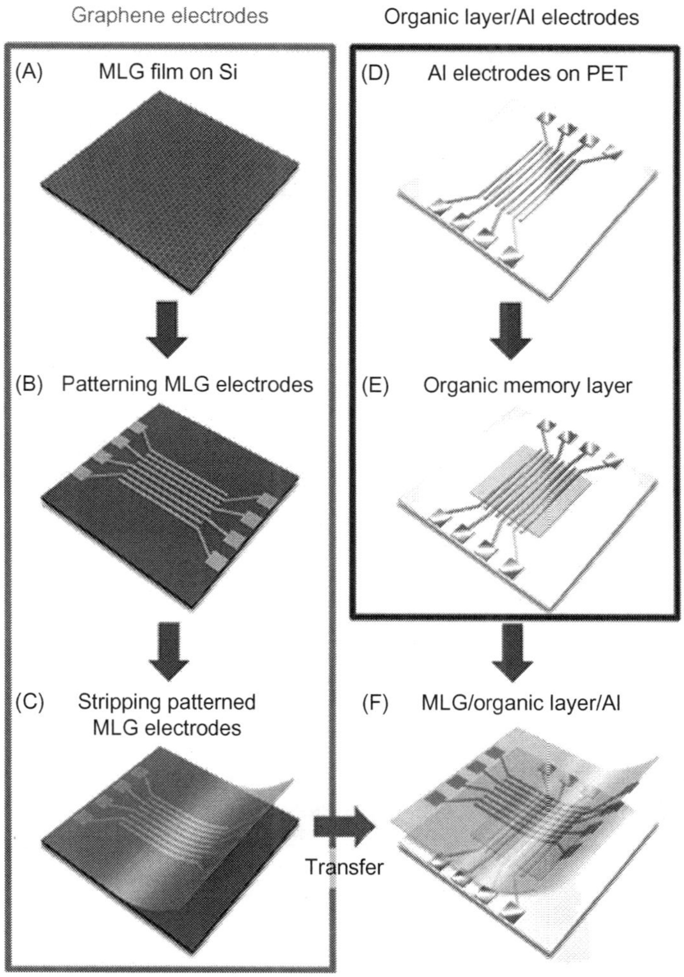

Figure 13.8

Schematic representation showing the fabrication process for the hybrid flexible memory device with transparent graphene top electrodes. *Source: Y. Ji, et al. (2011), Flexible organic memory devices with multilayer graphene electrodes, ACS Nano, 5(7), pp. 5995–6000. doi: 10.1021/nn201770s. Copyright © 2011 American Chemical Society.*

A stainless-steel foil can also serve as a conducting substrate for a nonvolatile information storage device because of its high conductivity and flexibility. Lee et al. [99] reported the resistive-switching characteristics of ZnO thin films grown on stainless-steel for flexible nonvolatile information storage devices. The device performance was not degraded upon bending, which indicates high potential for flexible memory applications. However, the introduction of a stainless-steel substrate may lead to an enhanced fabrication cost for commercialized information storage devices. In addition, a stainless-steel substrate might not be appropriate for flexible memory devices that require high transparency.

13.5.4 Electrical Characteristics of the Hybrid Information Storage Devices

In general, the electrical properties of nonvolatile information storage device based on hybrid nanocomposites can be investigated by using current−voltage (I−V) measurements. The applied voltage across the device is varied in a cycle from negative bias to positive bias and then to negative bias again. Two distinct conducting states, the ON-state and the OFF state, corresponding to the high-current and the low-current states, respectively, can be observed in the I−V curves. The electrical hysteresis behavior shown in the I−V characteristics is an essential feature of a resistive-switching information storage device [100]. The resistive transition (OFF−ON) occurs at a certain applied voltage, which is defined as the turn-on voltage. The transition is equivalent to the "writing" process in a digital memory cell. A high turn-on voltage means high power consumption and always reduces the reliability. Thus, the operating voltage for a hybrid nonvolatile memory should be only a few volts. After the transition is finished, the ON-state remains in the device even after the power is turned off, which reveals the nonvolatile nature of the hybrid information storage device [36].

Depending on the I−V characteristics, three types of nonvolatile information storage effects, write-once-read-many times, unipolar and bipolar electrical switching, can be observed for hybrid devices. The write-once-read-many-time memory exhibits an irreversible electrical transition upon external voltage stimulation. Both unipolar and bipolar memories can restore their original conducting state when a voltage with the same (for unipolar) or different (for bipolar) polarity is applied to the device. Typically, the maximum ON/OFF ratio in the I−V curves for hybrid nonvolatile information storage is less than 10^3, which distinguishes the two states and decreases the probability of misreading. In the case of a nonvolatile information storage device based on hybrid nanocomposites, the memory effect is strongly correlated with the presence of inorganic NMs, for example, ZnO QDs, Au NPs, CdSe NPs, and CNTs, embedded in a polymer matrix because the current difference between the ON and the OFF states for the polymer-only device is negligible [101]. Recently, for a hybrid device containing Au QDs embedded in a PI layer, Wu et al. reported an electrical bistability with an ON/OFF ratio of $\sim 10^8$, which is one of the largest values for a stable hybrid information storage device. In the case of a unipolar nonvolatile memory, due to the presence of a negative differential resistance in the I–V characteristics, the OFF state can be regained when a forward bias is given to the information storage device [102–105].

For bipolar memories, the OFF state of the hybrid device can be recovered by applying a reverse bias voltage, which is equivalent to the "erasing" process in a digital memory cell. For a nonvolatile rewritable hybrid memory, a "write-read-erase-read" sequence test is commonly carried out to investigate the endurance of the device. Before the ON and OFF states are no longer distinguishable, the number of cycles should be large enough to obtain an advantage over conventional semiconductor-based memories, which show a typical endurance of 10^3–10^7 cycles under ambient conditions. Besides the endurance, the retention time is also an important parameter for an information storage device. A test for the retention time can be performed by keeping the hybrid device in the ON or OFF state and observing the device current over time to identify the conductive state of the device. The retention time clearly reflects the environmental stability and capability of the memory to maintain its storage. Most commercial products claim to have retention times of less than 10 years [106].

As reported by Son et al. [90], some hybrid nonvolatile information storage, in which a hybrid nanocomposite of graphene poly (methyl methacrylate) was used as an active layer (Fig. 13.9), have been demonstrated to exhibit retention time comparable to those of commercial products. The switching speed is defined as the shortest time for programming or erasing a hybrid memory cell. Obviously, for practical applications, a high switching speed (<100 ns) is necessary to meet the requirement of ultra-high-density information storage. Ma et al. [34] reported that a hybrid memory cell could exhibit a switching speed as fast as 10 ns.

The device yield for hybrid memories should definitely be taken into account for future commercialization. Unfortunately, the yield is generally rather low due to the uncontrollability of the microstructure of the hybrid active layers. The uncertainty at the hybrid layer/electrode interface also contributes to low device reproducibility and cell-to-cell uniformity. Thus, the morphology of the hybrid active film should be precisely controlled because the carrier transport of a hybrid information storage device is strongly governed by the microstructure of the hybrid film. The growth parameters, such as the concentration of inorganic NPs, should be carefully adjusted to obtain a repeatable nonvolatile memory effect. It has been shown that the ON/OFF ratio of a hybrid memory based on a CdSe QD–poly (*N*-vinylcarbazole) structure can be maximized when the content of CdSe QDs is optimized [107]. Excessive CdSe leads to an aggregation of the poly (*N*–*N*-vinylcarbazole) matrix and to a degradation of the memory properties, including the device reproducibility.

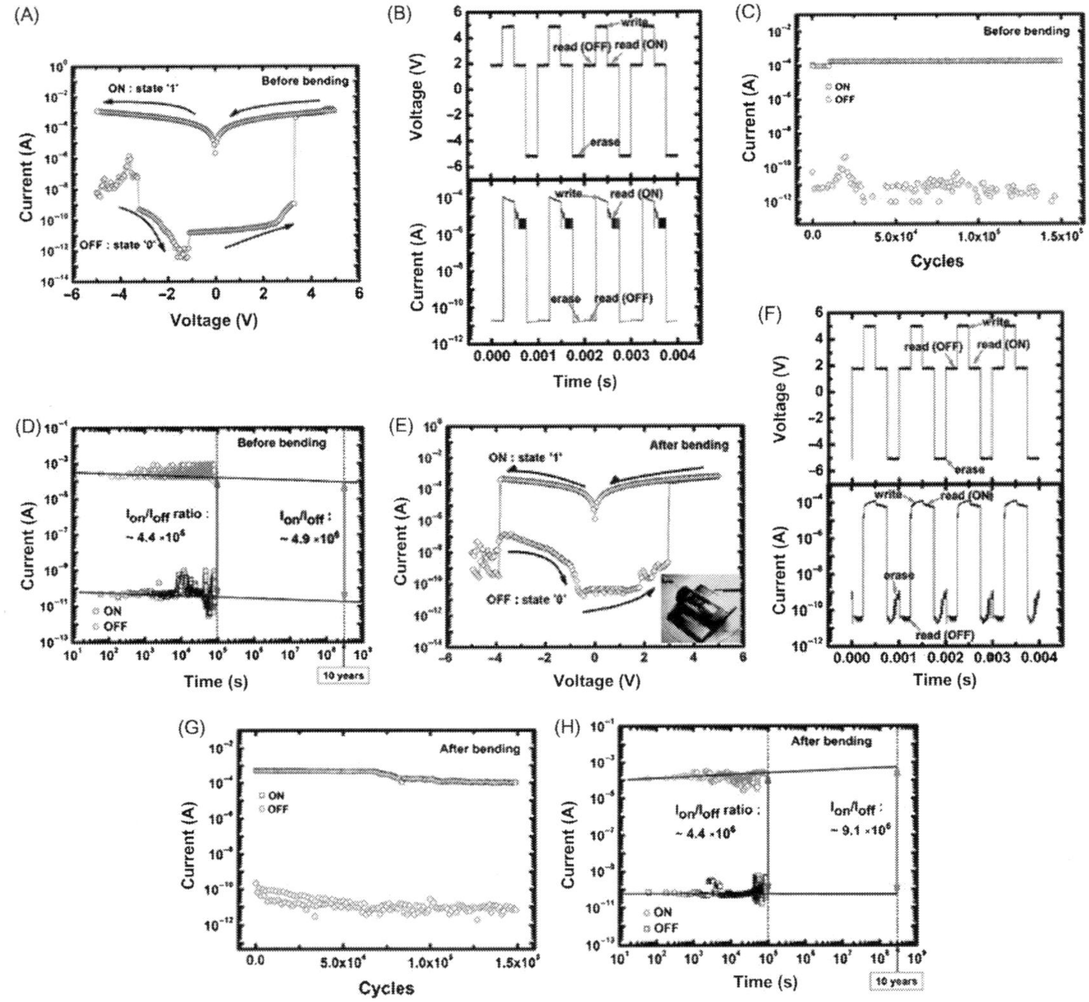

Figure 13.9

Memory characterization of the organic bistable devices (OBDs) before bending: (A) I–V curves for the Al/poly(methyl methacrylate) (PMMA)/graphene/PMMA/ITO/polyethylene terephthalate (PET) device, (B) operation of a write-read-erase-read (5/2/−5/2 V) sequence, (C) cycling stress test of the Al/PMMA/graphene/PMMA/ITO/PET device under a stress voltage of 2 V per 4 ms, and (D) retention test of OBDs in an ambient environment. (e) I–V curves for the Al/PMMA/graphene/PMMA/ITO/PET devices after bending, (F) operation of a write-read-erase-read (5/1.8/−5/1.8 V) sequence, (G) cycling stress test of the Al/PMMA/graphene/PMMA/ITO/PET device under a stress voltage of 1.8 V per 4 ms, and (H) retention test of the memory device in an ambient environment. *Source: D.I. Son, et al. (2010), Flexible organic bistable devices based on graphene embedded in an insulating poly(methyl methacrylate) polymer layer, Nano Letters, 10(7), pp. 2441–2447. doi: 10.1021/nl1006036. Copyright © 2010 American Chemical Society.*

13.5.5 Switching and Carrier Transport Mechanism

The switching mechanism has been under debate since the discovery of the memory phenomenon in hybrid devices. It has been argued that because the device characteristics are so similar across different material systems, one underlying mechanism might explain the phenomenon completely. However, a unifying explanation has been elusive due to inconclusive evidence. The lack of understanding of the switching mechanism has hindered progress toward the commercialization of these devices.

One approach to understanding the switching mechanism is to study the carrier transport mechanisms of the device at the two resistive states. Fittings to the I–V curves and their temperature dependences are often used to study the charge transport during device operation. Ohmic conduction, thermionic emission, space-charge-limited current (SPLC), and Fowler–Nordheim tunnelling are few of the more common mechanisms used to explain charge transport in organic devices. Even though this is not always explicitly acknowledged, the current distribution is almost always assumed to be uniformly distributed across the entire device.

In reality, this is difficult to achieve because of no uniformities in the devices. As a result, the parallel pathways with different transport mechanisms may exist in one device, which may complicate analysis. More recently, high-resolution transmission electron microscopy has been used to examine the change in the nanoscale structure of a device before and after switching. This may provide an important new evidence for a complete understanding of the switching phenomenon [108].

13.5.5.1 Filament formation

One of the earliest explanations for the large change in resistance is the formation of the highly conductive pathways in the composite layer. As in copper-ion memories [74,109,110], mobile metallic ions from the electrodes or the NPs can migrate to form a conductive filament between the two electrodes when a high enough voltage is applied to the device. It has been observed using a current-sensing atomic force microscope [33,72,76,77] and an infrared microscope [75] that only the switching occurred in localized areas of the devices.

Highly conductive spots are created after the device is switched to the low conductive state. In some cases, damage to the electrodes due to high-current densities at such spots is detected [73]. Because the filament is metallic in nature, the temperature dependence of the current at the low conductive state is very low. Because of the simplicity of the idea, many groups have attributed the switching phenomenon to the formation and the destruction of the conductive pathways despite the lack of direct evidence.

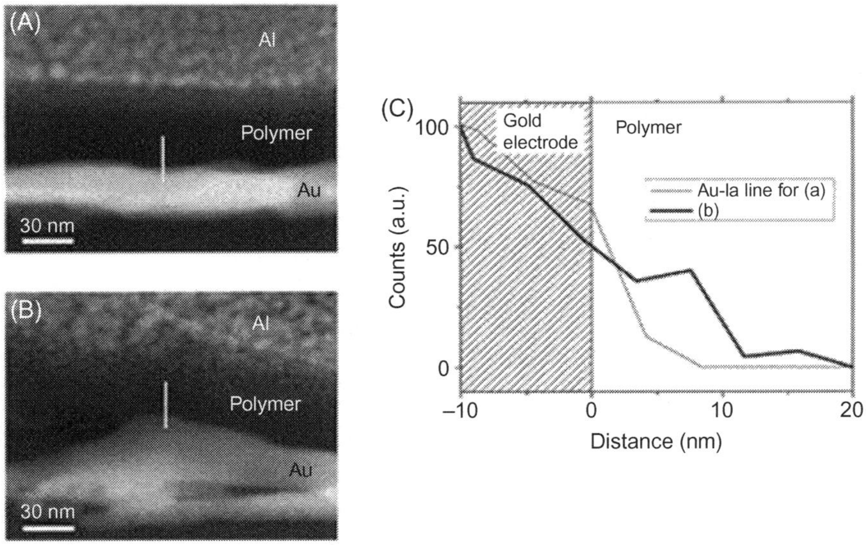

Figure 13.10

Cross-sectional scanning transmission electron microscope (STEM) images of devices whose bottom gold electrode was (A) flat and (B) irregular. The white lines denote the locations of the TEM-energy-dispersive X-ray spectroscopy (EDS) line scan. (C) TEM-EDS line scan data for (A) and (B). *Source: W.L. Kwan, et al. (2009), Direct observation of localized conduction pathways in photocross-linkable polymer memory, J. Appl. Phys., 105(12), pp. 11–16. doi: 10.1063/1.3153980. Copyright © 2009 American Institute of Physics.*

As the filament is usually nanoscale and sparsely distributed, probing the filament directly is very difficult. Only recently has direct observation of the filaments been possible, through the use of high-resolution transmission electron microscopy, for carefully prepared cross-sections of the device [41,79]. Fig. 13.10 shows that gold atoms from the electrode are injected into the polymer film after a high voltage is applied to the device ("forming" process). Fig. 13.11A and 13.11B show a highly conductive pathway between the top and the bottom electrodes after the device is switched to a low conductive state. The change in the temperature dependence of the current of the device (Fig. 13.11C) clearly shows that the device has changed from an electronic conduction mechanism to a filamentary one after the switching process. Although a device that switches with a filamentary mechanism has been shown to be able to switch more than a few thousand cycles [80], this is undesirable as it requires a large switching current and may damage the electrodes.

13.5.5.2 Space-charge-limited current (SPLC)

Usually a non-Ohmic contact is formed between the electrode and the organic/inorganic hybrid layer. When that happens, the current that flows through the device is injection

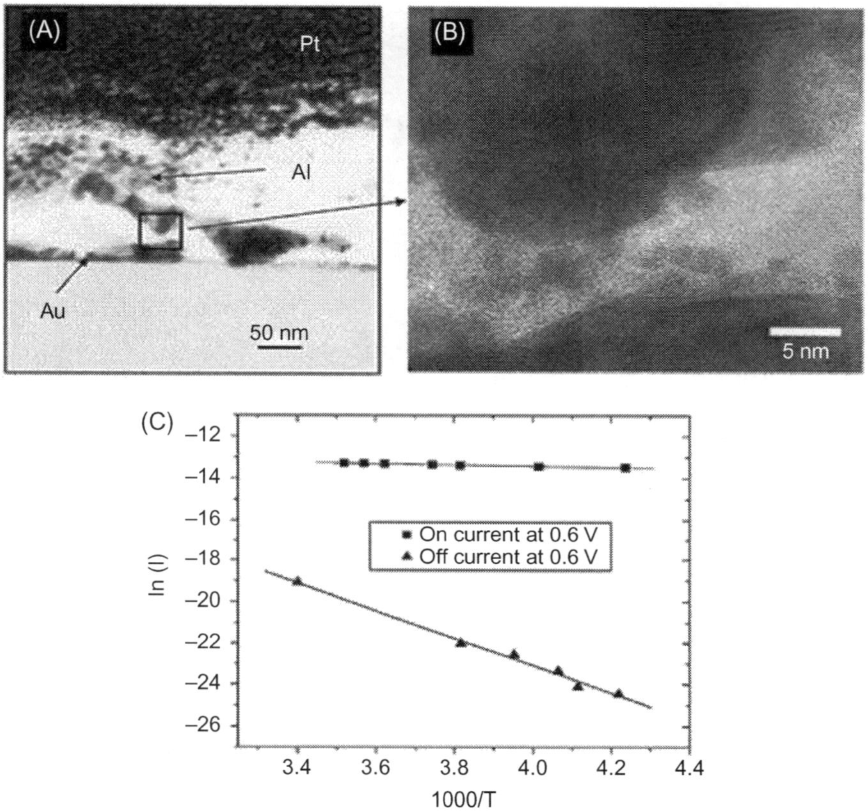

Figure 13.11
(A) TEM cross-sectional view of the device near the filament. Near the edge of the gold electrode, the top electrode was almost in contact with the bottom electrode. (B) Close-up TEM image of the boxed region in (A). (C) Arrhenius plots of the currents measured at 0.6 V at the ON and OFF states of the device. *Source: W.L. Kwan, et al. (2009), Direct observation of localized conduction pathways in photocross-linkable polymer memory, J. Appl. Phys., 105(12), pp. 11–16. doi: 10.1063/1.3153980. Copyright © 2009 American Institute of Physics.*

limited. I–V and temperature characteristics will, thus, reflect the injection mechanism of the interface. The two most common injection mechanisms found in organic devices are thermionic emission and Fowler–Nordheim tunnelling [83,111].

During the switching process, as more carriers are injected into the composites, traps that are formed from NPs are hypothesized to fill up, which eventually causes the current to increase rapidly, reaching an SPLC [81]. The NPs can act as traps as long as the energy level is within the lowest unoccupied molecular orbital and highest occupied molecular orbital levels of the organic matrix. Fig. 13.12 shows a transition from thermionic emission to an SPLC when the bias voltage is increased. Interestingly, the trap is able to hold the charges very stably, with the retention time being extrapolated to as long as 10 years [85].

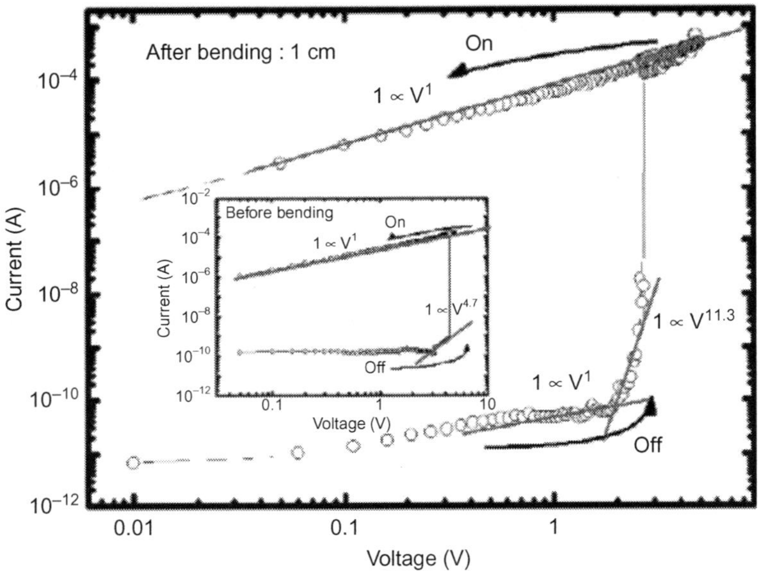

Figure 13.12

A log–log plot of the current as a function of the applied voltage for Al/PMMA/UGS/PMMA/ITO/PET devices. The curves are fitted for the Space charge limited current (SCLC) mechanism. The inset presents a log–log plot of the current as a function of the voltage for the bistable organic memory devices before bending. *Source: D.I. Son, et al. (2011), Polymer-ultrathin graphite sheet-polymer composite structured flexible nonvolatile bistable organic memory devices, Nanotechnology, 22(29). doi: 10.1088/0957-4484/22/29/295203. Copyright © 2011 IOP Publishing Ltd.*

13.5.5.3 Simmons and Verderder's model

In the Simmons and Verderder's model, when the traps are empty, the device is in the low resistive state, which is contrary to that of the SPLC model. The Simmons and Verderder's model argues that when the trap is not filled, the impurity band of the composite layer is free to conduct current. However, as more charges are introduced into the material through a larger bias voltage, the charges start to occupy the impurity states, thus lowering the impurity band. This prevents additional charge from being injected into the device. The trapped charge can be removed by applying an even larger voltage bias to the device [38,88]. This model is successful in explaining the negative differential resistance, whereby the current decreases with increasing voltage bias.

13.5.5.4 Electric field induced charge transfer

Because the nanocomposite layer is usually less than a few hundred nanometers, the electric field across the device can be very large, even though the applied bias is only a few volts. This high field is thought to cause charges to be transferred between donors and acceptors in the composite layer. Depending on the relative energy levels, the organic matrix and metallic/semiconducting NPs can act as donors and acceptors, respectively. The metal particles are usually surrounded by an insulating layer, so the charges are prevented from moving back even when the applied voltage is removed. As the electrons from the organic matrix are transferred from the highest occupied molecular orbital level to the NPs, additional holes are created after the charge transfer, thereby increasing the conductivity of the device [86].

In a competing theory on electric field transfer of charge, Ling et al. [87] suggest that a charge-transfer complex, which may be insulating or conducting, is formed after the threshold electric field has been reached. In the case of an electron-donating polymer and a GO composite, the electron from the polymer is transferred to the giant p-conjugated system of the GO, thereby reducing the insulating GO to a highly conductive graphene. This substantially enhances the conductivity [89].

13.6 Outcomes and Conclusive Aspect

Hybrid inorganic/organic nanocomposites provide a frontier for unique functions with promising applications in nonvolatile information storage devices. OBDs fabricated utilizing nanocomposites have emerged as excellent candidates for potential applications in the next-generation nonvolatile information storage devices. In this chapter, the NMs classification, nanocomposite, current bistability, operating mechanisms, memory mechanisms, and carrier transport mechanisms of OBDs fabricated utilizing nanocomposites were described on the basis of recently published experimental data. Currently, nonvolatile memories based on hybrid nanocomposites still have some problems. In comparison with Si-based memory device, the switching speed of hybrid memory device should be further improved, even though some hybrid systems have exhibited switching times of ~ 10 ns. Other problems that nonvolatile hybrid memory devices face are device-to-device uniformity and cell reproducibility. Solution processing not only simplifies the fabrication of nonvolatile hybrid information storage device but also leads to a high uncertainty in the microstructure. Thus, for practical applications, hybrid information storage should have a high device yield comparable to that of Si-based memories. Obviously, the information storage properties of hybrid devices should be improved, so new hybrid materials with superior memory effects should be explored. Moreover, the carrier transport mechanisms in hybrid nonvolatile information storage are still under debate and should be classified in the near future.

One of the important applications of nonvolatile information storage devices fabricated utilizing nanocomposites is flexible information storage device, which is necessary for fascinating applications of nonvolatile information storage devices in portable equipment. The device performance should be reproducible, regardless of repetitive bending. Because scaled-down experiments of memory devices with nanoscale electrodes are very important for enhancing storage density, the inherent problems due to the limitations of organic lithography for fabrication of OBDs with nanoscale electrodes based on inorganic/organic nanocomposites should be overcome. As an alternative, nonvolatile information storage with a 3-D architecture should be developed for the realization of an ultra-high-density information storage based on hybrid nanocomposites.

References

[1] V. Shukla, M. Singh, Synthesis of in-situ luminescent ZnS nanoparticles facile with CTAB micelles and their properties study, AIP Conf. Proc. (2016) 1724. Available from: https://doi.org/10.1063/1.4945246.

[2] V. Shukla, B. Raval, M. Singh, Impact of newly synthesized water soluble photoluminescent ZnS-L-Cysteine: Core-shell nanoparticles in defining the in-situ opto-electronic orbital model, Adv. Mater. Lett. 8 (2) (2016) 156–162. Available from: https://doi.org/10.5185/amlett.2017.7078.

[3] C.R. Martin, Welcome to nanomedicine, Nanomedicine 1 (1) (2006). Available from: https://doi.org/10.2217/17435889.1.1.5. 5–5.

[4] Swiss Reinsurance Company, Nanotechnology: small matter, many unknowns, Chem. Eng. News August (2004) 56. Available at: http://www.asse.org/nanotechnology/pdfs/govupdate_02-3-05_nanosafety.pdf.

[5] M.E. Kurtoglu, et al., Effect of calcination temperature and environment on photocatalytic and mechanical properties of ultrathin sol-gel titanium dioxide films, J. Am. Ceram. Soc. 94 (4) (2011) 1101–1108. Available from: https://doi.org/10.1111/j.1551-2916.2010.04218.x.

[6] Report, T. *et al.* (2016) 'The role of nanotechnology in electronic properties of Dr. Magy Mohamed Kandil', (July).

[7] Esc, P. (2007e) 'NATO lectures M. Meyyappan nanotechnology in aerospace applications', *Notes*, 2006 (2007), pp. 1–2.

[8] H. Presting, U. König, Future nanotechnology developments for automotive applications, Mater. Sci. Eng. C 23 (6–8) (2003) 737–741. Available from: https://doi.org/10.1016/j.msec.2003.09.120.

[9] R. Kamal, K. Khan, M.,F. Khan, Application of nanotechnology in medical diagnosis, Int. Conf. Innov. Eng. Technol. 2 (3) (2011) 22–34.

[10] K. Arivalagan, et al., Nanomaterials and its potential applications, Int. J. ChemTech Res. 3 (2) (2011) 534–538. Available from: https://doi.org/10.1021/la904014z.

[11] R. Roy, R.A. Roy, D.M. Roy, Alternative perspectives on "quasi-crystallinity": non-uniformity and nanocomposites, Mater. Lett. 4 (8–9) (1986) 323–328. Available from: https://doi.org/10.1016/0167-577X(86)90063-7.

[12] D. Schmidt, D. Shah, E.P. Giannelis, New advances in polymer/layered silicate nanocomposites, Curr. Opin. Solid State Mater. Sci. 6 (3) (2002) 205–212. Available from: https://doi.org/10.1016/S1359-0286(02)00049-9.

[13] L.M. Manocha, et al., Nanocomposites for structural applications, Indian J. Pure Appl. Phys. 44 (2) (2006) 135–142.

[14] F.E. Kruis, H. Fissan, A. Peled, Synthesis of nanoparticles in the gas phase for electronic, optical and magnetic applications—a review, J. Aerosol Sci. 29 (5–6) (1998) 511–535. Available from: https://doi.org/10.1016/S0021-8502(97)10032-5.

[15] S. Zhang, et al., Recent advances of superhard nanocomposite coatings: a review, Surf. Coat. Technol. 167 (2–3) (2003) 113–119. Available from: https://doi.org/10.1016/S0257-8972(02)00903-9.

[16] Hettich, B. and Materialforschung, G. (1991) 'Advanced a4 Aterials', 3(12), pp. 8–9.

[17] M. Birkholz, U. Albers, T. Jung, Nanocomposite layers of ceramic oxides and metals prepared by reactive gas-flow sputtering, Surf. Coat. Technol. 179 (2–3) (2004) 279–285. Available from: https://doi.org/10.1016/S0257-8972(03)00865-X.

[18] S.R. Bakshi, D. Lahiri, A. Agarwal, Carbon nanotube reinforced metal matrix composites—a review, Int. Mater. Rev. 55 (1) (2010) 41–64. Available from: https://doi.org/10.1179/095066009X12572530170543.

[19] Gash, A.E., Simpson, R.L. and Satcher, J.H. (2002) 'Energetic nanocomposites with sol-gel chemistry: synthesis, safety, and characterization'.

[20] Gash, A.E. et al. (2000) 'Making nanostructured pyrotechnics in a beaker', *Production*, April 10, 2000.

[21] K.R. Ryan, J.R. Gourley, S.E. Jones, Environmental anomalies at the World Trade Center: evidence for energetic materials, Environmentalist 29 (1) (2009) 56–63. Available from: https://doi.org/10.1007/s10669-008-9182-4.

[22] Araujo, P. De (2007) 'Stiffer by design', 6(January), pp. 9–11. doi: 10.2217/bmm.14.57.

[23] Y.-W. Mai, Z.-Z. Yu, s, Woodhead Publishing, Cambridge., Polymer Nanocomposite (2006). Available from: https://doi.org/10.1533/9781845691127.

[24] A. Zandiatashbar, C.R. Picu, N. Koratkar, Control of epoxy creep using graphene, Small 8 (11) (2012) 1676–1682. Available from: https://doi.org/10.1002/smll.201102686.

[25] K. O, Synthesis of nylon 6-clay hybrid, J. Mater. Res. 8 (5) (1993) 1179. Available from: https://doi.org/10.1557/JMR.1993.1179.

[26] A. Usuki, et al., Swelling behavior of montmorillonite cation exchanged for omega-amino acids by epsilon-caprolactam, J. Mater. Res. 8 (5) (1993) 1174–1178. Available from: https://doi.org/10.1557/jmr.1993.1174.

[27] Levchik, S.V. (2006) Introduction to flame retardancy and polymer flammability, flame retardant polymer nanocomposites, in: A. B. Morgan, C. A. Wilkie (Eds) Flame Retardant Polymer Nanocomposites, Wiley, Hoboken, pp. 1–29

[28] J.R. Heath, Molecular Electronics, Annual Review of Materials Research (2009). Available from: https://doi.org/10.1146/annurev-matsci-082908-145401.

[29] Waldner, J. (no date) 'Nanocomputers and Swarm Intelligence TS - RIS M4 - Citavi'.

[30] J.F. Kerrdge, I applied physics 1 Carbon Nanotubes for Next-, Science 278 (5335) (1997) 77–78.

[31] Y. Kuang, et al., Flexible single-component-polymer resistive memory for ultrafast and highly compatible nonvolatile memory applications, IEEE Electron Dev. Lett. 31 (7) (2010) 758–760. Available from: https://doi.org/10.1109/LED.2010.2048297.

[32] K.H. Lee, et al., High-mobility nonvolatile memory thin-film transistors with a ferroelectric polymer interfacing ZnO and pentacene channels, Adv. Mater. 21 (42) (2009) 4287–4291. Available from: https://doi.org/10.1002/adma.200900398.

[33] W. Lu, C.M. Lieber, Nanoelectronics from the bottom up, Nat. Mater. 6 (11) (2007) 841–850. Available from: https://doi.org/10.1038/nmat2028.

[34] L.P. Ma, J. Liu, Y. Yang, Organic electrical bistable devices and rewritable memory cells, Appl. Phys. Lett. 80 (16) (2002) 2997–2999. Available from: https://doi.org/10.1063/1.1473234.

[35] T. Sekitani, et al., Organic nonvolatile memory transistors for flexible sensor arrays, Science 326 (5959) (2009) 1516–1519. Available from: https://doi.org/10.1126/science.1179963.

[36] L. Ma, et al., Nonvolatile electrical bistability of organic/metal-nanocluster/organic system, Appl. Phys. Lett. 82 (9) (2003) 1419–1421. Available from: https://doi.org/10.1063/1.1556555.

[37] L.D. Bozano, et al., Organic materials and thin-film structures for cross-point memory cells based on trapping in metallic nanoparticles, Adv. Funct. Mater. 15 (12) (2005) 1933–1939. Available from: https://doi.org/10.1002/adfm.200500130.

[38] L.D. Bozano, et al., Mechanism for bistability in organic memory elements, Appl. Phys. Lett. 84 (4) (2004) 607–609. Available from: https://doi.org/10.1063/1.1643547.

[39] J.G. Park, et al., Multilevel nonvolatile small-molecule memory cell embedded with Ni nanocrystals surrounded by a NiO tunneling barrier, Nano Lett. 9 (4) (2009) 1713–1719. Available from: https://doi.org/10.1021/nl900429h.

[40] T.-Y. Chang, Y.-W. Cheng, P.-T. Lee, Electrical characteristics of an organic bistable device using an Al/Alq3/nanostructured MoO3/Alq3/p + -Si structure, Appl. Phys. Lett. 96 (4) (2010) 43309. Available from: https://doi.org/10.1063/1.3299265.

[41] W.L. Kwan, et al., Direct observation of localized conduction pathways in photocross-linkable polymer memory, J. Appl. Phys. 105 (12) (2009) 11–16. Available from: https://doi.org/10.1063/1.3153980.

[42] B. Lei, et al., Statistical characterization of the memory effect in polyfluorene based non-volatile resistive memory devices, Org. Electron. Phys. Mater. Appl. 10 (6) (2009) 1048–1053. Available from: https://doi.org/10.1016/j.orgel.2009.05.015.

[43] V.Y. Aristov, et al., Properties of hybrid organic-inorganic systems: Au nanoparticles embedded into an organic CuPc matrix, Appl. Phys. Lett. 97 (11) (2010). Available from: https://doi.org/10.1063/1.3488809.

[44] F. Li, D.I. Son, et al., Organic bistable devices based on core/shell CdSeZnS nanoparticles embedded in a conducting poly (N-vinylcarbazole) polymer layer, Appl. Phys. Lett. 91 (12) (2007) 1–4. Available from: https://doi.org/10.1063/1.2783189.

[45] T.R. Andersen, et al., Aqueous processing of low-band-gap polymer solar cells using roll-to-roll methods, ACS Nano 5 (5) (2011) 4188–4196. Available from: https://doi.org/10.1021/nn200933r.

[46] Y. Zhou, et al., Binaphthyl-containing green-and red-emitting molecules for solution-processable organic light-emitting diodes, Adv. Funct. Mater. 18 (20) (2008) 3299–3306. Available from: https://doi.org/10.1002/adfm.200800375.

[47] F.C. Chen, et al., Flexible polymer photovoltaic devices prepared with inverted structures on metal foils, IEEE Electron Dev. Lett. 30 (7) (2009) 727–729. Available from: https://doi.org/10.1109/LED.2009.2022345.

[48] G. Gelinck, et al., Organic transistors in optical displays and microelectronic applications, Adv. Mater. 22 (34) (2010) 3778–3798. Available from: https://doi.org/10.1002/adma.200903559.

[49] D. Kim, et al., Heterogeneous interfacial properties of ink-jet-printed silver nanoparticulate electrode and organic semiconductor, Adv. Mater. 20 (16) (2008) 3084–3089. Available from: https://doi.org/10.1002/adma.200702750.

[50] N.A. Azarova, et al., Fabrication of organic thin-film transistors by spray-deposition for low-cost, large-area electronics, Organ. Electron. Phys. Mater. Appl. 11 (12) (2010) 1960–1965. Available from: https://doi.org/10.1016/j.orgel.2010.09.008.

[51] Q.D. Ling, D.J. Liaw, et al., Polymer memories: bistable electrical switching and device performance, Polymer 48 (18) (2007) 5182–5201. Available from: https://doi.org/10.1016/j.polymer.2007.06.025.

[52] F. Li, T.W. Kim, et al., Formation and electrical bistability properties of ZnO nanoparticles embedded in polyimide nanocomposites sandwiched between two C60 layers, Appl. Phys. Lett. 92 (1) (2008) 1–4. Available from: https://doi.org/10.1063/1.2830617.

[53] J.H. Jung, et al., Nonvolatile organic bistable devices fabricated utilizing Cu2O nanocrystals embedded in a polyimide layer, Appl. Phys. Lett. 89 (12) (2006). Available from: https://doi.org/10.1063/1.2355465.

[54] J.H. Kim, et al., Formation and electrical properties of Ni1-x Fex nanocrystals embedded in a polyimide layers for applications as nonvolatile flash memories, Appl. Phys. Lett. 86 (3) (2005) 1–3. Available from: https://doi.org/10.1063/1.1850194.

[55] W.T. Kim, et al., Current bistability and carrier transport mechanisms of organic bistable devices based on hybrid Ag nanoparticle-polymethyl methacrylate polymer nanocomposites, Appl. Phys. Lett. 96 (25) (2010). Available from: https://doi.org/10.1063/1.3453661.

[56] W.L. Leong, et al., Micellar poly(styrene-b-4-vinylpyridine)-nanoparticle hybrid system for non-volatile organic transistor memory, J. Mater. Chem. 19 (39) (2009) 7354–7361. Available from: https://doi.org/10.1039/B911493a.

[57] N.G. Portney, A.A. Martinez-Morales, M.O. Ozkan, Nanoscale memory characterization of virus-templated semiconducting quantum dots, ACS Nano 2 (2) (2008) 191–196. Available from: https://doi.org/10.1021/nn700240z.

[58] C.Y. Huang, et al., Stability in the memory state of the silica nanoparticle-doped hybrid aligned nematic device, J. Appl. Phys. 109 (2) (2011) 2–6. Available from: https://doi.org/10.1063/1.3531993.

[59] P.Y. Lai, J.S. Chen, Electrical bistability and charge transport behavior in Au nanoparticle/poly(N-vinylcarbazole) hybrid memory devices, Appl. Phys. Lett. 93 (15) (2008) 153305. Available from: https://doi.org/10.1063/1.3002281.

[60] F. Li, D.I. Son, H.M. Cha, et al., Memory effect of CdSeZnS nanoparticles embedded in a conducting poly[2-methoxy-5-(2-ethylhexyloxy)-1,4-phenylene-vinylene] polymer layer, Appl. Phys. Lett. 90 (22) (2007) 2005–2008. Available from: https://doi.org/10.1063/1.2745219.

[61] M.F. Mabrook, et al., Memory effects in hybrid silicon-metallic nanoparticle-organic thin film structures, Org. Electron. Phys. Mater. Appl. 9 (5) (2008) 816–820. Available from: https://doi.org/10.1016/j.orgel.2008.05.023.

[62] S.H. Lee, et al., Variations in the memory capability of nonvolatile memory devices fabricated using hybrid composites of InP nanoparticles and a polystyrene layer due to the scale-down, J. Nanosci. Nanotechnol. 11 (1) (2011) 449–452. Available from: https://doi.org/10.1166/jnn.2011.3172.

[63] B. Cho, S. Song, et al., Organic resistive memory devices: performance enhancement, integration, and advanced architectures, Adv. Funct. Mater. 21 (15) (2011) 2806–2829. Available from: https://doi.org/10.1002/adfm.201100686.

[64] T.W. Kim, et al., One transistor-one resistor devices for polymer non-volatile memory applications, Adv. Mater. 21 (24) (2009) 2497–2500. Available from: https://doi.org/10.1002/adma.200803798.

[65] E.Y.H. Teo, et al., An organic-based diode-memory device with rectifying property for crossbar memory array applications, IEEE Electron Device Letters 30 (5) (2009) 487–489. Available from: https://doi.org/10.1109/LED.2009.2017387.

[66] F. Li, D.I. Son, J.H. Ham, et al., Memory effect of nonvolatile bistable devices based on CdSeZnS nanoparticles sandwiched between C60 layers, Appl. Phys. Lett. 91 (16) (2007) 10–13. Available from: https://doi.org/10.1063/1.2801357.

[67] C. Wu, F. Li, T. Guo, et al., Efficient nonvolatile rewritable memories based on three-dimensionally confined Au quantum dots embedded in ultrathin polyimide layers, Japanese J. Appl. Phys. 50 (3) (2011) 2–5. Available from: https://doi.org/10.1143/JJAP.50.030204.

[68] H.M. Koo, et al., Improvement of charge storage characteristics on floating gated nonvolatile memory devices with In2O3 nanoparticles embedded polyimide gate insulator, Appl. Phys. Lett. 91 (4) (2007) 43513. Available from: https://doi.org/10.1063/1.2764558.

[69] J.G. Yun, S. Cho, B.G. Park, Stacked-nanowire device with virtual source/drain (SD-VSD) for 3D NAND flash memory application, Solid-State Electron. 64 (1) (2011) 42–46. Available from: https://doi.org/10.1016/j.sse.2011.07.003.

[70] J. Kim, et al., A stacked memory device on logic 3D technology for ultra-high-density data storage, Nanotechnology 22 (25) (2011). Available from: https://doi.org/10.1088/0957-4484/22/25/254006.

[71] S. Song, et al., Three-dimensional integration of organic resistive memory devices, Adv. Mater. 22 (44) (2010) 5048–5052. Available from: https://doi.org/10.1002/adma.201002575.

[72] S. Choi, et al., High-performance programmable memory devices based on hyperbranched copper phthalocyanine polymer thin films, Adv. Mater. 20 (9) (2008) 1766–1771. Available from: https://doi.org/10.1002/adma.200702147.

[73] M. Cölle, M. Büchel, D.M. de Leeuw, Switching and filamentary conduction in non-volatile organic memories, Organ. Electron. Phys. Mater. Appl. 7 (5) (2006) 305–312. Available from: https://doi.org/10.1016/j.orgel.2006.03.014.

[74] W.J. Joo, et al., Study on threshold behavior of operation voltage in metal filament-based polymer memory, J. Phys. Chem. B 111 (27) (2007) 7756–7760. Available from: https://doi.org/10.1021/jp0684933.

[75] J. Kim, et al., Novel electrical properties of nanoscale thin films of a semiconducting polymer: quantitative current-sensing AFM analysis, Langmuir 23 (17) (2007) 9024–9030. Available from: https://doi.org/10.1021/la700785h.

[76] S. Baek, et al., Novel digital nonvolatile memory devices based on semiconducting polymer thin films, Adv. Funct. Mater. 17 (15) (2007) 2637–2644. Available from: https://doi.org/10.1002/adfm.200600892.

[77] T.W. Kim, et al., Reversible switching characteristics of polyfluorene-derivative single layer film for nonvolatile memory devices, Appl. Phys. Lett. 92 (25) (2008). Available from: https://doi.org/10.1063/1.2952825.

[78] W.L. Kwan, R.J. Tseng, Y. Yang, Multi-layer stackable polymer memory devices.', Philos. Trans. A, Math. Phys. Eng. Sci. 367 (1905) (2009) 4159–4167. Available from: https://doi.org/10.1098/rsta.2008.0263.

[79] B. Cho, J.M. Yun, et al., Direct observation of Ag filamentary paths in organic resistive memory devices, Adv. Funct. Mater. 21 (20) (2011) 3976–3981. Available from: https://doi.org/10.1002/adfm.201101210.

[80] Kwan, W.L. *et al.* (2007) 'Stackable resistive memory device using photo cross-linkable copolymer bametd Al electrod method csolubld could be ustedr to pattern tross-lin the rop', pp. 2007–2010.

[81] M.A. Lampert, R.B. Schilling, Current injection in solids: the regional approximation method, Semicond. Semimetals 6 (C) (1970) 1–96. Available from: https://doi.org/10.1016/S0080-8784(08)62630-7.

[82] J.G. Simmons, R.R. Verderber, New conduction and reversible memory phenomena in thin insulating films, Proc. Roy. Soc. A: Math. Phys. Eng. Sci. 301 (1464) (1967) 77–102. Available from: https://doi.org/10.1098/rspa.1967.0191.

[83] S.M. Sze, K.K. Ng, Physics of Semiconductor Devices, Third Edition, Wiley, 2006. Available at: https://www.wiley.com/en-us/Physics + of + Semiconductor + Devices%2C + 3rd + Edition-p-9780471143239.

[84] H.Y. Jeong, et al., Graphene oxide thin films for flexible nonvolatile memory applications, Nano Lett. 10 (11) (2010) 4381–4386. Available from: https://doi.org/10.1021/nl101902k.

[85] D.I. Son, et al., Polymer-ultrathin graphite sheet-polymer composite structured flexible nonvolatile bistable organic memory devices, Nanotechnology 22 (29) (2011). Available from: https://doi.org/10.1088/0957-4484/22/29/295203.

[86] Q. Ling, et al., Non-volatile polymer memory device based on a novel copolymer of N-vinylcarbazole and Eu-complexed vinylbenzoate, Adv. Mater. 17 (4) (2005) 455–459. Available from: https://doi.org/10.1002/adma.200401048.

[87] Q.D. Ling, S.L. Lim, et al., Nonvolatile polymer memory device based on bistable electrical switching in a thin film of poly(N-vinylcarbazole) with covalently bonded C60, Langmuir 23 (1) (2007) 312–319. Available from: https://doi.org/10.1021/la061504z.

[88] J. Ouyang, et al., Programmable polymer thin film and non-volatile memory device, Nat. Mater. 3 (12) (2004) 918–922. Available from: https://doi.org/10.1038/nmat1269.

[89] X.D. Zhuang, et al., Conjugated-polymer-functionalized graphene oxide: Synthesis and nonvolatile rewritable memory effect, Adv. Mater. 22 (15) (2010) 1731–1735. Available from: https://doi.org/10.1002/adma.200903469.

[90] D.I. Son, et al., Flexible organic bistable devices based on graphene embedded in an insulating poly (methyl methacrylate) polymer layer, Nano Letters 10 (7) (2010) 2441–2447. Available from: https://doi.org/10.1021/nl1006036.

[91] C. Wu, F. Li, Y. Zhang, et al., Highly reproducible memory effect of organic multilevel resistive-switch device utilizing graphene oxide sheets/polyimide hybrid nanocomposite, Appl. Phys. Lett. 99 (4) (2011) 2011–2014. Available from: https://doi.org/10.1063/1.3619815.

[92] S.J. Kim, J.S. Lee, Flexible organic transistor memory devices, Nano Lett. 10 (8) (2010) 2884–2890. Available from: https://doi.org/10.1021/nl1009662.

[93] Q. Lai, et al., An organic/Si nanowire hybrid field configurable transistor, Nano Lett. 8 (3) (2008) 876–880. Available from: https://doi.org/10.1021/nl073112y.

[94] D. Son, et al., Nonvolatile flexible organic bistable devices fabricated utilizing CdSe/ZnS nanoparticles embedded in a conducting poly N-vinylcarbazole polymer layer, Nanotechnology 19 (2008) 55204. Available from: https://doi.org/10.1088/0957-4484/19/5/055204.

[95] Z.L. Tseng, et al., Electrical bistability in hybrid ZnO nanorod/polymethylmethacrylate heterostructures, Appl. Phys. Lett. 97 (21) (2010). Available from: https://doi.org/10.1063/1.3511756.

[96] B. Yuan, et al., Electrical bistability in self-assembled hybrid multilayers of phospholipid and nanoparticles, Nanotechnology 22 (31) (2011). Available from: https://doi.org/10.1088/0957-4484/22/31/315303.

[97] D.I. Son, et al., Single active-layer structured dual-function devices using hybrid polymer-quantum dots, Nanotechnology 19 (39) (2008). Available from: https://doi.org/10.1088/0957-4484/19/39/395201.

[98] Y. Ji, et al., Flexible organic memory devices with multilayer graphene electrodes, ACS Nano 5 (7) (2011) 5995–6000. Available from: https://doi.org/10.1021/nn201770s.

[99] S. Lee, et al., Resistive switching characteristics of ZnO thin film grown on stainless steel for flexible nonvolatile memory devices, Appl. Phys. Lett. 95 (26) (2009) 262113. Available from: https://doi.org/10.1063/1.3280864.

[100] Y. Yang, et al., Electrical switching and bistability in organic/polymeric thin films and memory devices, Adv. Funct. Mater. 16 (8) (2006) 1001–1014. Available from: https://doi.org/10.1002/adfm.200500429.

[101] F. Li, D.I. Son, et al., Nonvolatile electrical bistability and operating mechanism of memory devices based on CdSe/ZnS nanoparticle/polymer hybrid composites, Appl. Phys. Lett. 93 (2) (2008) 32–35. Available from: https://doi.org/10.1063/1.2959786.

[102] X.N. Xie, et al., Aggregates-induced dynamic negative differential resistance in conducting organic films, Appl. Phys. Lett. 95 (20) (2009) 2007–2010. Available from: https://doi.org/10.1063/1.3266854.

[103] L.Z. Yu, C.T. Lee, Investigation of three-terminal organic-based devices with memory effect and negative differential resistance, Appl. Phys. Lett. 95 (10) (2009) 2007–2010. Available from: https://doi.org/10.1063/1.3224201.

[104] J. Chen, et al., Negative differential resistance effect in organic devices based on an anthracene derivative, Appl. Phys. Lett. 89 (8) (2006). Available from: https://doi.org/10.1063/1.2338513.

[105] M. Vilkman, et al., Negative differential resistance in polymeric memory devices containing disordered block copolymers with semiconducting block, Org. Electron. Phys. Mater. Appl. 10 (8) (2009) 1478–1482. Available from: https://doi.org/10.1016/j.orgel.2009.08.012.

[106] M.J. Lee, et al., A fast, high-endurance and scalable non-volatile memory device made from asymmetric Ta2O5-xx/TaO2-xbilayer structures, Nat. Mater. 10 (8) (2011) 625–630. Available from: https://doi.org/10.1038/nmat3070.

[107] F. Li, H. Cha, et al., Investigation of carrier transport mechanism in CDSE/PVK hybrid nanocomposites, J. Korean Phys. Soc. 53 (1) (2008) 376–379. Available from: https://doi.org/10.3938/jkps.53.376.

[108] S. Möller, et al., A polymer/semiconductor write-once read-many-times memory, Nature 426 (6963) (2003) 166–169. Available from: https://doi.org/10.1038/nature02070.

[109] W.J. Joo, et al., Metal filament growth in electrically conductive polymers for nonvolatile memory application, J. Phys. Chem. B 110 (47) (2006) 23812–23816. Available from: https://doi.org/10.1021/jp0649899.

[110] L. Ma, Q. Xu, Y. Yang, Organic nonvolatile memory by controlling the dynamic copper-ion concentration within organic layer, Appl. Phys. Lett. 84 (24) (2004) 4908–4910. Available from: https://doi.org/10.1063/1.1763222.

[111] I.D. Parker, Carrier tunneling and device characteristics in polymer light-emitting diodes, J. Appl. Phys. 75 (3) (1994) 1656–1666. Available from: https://doi.org/10.1063/1.356350.

Index

Note: Page numbers followed by "*f*" and "*t*" refer to figures and tables, respectively.

D

G

H

N

P

S

U

V

W

Made in the USA
Monee, IL
13 September 2026

63138391R00262